Probability

Bayes' theorem

$$P(B_i|A) = \frac{P(B_i) \cdot P(A|B_i)}{P(B_1) \cdot P(A|B_1) + P(B_2) \cdot P(A|B_2) + \cdots + P(B_k) \cdot P(A|B_k)}$$

Conditional probability

$$P(A|B) = \frac{P(A \cap B)}{P(B)}$$

General addition rule

$$P(A \cup B) = P(A) + P(B) - P(A \cap B)$$

General multiplication rule

$$P(A \cap B) = P(B) \cdot P(A|B) \qquad \text{or} \qquad P(A \cap B) = P(A) \cdot P(B|A)$$

Mathematical expectation

$$E = a_1 p_1 + a_2 p_2 + \cdots + a_k p_k$$

Probability Distributions

Binomial distribution

$$f(x) = \binom{n}{x} p^x (1 - p)^{n-x}$$

Mean of probability distribution

$$\mu = \sum x \cdot f(x)$$

Standard deviation of probability distribution

$$\sigma = \sqrt{\sum (x - \mu)^2 \cdot f(x)}$$

Continued inside back cover

Eighth Edition

MODERN ELEMENTARY STATISTICS

John E. Freund
Arizona State University

Gary A. Simon
New York University

This Instructor's Edition contains a full answer key to all of the exercises in the text. In all other respects this edition is identical to your student's textbook.

When ordering the text for your students, please be sure to use the ISBN for the student text, 0-13-602699-0.

PRENTICE HALL
Englewood Cliffs, New Jersey 07632

INSTRUCTOR'S EDITION

Editor-in-chief: Tim Bozik
Acquisition editor: Steve Conmy
Editorial/production: Nick Romanelli
Interior/cover design: Amy Rosen
Design director: Florence Dara Silverman
Cover art: Salem Krieger
Pre-press buyer: Paula Massenaro
Manufacturing buyer: Lori Bulwin
Editorial assistant: Joan Dello Stritto

© 1992, 1988, 1984, 1979, 1973, 1967, 1960, 1952 by Prentice-Hall, Inc.
A Simon & Schuster Company
Englewood Cliffs, New Jersey 07632

Printed in the United States of America
10   9   8   7   6   5   4   3   2   1

ISBN   0-13-602707-5

Prentice-Hall International (UK) Limited, *London*
Prentice-Hall of Australia Pte. Limited, *Sydney*
Prentice-Hall Canada, Inc. *Toronto*
Prentice-Hall Hispanoamericana, S.A., *Mexico*
Prentice-Hall of India Private Limited, *New Delhi*
Prentice-Hall of Japan, Inc. *Tokyo*
Simon & Schuster Asia Pte. Ltd., *Singapore*
Editora Prentice-Hall do Brasil, Ltda., *Rio de Janeiro*

CONTENTS

10 SAMPLING AND SAMPLING DISTRIBUTIONS 250

11 INFERENCES ABOUT MEANS 281

15 REGRESSION

16 CORRELATION

17 NONPARAMETRIC TESTS

PREFACE

The main objective of this edition, like that of all previous editions of this book, is to introduce beginning students to the fundamentals of modern statistics. The basic organization has remained unchanged, but there are many changes in content and emphasis. These changes reflect current trends in the teaching of statistics, brought about by the general availability of computers, by the realization that there are statistical problems which do not require that we make an inference, and by the desire to give students a better understanding of what is really involved in the various statistical techniques.

The eighth edition differs from the seventh primarily in revisions and additions to the exercise sets. Many of these changes ask the students to think about the nature of the statistical issues, and other changes bring in concepts which are extensions of the material in the text. The eighth edition has about 1,400 exercises, involving both computational and conceptual types. They are designed to make the reader think about the various techniques and not merely familiarize him or her with the mechanics of performing the necessary calculations. These exercises affirm the merits of the "bread and butter techniques" while making the student aware of their limitations.

The p-value concept has been formally integrated into the testing of statistical hypotheses. In addition, many small changes in the presentation have been made throughout the book. The revisions for the eighth edition are entirely the work of Gary Simon.

Although most of the calculations required of the reader can be done with a small hand-held calculator, he or she will be made aware that most of the calculations that we do in statistics can be done through commonly available computer software. It is also shown that computers can be used to great advantage in simulating values of random variables (say, in connection with the law of large numbers or the sampling distribution of the mean); that computers can provide information that is otherwise not readily accessible (for instance, in determining tail probabilities for non-normal distributions or binomial probabilities for values of n and p not found in most tables); that they can be used to judge the closeness of approximations (say, the Poisson approximation to the binomial distribution); and, of course, that they are indispensable (as in multiple regression) when the computational load would otherwise be prohibitive.

As in previous editions, the authors shall resist the temptation to tell anyone specifically what chapters, sections, or subjects to study or teach, but it should be observed that there are topics which may be omitted without loss of continuity.

Some sections are thus marked ★, meaning that they are optional, although what is optional to one person may be essential to another. For instance, many instructors will not cover the material on the description of grouped data, yet a person who has to work with published government data may well consider it essential. Although quite a few colleagues indicated that they do not cover the material on subjective probability, to mark it optional would be an affront to those who hold the subjectivist point of view. However, risking their wrath, the sections dealing with Bayes' theorem and Bayesian estimation are marked optional.

Exercises which are marked ★ are based on material from sections marked ★ or are based on extensions of the ideas beyond what is directly presented in the text. Exercises which are marked with the computer icon furnish useful practice in the use of computer software. We do not endorse any particular computer software, and the use of MINITAB in our illustrations simply reflects our observation that it is a widely used program.

As indicated in previous editions, the study of statistics may not only be directed toward application in various specialized fields of inquiry, but it may also be presented at various levels of mathematical difficulty and in almost any balance between theory and application. As it is more important, in the authors' opinions, to understand the meaning and implications of basic ideas than it is to memorize an impressive list of formulas, some of the details that are sometimes included in introductory courses in statistics have been sacrificed. This many be unfortunate in some respects, but it should prevent the reader from getting lost in an excessive amount of detail which could easily obscure the more important issues. It is hoped that this will avoid some of the unfortunate consequences which often result from the indiscriminate application of so-called standard techniques without a thorough understanding of the basic ideas that are involved.

It cannot be denied that a limited amount of mathematics is a prerequisite for any course in statistics, and that a thorough study of the theoretical principles of statistics would require a knowledge of mathematical subjects taught ordinarily only on the graduate level. Since this book is designed for students with relatively little background in mathematics, the aims, and therefore also the prerequisites, are considerably more modest. Actually, the mathematical background needed for this study of statistics is amply covered in a course of college algebra; in fact, even a good knowledge of high school algebra provides a sufficient foundation.

The authors would like to express appreciation to the many colleagues and students whose helpful suggestions and criticisms contributed greatly to previous editions of this text as well as to this eighth edition. In particular, the authors would like to thank Mrs. Rita Ewer for reading various drafts of the manuscript and helping with the proofreading. The authors would also like to express their appreciation to Steve Conmy, mathematics editor, and to Nicholas Romanelli, production editor, for their untiring help in the production of this book.

The authors are also indebted to Prentice Hall for permission to reproduce part of Table 2 of R. A. Johnson and D. W. Wichern's *Applied Multivariate Statistical Analysis*; to Professor E. S. Pearson and the *Biometrika* trustees to reproduce the material in Tables III and IV; to the Addison-Wesley Publishing Company to base Table VII on Table 11.4 of D. B. Owen's *Handbook of Statistical*

Tables; to the American Cynamid Company to reproduce the material in Table VI; to the editor of the *Annals of Mathematical Statistics* to reproduce the material in Table VIII; to the RAND Corporation to reproduce the sample pages of random normal numbers shown in Table XII; and to MINITAB to reproduce the computer printouts shown in the text.

Gary A. Simon

INTRODUCTION

The collection, processing, interpretation, and presentation of numerical data all belong to the domain of statistics. These tasks include the calculation of baseball batting averages, collecting data on births and deaths, evaluating the effectiveness of commercial products, and forecasting the weather. Statistical information is presented to us constantly on radio and television. Our enthusiasm for statistical facts is encouraged by national newspapers such as *The Wall Street Journal* and *USA Today*.

The word "statistics" is used in several ways. It can refer not only to the mere tabulation of numeric information, as in reports of stock market transactions, but also to the body of techniques used in processing or analyzing data.

The word "statistician" is also used in several ways. The term can be applied to those who simply collect information, as well as to those who prepare analyses or interpretations, and it is also applied to scholars who develop the mathematical theory on which statistics is based.

In Sections 1.1 and 1.2 we discuss the recent growth of statistics and its ever-widening range of applications. In Section 1.3 we explain the distinction between the two major branches of statistics, descriptive statistics and statistical inference, and in the optional Section 1.4 we discuss the nature of various kinds of data and in connection with this warn the reader against the indiscriminate mathematical treatment of statistical data.

1.1

THE GROWTH OF MODERN STATISTICS

There are several reasons why the scope of statistics and the need to study statistics have grown enormously in the last fifty or so years. One reason is the increasingly quantitative approach employed in all the sciences, as well as in business and many other activities which directly affect our lives. This includes the use of mathematical techniques in the evaluation of anti-pollution controls, in inventory planning, in the analysis of traffic patterns, in the study of the effects of various kinds of medications, in the evaluation of teaching techniques, in the analysis of competitive behavior of businessmen and governments, in the study of diet and longevity, and so forth. The availability of powerful computers has greatly increased our ability to deal with numerical information. Many types of computers are also inexpensive, so that sophisticated statistical work can be done by small businesses, college students, and even high-school students.

The other reason is that the amount of data that is collected, processed, and disseminated to the public for one reason or another has increased almost beyond comprehension, and what part is "good" statistics and what part is "bad" statistics is anybody's guess. To act as watchdogs, more and more persons with some knowledge of statistics are needed to take an active part in the collection of the data, in the analysis of the data, and, what is equally important, in all of the preliminary planning. Without the latter, it is frightening to think of all the things that can go wrong in the compilation of statistical data. The results of costly surveys can be useless if questions are ambiguous or asked in the wrong way, if they are asked of the wrong persons, in the wrong place, or at the wrong time. Much of this is just common sense, as is illustrated by the following examples:

EXAMPLE To determine public sentiment about the continuation of a certain government program, an interviewer asks: "Do you feel that this wasteful program should be stopped?" Explain why this will probably not yield the desired information.

Solution The interviewer is "begging the question" by suggesting, in fact, that the program is wasteful.

EXAMPLE To study consumer reaction to a new convenience food, a house-to-house survey is conducted during weekday mornings, with no provisions for return visits in case no one is home. Explain why this may well yield misleading information.

Solution This survey will fail to reach those who are most likely to use the product: single persons and married couples with both spouses employed.

Although much of the above-mentioned growth of statistics began prior to the "computer revolution," the widespread availability and use of computers have greatly accelerated the process. In particular, computers enable us to handle, analyze, and dissect large masses of data, and they enable us to perform calculations which previously had been too cumbersome even to contemplate. Our objective in this book will be your gaining an understanding of the ideas of statistics. Access to a computer is not critical for this objective. Computer uses are occasionally illustrated in this textbook, but nearly all the exercises can be done with nothing more than a four-function calculator.

1.2

THE STUDY OF STATISTICS

The subject of statistics can be presented at various levels of mathematical difficulty, and it may be directed toward applications in various fields of inquiry. Accordingly, many textbooks have been written on business statistics, educational statistics, medical statistics, psychological statistics, ..., and even on statistics for historians. Although problems arising in these various disciplines will sometimes require special statistical techniques, none of the basic methods discussed in this text is restricted to any particular field of application. In the same way in which $2 + 2 = 4$ regardless of whether we are adding dollar amounts, horses, or trees, the methods we shall present provide **statistical models** which apply regardless of whether the data are IQ's, tax payments, reaction times, humidity readings, test scores, and so on. To illustrate this further, consider Exercise 13.96 on page 388, made up by the author.

> **13.96** In a random sample of 200 retired persons, 137 stated that they prefer living in an apartment to living in a one-family home. At the 0.05 level of significance does this refute the claim that 60 percent of all retired persons prefer living in an apartment to living in a one-family home?

The question asked here should be clear, and it should also be clear that the answer would be of interest mainly to social scientists or to persons in the construction industry. However, if we wanted to cater to the special interests of students of biology, engineering, education, or ecology, we might rephrase the exercise as follows:

> **13.96** In a random sample of 200 citrus trees exposed to a 20° frost, 137 showed some damage to their fruit. At the 0.05 level of significance does this refute the claim that 60 percent of all citrus trees exposed to a 20° frost will show some damage to their fruit?

> **13.96** In a random sample of 200 transistors made by a given manufacturer, 137 passed an accelerated performance test. At the 0.05 level of significance does this refute the claim that 60 percent of all the transistors made by the manufacturer will pass the test?

> **13.96** In a random sample of 200 high school seniors in a large city, 137 said that they will go on to college. At the 0.05 level of significance does this refute the claim that 60 percent of all the high school seniors in this city will go on to college?

> **13.96** In a random sample of 200 cars tested for the emission of pollutants, 137 failed to meet a state's legal standards. At the 0.05 level of significance does this refute the claim that 60 percent of all cars tested in this state will fail to meet its legal emission standards?

So far as the work in this book is concerned, the statistical treatment of all these versions of Exercise 13.96 is the same, and with some imagination the reader should be able to rephrase it for virtually any field of specialization. As some authors do, we could present, and so designate, special problems for readers with special interests, but this would defeat our goal of impressing upon the reader the importance of statistics in all of science, business, and everyday life. To attain this goal, we have included in this text exercises covering a wide spectrum of interests.

To avoid the possibility of misleading anyone with our various versions of Exercise 13.96, let us make it clear that we cannot squeeze all statistical problems into the same mold. Although the methods we shall study in this book are all widely applicable, it is always important to make sure that the statistical model we are using is the right one.

EXERCISES

1.1 Rephrase Exercise 13.96 referred to on page 4 so that it will be of special interest to
   (a) a cosmetics salesperson;
   (b) a musician;
   (c) a traffic engineer.

1.2 "Bad" statistics may well result from asking questions in the wrong way or of the wrong persons. Explain why the following may lead to useless data:
   (a) To determine public sentiment about a certain foreign trade restriction, an interviewer asks voters: "Do you feel that this unfair practice should be stopped?"
   (b) In order to predict a municipal election, a public opinion poll telephones persons selected haphazardly from the city's telephone directory.

1.3 "Bad" statistics may well result from asking questions in the wrong place or at the wrong time. Explain why the following may lead to useless data:
   (a) To predict an election, a poll taker interviews persons coming out of the building which houses the national headquarters of a political party.

   (b) To study the spending patterns of families in a certain income group, a survey is conducted during the first three weeks of December.

1.4 Explain why each of the following studies may fail to yield the desired information:
   (a) To determine the proportion of improperly sealed cans of coffee, a quality-control inspector examines every 50th can coming off a production line.
   (b) To determine the average annual income of its graduates 10 years after graduation, a college's alumni office sent questionnaires in 1990 to all members of the class of 1980, and the estimate was based on the questionnaires returned.
   (c) To ascertain facts about tooth-brushing habits, a sample of the residents of a community are asked how many times they brush their teeth each day.

1.3
DESCRIPTIVE STATISTICS AND STATISTICAL INFERENCE

The origin of modern statistics can be traced to two areas of interest which, on the surface, have very little in common: government (political science) and games of chance.

Governments have long used censuses to count persons and property, and the problem of describing, summarizing, and analyzing census data has led to the development of methods which, until recently, constituted about all there was to the subject of statistics. These methods, which at first consisted primarily of presenting data in the form of tables and charts, make up what we now call **descriptive statistics**. This includes anything done to data which is designed to summarize, or describe, without going any further; that is, without attempting to infer anything that goes beyond the data, themselves. For instance, if tests performed on six small cars imported in 1986 showed that they were able to accelerate from 0 to 60 miles per hour in 18.7, 19.2, 16.2, 12.3, 17.5, and 13.9 seconds, and we report that half of them accelerated from 0 to 60 mph in less than 17.0 seconds, our work belongs to the domain of descriptive statistics. This would also be the case if we claim that these six cars averaged

$$\frac{18.7 + 19.2 + 16.2 + 12.3 + 17.5 + 13.9}{6} = 16.3 \text{ seconds}$$

but not if we conclude that half of *all* cars imported that year could accelerate from 0 to 60 mph in less than 17.0 seconds.

Although descriptive statistics is an important branch of statistics and it continues to be widely used, statistical information usually arises from samples (from observations made on only part of a large set of items), and this means that its analysis requires generalizations which go beyond the data. As a result, the most important feature of the recent growth of statistics has been a shift in emphasis from methods which merely describe to methods which serve to make generalizations; that is, a shift in emphasis from descriptive statistics to the methods of **statistical inference**.

Such methods are required, for instance, to predict the operating life span of a hand-held calculator (on the basis of the performance of several such calculators); to estimate the 1995 assessed value of all privately owned property in Orange County, California (on the basis of business trends, population projections, and so forth); to compare the effectiveness of two reducing diets (on the basis of the weight losses of persons who have been on the diets); to determine the most effective dose of a new medication (on the basis of tests performed with volunteer patients from selected hospitals); or to predict the flow of traffic on a freeway which has not yet been built (on the basis of past traffic counts on alternative routes).

In each of the situations described in the preceding paragraph, there are uncertainties because there is only partial, incomplete, or indirect information; therefore, the methods of statistical inference are needed to judge the merits of our results, to choose a "most promising" prediction, or to select a "most reasonable" (perhaps, a "potentially most profitable") course of action.

In view of the uncertainties, we handle problems like these with statistical methods which find their origin in games of chance. Although the mathematical study of games of chance dates back to the seventeenth century, it was not until the early part of the nineteenth century that the theory developed for "heads or tails," for example, or "red or black" or "even or odd," was applied also to real-life situations where the outcomes were "boy or girl," "life or death," "pass or fail," and so forth. Thus, **probability theory** was applied to many problems in the behavioral, natural, and social sciences, and nowadays it provides an important tool for the analysis of any situation (in science, in business, or in everyday life) which in some way involves an element of uncertainty or chance. In particular, it provides the basis for the methods which we use when we generalize from observed data, namely, when we use the methods of statistical inference.

In recent years, it has been suggested that the emphasis has swung too far from descriptive statistics to statistical inference, and that more attention should be paid to the treatment of problems requiring only descriptive techniques. To accommodate these needs, some new descriptive methods have recently been developed under the general heading of **exploratory data analysis**. Two of these will be presented in Sections 2.3 and 4.5.

1.4

THE NATURE OF STATISTICAL DATA ★[†]

Statistical data are the raw material of statistical investigations—they arise whenever measurements are made or observations are recorded. They may be weights of animals, measurements of personality traits, or earthquake intensities, and they may be simple "yes or no" answers or descriptions of persons' marital status as single, married, widowed, or divorced. Since we said on page 2 that statistics deals with numerical data, this requires some explanation, because "yes or no" answers and descriptions of marital status would hardly seem to qualify as being numerical. Observe, however, that we can record "yes or no" answers to a question as 0 and 1 (or as 1 and 2, or perhaps as 29 and 30 if we are referring to the 15th "yes or no" question of a test), and that we can record a person's marital status as 1, 2, 3, or 4, depending on whether the person is single, married, widowed, or divorced. In this artificial, or nominal way, categorical (qualitative, or descriptive) data can be made into numerical data, and if we thus code the various categories, we refer to the numbers we record as **nominal data**.

Nominal data are numerical in name only, because they do not share any of the properties of the numbers we deal with in ordinary arithmetic. For instance, if we record marital status as 1, 2, 3, or 4 as suggested above, we cannot write $3 > 1$ or $2 < 4$, and we cannot write $2 - 1 = 4 - 3$, $1 + 3 = 4$, or $4 \div 2 = 2$. It is important, therefore, always to check whether mathematical calculations performed in a statistical analysis are really legitimate.

Let us now consider some examples where data share some, but not necessarily all, of the properties of the numbers we deal with in ordinary arithmetic. For instance, in mineralogy the hardness of solids is sometimes determined by observing "what scratches what." If one mineral can scratch another it receives a higher hardness number, and on Mohs' scale the numbers from 1 to 10 are assigned, respectively, to talc, gypsum, calcite, fluorite, apatite, feldspar, quartz, topaz, sapphire, and diamond. With these numbers we can write $6 > 3$, for example, or $7 < 9$, since feldspar is harder than calcite and quartz is softer than sapphire. On the other hand, we cannot write $10 - 9 = 2 - 1$, for

---

[†] As is explained in the Preface, all sections marked ★ are optional. Although the material in this section is meant to serve as a warning against the indiscriminate mathematical treatment of statistical data, it is most relevant to students of the behavioral and social sciences, where artificial scales serve to measure, say, neurotic tendencies, happiness, or conformity to social standards.

example, because the difference in hardness between diamond and sapphire is actually much greater than that between gypsum and talc. Also, it would be meaningless to say that topaz is twice as hard as fluorite simply because their respective hardness numbers on Mohs' scale are 8 and 4.

If we cannot do anything except set up inequalities, as was the case in the preceding example, we refer to the data as **ordinal data**. In connection with ordinal data, > does not necessarily mean "greater than;" it may be used to denote "happier than," "preferred to," "more difficult than," "tastier than," and so forth.

If we can also form differences, but not multiply or divide, we refer to the data as **interval data**. To give an example, suppose we are given the following temperature readings in degrees Fahrenheit: 63°, 68°, 91°, 107°, 126°, and 131°. Here we can write 107° > 68° or 91° < 131°, which simply means that 107° is warmer than 68° and that 91° is colder than 131°. Also, we can write 68° − 63° = 131° − 126°, since equal temperature differences are equal in the sense that the same amount of heat is required to raise the temperature of an object from 63° to 68° as from 126° to 131°. On the other hand, it would not mean much if we say that 126° is twice as hot as 63°, even though 126 ÷ 63 = 2. To show why, we have only to change to the Celsius scale, where the first temperature becomes $\frac{5}{9}(126 - 32) = 52.2°$, the second temperature becomes $\frac{5}{9}(63 - 32) = 17.2°$, and the first figure is now more than three times the second. This difficulty arises because the Fahrenheit and Celsius scales both have artificial origins (zeros); in other words, the number 0 of neither scale is indicative of the absence of whatever quantity we are trying to measure.

If we can also form quotients, we refer to the data as **ratio data**, and such data are not difficult to find. They include all the usual measurements (or determinations) of length, height, money amounts, weight, volume, area, pressure, elapsed time (though not calendar time), sound intensity, density, brightness, velocity, and so on.

The distinction we have made here between nominal, ordinal, interval, and ratio data is important, for as we shall see, the nature of a set of data may suggest the use of particular statistical techniques. To emphasize the point that what we can and cannot do arithmetically with a given set of data depends on the nature of the data, consider the following scores which four students obtained in the three parts of a comprehensive history test

	American history	European history	Ancient history
Linda	89	51	40
Tom	61	56	54
Henry	40	70	55
Rose	13	77	72

The totals for the four students are 180, 171, 165, and 162, so that Linda scored highest, followed by Tom, Henry, and Rose.

Suppose now that somebody proposes that we compare the overall performance of the four students by ranking their scores from high to low for each part of the test, and then average their ranks. What we get is shown in the following table:

	American history	European history	Ancient history	Average rank
Linda	1	4	4	3
Tom	2	3	3	$2\frac{2}{3}$
Henry	3	2	2	$2\frac{1}{3}$
Rose	4	1	1	2

Here Linda's average rank was calculated as $\frac{1 + 4 + 4}{3} = \frac{9}{3} = 3$, Tom's as $\frac{2 + 3 + 3}{3} = \frac{8}{3} = 2\frac{2}{3}$, and so forth.

Now, if we look at the average ranks, we find that Rose came out best, followed by Henry, Tom, and Linda, so that the order has been reversed from what it was before. How can this be? Well, strange things can happen when we average ranks. For instance, when it comes to their ranks, Linda's outscoring Tom by 28 points in American history counts just as much as Tom's outscoring her by 5 points in European history, and Tom's outscoring Henry by 21 points in American history counts just as much as Henry's outscoring him by a single point in ancient history. We conclude that, perhaps, we should not have averaged the ranks, but it might also be pointed out that, perhaps, we should not even have totaled the original scores. The variation of the American history scores, which go from 13 to 89, is much greater than that of the other two kinds of scores, and this strongly affects the total scores and suggests a possible shortcoming of the procedure. We shall not go into this here, as it has been our goal merely to alert the reader against the indiscriminate use of statistical techniques.

EXERCISES†

1.5 In five biology tests a student received grades of 46, 61, 74, 79, and 88. Which of the following conclusions can be obtained from these figures by purely descriptive methods and which require generalizations? Explain your answers.
   (a) Only two of the grades exceeded 75.
   (b) The student's grades increased from each test to the next.

(c) The student must have studied harder for each successive test.
(d) The difference between the highest and lowest grades is 42.

1.6 Mary and Jean are real estate salespersons. In the first three months of 1990 Mary sold 3, 6, and 2 one-family homes and Jean sold 4, 0, and 5 one-family homes. Which of the following conclusions can be obtained

---

† Exercise marked ★ pertain to optional material.

from these figures by purely descriptive methods and which require generalizations? Explain your answers.

(a) During the three months Mary sold more one-family homes than Jean.

(b) Mary is a better real estate salesperson than Jean.

(c) Mary sold at least two one-family homes during each of the three months.

(d) Jean probably took her annual vacation during the second month.

1.7 On three consecutive days, a traffic policeman issued 9, 14, and 10 speeding tickets, and 5, 10, and 12 tickets for going through red lights. Which of the following conclusions can be obtained from these data by purely descriptive methods and which require generalizations? Explain your answers.

(a) Altogether on these three days, the policeman issued more speeding tickets than tickets for going through red lights.

(b) On two of the three days, the policeman issued more speeding tickets than tickets for going through red lights.

(c) The policeman issued the smallest number of tickets on the first day because he was new on the job.

(d) This policeman will seldom give more than 15 speeding tickets on any one day.

1.8 The three lemons which a person bought at a supermarket weighed 7, 8, and 12 ounces. Which of the following conclusions can be obtained from these data by purely descriptive methods and which require generalizations? Explain your answers.

(a) The average weight of the three lemons is 9 ounces.

(b) The average weight of lemons sold at that supermarket is 9 ounces.

★ 1.9 Will we get nominal data or ordinal data if

(a) mechanics have to say whether changing the spark plugs on a new model car is very difficult, difficult, easy, or very easy;

(b) the religion of persons attempting suicide is coded 1, 2, 3, 4, or 5, representing Protestant, Catholic, Jewish, other, and none;

(c) consumers must say whether they prefer brand A to brand B, like them equally, or prefer brand B to brand A;

(d) consumers must say whether they prefer brand A to brand B, like them equally, prefer brand B to brand A, or have no opinion.

★ 1.10 Are the following nominal, ordinal, interval, or ratio data? Explain your answers.

(a) Social Security numbers.

(b) The number of passengers on buses from Los Angeles to San Diego.

(c) Vocational interest scores consisting of the total number of "yes" answers given to a set of questions, if it can be assumed that each "yes" answer represents the same increment of vocational interest.

(d) Military ranks.

★ 1.11 IQ scores are sometimes looked upon as interval data. What assumption would this entail about the differences in intelligence of three persons with IQ's of 95, 105, and 135? Is this assumption reasonable?

★ 1.12 On page 8 we indicated that data pertaining to calendar time (for instance, the years in which Army beat Navy in football) are not ratio data. Explain why. What kind of time measurements do constitute ratio data?

1.5

CHECKLIST OF KEY TERMS[†]
(with page references to their definitions)

Descriptive statistics, 5
Exploratory data analysis, 7
★ Interval data, 8
★ Nominal data, 7

★ Ordinal data, 8
Probability theory, 6
★ Ratio data, 8
Statistical inference, 6
Statistical model, 4

—————

[†] Terms marked ★ pertain to optional material.

1.6
REVIEW EXERCISES[†]

1.13 The paid attendance at a small college's home football games was 12,305, 10,984, 6,850, 11,733, and 10,641. Which of the following conclusions can be obtained from these figures by purely descriptive methods and which require generalizations? Explain your answers.

(a) The attendance at the third home game was so low because it rained.

(b) Among the five games, the paid attendance was highest at the first game.

(c) The paid attendance exceeded 11,000 at two of the five games.

(d) The paid attendance increased from the third home game to the fourth home game because the college's football team had been winning.

★ 1.14 Are the following nominal, ordinal, interval, or ratio data? Explain your answers.

(a) Elevations above sea level.

(b) Responses to the question whether (in the downtown area of a large city) living conditions are "getting much worse," "getting a little worse," "staying the same," "getting a little better," or "getting much better."

(c) Ages of secondhand cars.

(d) Responses on drivers' licenses with regard to eye color.

1.15 Explain why each of the following data may well fail to yield the desired information:

(a) To predict a municipal election, a public opinion poll questions persons walking in front of the municipal government office.

(b) To determine public sentiment about certain import restrictions, an interviewer asks voters: "Do you feel that American consumers should be denied the use of these products?"

★ 1.16 If students calculate their grade-point indexes (that is, average their grades) by counting A, B, C, D, and F as 4, 3, 2, 1, and 0, what does this assume about the nature of the grades?

1.17 Explain why each of the following data may well fail to yield the desired information:

(a) To see how the public feels about imports from India, selected persons are asked whether they like Indian art.

(b) To ascertain facts about their bathing habits, a sample of the citizens of a European country are asked how many times on the average they bathe per week.

1.18 Using the same model car, five drivers averaged 23.4, 22.5, 24.0, 23.4, and 22.7 miles per gallon. Which of the following conclusions can be obtained by purely descriptive methods and which require generalizations? Explain your answers.

(a) More often than any of the other figures, the drivers averaged 23.4 miles per gallon.

(b) More often than any of the other figures, drivers of this kind of car will average 23.4 miles per gallon.

(c) None of the averages differs from 23.5 by more than 1.0 mile.

(d) If the whole experiment is repeated, none of the drivers will average less than 22.5 or more than 24.5 miles per hour.

★ 1.19 In two major golf tournaments one professional golfer finished second and ninth, while another finished sixth and fifth. Comment on the argument that since 2 + 9 = 6 + 5, the overall performance of the two golfers in these two tournaments was equally good.

1.20 Rephrase the exercise referred to on page 4 so that it would be of special interest to

(a) a lawyer;

(b) a travel agent;

(c) an author.

[†] Review exercises marked ★ pertain to optional material.

1.7
REFERENCES

Brief and informal discussions of what statistics is and what statisticians do may be found in pamphlets titled *Careers in Statistics* and *Statistics as a Career: Women at Work*, which are published by the American Statistical Association. They may be obtained by writing to this organization at 1429 Duke Street, Alexandria, VA, 22314. Among the few books on the history of statistics, there is on the elementary level

WALKER H. M., *Studies in the History of Statistical Method*. Baltimore: The Williams & Wilkins Company, 1929.

and on the more advanced level

PEARSON E. S., and KENDALL, M. G., eds., *Studies in the History of Statistics and Probability*. New York: Hafner Press, 1970.

KENDALL, M. G., and PLACKETT, R. L., eds., *Studies in the History of Statistics and Probability, Vol. II*. New York: Macmillan Publishing Co., Inc., 1977.

STIGLER, S. M., *The History of Statistics*. Cambridge, Mass.: Harvard University Press, 1986.

A more detailed discussion of the nature of statistical data and the general problem of **scaling** (namely, the problem of constructing scales of measurement) may be found in

HILDEBRAND, D. K., LAING, J. D., and ROSENTHAL, H., *Analysis of Ordinal Data*. Beverly Hills, Calif.: Sage Publications, Inc., 1977.

REYNOLDS, H. T., *Analysis of Nominal Data*. Beverly Hills, Calif.: Sage Publications, Inc., 1977.

SIEGEL, S., *Nonparametric Statistics for the Behavioral Sciences*. New York: McGraw-Hill Book Company, 1956.

The following are some titles from the ever-growing list of books on statistics which are written for the layman.

BROOK, R. J., ARNOLD, G. C., HASSARD, T. H., and PRINGLE, R. M., eds., *The Fascination of Statistics*. New York: Marcel Dekker, Inc., 1986.

CAMPBELL, S. K., *Flaws and Fallacies in Statistical Thinking*. Englewood Cliffs, N.J.: Prentice-Hall, Inc., 1974.

HOLLANDER, M., and PROSCHAN, F., *The Statistical Exorcist: Dispelling Statistics Anxiety*. New York: Marcel Dekker, Inc., 1984.

HOOKE, R., *How to Tell the Liars from the Statisticians*. New York: Marcel Dekker, Inc., 1983.

KIMBLE, G. A., *How to Use (and Misuse) Statistics*. Englewood Cliffs, N. J.: Prentice-Hall, Inc., 1978.

LARSEN, R. J., and STROUP, D. F., *Statistics in the Real World.* New York: Macmillan Publishing Co., Inc., 1976.

RUNYON, R. P., *Winning with Statistics.* Reading, Mass.: Addison-Wesley Publishing Company, Inc., 1977.

TANUR, J. M. ed., *Statistics*: *A Guide to the Unknown.* San Francisco: Holden-Day, Inc., 1972.

SUMMARIZING DATA: FREQUENCY DISTRIBUTIONS

In recent years the collection of statistical data has grown at such a rate that it would be impossible to keep up with even a small part of the things which directly affect our lives unless this information is disseminated in "predigested" or summarized form. The whole matter of putting large masses of data into a usable form has always been important, but it has multiplied greatly in the last few decades. This has been due partly to the development of computers which now make it possible to accomplish in minutes what was previously left undone because it would have taken months or years, and partly to the deluge of data generated by the increasingly quantitative approach of the sciences, especially the behavioral and social sciences, where nearly every aspect of human life is nowadays measured in one way or another.

The most common method of summarizing data is to present them in condensed form in tables or charts, and at one time this took up the better part of an elementary course in statistics. Nowadays, there is so much else to learn in statistics that very little time is devoted to this kind of work. In a way this is unfortunate, because one does not have to look far in newspapers, magazines, and even professional journals to find unintentionally or intentionally misleading statistical charts.

Section 2.1 deals with the task of listing numeric values and presents a new technique, the stem-and-leaf display. Section 2.2 deals with frequency distributions, the standard method for grouping data. Section 2.3 shows some graphical methods.

2.1
LISTING NUMERIC VALUES

Organizing and presenting a set of numeric information is one of the first tasks in understanding a problem. As a typical situation, consider the values below, which represent the travel time to work of 100 employees in a large downtown office building. The times are given in minutes, and each value represents an employee's average time over five consecutive work days. The mere gathering together of this information is no small task, but it is clear that more must be done to make the numbers comprehensible.

44.0	35.4	28.4	37.0	46.0	35.4	19.4	20.4	56.4	43.2
36.2	38.4	49.2	31.8	86.4	12.6	27.4	14.0	39.4	39.4
15.8	28.8	38.0	44.0	38.4	74.0	23.0	11.4	39.8	30.2
29.2	40.6	49.6	30.4	12.2	123.8	42.0	47.0	32.4	39.2
35.2	56.4	31.0	45.0	90.2	100.0	39.0	37.0	49.4	28.2
12.6	27.0	47.8	52.6	41.0	40.0	28.0	23.6	37.6	37.8
30.0	45.8	18.0	41.0	22.6	24.2	89.6	90.4	43.0	29.8
56.2	24.8	12.6	53.6	125.4	16.2	39.0	40.8	33.6	39.4
45.6	37.4	18.0	50.6	103.4	52.4	20.2	64.6	22.2	60.0
42.2	42.0	16.2	108.2	44.0	42.6	39.4	37.6	41.4	40.4

What can be done to make this mass of information more usable? Some people find it interesting to locate the extreme values.

For this list, the smallest value is 11.4 minutes, and the largest is 125.4 minutes. It is occasionally useful to sort the data values into increasing or decreasing order. The list below gives these sorted values, and it is now possible to learn more about these travel times. For instance, we see now that there are many values near 40 minutes.

11.4	12.2	12.6	12.6	12.6	14.0	15.8	16.2	16.2	18.0
18.0	19.4	20.2	20.4	22.2	22.6	23.0	23.6	24.2	24.8
27.0	27.4	28.0	28.2	28.4	28.8	29.2	29.8	30.0	30.2
30.4	31.0	31.8	32.4	33.6	35.2	35.4	35.4	36.2	37.0
37.0	37.4	37.6	37.6	37.8	38.0	38.4	38.4	39.0	39.0
39.2	39.4	39.4	39.4	39.4	39.8	40.0	40.4	40.6	40.8
41.0	41.0	41.4	42.0	42.0	42.2	42.6	43.0	43.2	44.0
44.0	44.0	45.0	45.6	45.8	46.0	47.0	47.8	49.2	49.4
49.6	50.6	52.4	52.6	53.6	56.2	56.4	56.4	60.0	64.6
74.0	86.4	89.6	90.2	90.4	100.0	103.4	108.2	123.8	125.4

Sorting a large set of numbers into increasing or decreasing order is a surprisingly difficult task.

This list of numbers, even in sorted form, is still a ponderous piece of information. It will be helpful to have other ways of dealing with these values. A recently developed technique, the **stem-and-leaf** display, will give a good overall impression of the data.

To illustrate this technique, consider the following scores on a test of physical coordination given to 20 students who had consumed an amount of alcohol equal to 0.1% of their weight:

69	84	52	93	61	74	79	65	88	63
57	64	67	72	74	55	82	61	68	77

Now break each number into its tens and units digits, tallying together values which share the tens digit. That is, we will think of the number 69 as 6 | 9. The tens digits will then be aligned vertically with the units digits displayed to the side. For the set of 20 physical coordination scores, the picture is this:

```
5 | 2 7 5
6 | 9 1 5 3 4 7 1 8
7 | 4 9 2 4 7
8 | 4 8 2
9 | 3
```

The first row of the display, namely 5 | 2 7 5, tells us that the list contains values 52, 57, and 55. The second row tells us that the list contains eight values in the 60s.

This table is referred to as a **stem-and-leaf** display, since each row represents a **stem** position, and each digit to the right of the vertical line can be thought of as a **leaf**. To make this stem-and-leaf display, begin with the stems only, as

```
5 |
6 |
7 |
8 |
9 |
```

This step need not be perfect; after all, it's easy to stick on additional stem positions at the top or bottom. Then mark the leaves by going through the individual data items sequentially. After the first three values (69, 84, 52) the stem-and-leaf display will look like this:

```
5 | 2
6 | 9
7 |
8 | 4
9 |
```

After just one pass through the data, the complete stem-and-leaf display will be done.

The stem-and-leaf display conveys the same information as the original list, but it is much more compact. The stem-and-leaf display highlights the important aspects of the data. For instance, in this case it reveals immediately that most of the values are in the 60s.

The stem-and-leaf display accomplishes most of the task of sorting the values. Most people like to complete the sorting by ordering the leaves as well. In this example, it gives the following:

```
5 | 2 5 7
6 | 1 1 3 4 5 7 8 9
7 | 2 4 4 7 9
8 | 2 4 8
9 | 3
```

Not every set of values can be placed into a stem-and-leaf display this easily. The treatment of more complicated situations is a matter of taste. Let's consider the 100 travel times noted before. These numbers were given to *tenths* of minutes. In making the stem-and-leaf display, we recommend ignoring the tenths rather than rounding the values to the nearest minute. This introduces a half-minute bias to the display, but produces with less effort nearly the same result as proper rounding.[†] We will use the tens digit to label the stem, and this produces the display below:

```
 1 | 122224566889
 2 | 0022334477888899
 3 | 000112355567777788899999999
 4 | 00001112222334445556677999
 5 | 0223666
 6 | 04
 7 | 4
 8 | 69
 9 | 00
10 | 038
11 |
12 | 35
13 |
```

NOTE: 7 | 4   means   74 minutes

12 | 3   means   123 minutes

This stem-and-leaf display requires just looking through the original list once. The leaves can be sorted. Indeed, this picture conveys the message of the original list in a very clean pictorial form. It is helpful to attach a note as above to help people read the display. This is certainly helpful when the stem labels are not the tens digits.

There are various ways in which stem-and-leaf displays can be modified to meet particular needs. If we want to construct a stem-and-leaf display with more stems than there would be otherwise, we can divide each stem position in two. Use the first stem position to hold leaves 0, 1, 2, 3, and 4, and use the second stem position to hold leaves 5, 6, 7, 8, and 9. For the second set of data on page 16, we would thus get the **double-stem display:**

```
5· | 2
5* | 5  7
6· | 1  1  3  4
6* | 5  7  8  9
7· | 2  4  4
7* | 7  9
8· | 2  4
8* | 8
9· | 3
```

In this display, we doubled the number of stem positions by cutting in half the interval covered by each tens digit.

Also, for some further descriptions (see Section 3.4) it is convenient to have the leaves on each stem arranged according to size, as in Figure 2.1, where we used a computer to construct the double-stem display shown above.

[†] Dropping off digits has a number of advantages over the process of careful rounding, and it is the procedure recommended by John Tukey, the inventor of the stem-and-leaf display. His book appears in the reference list at the end of this chapter.

```
MTB > SET C1
DATA> 69  84    52    93    61    74    79    65    88    63
DATA> 57  64    67    72    74    55    82    61    68    77
MTB > STEM C1

  STEM-AND-LEAF DISPLAY OF C1
  LEAF DIGIT UNIT =    1.000
  1 2 REPRESENTS 12.


      1       5*  2
      3       5.  57
      7       6*  1134
     (4)      6.  5789
      9       7*  244
      6       7.  79
      4       8*  24
      2       8.  8
      1       9*  3
```

FIGURE 2.1 *Computer printout for the construction of a double-stem display.*

We shall not treat stem-and-leaf displays in any great detail, as it has been our objective mainly to present one of the relatively new techniques which come under the general heading of **exploratory data analysis**. These techniques are used primarily to explore data without, or before, using the more traditional methods of statistical analysis.

EXERCISES

2.1 The following are the heights in centimeters of sixteen high-school students: 172, 182, 177, 174, 166, 158, 170, 178, 163, 161, 191, 167, 171, 201, 166, 172. Construct a stem-and-leaf display with the stem labels 15, 16, 17, 18, 19, and 20.

2.2 The following are the weights in pounds of twenty applicants for jobs with a city's fire department: 225, 182, 194, 210, 205, 172, 181, 198, 164, 176, 180, 193, 178, 193, 208, 186, 183, 170, 186, 188. Construct a stem-and-leaf display with the stem labels 16, 17, 18, 19, 20, 21, and 22.

2.3 The following are the weekly earnings in dollars of fifteen salespersons: 425, 440, 610, 518, 324, 482, 624, 390, 468, 457, 509, 561, 482, 480, 520. Construct a stem-and-leaf display with the stem labels 3, 4, 5, and 6; the tens digits should be used as the leaves.

2.4 List the data values which are given in the stem-and-leaf display below.

```
4 | 0  2  3
5 | 1  1  8  9
6 | 2  3  3  7  7  9
7 | 0
```

2.5 Construct a double-stem display for the data of Exercise 2.2.

2.6 The following are the ages of thirty-two heads of household in a retirement community: 68, 81, 62, 61, 76, 65, 69, 73, 66, 68, 71, 74, 64, 70, 68, 73, 82, 79, 63, 69, 68, 66, 73, 74, 77, 80, 73, 66, 67, 81, 77, and 66. Construct a double-stem display (see page 18) for these values.

2.2

FREQUENCY DISTRIBUTIONS

When we deal with large sets of data, a good overall picture and all the information we need can often be conveyed by grouping the data into a number of classes, intervals, or categories. For instance, 1983 data on the size of cable television systems in the United States may be summarized as follows:

Number of subscribers	Cable television systems
Less than 1,000	2,444
1,000– 3,499	1,573
3,500– 9,999	958
10,000–19,999	369
20,000–49,999	239
50,000 or more	57
Total	5,640

A table like this is called a **frequency distribution** (or simply a **distribution**)—it shows how the cable television systems are distributed among the six classes. When data are thus grouped according to size, we refer to a table like the one above as a **numerical** (or **quantitative**) **distribution**.

In the cable-television example, each class covered a wide range of values, but there are also numerical distributions where each class covers only a single value. This is illustrated by the following example, based on a study in which 200 persons were asked how many times they had visited the local zoo during the last twelve months:

Number of visits to local zoo	Number of persons
0	90
1	72
2	26
3	8
4	3
5	0
6	1
Total	200

If data are grouped into nonnumerical categories, the resulting table is called a **categorical** (or **qualitative**) **distribution**. This kind of distribution is illustrated by the following table pertaining to 2,439 complaints about comfort-related characteristics of an airline's planes:

Nature of complaint	Number of complaints
Inadequate leg room	719
Uncomfortable seats	914
Narrow aisles	146
Insufficient carry-on facilities	218
Insufficient restrooms	58
Miscellaneous other complaints	384
Total	2,439

We could convert a distribution like this into a numerical distribution by coding the data, say, by assigning the six alternatives the numbers 1, 2, 3, 4, 5, and 6, but this would give us nominal data, which are numerical only in a trivial sense.

Frequency distributions present data in a relatively compact form, give a good overall picture, and contain adequate information for many purposes, but there are usually some things which cannot be determined without referring to the original data. For instance, from the first table of this section we can find neither the size of the smallest of the 5,640 cable television systems, nor the average number of subscribers for the ten largest systems. Similarly, from the third table we cannot tell how many of the complaints about uncomfortable seats were over the width of the seats, or how many of the complaints about insufficient carry-on facilities were over space for suit carriers. Nevertheless, frequency distributions present **raw** (unprocessed) **data** in a more readily usable form, and the price we have to pay for this—the loss of some information—is usually worthwhile.

The construction of a frequency distribution consists essentially of three steps: (1) choosing the **classes** (intervals or categories), (2) sorting or tallying the data into these classes, and (3) counting the number of items in each class. Since the second and third steps are purely mechanical, we shall concentrate here on the first, namely, that of choosing a suitable classification.

For numerical distributions, this consists of deciding how many classes we are going to use and from where to where each class should go. Both of these choices are essentially arbitrary, but the following rules are usually observed:

> **We seldom use fewer than 6 or more than 15 classes; the exact number we use in a given situation will depend largely on how many measurements or observations there are.**

Clearly, we would lose more than we gain if we group five observations into twelve classes with most of them empty, and we would probably discard too much information if we group a thousand measurements into three classes.

We always make sure that each item (measurement or observation) goes into one and only one class.

To this end we must make sure that the smallest and largest values fall within the classification, that none of the values can fall into a gap between successive classes, and that the classes do not overlap, namely, that successive classes have no values in common.

Whenever possible, we make the classes cover equal ranges of values.

Also, if we can, we make these ranges multiples of numbers that are easy to work with, such as 5, 10, or 100, since this will tend to facilitate the construction and the use of a distribution.

In connection with these rules, the cable-television distribution on page 15 satisfies the first two, but violates the third. Actually, the third rule is violated in several ways, since the classes 1,000–3,499, 3,500–9,999, 10,000–19,999, and 20,000–49,999 all cover different ranges of values, the first class has no specified lower limit, and the last class has no specified upper limit. Presumably, the government statisticians responsible for this table had good reasons for choosing the classes as they did.

Classes of the "less than," "or less," "more than," or "or more" variety are referred to as **open classes**, and they are used to reduce the number of classes that are needed when some of the values are much smaller than or much greater than the rest. Generally, open classes should be avoided, however, because they make it impossible to calculate certain values of interest, such as averages or totals (see Exercise 3.52 on page 65).

So far as the second rule is concerned, we have to watch whether the data are given to the nearest dollar or to the nearest cent, whether they are given to the nearest inch or to the nearest tenth of an inch, whether they are given to the nearest ounce or to the nearest hundreth of an ounce, and so on. For instance, if we want to group the weights of certain animals, we might use the first of the following classifications when the weights are given to the nearest kilogram, the second when the weights are given to the nearest tenth of a kilogram, and the third when the weights are given to the nearest hundreth of a kilogram:

Weight (kilograms)	Weight (kilograms)	Weight (kilograms)
10–14	10.0–14.9	10.00–14.99
15–19	15.0–19.9	15.00–19.99
20–24	20.0–24.9	20.00–24.99
25–29	25.0–29.9	25.00–29.99
30–34	30.0–34.9	30.00–34.99
etc.	etc.	etc.

To illustrate what we have been discussing in this section, let us now go through the actual steps of grouping a set of data into a frequency distribution.

EXAMPLE Construct a distribution of the following amounts of sulfur oxides (in tons) emitted by an industrial plant on 80 days:

15.8	26.4	17.3	11.2	23.9	24.8	18.7	13.9	9.0	13.2
22.7	9.8	6.2	14.7	17.5	26.1	12.8	28.6	17.6	23.7
26.8	22.7	18.0	20.5	11.0	20.9	15.5	19.4	16.7	10.7
19.1	15.2	22.9	26.6	20.4	21.4	19.2	21.6	16.9	19.0
18.5	23.0	24.6	20.1	16.2	18.0	7.7	13.5	23.5	14.5
14.4	29.6	19.4	17.0	20.8	24.3	22.5	24.6	18.4	18.1
8.3	21.9	12.3	22.3	13.3	11.8	19.3	20.0	25.7	31.8
25.9	10.5	15.9	27.5	18.1	17.9	9.4	24.1	20.1	28.5

Solution Since the smallest value is 6.2 and the largest value is 31.8, we might choose the six classes 5.0–9.9, 10.0–14.9,..., and 30.0–34.9, the seven classes 5.0–8.9, 9.0–12.9,..., and 29.0–32.9, the nine classes 5.0–7.9, 8.0–10.9,..., and 29.0–31.9, or numerous other classifications. Note that in each of the classifications cited, the classes accommodate all the data, they do not overlap, and they are all of the same size.

Essentially, the choice among these classifications is arbitrary, but supposing that for some legal reasons (say, government regulations) it is necessary to keep all values greater than 16.9 but less than 21.0 in one class, we choose the second one. Then, tallying the 80 measurements into the seven classes, we get the following table:

Tons of sulfur oxides	Tally	Frequency
5.0– 8.9	///	3
9.0–12.9	ﬀﬀ ﬀﬀ	10
13.0–16.9	ﬀﬀ ﬀﬀ ////	14
17.0–20.9	ﬀﬀ ﬀﬀ ﬀﬀ ﬀﬀ ﬀﬀ	25
21.0–24.9	ﬀﬀ ﬀﬀ ﬀﬀ //	17
25.0–28.9	ﬀﬀ ////	9
29.0–32.9	//	2
	Total	80

In the final presentation of such a table, the tally is usually omitted.

The numbers given in the right-hand column of the table above, which show how many items fall into each class, are called the **class frequencies**. The smallest and largest values that can go into any given class are called its **class limits**, and for the distribution of the emission data they are 5.0 and 8.9, 9.0 and 12.9, 13.0 and 16.9,..., and 29.0 and 32.9. More specifically, 5.0, 9.0, 13.0,..., and 29.0 are called the **lower class limits**, and 8.9, 12.9, 16.9,..., and 32.9 are called the **upper class limits**.

The amounts which we grouped in our example were all given to the nearest tenth of a ton, so that 5.0 actually includes everything from 4.95 to 5.05, 8.9 includes everything from 8.85 to 8.95, and the class 5.0–8.9 includes everything from 4.95 to 8.95. Similarly, the second class includes everything from 8.95 to 12.95,..., and the seventh class includes everything from 28.95 to 32.95. It is customary to refer to 4.95, 8.95, 12.95,..., and 32.95 as the **class boundaries** or the **real class limits**. Although 8.95 is the upper boundary of the first class and also the lower boundary of the second class,..., and 28.95 is the upper boundary of the sixth class and also the lower boundary of the seventh class, there is no cause for alarm. The class boundaries are, by their very nature, impossible values which cannot occur among the data being grouped. For instance, in the cable-television example on page 20, the class boundaries are the impossible values 999.5, 3,499.5, 9,999.5, 19,999.5, and 49,999.5. They are impossible values because a cable television system cannot very well have half a subscriber.

To avoid gaps in the continuous number scale, some statistics texts and some widely-used computer programs (*Minitab*, for example) include in each class its lower boundary. They would include 4.95, but not 8.95, in the first class of the distribution of the sulfur oxides emision data. Similarly, they would include 8.95, but not 12.95, in the second class, and so forth. All this is totally immaterial, of course, so long as the class boundaries are impossible values which cannot occur among the data being grouped. Especially for this reason, the use of impossible class boundaries cannot be overemphasized (see discussion of Figure 10.3 on page 266).

Numerical distributions also have what we call **class marks** and **class intervals**. Class marks are simply the midpoints of the classes, and they are found by adding the lower and upper limits of a class (or its lower and upper boundaries) and dividing by 2. A class interval is merely the length of a class, or the range of values it can contain, and it is given by the difference between its boundaries. If the classes of a distribution are all equal in length, their common class interval, which we call the **class interval of the distribution**, is also given by the difference between any two successive class marks.

EXAMPLE Find the class marks and the class interval of the distribution of the sulfur oxides emission data.

Solution The class marks are $\dfrac{5.0 + 8.9}{2} = 6.95$, $\dfrac{9.0 + 12.9}{2} = 10.95$, $\dfrac{13.0 + 16.9}{2} = 14.95$,

$\dfrac{17.0 + 20.9}{2} = 18.95$, $\dfrac{21.0 + 24.9}{2} = 22.95$, $\dfrac{25.0 + 28.9}{2} = 26.95$, and

$\dfrac{29.0 + 32.9}{2} = 30.95$. Therefore, since the class intervals are $8.95 - 4.95 = 4$,

12.95 − 8.95 = 4, . . . , and 32.95 − 28.95 = 4, and they are all equal, the class interval of the distribution is 4. Note that if we had taken the differences between the class limits instead of the differences between the class boundaries, we would have obtained 3.9 instead of 4, and 3.9 is not the class interval.

There are essentially two ways in which frequency distributions can be modified to suit particular needs. One way is to convert a distribution into a **percentage distribution** by dividing each class frequency by the total number of items grouped, and then multiplying by 100%.

EXAMPLE   Convert the distribution of the sulfur oxides emission data into a percentage distribution.

Solution   The first class contains $\frac{3}{80} \cdot 100\% = 3.75\%$ of the data, the second class contains $\frac{10}{80} \cdot 100\% = 12.50\%$ of the data, . . . , and the seventh class contains $\frac{2}{80} \cdot 100\% = 2.50\%$ of the data. These results are shown in the following table:

Tons of sulfur oxides	Percentage
5.0– 8.9	3.75
9.0–12.9	12.50
13.0–16.9	17.50
17.0–20.9	31.25
21.0–24.9	21.25
25.0–28.9	11.25
29.0–32.9	2.50
	100.00

Percentage distributions are often used when we want to compare two or more distributions; for instance, if we want to compare oxides emission of the plant considered in our example with that of a plant at a different location.

The other way of modifying a frequency distribution is to convert it into a "less than," "or less," "more than," or "or more" **cumulative distribution**. To construct a cumulative distribution, we simply add the class frequencies, starting either at the top or at the bottom of the distribution.

EXAMPLE   Convert the distribution on page 23 into a "less than" cumulative distribution.

Solution   Since none of the values is less than 5.0, 3 of the values are less than 9.0, 3 + 10 = 13 of the values are less than 13.0, 3 + 10 + 14 = 27 of the values are less than 17.0, and so on, we get the results shown in the following table:

Tons of sulfur oxides	Cumulative frequency
Less than 5.0	0
Less than 9.0	3
Less than 13.0	13
Less than 17.0	27
Less than 21.0	52
Less than 25.0	69
Less than 29.0	78
Less than 33.0	80

Note that instead of "less than 5.0," "less than 9.0," "less than 13.0," ..., we could have written "4.9 or less," "8.9 or less," "12.9 or less," ..., or "less than 4.95," "less than 8.95," "less than 12.95," ....

In the same way we can also convert a percentage distribution into **cumulative percentage distributions**. We simply add the percentages instead of the frequencies, starting either at the top or at the bottom of the distribution.

So far we have discussed only the construction of numerical distributions, but the general problem of constructing categorical (or qualitative) distributions is about the same. Here again we must decide how many categories (classes) to use and what kind of items each category is to contain, making sure that all the items are accommodated and that there are no ambiguities. Since the categories must often be chosen before any data are actually collected, it is usually prudent to include a category labeled "others" or "miscellaneous."

For categorical distributions, we do not have to worry about such mathematical details as class limits, class boundaries, and class marks. On the other hand, there is often a serious problem with ambiguities and we must be very careful and explicit in defining what each category is to contain. For instance, if we had to classify items sold at a supermarket into "meats," "frozen foods," "baked goods," and so forth, it would be difficult to decide, for example, where to put frozen beef pies. Similarly, if we had to classify occupations, it would be difficult to decide where to put a farm manager, if our table contained (without qualification) the two categories "farmers" and "managers." For this reason, it is advisable, where possible, to use standard categories developed by the Bureau of the Census and other government agencies. References to lists of such categories may be found in the book by P. M. Hauser and W. R. Leonard listed among the references on page 38.

EXERCISES

2.7 The weights of the members of a university's football team vary from 168 to 266 pounds. Indicate the limits of eleven classes into which these weights might be grouped.

2.8 Measurements of the boiling point of a flavoring extract, measured to the nearest tenth of a Celsius degree, vary from 148.2° to 160.6°. Indicate the limits of seven classes into which these measurements might be grouped.

2.9 The weekly earnings of the piecework employees of a furniture factory vary from $227.82 to $396.05. Indicate the limits of seven classes with an interval of $25 into which these values might be grouped.

2.10 The numbers of empty seats on bus runs from Philadelphia to Baltimore are grouped into a table with classes 0–4, 5–9, 10–14, 15–19, 20–24, and 25 or more. Will it be possible to determine exactly from this table the number of bus runs on which there were
   (a) at least 10 empty seats;
   (b) more than 10 empty seats;
   (c) more than 14 empty seats;
   (d) at least 14 empty seats;
   (e) exactly 9 empty seats?

2.11 The postmaster at a suburban post office has grouped the values of stamp purchases into a frequency distribution with the classes $0.00–4.99, $5.00–9.99, $10.00–14.99, $15.00–19.99, $20.00–24.99, $25.00–29.99, and $30.00 and over. Is it possible to determine from this distribution the proportion of purchases valued at
   (a) less than $10.00;
   (b) $10.00 or less;
   (c) more than $25.00;
   (d) $25.00 or more?

2.12 The following is the distribution of the weights of 125 mineral specimens collected on a field trip:

Weight in grams	Number of specimens
0.0– 19.9	16
20.0– 39.9	38
40.0– 59.9	35
60.0– 79.9	20
80.0– 99.9	11
100.0–119.9	4
120.0–139.9	1
Total	125

If possible, find how many of the specimens weigh
   (a) at most 59.9 grams;
   (b) more than 59.9 grams;
   (c) more than 80.0 grams;
   (d) 80.0 grams or less;
   (e) exactly 70.0 grams;
   (f) anywhere from 60.0 grams to 100.0 grams.

2.13 The numbers of nurses on duty each day at a hospital are grouped into a distribution having the classes 15–29, 30–44, 45–59, 60–74, and 75–89. Find
   (a) the class limits;
   (b) the class boundaries;
   (c) the class marks;
   (d) the class interval of the distribution.

2.14 The declared values of packages mailed from Great Britain to the United States are grouped into a distribution with classes $0.00–49.99, $50.00–99.99, $100.00–149.99, $150.00–199.99, $200.00–249.99, $250.00–299.99, and $300.00 and over. Find
   (a) the class limits;
   (b) the class boundaries;
   (c) the class marks;
   (d) the class interval of the distribution.

2.15 The class marks of a distribution of the daily number of burglaries reported by a police precinct are 4, 13, 22, 31, and 40. If the class intervals are all equal, find
   (a) the class boundaries;
   (b) the class limits.

2.16 The class marks of a distribution of the daily number of VCRs (video cassette recorders) repaired by a manufacturer's regional service center are 6, 19, 32, and 45. If the class intervals are all equal, find
   (a) the class boundaries;
   (b) the class limits.

2.17 To group data on the number of rainy days reported by a weather station for the month of August during the last sixty years, a meteorologist uses the classes 0–5, 6–11, 12–16, 18–24, and 24–30. Explain where difficulties might arise.

2.18 To group sales receipts ranging from $10.00 to $60.00, a clerk uses the following classifications: $10.00–19.99, $20.00–35.99, $35.00–49.90, and $50.00–59.99. Explain where difficulties might arise.

2.19 The following are the grades which 40 students obtained in a psychology test:

75	89	66	52	90	68	83	94	77	60
38	47	87	65	97	49	65	72	73	81
63	77	31	88	74	37	85	76	74	63
69	72	91	87	76	58	63	70	72	65

Group these grades into a distribution having the classes 20–29, 30–39, 40–49, 50–59, 60–69, 70–79, 80–89, and 90–99.

2.20 Convert the distribution obtained in the preceding exercise into a percentage distribution.

2.21 Convert the distribution obtained in Exercise 2.19 into a "less than" cumulative distribution, beginning with "less than 20."

2.22 The following are the body weights (in grams) of 50 rats used in a study of vitamin deficiencies:

136	92	115	118	121	137	132	120	104
125	119	115	101	129	87	108	110	133
135	126	127	103	110	126	118	82	104
137	120	95	146	126	119	119	105	132
126	118	100	113	106	125	117	102	146
129	124	113	95	148				

Group these weights into a distribution having the classes 80–89, 90–99, 100–109, . . . , and 140–149.

2.23 Convert the distribution obtained in the preceding exercise into a percentage distribution.

2.24 Convert the distribution obtained in Exercise 2.22 into an "or more" cumulative distribution, beginning with "80 or more."

2.25 The following are the numbers of customers a restaurant served for lunch on 120 weekdays:

50	64	55	51	60	41	71	53	63	64
46	59	66	45	61	57	65	62	58	65
55	61	50	55	53	57	58	66	53	56
64	46	59	49	64	60	58	64	42	47
59	62	56	63	61	68	57	51	61	51
60	59	67	52	52	58	64	43	60	62
48	62	56	63	55	73	60	69	53	66
54	52	56	59	65	60	61	59	63	56
62	56	62	57	57	52	63	48	58	64
59	43	67	52	58	47	63	53	54	67
57	61	76	78	60	66	63	58	60	55
61	59	74	62	49	63	65	55	61	54

Group these figures into a distribution having the classes 40–44, 45–49, 50–54, 55–59, 60–64, 65–69, 70–74, and 75–79.

2.26 Convert the distribution obtained in the preceding exercise into a
(a) percentage distribution:
(b) "less than" cumulative percentage distribution.

2.27 The following are the miles per gallon obtained with 40 tanksful of gas:

24.5	23.6	24.1	25.0	22.9	24.7	23.8	25.2
23.7	24.4	24.7	23.9	25.1	24.6	23.3	24.3
24.6	23.9	24.1	24.4	24.5	25.7	23.6	24.0
23.9	24.2	24.7	24.9	25.0	24.8	24.5	23.4
24.9	24.8	24.7	24.1	22.8	23.1	25.3	24.6

Group these figures into a distribution having the classes 22.5–22.9, 23.0–23.4, 23.5–23.9, 24.0–24.4, 24.5–24.9, 25.0–25.4, and 25.5–25.9.

2.28 Convert the distribution obtained in the preceding exercise into a
(a) "more than" cumulative distribution;
(b) "more than" cumulative percentage distribution.

2.29 On a certain freeway there were 1, 0, 3, 2, 3, 5, 3, 0, 2, 7, 0, 4, 3, 1, 3, 1, 5, 3, 4, 4, 2, 1, 3, 1, 2, 2, 1, 0, 2, 0, 0, 1, 1, 2, 4, 5, 3, 4, 3, 4, 3, 3, 5, 2, 1, 6, 1, 2, 4, and 6 traffic accidents during fifty afternoon rush hours. Construct a distribution showing on how many afternoons there were 0, 1, 2, 3, 4, 5, 6, or 7 accidents.

2.30 An audit of 60 sales invoices revealed 0, 0, 2, 0, 2, 1, 0, 1, 3, 1, 2, 0, 1, 1, 0, 1, 0, 3, 1, 4, 1, 1, 0, 0, 0, 0, 1, 0, 3, 2, 0, 1, 0, 0, 1, 2, 2, 1, 0, 0, 3, 1, 0, 0, 2, 0, 2, 1, 1, 2, 0, 4, 2, 0, 1, 0, 0, 1, 0, and 1 mistakes in quoting prices. Construct a distribution showing on how many of the invoices there were 0, 1, 2, 3, or 4 mistakes.

2.31 Convert the distribution obtained in the previous exercise into a cumulative "less than" distribution.

2.32 A survey taken at a New York hotel indicated that 40 guests arrived by the following means of transportation: car, car, bus, plane, train, bus, bus, plane, car, plane, plane, bus, plane, car, car, train, train, car, car, car, car, plane, plane, car, bus, car, bus, car, plane, car, plane, plane, car, car, car, bus, train, car, bus, and car. Construct a categorical distribution showing the frequencies corresponding to the different means of transportation.

2.33 In a categorical distribution, women's dresses are classified according to whether they are made of wool, cotton, silk, or synthetic fibers. Explain where difficulties might arise.

2.3

GRAPHICAL PRESENTATIONS

When frequency distributions are constructed primarily to condense large sets of data and display them in an "easy to digest" form, it is usually best to present them graphically. A picture speaks louder than a thousand words, and this was true even before the current popularity of computer graphics, where each software package strives to outdo its competitors by means of elaborate pictorial presentations of statistical data.

For frequency distributions, the most common form of graphical presentation is the **histogram**, like the one shown in Figure 2.2. Histograms are constructed by representing the measurements or observations that are grouped (in Figure 2.2 the sulfur oxides emission data) on a horizontal scale, the class frequencies on a vertical scale, and drawing rectangles whose bases equal the class intervals and whose heights are the corresponding class frequencies. (There is nothing sacred about this arrangement and, if convenient, the scale of measurement and the class frequencies may be represented as in the computer printout of Figure 2.3.)

The markings on the horizontal scale of a histogram can be the class limits as in Figure 2.2, the class boundaries, the class marks, or arbitrary key values. For easy readability, it is usually preferable to show the class limits, although the rectangles actually go from one class boundary to the next. Note that histograms cannot be drawn for distributions with open classes, and that they require special care when the class intervals are not all equal (see Exercise 2.39 on page 34).

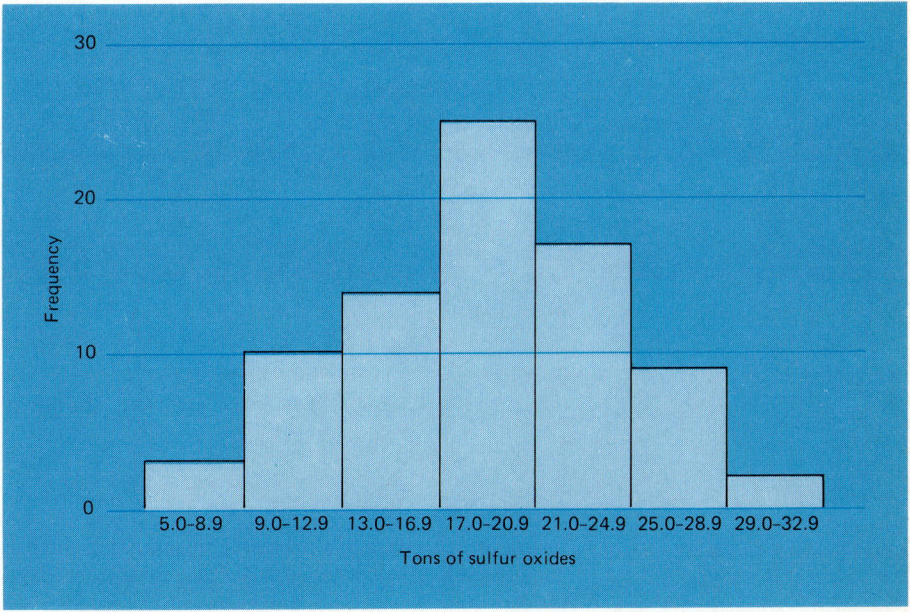

FIGURE 2.2 *Histogram of the distribution of the sulfur oxides emission data.*

```
MTB > SET C1
DATA> 15.8    26.4    17.3    11.2    23.9    24.8    18.7    13.9     9.0    13.2
DATA> 22.7     9.8     6.2    14.7    17.5    26.1    12.8    28.6    17.6    23.7
DATA> 26.8    22.7    18.0    20.5    11.0    20.9    15.5    19.4    16.7    10.7
DATA> 19.1    15.2    22.9    26.6    20.4    21.4    19.2    21.6    16.9    19.0
DATA> 18.5    23.0    24.6    20.1    16.2    18.0     7.7    13.5    23.5    14.5
DATA> 14.4    29.6    19.4    17.0    20.8    24.3    22.5    24.6    18.4    18.1
DATA>  8.3    21.9    12.3    22.3    13.3    11.8    19.3    20.0    25.7    31.8
DATA> 25.9    10.5    15.9    27.5    18.1    17.9     9.4    24.1    20.1    28.5
MTB > HIST C1 6.95 4.0

 C1

   MIDDLE OF     NUMBER OF
   INTERVAL      OBSERVATIONS
     6.95          3      ***
    10.95         10      **********
    14.95         14      **************
    18.95         25      *************************
    22.95         17      *****************
    26.95          9      *********
    30.95          2      **
```

FIGURE 2.3 *Computer printout for the construction of the histogram of the sulfur oxides emission data.*

The data which led to Figure 2.2 were easy to group because there were only 80 values in the sample. For really large sets of data, it may be convenient to construct histograms directly from raw data by using a suitable computer package. For instance, Figure 2.3 shows a computer-generated histogram of the sulfur oxides emission data.[†] As can be seen from the command "HIST C1 6.95 4.0," the first class mark is 6.95 and the class interval is 4, so that the classes are as in the table on page 23. Compared to Figure 2.2, the diagram is on its side, and it is not really a histogram in accordance with the definition which we gave above. However, it combines some of the features of Figures 2.4 and 2.7 and conveys the same idea.

Similar to histograms are **bar charts**, like the one shown in Figure 2.4. The heights of the rectangles, or bars, represent the class frequencies as in a histogram, but there is no pretense of having a continuous horizontal scale.

Another, less widely used form of graphical presentation is the **frequency polygon**, like the one shown in Figure 2.5. Here the class frequencies are plotted at the class marks and the successive points are connected by straight lines. Note that

---

[†] In the printout of Figure 2.3 and in others appearing in this text, words and numbers appear which relate to the technical aspects of operating the particular computer program employed. If a computer is available, the reader should refer to the appropriate manuals for operating instructions and for a list of problems which can be solved with existing programs. Trained users can create additional programs as needed.

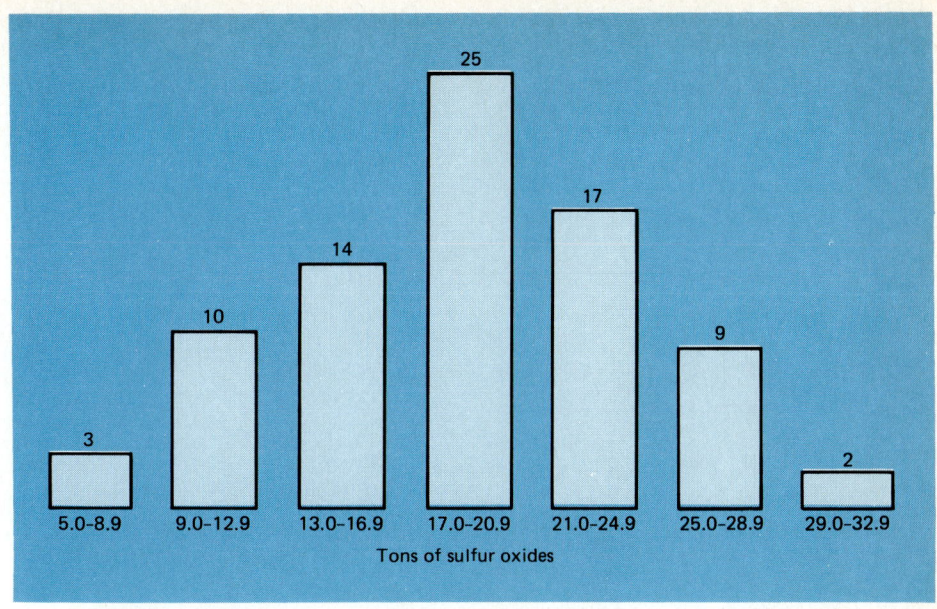

FIGURE 2.4 *Bar chart of the distribution of the sulfur oxides emission data.*

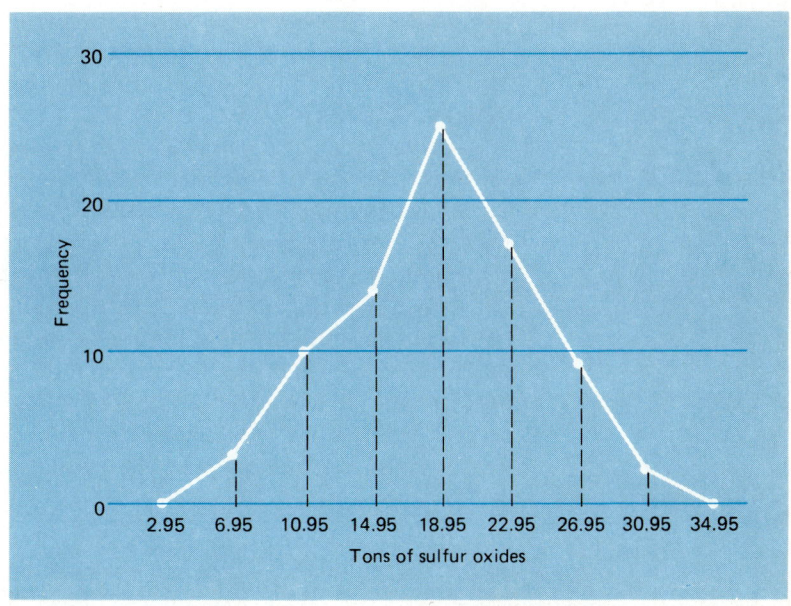

FIGURE 2.5 *Frequency polygon of the distribution of the sulfur oxides emission data.*

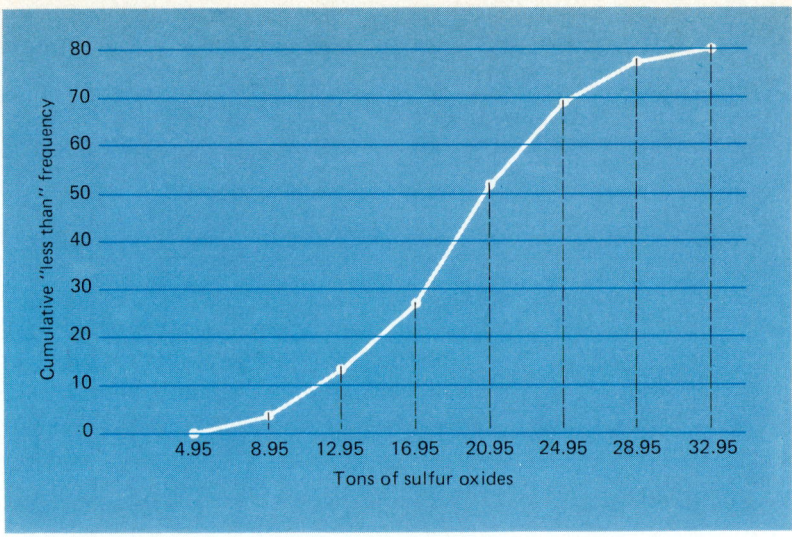

FIGURE 2.6 *Ogive of the cumulative distribution of the sulfur oxides emission data.*

we added classes with zero frequencies at both ends of the distribution to "tie down" the graph to the horizontal scale. If we apply a similar technique to a cumulative distribution, we obtain what is called an **ogive**. However, in an ogive the cumulative frequencies are plotted at the class boundaries instead of the class marks—it stands to reason that the cumulative frequency corresponding to, say, "less than 13.0" should be plotted at the class boundary 12.95, since "less than 13.0" actually includes everything up to 12.95. Figure 2.6 shows an ogive of the "less than" cumulative distribution obtained on page 26 for the sulfur oxides emission data.

Although the visual appeal of histograms, bar charts, frequency polygons, and ogives is a marked improvement over that of mere tables, there are various ways in which distributions can be presented even more dramatically and often more effectively. An example of such pictorial presentations (often seen in newpapers, magazines, and reports of various sorts) is the **pictogram** shown in Figure 2.7.

Categorical (or qualitative) distributions are often presented graphically as **pie charts**, like the one shown in Figure 2.8, where a circle is divided into sectors—pie-shaped pieces—which are proportional in size to the corresponding frequencies or percentages. To construct a pie chart, we first convert the distribution into a percentage distribution. Then since a complete circle corresponds to 360 degrees, we obtain the central angles of the various sectors by multiplying the percentages by 3.6.

Many computers are preprogrammed so that, once the data have been entered, a simple command will produce a pie chart, or variations thereof. Some computer-generated pie charts use color, some are three dimensional, some cut out sectors (like pieces of pie) for emphasis, and some shade or tint the different sectors.

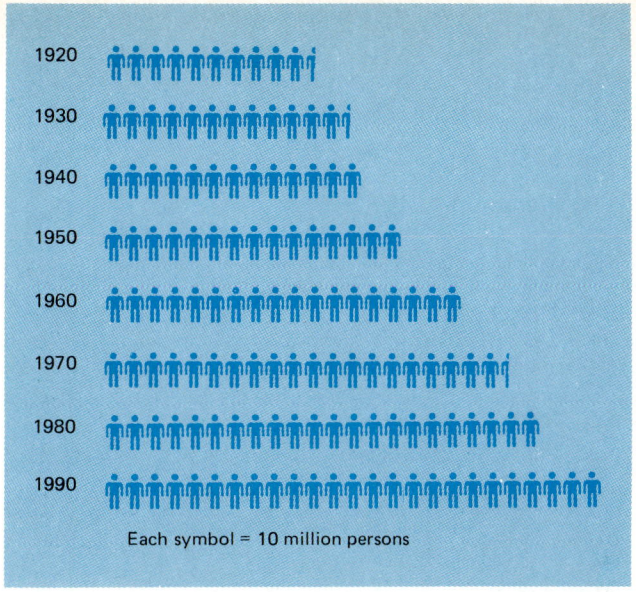

FIGURE 2.7 *Pictogram of the population of the United States.*

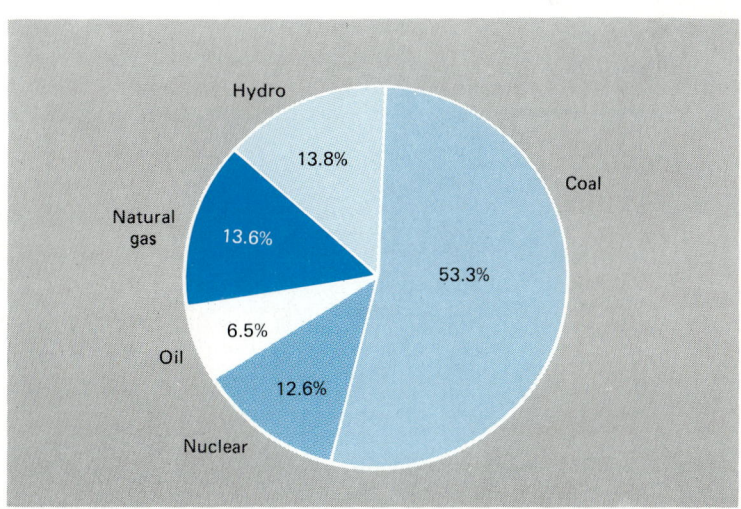

FIGURE 2.8 *Electric energy production of the United States in 1982.*

Research on ability to read graphical displays has suggested that untrained users have trouble interpreting pie graphs. Comparing information from two adjacent pie charts is particularly difficult. The comparison of bar heights in histograms is rather easy. It is recommended that pie charts be used only rarely and with great care.

EXERCISES

2.34 The following is the distribution of the monthly bills for 200 charge accounts at a department store:

Amount in dollars	Frequency
0.00– 19.99	22
20.00– 39.99	47
40.00– 59.99	66
60.00– 79.99	35
80.00– 99.99	21
100.00–119.99	9

(a) Draw a histogram of this distribution.
(b) Draw a bar chart of this distribution.

2.35 Convert the distribution of the preceding exercise into a cumulative "less than" distribution and draw an ogive.

2.36 The following is the distribution of the weights of the 150 freshmen women entering a certain college:

Weight in pounds	Frequency
90– 99	6
100–109	25
110–119	46
120–129	37
130–139	22
140–149	7
150–159	3
160–169	3
170–179	0
180–189	1

(a) Draw a histogram of this distribution.
(b) Draw a frequency polygon of this distribution.

2.37 Convert the distribution of the preceding exercise into a cumulative "less than" distribution and draw an ogive.

2.38 The following table shows how workers in Denver, Colorado, get to work:

Means of Transportation	Percentage
Ride alone	82
Car pool	13
Ride bus	2
Varies or work at home	3

Construct a pie chart of this percentage distribution.

2.39 Figure 2.9 shows the distribution of the scores of 80 entering college freshmen on a foreign language placement examination. Explain why it might easily give a misleading impression and indicate how it might be improved.

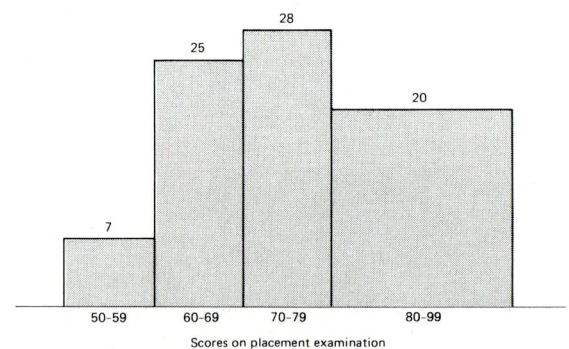

FIGURE 2.9 *Distribution of scores on foreign language placement examination.*

2.40 The pictogram of Figure 2.10 is intended to illustrate that in a certain region average family income has doubled from $7,000 in 1970 to $14,000 in 1982. Explain why this pictogram conveys a misleading impression and indicate how it might be modified.

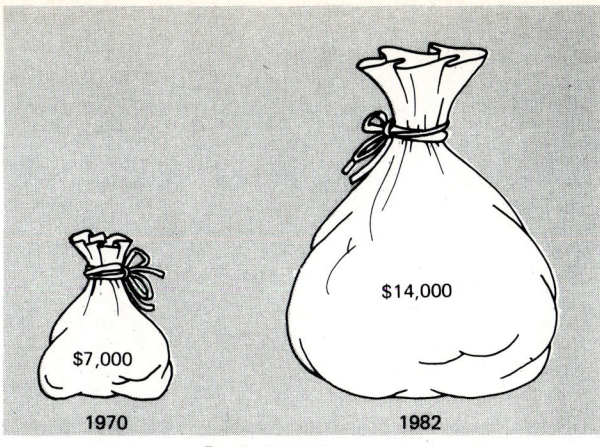

$7,000   1970

$14,000   1982

Family income

FIGURE 2.10 *Pictogram for Exercise 2.40*

2.41 Construct a pie chart of the following distribution, which shows the numbers of motor vehicle registrations at a state office:

Type of vehicle	Number
Passenger car	248
Minivan	62
Two-axle truck	42
Multi-axle truck	12
Motorcycle	55
Motorboat	9
Total	

2.42 Here, again, are the grades (from Exercise 2.19) which 40 students obtained in a psychology test:

75	89	66	52	90	68	83	94	77	60
38	47	87	65	97	49	65	72	73	81
63	77	31	88	74	37	85	76	74	63
69	72	91	87	76	58	63	70	72	65

Use a computer package to construct a histogram with the classes 20–29, 30–39, 40–49, 50–59, 60–69, 70–79, 80–89, and 90–99. Also construct a histogram with the classes 20–39, 40–59, 60–79, and 80–99. Which histogram do you prefer?

2.43 Use a computer package to construct a histogram with the classes 40–44, 45–49, 50–54, 55–59, 60–64, 65–69, 70–74, and 75–79 for the number of lunch customers given in Exercise 2.25.

2.44 Use a computer package to construct a histogram with the classes 22.5–22.9, 23.0–23.4, 23.5–23.9, 24.0–24.4, 24.5–24.9, 25.0–25.4, and 25.5–25.9 for the miles-per-gallon data of Exercise 2.27 on page 28.

2.4
CHECKLIST OF KEY TERMS[†]
(with page references to their definitions)

Bar chart, 30
Categorical distribution, 20
Class, 22
Class boundary, 24

Class frequency, 23
Class interval, 24
Class limit, 23
Class mark, 24

[†] Terms introduced in exercises are given in italics.

2.5
REVIEW EXERCISES

2.45 The class marks of a distribution of the daily number of calls received by a small cab company are 18, 25, 32, 39, 46, and 53. If the class intervals are all equal, what are the class limits?

2.46 The following are the numbers of deer observed in 72 sections of land in a wildlife count:

18	8	9	22	12	16	20	33	15	21	18	13
13	19	0	2	14	17	11	18	16	13	12	6
8	12	13	21	8	11	19	1	14	4	19	16
2	16	11	18	10	28	15	24	8	20	6	7
21	0	16	12	20	17	13	20	10	16	5	10
15	10	16	14	29	17	4	18	21	10	16	9

Group these data into a distribution having the classes 0–4, 5–9, 10–14, 15–19, 20–24, 25–29, and 30–34.

2.47 Draw a histogram of the distribution obtained in the preceding exercise.

2.48 Convert the distribution of Exercise 2.46 into an "or less" cumulative distribution and draw an ogive.

2.49 The ages of a company's employees are to be grouped into the following classes: under 19, 20–24, 25–29, 30–34, 34–39, and over 39. Explain where difficulties might arise.

2.50 The daily number of persons attending an art exhibit are grouped into a distribution with the classes 0–39, 40–79, 80–119, and 120–159. Is it possible to determine from this distribution on how many days
  (a) at least 79 persons attended the exhibit;
  (b) more than 79 persons attended the exhibit;
  (c) 40 or more persons attended the exhibit;
  (d) at most 120 persons attended the exhibit?

2.51 The following is the distribution of the number of meals which 60 real estate salespersons charged as business expenses in a given week:

Number of meals	Frequency
0–1	16
2–3	25
4–5	13
6–7	4
8–9	2

Find
  (a) the class marks;
  (b) the class boundaries;
  (c) the class interval of the distribution.

2.52 Convert the distribution of the preceding exercise into an "or more" cumulative percentage distribution and draw a bar chart.

2.53 In 1982, the 1,572 symphony orchestras in the United States included 385 college orchestras, 919 community orchestas, 94 urban orchestras, 110 metropolitan orchestras, and 64 others. Present this information in the form of a
  (a) bar chart;
  (b) pic chart.

2.54 In 1986, the annual salaries paid to teachers in a certain school district varied from $18,400 to $32,600. Indicate the limits of six classes, each with an interval of $2,500, into which these salaries might be grouped.

★ **2.55** Among histograms, bar charts, and pie charts, which ones can be used to represent
  (a) nominal data;
  (b) ordinal data;
  (c) interval data?

2.56 Measurements of the lengths of fish to the nearest tenth of an inch are grouped into a table whose classes have the boundaries 5.95, 7.95, 9.95, 11.95, 13.95, and 15.95. What are the lower and upper limits of each class?

2.57 Asked whether they ever attend meetings of their town council, fifty residents of a community replied: never, occasionally, rarely, rarely, never, rarely, occasionally, frequently, never, rarely, rarely, rarely, occasionally, rarely, occasionally, never, never, rarely, frequently, never, never, rarely, occasionally, occasionally, rarely, rarely, never, never, rarely, rarely, frequently, occasionally, occasionally, never, rarely, never, rarely, rarely, occasionally, rarely, never, never, rarely, occasionally, never, rarely, rarely, occasionally, rarely, and never. Construct a categorical distribution and draw a pie chart.

2.58 List the data which correspond to the following stems of stem-and-leaf displays:
  (a) 125 | 3 0 4 8 7 6 6 5
  (b) 34 | 67 05 19 48
  (c) 1∗ | 1 0
     1· | 8 6 7 7
     2∗ | 2 4 0 1 1 3
     2· | 9 6 7 7
     3∗ | 3 1 2
     3· | 8 5
     4∗ | 2

2.59 The following are the numbers of false alarms (tripped accidentally or by equipment malfunction) which a security monitoring service received on thirty days: 3, 6, 2, 4, 5, 8, 2, 5, 6, 3, 4, 7, 4, 6, 5, 5, 5, 4, 3, 7, 4, 4, 6, 3, 9, 5, 7, 4, 4, and 6. Construct a frequency distribution.

2.60 Construct a histogram of the distribution obtained in the preceding exercise.

2.61 The following are the systolic blood pressures of twenty hospital patients: 165, 135, 151, 153, 155, 182, 142, 158, 146, 149, 124, 162, 173, 204, 159, 130, 177, 162, 141, and 156. Construct a stem-and-leaf display with one-digit leaves.

2.62 In a survey, persons are asked whether they
  (1) finished high school;
  (2) finished college;
  (3) have an advanced degree.
Explain where difficulties might arise.

2.6
REFERENCES

Detailed information about statistical charts may be found in

CLEVELAND, W. S., *The Elements of Graphing Data.* Monterey, Calif.: Wadsworth Advanced Books and Software, 1985.
SCHMID, C. F., *Statistical Graphics: Design Principles and Practices.* New York: John Wiley & Sons, Inc., 1983.

TUFTE, E. R., *The Visual Display of Quantitative Information*, Chesshire, Conn.: Graphics Press, 1985.

and some interesting information about the history of the graphical presentation of statistical data is given in an article by E. Royston in

PEARSON, E. S. and KENDALL, M. G., eds., *Studies in the History of Statistics and Probability*. New York: Hafner Press, 1970.

Discussions of what not to do in the presentation of statistical data may be found in

CAMPBELL, S. K., *Flaws and Fallacies in Statistical Thinking*. Englewood Cliffs, N.J.: Prentice-Hall, Inc., 1974.

HUFF, D., *How to Lie with Statistics*. New York: W. W. Norton & Company, Inc., 1954.

REICHMAN, W. J., *Use and Abuse of Statistics*. New York: Penguin Books, 1971.

Useful references to lists of standard categories are given in

HAUSER, P. M., and LEONARD, W. R., *Government Statistics for Business Use, 2nd ed.* New York: John Wiley & Sons, Inc., 1956.

For further information about exploratory data analysis and stem-and-leaf displays in particular, see

HARTWIG, F., and DEARING, B. E., *Exploratory Data Analysis*. Beverly Hills, Calif.: Sage Publications, Inc., 1979.

HOAGLIN, D. C., MOSTELLER, F., and TUKEY, J. W., *Understanding Robust and Exploratory Data Analysis*. New York: John Wiley & Sons, Inc., 1983.

KOOPMANS, L. H., *An Introduction to Contemporary Statistics*. Boston: Duxbury Press, 1981.

TUKEY, J. W., *Exploratory Data Analysis*. Reading, Mass.: Addison-Wesley Publishing Company, Inc., 1977.

VELLEMAN, P. F., and HOAGLIN, D. C., *Applications, Basics, and Computing for Exploratory Data Analysis*. North Scituate, Mass.: Duxbury Press, 1980.

3 SUMMARIZING DATA: MEASURES OF LOCATION

When we describe a set of data, we try to say neither too little nor too much. Statistical descriptions can be brief or elaborate, depending on the purposes they are to serve. Sometimes we present data in raw form and let them speak for themselves. On other occasions we present data as frequency distributions or as graphs. Most of the time, however, we must describe data by one or two carefully chosen numbers.

It is often necessary to summarize data by means of a single number which, in its way, is descriptive of the entire set. Exactly what sort of number we choose depends on the particular characteristic we want to describe. In one study we may be interested in the value which is exceeded by only 25 percent of the data; in another, in the value which exceeds the lowest 10 percent of the data; and in still another, in a value which somehow describes the center or middle of the data. The statistical measures which describe such characteristics are called **measures of location**; among them, the ones which describe the center or middle of the data are called **measures of central location**.

In Sections 3.2, 3.3, 3.4, and 3.6 we present four of the most widely used measures of central location; some measures of location other than central location and the description of grouped data are discussed in Sections 3.5 and 3.7, which are optional.

3.1

POPULATIONS AND SAMPLES

Before we study particular statistical descriptions, let us make the following distinction:

> **If a set of data consists of all conceivably possible (or hypothetically possible) observations of a given phenomenon, we call it a population; if a set of data consists of only a part of these observations, we call it a sample.**

Here we added the phrase "hypothetically possible" to take care of such clearly hypothetical situations as where we look at the outcomes (heads or tails) of 12 flips of a coin as a sample from the potentially unlimited number of flips of the coin, where we look at the weights of ten 30-day-old lambs as a sample of the weights of all (past, present, and future) 30-day-old lambs raised at a certain farm, or where we look at four determinations of the uranium content of an ore as a sample of the many determinations that could conceivably be made. In fact, we often look at the results of an experiment as a sample of what we might get if the experiment were repeated over and over again.

Originally, statistics dealt with the description of human populations, census counts and the like (see page 5), but as it grew in scope, the term "population" took on the much wider connotation given to it above. Whether or not it sounds strange to refer to the heights of all the trees in a forest or the speeds of all the cars passing a checkpoint as populations is beside the point—in statistics, "population" is a technical term with a meaning of its own.

Although we are free to call any group of items a population, what we do in practice depends on the context in which the items are to be viewed. Suppose, for instance, that we are offered a lot of 400 ceramic tiles, which we may or may not buy depending on their strength. If we measure the breaking strength of 20 of these tiles in order to estimate the average breaking strength of all the tiles, these 20 measurements are a sample from the population which consists of the breaking strengths of the 400 tiles. In another context, however, if we consider entering into a long-term contract calling for the delivery of tens of thousands of such tiles, we would look upon the breaking strengths of the original 400 tiles only as a sample. Similarly, the complete figures for a recent year, giving the elapsed times between the filing and disposition of divorce suits in San Diego County, can be looked upon as either a population or a sample. If we are interested only in San Diego County and that particular year, we would look upon the data as a population; on the other hand, if we want to generalize about the time that is required for the disposition of divorce suits in the entire United States, in some other county, or in some other year, we would look upon the data as a sample.

As we have used it here, the word "sample" has very much the same meaning as it has in everyday language. A newspaper considers the attitudes of 150 readers toward a proposed school bond to be a sample of the attitudes of all its readers toward the bond. A consumer considers a box of Mrs. See's candy a sample of the firm's product. Later, we shall use the word "sample" only when referring to data which can reasonably serve as the basis for valid generalizations about the populations from which they came; in this more technical sense, many sets of data which are popularly called samples are not samples at all.

In this chapter and in Chapter 4 we shall describe things statistically without making any generalizations. For future reference, though, it is important even here to distinguish between populations and samples. Thus, we shall use different symbols depending on whether we are describing populations or samples.

3.2
THE MEAN

The most popular measure of central location is what the layman calls an "average" and what the statistician calls an **arithmetic mean**, or simply a **mean**.[†] It is defined as follows:

The mean of *n* numbers is their sum divided by *n*.

———————

[†] The term "arithmetic mean" is used mainly to distinguish the mean from the **geometric mean** and the **harmonic mean**, two other kinds of averages used only in very special situations (see Exercises 3.17 and 3.18).

It is alright to use the word "average" and on occasion we shall use it ourselves, but there are other kinds of averages in statistics and we cannot afford to speak loosely when there is any danger of ambiguity.

EXAMPLE   For the 12 months of 1990, a police department reported 4, 3, 5, 5, 10, 8, 9, 6, 3, 4, 8, and 7 armed robberies. Find the mean, namely, the average number of armed robberies per month.

Solution   The total for the 12 months is $4 + 3 + 5 + 5 + 10 + 8 + 9 + 6 + 3 + 4 + 8 + 7 = 72$, and therefore

$$\text{mean} = \frac{72}{12} = 6$$

EXAMPLE   A supermarket manager, who wants to study the "traffic" in her store, finds that 295, 1002, 941, 768, and 1283 persons entered the store during the past five days. Find the mean number of persons who entered the supermarket during these five days.

Solution   Altogether the number of persons who entered the supermarket during the past five days is $295 + 1002 + 941 + 768 + 1283 = 4289$. Since $\frac{4289}{5} = 857.8$, this is the mean (or average) number of persons who entered the store per day.

Since we shall have occasion to calculate the means of many different sets of sample data, it will be convenient to have a simple formula which is always applicable. This requires that we represent the figures to be averaged by some general symbol such as x, y, or z; the number of values in a sample, the **sample size**, is usually denoted by the letter n. Choosing the letter x, we can refer to the n values in a sample as $x_1, x_2, \ldots,$ and x_n (which are read "x sub-one," "x sub-two," $\ldots,$ and "x sub-n"), and write

$$\text{sample mean} = \frac{x_1 + x_2 + x_3 + \cdots + x_n}{n}$$

This formula will take care of any set of sample data, but it can be made more compact by assigning the sample mean the symbol $\bar{x}$ (which is read "x bar") and using the $\sum$ notation. The symbol $\sum$ is capital *sigma*, the Greek letter for S. In this notation we let $\sum x$ stand for "the sum of the x's" (that is, $\sum x = x_1 + x_2 + \cdots + x_n$), and we can write

Sample mean

$$\bar{x} = \frac{\sum x}{n}$$

If we refer to the measurements as y's or z's, we write their mean as $\bar{y}$ or $\bar{z}$. In the formula for $\bar{x}$, the term $\sum x$ does not state explicitly which values of x are added; let it be understood, however, that $\sum x$ always refers to the sum of all the x's under consideration in a given situation. In the technical note of Section 3.8 the use of the sigma notation is discussed in more detail.

The number of values in a population, the **population size**, is usually denoted by N. The mean of a population of N items is defined in the same way. It is the sum of the N items, $x_1 + x_2 + x_3 + \cdots + x_N$, or $\sum x$, divided by N.

Assigning the population mean the symbol μ (*mu*, the Greek letter for lower-case *m*), we write

Population mean

$$\mu = \frac{\sum x}{N}$$

with the reminder that $\sum x$ is now the sum of all N values of x which constitute the population.[†]

Also, to distinguish between descriptions of populations and descriptions of samples, we not only use different symbols such as μ and $\bar{x}$, but we refer to a description of a population as a **parameter** and a description of a sample as a **statistic**. Parameters are usually denoted by Greek letters.

To illustrate the terminology and notation introduced in this section, suppose we are interested in the mean lifetime of a production lot of $N = 40,000$ light bulbs. Obviously, we cannot test all of the light bulbs for there would be none left to use or sell, so we take a sample, calculate $\bar{x}$, and use this quantity to estimate μ. If $n = 5$ and the light bulbs in the sample last 967, 949, 940, 952, and 922 hours, we have

$$\bar{x} = \frac{967 + 949 + 940 + 952 + 922}{5} = 946 \text{ hours}$$

If these lifetimes constitute a sample in the technical sense (that is, a set of data from which valid generalizations can be made), we can estimate the mean lifetime μ of all the 40,000 light bulbs as 946 hours.

For non-negative data, the mean not only describes the middle of a set of data, but it also puts some limitation on their size. If we multiply by n on both sides of the equation $\bar{x} = \dfrac{\sum x}{n}$, it follows that $\sum x = n \cdot \bar{x}$ and, hence, no single x-value can exceed $n \cdot \bar{x}$.

EXAMPLE   If the mean annual salary paid to the three top executives of a firm is \$156,000, can one of them receive an annual salary of \$500,000?

Solution   Since $n = 3$ and $\bar{x} = \$156,000$, we get $\sum x = 3 \cdot 156,000 = \$468,000$, and it is impossible for any of the executives to receive more than that.

[†] In cases in which the population size is unlimited, as discussed in Section 3.1, the population mean cannot be defined in this manner. The mean of an infinite population is discussed in the references in Section 3.11.

EXAMPLE If nine high school juniors averaged 41 on the verbal part of the PSAT/NMSQT test, at most how many of them can have scored 65 or more?

Solution Since $n = 9$ and $\bar{x} = 41$, we get $\sum x = 9 \cdot 41 = 369$, and since 65 goes into 369 five times $(369 = 5 \cdot 65 + 44)$, it follows that at most five of the nine high school juniors can have scored 65 or more.

The popularity of the mean as a measure of the "middle" or "center" of a set of data is not accidental. Any time we use a single number to describe some aspect of a set of data, there are certain requirements, or desirable features, that should be kept in mind. Aside from the fact that the mean is a simple and familiar measure, the following are some of its noteworthy properties:

It can be calculated for any set of numerical data, so it always exists.

A set of numerical data has one and only one mean, so it is always unique.

It lends itself to further statistical treatment; as we shall see, for example, the means of several sets of data can always be combined into the overall mean of all the data.

It is relatively reliable in the sense that means of many samples drawn from the same population usually do not fluctuate, or vary, as widely as other statistical measures used to estimate the mean of a population.

The last of these properties is of fundamental importance in statistical inference, and we shall study it in some detail in Chapter 10.

There is another property of the mean which, on the surface, seems desirable:

It takes into account every item of a set of data.

However, sometimes samples contain very small or very large values which are so far removed from the main body of the data that the appropriateness of including them in a sample is questionable. Such values may be due to chance, or they may be due to gross errors in recording the data, gross errors in calculations, malfunctioning of equipment, or other identifiable sources of contamination. In any case, when such values are averaged in with the other values, they can affect the mean to such an extent that it is debatable whether it really provides a useful description of the "middle" of the data.

EXAMPLE With reference to the illustration dealing with the light bulbs on page 43, suppose that the second value is recorded incorrectly as 499 instead of 949. Find the error this would cause in calculating the mean lifetime of the five light bulbs.

Solution The mean of 967, 499, 940, 952, and 922 is

$$\bar{x} = \frac{967 + 499 + 940 + 952 + 922}{5} = 856$$

and this differs from 946, the mean we got on page 43, by $946 - 856 = 90$ hours.

EXAMPLE The ages of six students who went on a geology field trip are 18, 19, 20, 17, 19, and 18, and the age of the instructor who went with them is 50. Find the mean age of these seven persons.

Solution The mean is

$$\bar{x} = \frac{18 + 19 + 20 + 17 + 19 + 18 + 50}{7} = 23$$

but any statement to the effect that the mean age of the group is 23 could easily be misinterpreted. We might well infer incorrectly that the persons who went on the field trip are all in their low twenties.

To avoid the possibility of being misled by a mean affected by a very small value or a very large value, we sometimes find it preferable to describe the middle or center of a set of data with a statistical measure other than the mean; perhaps, with the **median**, which we shall discuss in Section 3.4.

3.3
THE WEIGHTED MEAN

When we calculate an average, we may be making a serious mistake if we overlook the fact that the quantities are not all of equal importance with references to the phenomenon being described. Consider, for example, the following information on the percentage of housing units that were owner occupied in three California cities in 1980:

	Percent owner occupied
Los Angeles	40.3
Sacramento	56.4
San Jose	62.1

The mean of these three percentages is $\dfrac{40.3 + 56.4 + 62.1}{3} = 52.9$, but we cannot very well say that this is the average owner occupancy rate for the three cities. The three figures do not carry equal weight because there are great differences in the size of the three cities.

In order to give quantities being averaged their proper degree of importance, it is necessary to assign them (relative importance) **weights**, and then calculate a **weighted mean**. In general, the weighted mean $\bar{x}_w$ of a set of numbers $x_1, x_2, x_3, \ldots$, and x_n, whose relative importance is expressed numerically by a corresponding set of numbers $w_1, w_2, w_3, \ldots,$ and w_n, is given by

Weighted mean

$$\bar{x}_w = \frac{w_1 x_1 + w_2 x_2 + \cdots + w_n x_n}{w_1 + w_2 + \cdots + w_n} = \frac{\sum w \cdot x}{\sum w}$$

Here $\sum w \cdot x$ is the sum of the products obtained by multiplying each x by the corresponding weight, and $\sum w$ is simply the sum of the weights. Note that when the weights are all equal, the formula for the weighted mean reduces to that for the ordinary (arithmetic) mean.

EXAMPLE Given that there were 1,135 thousand housing units in Los Angeles, 113 thousand in Sacramento, and 210 thousand in San Jose, use these figures and the percentages in the text above to determine the average owner occupancy rate (percentage) for the three cities.

Solution Substituting $x_1 = 40.3$, $x_2 = 56.4$, $x_3 = 62.1$, $w_1 = 1,135$, $w_2 = 113$, and $w_3 = 210$ into the formula for $\bar{x}_w$, we get

$$\bar{x}_w = \frac{(1,135)(40.3) + (113)(56.4) + (210)(62.1)}{1,135 + 113 + 210}$$

$$= \frac{65,154.7}{1,458}$$

$$= 44.7$$

Note that the value we obtained for $\bar{x}_w$ is much smaller than that of $\bar{x}$, 44.7 compared to 52.9, and this is due entirely to the large size of Los Angeles and its low owner-occupancy rate.

A special application of the formula for the weighted mean arises when we must find the overall mean, or **grand mean**, of k sets of data having the means $\bar{x}_1, \bar{x}_2, \bar{x}_3, \ldots,$ and $\bar{x}_k$, and consisting of $n_1, n_2, n_3, \ldots,$ and n_k measurements or observations. The result is given by

Grand mean
of combined data

$$\bar{\bar{x}} = \frac{n_1 \bar{x}_1 + n_2 \bar{x}_2 + \cdots + n_k \bar{x}_k}{n_1 + n_2 + \cdots + n_k} = \frac{\sum n \cdot \bar{x}}{\sum n}$$

where the weights are the sizes of the respective sets of data, the numerator is the total of all the measurements or observations, and the denominator is the total number of items in the combined data.

EXAMPLE In a biology class there are 20 freshmen, 18 sophomores, and 12 juniors. If the freshmen averaged 68 in an examination, the sophomores averaged 75, and the juniors averaged 86, find the mean grade for the entire class.

Solution Substituting $n_1 = 20$, $n_2 = 18$, $n_3 = 12$, $\bar{x}_1 = 68$, $\bar{x}_2 = 75$, and $\bar{x}_3 = 86$ into the formula for the grand mean of combined data, we get

$$\bar{\bar{x}} = \frac{20 \cdot 68 + 18 \cdot 75 + 12 \cdot 86}{20 + 18 + 12}$$

$$= \frac{3,742}{50}$$

$$= 74.84$$

or 75 rounded to the nearest integer.

EXERCISES

3.1 Suppose that we are given complete information about the travel expenses which the administrators of a business's computer support staff charged to their expense accounts during 1990. Give one illustration each of a situation where these data would be looked upon as
  (a) a population;
  (b) a sample.

3.2 The final election returns from a county show that the three candidates for a certain office received 14,276, 10,210, and 2,873 votes. Indicate an office that these candidates might be running for so that these figures would constitute
  (a) a population;
  (b) a sample.

3.3 Suppose that we have the complete information on the number of purchase returns at each of 23 department stores. Give one illustration each of a situation where these data would be looked upon as
  (a) a population;
  (b) a sample.

3.4 Suppose that we have just obtained the results of a questionnaire given to 848 students of a particular college. Indicate a situation in which these data would be looked upon as
  (a) a population;
  (b) a sample.

3.5 The following are the ages of thirty persons empaneled for jury duty by a court: 42, 45, 51, 39, 32, 61, 27, 62, 53, 51, 48, 40, 34, 37, 28, 58, 55, 43, 29, 39, 40, 22, 58, 28, 31, 31, 52, 44, 38, and 36. Find their mean age.

3.6 The following are the numbers of stray dogs captured or turned in to a town animal shelter on twenty working days: 4, 6, 8, 4, 2, 6, 4, 3, 4, 9, 5, 8, 5, 3, 5, 7, 6, 3, 8, and 6. Find the mean.

3.7 At a roadblock, twelve drivers cited for speeding were going 8, 11, 14, 6, 8, 10, 20, 11, 13, 18, 9, and 15 miles per hour over the speed limit.
  (a) On the average, by how many miles per hour were these drivers exceeding the speed limit?
  (b) If a driver exceeding the speed limit by less than 15 miles per hour was fined $60 and the others were fined $88, find the mean of the fines which these drivers had to pay.

3.8 At their first inaugurations, the first ten presidents of the United States were 57, 61, 57, 57, 58, 57, 61, 54, 68, and 51 years old. Find the mean age of those presidents at their first inauguration.

3.9 As part of a laboratory assignment in nutrition, fifteen students determined the number of calories in a serving of lasagna. They obtained the values 329, 335, 347, 318, 322, 330, 351, 362, 315, 342, 346, 353, 316, 327, and 333.

(a) Find the mean.

(b) Subtract 300 from each value and then find the mean of the numbers thus obtained. Add 300 to the result. Does this suggest a simplification for the calculation of a mean?

3.10 A bridge is designed to carry a maximum load of 150,000 pounds. Is this bridge overloaded if it is carrying 18 vehicles having a mean weight of 4,630 pounds?

3.11 An elevator in an office building is designed to carry a maximum load of 2,000 pounds. Is it overloaded if it is carrying nine women whose mean weight is 123 pounds and five men whose mean weight is 174 pounds?

3.12 A clerk lost one of ten sales slips for purchases made in the last hour. The mean value of all ten slips was $7.20, and the remaining nine slips had the values $4.80, $7.10, $7.90, $9.55, $4.45, $5.72, $7.54, $8.34, and $9.70. What is the value of the slip that was lost?

3.13 Careful measurements show that the actual amounts of coffee in six four-ounce jars of instant coffee are 4.02, 3.98, 4.01, 4.05, 3.97, and 4.03 ounces.

(a) Find the mean coffee content of these six jars.

(b) What would be the error in calculating the mean coffee content of the six jars if the fourth value is recorded incorrectly as 4.50 instead of 4.05?

3.14 The mean weight of the 45 members of a football team is 215 pounds. If none of the players weighs less than 170 pounds, at most how many of them can weigh 250 pounds or more?

3.15 The argument of the examples on page 43 can be generalized. For any set of non-negative data with the mean $\bar{x}$, the fraction of the data that are greater than or equal to any positive value k cannot exceed $\bar{x}/k$. Use this result, called **Markov's theorem**, in solving the following:

(a) If the mean adult weight of a certain breed of dog is 35 pounds, at most what fraction can have a weight exceeding 40 pounds?

(b) If the citrus trees in a certain orchard have a mean diameter of 16.0 cm, at most what fraction of the trees can have a diameter of 24 cm or more?

3.16 Records show that in Phoenix, Arizona, the normal daily maximum temperature for each month is 65, 69, 74, 84, 93, 102, 105, 102, 98, 88, 74, and 66 degrees Fahrenheit. Verify that the mean of these figures is 85, and comment on the claim that the average daily maximum temperature in Phoenix is a very comfortable 85 degrees.

3.17 The **geometric mean** of n positive numbers is the nth root of their product. For example, the geometric mean of 3 and 12 is

$$\sqrt{(3)(12)} = \sqrt{36} = 6$$

The geometric mean of 2, 3, and 36 is

$$\sqrt[3]{(2)(3)(36)} = \sqrt[3]{216} = 6$$

The geometric mean is used mainly in averaging ratios, rates of change, and economic indices.

(a) Find the geometric mean of 0.8 and 3.2.

(b) Find the geometric mean of 1, 2, 8, and 16.

(c) During a flu epidemic, 12 cases were reported on the first day, 18 on the second day, and 48 on the third day. From the first day to the second day, the number of cases was multiplied by $\frac{18}{12}$, and from the second day to the third day the number of cases was multiplied by $\frac{48}{18}$. Find the geometric mean of these two growth rates and (assuming that the growth pattern continues) predict the numbers of cases that will be reported on the fourth and fifth day.

3.18 The **harmonic mean** of n numbers $x_1, x_2, \ldots,$ and x_n is defined as n divided by the sum of the reciprocals of the n numbers. The calculation required is $\dfrac{n}{\sum 1/x}$. The harmonic mean is used in dealing with musical frequencies and in some other special situations. For instance, if a commuter drives 10 miles on a freeway at 60 miles per hour and then the next 10 miles off the freeway at 30 miles per hour, his average speed is *not* 45 miles an hour. He will have driven a total of 20 miles in 30 minutes, so that his correct average speed is 40 miles per hour.

(a) Verify that the harmonic mean of 60 and 30 is 40, so that it is the appropriate "average" for this example.

(b) If an investor buys $18,000 worth of a company's stock at $45 a share and then buys $18,000 worth at $36 a share, find the average price that the investor has paid per share. Verify that this price is the harmonic mean of $45 and $36.

(c) If a bakery buys $36 worth of an ingredient at 60 cents per pound, $36 worth at 72 cents per pound, and $36 worth at 90 cents per pound, what is the average cost per pound?

3.19 An instructor counts the final examination in a course three times as much as each of the three one-hour examinations. What is the average grade of a student who received grades of 72, 86, and 80 in the three one-hour examinations and 90 in the final examination?

3.20 In a recent year, the average salaries of elementary school teachers in three cities were $28,300, $34,500, and $31,000. Given that there were 800, 640, and 450 elementary school teachers in these cities, find the average salary of all the elementary school teachers in the three cities.

3.21 A sample survey conducted by a public health organization yielded the following data on the average number of times that persons in various age groups visit a dentist:

Age group	Number of persons in the sample	Mean number of visits
Under 6	55	0.6
6–24	112	1.9
25–64	145	1.8
65 and over	88	1.5
Total	400	

What is the mean for all persons in the sample?

3.22 In a recent season, an amateur baseball team's five best hitters had batting averages of 0.381, 0.367, 0.321, 0.312, and 0.293. If these players had, respectively, 223, 180, 274, 125, and 191 at bats, find their combined batting average.

3.4
THE MEDIAN

To avoid the possibility of being misled by very small or very large values, we sometimes describe the "middle" or "center" of a set of data with statistical measures other than the mean. One of these, the **median** of n values, requires that we arrange the data according to size, and it is defined as follows:

> **The median is the value of the middle item when n is odd, and the mean of the two middle items when n is even.**

In either case, when no two values are alike, the median is exceeded by as many values as it exceeds. When some of the values are alike, this may not be the case.

EXAMPLE In a recent month, a state's Game and Fish Department reported 53, 31, 67, 53, and 36 hunting or fishing violations for five different regions. Find the median number of violations for these months.

Solution The median is not 67, the third (or middle) item, because the figures must first be arranged according to size. Thus, we get

$$31 \quad 36 \quad 53 \quad 53 \quad 67$$

and it can be seen that the median is 53.

Note that in this example there are two 53's among the data and that we do not refer to either of them as *the* median—the median is a number and not necessarily a particular measurement or observation.

EXAMPLE In some areas, persons cited for certain minor traffic violations can attend a class on defensive driving instead of paying a fine. If twelve such classes were attended by 40, 32, 37, 30, 24, 40, 38, 35, 40, 28, 32, and 37 persons, find the median attendance.

Solution Arranging these figures according to size, we get

$$24 \quad 28 \quad 30 \quad 32 \quad 32 \quad 35 \quad 37 \quad 37 \quad 38 \quad 40 \quad 40 \quad 40$$

and we find that the median is $\dfrac{35 + 37}{2} = 36$, namely, the mean of the two values nearest the middle.

Some of the values were alike in the preceding example, but this did not affect the median, which exceeds six of the values and is exceeded by the other six. The situation is quite different, though, in the example which follows:

EXAMPLE On the third hole of a certain golf course, nine golfers scored 4, 3, 4, 5, 4, 3, 3, 4, and 3. Find the median.

Solution Arranging these figures according to size, we get

$$3 \quad 3 \quad 3 \quad 3 \quad 4 \quad 4 \quad 4 \quad 4 \quad 5$$

and it can be seen that the median, the fifth value, is 4.

This time the median exceeds four of the values but is exceeded by only one, and it would be misleading to think of it as the "middle" of the scores—It is not exceeded by as many values as it exceeds.

The symbol which we use for the median of n sample values $x_1, x_2, x_3, \ldots,$ and x_n is $\tilde{x}$ (and, hence, $\tilde{y}$ or $\tilde{z}$ if we refer to the values of y's or z's). The symbol $\tilde{x}$ is usually read as "x tilde." If a set of data constitutes a population, we denote its median by $\tilde{\mu}$.

Thus, we have a symbol for the median, but no formula; there is only a formula for the **median position**. Referring again to data arranged according to size, usually ranked from low to high, we can write

Median position

> *The median is the value of the $\dfrac{n + 1}{2}$ th item.*

When n is odd, $\dfrac{n + 1}{2}$ is an integer and it gives the position of the median; when n is even, $\dfrac{n + 1}{2}$ is halfway between two integers and the median is the mean of the values of the corresponding items.

EXAMPLE   Find the median position for **(a)** $n = 15$ and **(b)** $n = 45$.

Solution   With the data arranged according to size (and counting from either end)

(a) $\dfrac{n+1}{2} = \dfrac{15+1}{2} = 8$, so that the median is the value of the 8th item;

(b) $\dfrac{n+1}{2} = \dfrac{45+1}{2} = 23$, so that the median is the value of the 23rd item.

EXAMPLE   Find the median position for **(a)** $n = 20$ and **(b)** $n = 48$.

Solution   With the data arranged according to size (and counting from either end)

(a) $\dfrac{n+1}{2} = \dfrac{20+1}{2} = 10.5$, so that the median is the mean of the values of the 10th and 11th items;

(b) $\dfrac{n+1}{2} = \dfrac{48+1}{2} = 24.5$, so that the median is the mean of the values of the 24th and 25th items.

It is important to remember that $\dfrac{n+1}{2}$ is a formula for the median position, and not a formula for the median, itself.

The determination of a median can sometimes be simplified, particularly for larger sets of data, by utilizing the grouping provided by a stem-and-leaf display.

EXAMPLE   The following are the numbers of passengers on 50 runs of a ferryboat: 61, 52, 65, 84, 35, 57, 58, 95, 82, 64, 50, 53, 103, 40, 62, 77, 78, 66, 60, 41, 58, 92, 51, 65, 71, 75, 89, 37, 54, 67, 59, 79, 80, 73, 49, 71, 97, 62, 68, 53, 43, 80, 75, 70, 45, 91, 50, 64, 56, and 86. Construct a stem-and-leaf display with one-digit leaves and use it to find the median.

Solution   First constructing the stem-and-leaf display, we get

```
 3 | 5 7
 4 | 0 1 9 3 5
 5 | 2 7 8 0 3 8 1 4 9 3 0 6
 6 | 1 5 4 2 6 0 5 7 2 8 4
 7 | 7 8 1 5 9 3 1 5 0
 8 | 4 2 9 0 0 6
 9 | 5 2 7 1
10 | 3
```

Since the median position is $\dfrac{50+1}{2} = 25.5$ and nineteen of the values fall on the first three stems, we must find the mean of the 6th and 7th values on the fourth stem (counting from low to high). Arranging the leaves on the fourth stem according to size, we get 0, 1, 2, 2, 4, 4, 5, 5, 6, 7, and 8, so that the 6th and 7th leaves are 4 and 5, and the median is $\dfrac{64 + 65}{2} = 64.5$.

Besides the median and the mean there are many other measures of central location (see, for example, the **midrange** described in Exercise 3.36 and the **midquartile** defined on page 56). Each describes the "middle" or "center" of a set of data in its own way and it should not come as a surprise that their values may not be the same. For instance, for the example where the seven students read 16, 10, 14, 13, 20, 11, and 17 of the books that were assigned, the median is 14 and the mean is

$$\frac{16 + 10 + 14 + 13 + 20 + 11 + 17}{7} = \frac{101}{7} = 14.4$$

(rounded to one decimal). The median is average in the sense that it splits the data into two parts so that, unless there are duplicates, as many values are below the median as there are above it. The mean, on the other hand, is average in the sense that if each value in a set of data is replaced by some number k while the total remains unchanged, this number k will have to be the mean. This follows directly from the relationship $n \cdot \bar{x} = \sum x$.

The median shares some, but not all of the properties of the mean which we listed on page 44. Like the mean, the median always exists and it is unique for any set of data. Also like the mean, the median is simple enough to find once the data have been arranged according to size, but ordering a large set of data manually can be a fairly tedious task.

Unlike the mean, the medians of several sets of data cannot generally be combined into an overall median of all the data, and in problems of statistical inference the median is usually less reliable than the mean. This is meant to say that the medians of many samples drawn from the same population will usually vary more widely than the corresponding sample means (see Exercises 3.33 on page 58 and 10.53 on page 276). On the other hand, in some situations the median may be preferable to the mean because it is not so easily affected by extreme (very small or very large) values. For instance, on page 43 we showed that the mean of 967, 949, 940, 952, and 922 (the lifetimes of five light bulbs) is 946, and on page 44 we showed that if 949 is misread as 499, the mean becomes 856. Thus, the error of this mistake is $946 - 856 = 90$. Had we used the median instead of the mean, we would have obtained 949 and 940, and the error would have been only $949 - 940 = 9$.

Finally, the median can be used to define the middle of a number of objects, properties, or qualities which can be ranked, namely, when we deal with ordinal data. For instance, we might rank a number of tasks according to their difficulty and then describe the middle (or median) one as being of "average" difficulty; also, we might rank samples of chocolate fudge according to their consistency and then describe the middle (or median) one as having "average" consistency. Since numbers were not used, the mean cannot be calculated for these situations.

3.5

OTHER FRACTILES ★

The median is but one of many **fractiles** which divide data into two or more parts, as nearly equal as they can be made. Among them we also find **quartiles, deciles,** and **percentiles**, which are intended to divide data into four, ten, and a hundred parts. Until recently, fractiles were determined mainly for distributions of large sets of data, and in this connection we shall study them in Section 3.7.

In this section we shall concern ourselves mainly with a problem that has arisen in **exploratory data analysis**—in the preliminary analysis of relatively small sets of data. It is the problem of dividing such data into four nearly equal parts, where we say "nearly equal" because there is no way in which we can divide a set of data into four equal parts for, say, $n = 27$ or $n = 33$. Statistical measures designed for this purpose have traditionally been referred to as the three **quartiles**, Q_1, Q_2, and Q_3, and there is no argument about Q_2, which is simply the median. On the other hand, there is ample room for arbitrariness in the definition of Q_1 and Q_3.

The following are some of the desirable properties we would like the quartiles Q_1 and Q_3 to have:

1. **Q_1 is exceeded by three times as many values as it exceeds, and it is the other way around for Q_3.**

2. **There are as many values less than Q_1 as there are between Q_1 and Q_2, between Q_2 and Q_3, and greater than Q_3.**

3. **Half the data fall between Q_1 and Q_3.**

It is assumed here that no two values are alike; otherwise, the wording would have to be changed as indicated on page 55.

To illustrate, consider the following high temperature readings in twelve European cities on a June day: 90, 75, 86, 77, 85, 72, 78, 79, 94, 82, 74, and 93. Arranging these figures according to size, we get

$$72 \quad 74 \quad 75 \quad 77 \quad 78 \quad 79 \quad 82 \quad 85 \quad 86 \quad 90 \quad 93 \quad 94$$

and it can be seen that the dashed lines in the upper part of Figure 3.1 divide the data into four equal parts. If we let the midpoints between 75 and 77, 79 and 82, and 86 and 90 be the three quartiles, we get

$$Q_1 = \frac{75 + 77}{2} = 76, \quad Q_2 = \frac{79 + 82}{2} = 80.5, \quad \text{and} \quad Q_3 = \frac{86 + 90}{2} = 88.$$

Of course, $Q_2 = 80.5$ is also the median and it can easily be verified that all three of the desirable properties listed above for the quartiles are satisfied.

Everything worked nicely in this example because 12, the sample size, happened to be a multiple of 4. What can we do, though, when this is not the case? Suppose, for instance, that the city where the high temperature was 77 failed

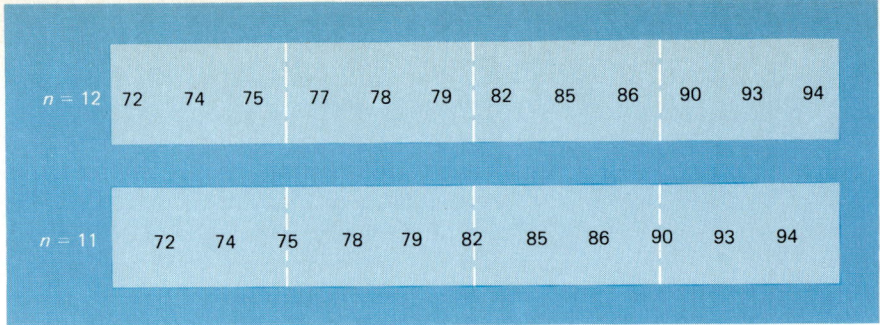

FIGURE 3.1 *Quartiles.*

to report, so that we are left with the following eleven numbers arranged according to size:

$$72 \quad 74 \quad 75 \quad 78 \quad 79 \quad 82 \quad 85 \quad 86 \quad 90 \quad 93 \quad 94$$

The median, or Q_2, is now 82, but what can we do about the other two dividing lines? If it is felt that the second of the three properties listed on page 53 is most relevant, the dividing lines can be drawn as in the lower part of Figure 3.1, so that $Q_1 = 75$, $Q_2 = 82$, and $Q_3 = 90$. There are two values less than Q_1, two values between Q_1 and Q_2, two values between Q_2 and Q_3, and two values greater than Q_3, but Q_1 is exceeded by four times as many values as it exceeds and only five of the eleven values fall between Q_1 and Q_3.

This procedure for finding the quartiles can be given a simple definition. Assuming that no two values are alike (but see the first paragraph on the next page), we write

The lower quartile is the median of all values less than the median of the whole set of data.

The upper quartile is the median of all values greater than the median of the whole set of data.

EXAMPLE The following are the scores of nine students on a history test: 86, 82, 73, 94, 88, 66, 79, 90, and 74. Find the median and the two quartiles.

Solution For $n = 9$, the median position is $\dfrac{9 + 1}{2} = 5$. The lower quartile is the median of the four values below the median, and the upper quartile is the median of the four values above the median. Arranging the data according to size, we get

$$66 \quad 73 \quad 74 \quad 79 \quad 82 \quad 86 \quad 88 \quad 90 \quad 94$$

and it can be seen that the median is 82, the lower quartile is $\dfrac{73 + 74}{2} = 73.5$, and the upper quartile is $\dfrac{88 + 90}{2} = 89$.

If some of the values are alike, we modify the definition of quartiles by replacing "less than the median" by "to the left of the median position" and "greater than the median" by "to the right of the median position." For instance, the nine golfers in the example on page 50 scored

$$3 \quad 3 \quad 3 \quad 3 \quad 4 \quad 4 \quad 4 \quad 4 \quad 5$$

on the third hole of a certain course. The lower quartile, the mean of the second and third values, is 3. The median, the fifth value, is 4. The upper quartile, the mean of the second and third values from the right, is 4.

Other definitions for quartiles are given in the exercises.

In exploratory data analysis we look at the process of finding values which divide a set of data into four parts in a different way—as a process of **folding**. With reference to the first of our two examples, where the sample size was twelve, suppose that in the upper part of Figure 3.1, or that of Figure 3.2, we fold the page along the dashed line on the left, along the dashed line on the right, and then along the dashed line in the middle. If we do this, the four parts into which we have divided the data will overlap. If we do this for our second example, where the sample size was eleven, the four parts will not overlap if we fold the page along the dashed lines in the lower part of Figure 3.1. However, they will overlap if we move the dashed lines as in the lower part of Figure 3.2.

The new dividing lines on the left and on the right are at the midpoints between 75 and 78, and 86 and 90, and we could write

$$Q_1 = \frac{75 + 78}{2} = 76.5 \quad \text{and} \quad Q_3 = \frac{86 + 90}{2} = 88.$$

Actually, in exploratory data analysis these two values are referred to as **hinges**—the **lower hinge** is 76.5 and the **upper hinge** is 88. This terminology reflects the process of folding, which we used to divide the data into four parts.

Having introduced the concept of a hinge by means of an example, let us now give a formal definition. Assuming that no two values are alike, but see below, we

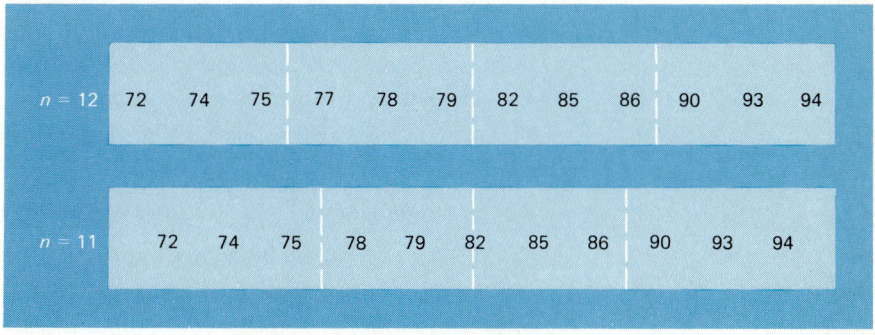

FIGURE 3.2 *Hinges.*

write

The lower hinge is the median of all the values less than or equal to the median of the whole set of data; the upper hinge is the median of all the values greater than or equal to the median of the whole set of data.

In practice, we first find the position of one hinge, and then count as many places from the other end to find the position of the other hinge.

EXAMPLE The following are the pulse rates of nine persons after they have done some fairly strenuous exercise: 104, 100, 98, 111, 91, 94, 103, 96, and 108. Find the median and the two hinges.

Solution For $n = 9$, the median position is $\dfrac{9 + 1}{2} = 5$. Thus, the position of the lower hinge is $\dfrac{5 + 1}{2} = 3$, and the upper hinge is the third value from the other end. Arranging the data according to size, we get

$$91 \quad 94 \quad 96 \quad 98 \quad 100 \quad 103 \quad 104 \quad 108 \quad 111$$

and it can be seen that the lower hinge is 96, the median is 100, and the upper hinge is 104. Also, if we imagine dashed lines drawn through these values as in Figure 3.2, we will find that the four parts will, indeed, overlap.

If some of the values are alike, we may have to modify the definition of the hinges by replacing "less than or equal to the median" by "to the left of or at the median position" and "greater than or equal to the median" by "to the right of or at the median position." Otherwise, the procedure is exactly the same. For instance, in the example on page 50, nine golfers scored

$$3 \quad 3 \quad 3 \quad 3 \quad 4 \quad 4 \quad 4 \quad 4 \quad 5$$

on the third hole of a certain course. Since the sample size is the same as in the preceding example, we find that the lower hinge, the third value, is 3, the median, the fifth value, is 4, and the upper hinge, the third value from the right, is 4.

In practice, quartiles and hinges are often used interchangeably.

Quartiles and hinges are not intended to be descriptive of the "middle" or "center" of a set of data, and we have given them here mainly because, like the median, they are fractiles and they are determined in more or less the same way. The **midquartile**, $\dfrac{Q_1 + Q_3}{2}$, has been used on occasion as a measure of central location, and presumably the mean of the two hinges can be used in the same way.

The information provided by the median, the two quartiles, and the smallest and largest values is sometimes presented in the form of a **box-and-whisker plot**, often just called a **box plot**. Such a plot is shown in Figure 3.3 for the data on the

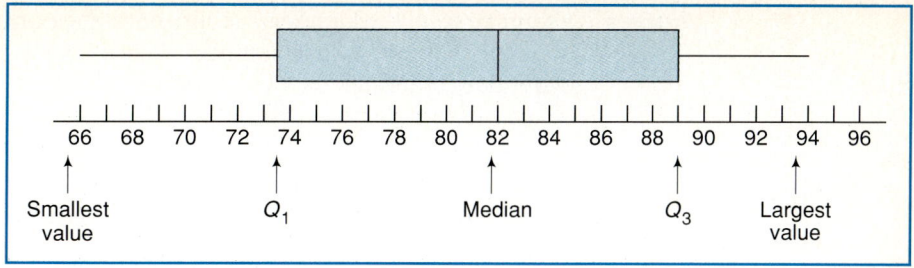

FIGURE 3.3 *Box-and-whisker plot for data on history test scores.*

history test scores on page 54. In practice, the two hinges are sometimes used instead of the two quartiles. This figure is sometimes embellished with other features of the data, but the simple form given here is adequate for most purposes.

3.6
THE MODE

Another measure that is sometimes used to describe the "middle" or "center" of a set of data is the **mode**, which is defined simply as the value which occurs with the highest frequency and more than once. Its two main advantages are that it requires no calculation, only counting, and it can be determined even for qualitative, or nominal, data.

EXAMPLE   The 20 meetings of a square dance club were attended by 26, 25, 28, 23, 25, 24, 24, 21, 23, 26, 28, 26, 24, 32, 25, 27, 24, 23, 24, and 22 of its members. Find the mode.

Solution   Among the twenty numbers, 21, 22, 27, and 32 each occurs once; 28 occurs twice; 23, 25, and 26 each occurs three times; and 24 occurs five times. Thus, 24 is the **modal attendance**.

Also, if more visitors to California want to see Disneyland than any other tourist attraction, we say that Disneyland is their **modal choice**.

Aside from the fact that the mode is seldom of any use in statistical inference, it also has the disadvantage that it may not exist (which is the case when no two values are alike) or that it may not be unique.

EXAMPLE   A sample of the records of a motor vehicle bureau shows that 18 drivers in a certain age group received 3, 2, 0, 0, 2, 3, 3, 1, 0, 1, 0, 3, 4, 0, 3, 2, 3, and 0 traffic tickets during the last three years. Find the mode.

Solution As can be seen, the number 4 occurs once, the number 1 occurs twice, the number 2 occurs three times, and the numbers 0 and 3 each occurs six times. So, there are the two modes 0 and 3.

An additional difficulty with the mode is that it behaves erratically when data values are rounded (see Exercise 3.51 on page 60).

There are many other measures of central location besides the ones we have mentioned, and at least one more, the **midrange**, is given in Exercise 3.36. What particular "average" should be used in a given situation can depend on many things (see Section 7.3, for example) and the choice may be difficult to make. Since the selection of statistical descriptions often contains an element of arbitrariness, there are persons who believe that the magic of statistics can be used to prove almost anything. Indeed, a famous nineteenth-century British statesman said that there are three kinds of lies: *lies, damned lies, and statistics,* and Exercises 3.35 and 3.36 on page 57 describe a situation where this criticism would be well justified.

EXERCISES

3.23 Find the median position for
  (a) $n = 25$;
  (b) $n = 32$.

3.24 Find the median position for
  (a) $n = 37$;
  (b) $n = 64$.

3.25 On fifteen days, a restaurant served breakfast to 40, 52, 55, 38, 40, 48, 56, 56, 60, 37, 58, 63, 46, 50, and 61 customers. Find the median.

3.26 In 1991, twelve used-car salespersons sold, respectively, 58, 70, 85, 42, 64, 46, 66, 89, 44, 93, 58, and 79 used cars. Find the median.

3.27 Twenty power failures lasted 18, 125, 44, 96, 31, 26, 80, 49, 125, 63, 45, 33, 89, 12, 103, 75, 40, 80, 61, and 28 minutes. Find the median.

3.28 Find the median number of traffic tickets for the 18 drivers given in the example on page 57.

3.29 On the nineteen pages of a report, a typist made 0, 0, 1, 2, 0, 3, 1, 0, 0, 0, 0, 1, 0, 0, 4, 1, 0, 0, and 2 mistakes. Find
  (a) the mean;
  (b) the median.

3.30 The following values are the times in minutes of twenty-five National Basketball Association (NBA) games:

138	142	113	126	135
142	159	157	140	157
121	128	142	164	155
139	143	158	140	118
142	146	123	130	137

  (a) Find the median directly by arranging the data according to size.
  (b) Find the median by first constructing a stem-and-leaf display.

3.31 Use the stem-and-leaf display on page 16 to find the median of the scores which the twenty students got on the physical coordination test.

3.32 In a certain month, fifteen salespersons reached 107, 90, 80, 92, 86, 109, 102, 92, 353, 78, 74, 102, 106, 95, and 91 percent of their sales quotas. Calculate the mean and the median of these percentages, and indicate which of the two measures gives a better indication of these salespersons' "average" performance.

3.33 To verify the claim that the mean is generally more reliable than the median (namely, that it is subject to smaller chance fluctuations), a student conducted an experiment consisting of 12 tosses of three dice. The following are his results: 2, 4, and 6; 5, 3, and 5; 4, 5, and 3; 5, 2, and 3; 6, 1, and 5; 3, 2, and 1; 3, 1, and 4; 5, 5, and 2; 3, 3, and 4; 1, 6, and 2; 3, 3, and 3; 4, 5, and 3.

(a) Calculate the twelve medians and the twelve means.

(b) Group the medians and the means obtained in part (a) into separate distributions having the classes 1.5–2.5, 2.5–3.5, 3.5–4.5, and 4.5–5.5. (Note that there will be no ambiguities since the medians of three whole numbers and the means of three whole numbers cannot equal 2.5, 3.5, or 4.5.)

(c) Draw histograms of the two distributions obtained in part (b) and explain how they illustrate the claim that the mean is generally more reliable than the median.

3.34 Repeat the preceding exercise with your own data by repeatedly rolling three dice (or one die three times) and constructing corresponding distributions of the medians and the means. (If no dice are available, simulate the experiment mentally, with the use of a computer, or by drawing numbered slips of paper out of a hat.)

3.35 A consumer testing service obtained the following miles per gallon in five test runs performed with each of three compact cars:

Car A:	27.9	30.4	30.6	31.4	31.7
Car B:	31.2	28.7	31.3	28.7	31.3
Car C:	28.6	29.1	28.5	32.1	29.7

(a) If the manufacturers of car A want to advertize that their car performed best in this test, which of the "averages" discussed in this text could they use to substantiate their claim?

(b) If the manufacturers of car B want to advertize that their car performed best in this test, which of the "averages" discussed in this text could they use to substantiate their claim?

3.36 Suppose that the manufacturers of car C of the preceding exercise hire an unscrupulous statistician and instruct him to find some kind of "average" which will show that their car performed best in the test. Show that the **midrange**, the mean of the smallest and largest values, will serve their purpose.

★ 3.37 The records of the library of a large university showed that the twenty-two senior philosophy majors checked out these numbers of books during the academic year:

62	73	40	72	79
88	35	51	48	42
75	65	69	82	50
66	103	68	54	38
52	76			

(a) Find the median.

(b) Find the two hinges.

★ 3.38 In a study of the stopping ability of a car with a new braking system, twenty-one drivers going at 30 miles per hour were able to stop in the distances below, expressed in feet:

69	58	70	80	46
61	65	74	75	55
67	56	70	72	61
66	58	68	70	68
58				

(a) Find the median.

(b) Find the two hinges.

★ 3.39 Find the two hinges for the lengths of the NBA games gives in Exercise 3.30.

★ 3.40 If n sample values are arranged according to size, find the positions of the median, the two hinges, and the quartiles:

(a) when $n = 40$;

(b) when $n = 41$;

(c) when $n = 42$;

(d) when $n = 43$.

★ 3.41 Find the lower and upper quartiles for the lengths of the NBA games given in Exercise 3.30.

★ 3.42 Find the quartiles and the hinges for the power failure data of Exercise 3.27.

★ 3.43 In this problem we will give a procedure for finding general **fractiles** or **percentiles**. Let p be a fraction between 0 and 1. Find the fractile corresponding to p as follows:

Compute pn. If this is not an integer, use the next highest integer for the pth fractile position; if it is an integer, use the mean of the values in positions pn and $pn + 1$ as the pth fractile.

For example, if you wanted fractile $p = 0.6$ in a list with $n = 103$, you would find $pn = 61.8$ and use the value in position 62. If you wanted fractile $p = 0.6$ in a list with $n = 110$, you would find $pn = 66$ and use the mean of the values in positions 66 and 67. Fractiles are often expressed as percentiles; for example, the 0.6 fractile is called the 60th percentile. Find the 60th percentile for the lengths of the NBA games given in Exercise 3.30.

★ **3.44** Some statisticians and some computer packages use **interpolation** to determine the position of quartiles and other fractiles. This technique begins by finding the position for the fractile corresponding to p as $p(n + 1)$. For instance, to find the position of the lower quartile of 30 values we substitute $p = 0.25$ and $n = 30$ to get $0.25(30 + 1) = 7.75$. This means that we must go three-quarters of the way between the seventh and eight values. If these values happen to be 146 and 148, then the lower quartile is

$$146 + \frac{3}{4}(148 - 146) = 147.5$$

A college had 8, 3, 20, 5, 2, 8, 14, 2, 6, 10, 7, and 15 applicants for twelve different teaching positions. Find the hinges and also find the quartiles using the technique of Exercise 3.43. Then use the interpolation method to obtain the quartiles.

★ **3.45** Use the results of Exercises 3.27 and 3.42 to construct a box-and-whisker plot for the lengths of the power failures.

★ **3.46** Use the results of Exercises 3.30 and 3.39 to construct a box-and-whisker plot for the lengths of the NBA games.

3.47 The following are the number of days in advance that seventeen persons purchased tickets for a sporting event: 7, 3, 4, 12, 18, 3, 8, 14, 6, 16, 7, 6, 11, 7, 9, 5, and 2. Find the mode.

3.48 On fifty days, these were the numbers of students absent from an algebra class:

```
1 3 0 0 1 0 4 1 1 0
1 2 6 0 1 0 0 1 0 0
0 0 1 3 3 2 5 0 1 1
1 0 3 0 0 0 4 1 1 2
1 2 0 1 0 1 0 0 3 2
```

Find the mode.

3.49 Find the mode (if it exists) of each of the following sets of blood pressure readings:
(a) 144, 145, 146, 146, 148, 146, 146, 145, 147, 145, 144;
(b) 146, 149, 146, 141, 146, 149, 147, 147, 149, 149, 145;
(c) 167, 151, 175, 144, 152, 148, 156, 169, 143, 177, 161.

3.50 Thirty persons were asked for their favorite color. These are their responses:

red	blue	blue	green	yellow
blue	brown	red	red	red
green	white	blue	red	yellow
blue	blue	red	green	yellow
blue	blue	orange	green	blue
blue	green	red	purple	blue

What is their modal choice?

3.51 The figures below are a local bakery's daily flour utilization, in pounds, for twenty consecutive weekdays:

440	677	481	690	707
514	671	488	483	554
611	638	572	514	623
664	631	570	484	612

(a) Find the mean, median, and mode for this set of values.
(b) Round the twenty values to the nearest ten pounds. Then again give the mean, median, and mode.
(c) Round the twenty values to the nearest hundred pounds. Then again give the mean, median, and mode.
(d) State a conclusion about the effect of rounding on the mean, median, and mode.

3.7

THE DESCRIPTION OF GROUPED DATA★

In the past, considerable attention was paid to the description of grouped data because it was generally advantageous to group data before calculating various statistical descriptions. This is no longer the case, since the necessary calculations can now be made in a matter of seconds by using computers or even hand-held calculators. Nevertheless, we shall devote this section and Section 4.4 to the description of grouped data, since some data (published government figures, for example) are available only in the form of frequency distributions.

As we have already seen, the grouping of data entails some loss of information. Each item loses its identity, so to speak; we only know how many items there are in each class, so we must be satisfied with approximations. To determine the mean, we can usually get a good approximation by **assigning to each item falling into a class the value of the corresponding class mark**. For instance, to calculate the mean of the grouped sulfur oxides emission data on page 23, we treat the three values in the first class as if they were all 6.95, the ten values in the second class as if they were all 10.95, ..., and the two values falling into the seventh class as if they were all 30.95. This procedure is usually quite satisfactory, since the errors which are thus introduced into the calculations will tend to "average out."

To give a general formula for the mean of a distribution with k classes, let us denote the successive class marks by $x_1, x_2, \ldots,$ and x_k, and the corresponding class frequencies by $f_1, f_2, \ldots,$ and f_k. Then, the sum of all the measurements is approximated by

$$x_1 \cdot f_1 + x_2 \cdot f_2 + \cdots + x_k \cdot f_k = \sum x \cdot f,$$

and the mean of the distribution is given by

Mean of grouped data

$$\bar{x} = \frac{\sum x \cdot f}{n}$$

Here n is the size of the sample, $f_1 + f_2 + \cdots + f_k$, and to write a corresponding formula for the mean of a population we substitute μ for $\bar{x}$ and N for n.

EXAMPLE Calculate the mean of the distribution of the sulfur oxides emission data on page 23.

Solution To get $\sum x \cdot f$, we perform the calculations shown in the following table, where the first column contains the class marks, the second column is copied from the original distribution, and the third column contains the products $x \cdot f$:

Class mark x	Frequency f	$x \cdot f$
6.95	3	20.85
10.95	10	109.50
14.95	14	209.30
18.95	25	473.75
22.95	17	390.15
26.95	9	242.55
30.95	2	61.90
	80	1,508.00

Then, substitution into the formula yields

$$\bar{x} = \frac{1{,}508.00}{80} = 18.85$$

To check on the **grouping error**, namely, the error introduced by replacing each value in a class by the class mark, let us refer to the computer printout of Figure 3.4, which shows that the mean of the original ungrouped data is 18.896. Thus, the error is only $18.85 - 18.896 = -0.046$, which is very small.

Once a set of data has been grouped, we can still determine most other statistical measures besides the mean, but we may have to make special assumptions or modify the definitions. For instance, we define the median of a distribution in the following way:

> **The median of a distribution is such that half the total area of the rectangles of the histogram of the distribution lies to its left and the other half lies to its right.**

This definition, which is illustrated by Figure 3.5, amounts to the assumption that the values in the class containing the median of the ungrouped data are distributed evenly—that is, spread out evenly—throughout that class.

To find the dividing line between the two halves of a histogram (each of which represents $\frac{n}{2}$ of the items grouped), we must count $\frac{n}{2}$ of the items starting at either end of the distribution. How this is done is illustrated by the following example:

EXAMPLE   Find the median of the distribution of the sulfur oxides emission data.

Solution   Since $\frac{n}{2} = \frac{80}{2} = 40$, we must count 40 of the items starting at either end. Starting at the bottom of the distribution (that is, beginning with the smallest values), we find

```
MTB > SET C1
DATA> 15.8    26.4    17.3    11.2    23.9    24.8    18.7    13.9     9.0    13.2
DATA> 22.7     9.8     6.2    14.7    17.5    26.1    12.8    28.6    17.6    23.7
DATA> 26.8    22.7    18.0    20.5    11.0    20.9    15.5    19.4    16.7    10.7
DATA> 19.1    15.2    22.9    26.6    20.4    21.4    19.2    21.6    16.9    19.0
DATA> 18.5    23.0    24.6    20.1    16.2    18.0     7.7    13.5    23.5    14.5
DATA> 14.4    29.6    19.4    17.0    20.8    24.3    22.5    24.6    18.4    18.1
DATA>  8.3    21.9    12.3    22.3    13.3    11.8    19.3    20.0    25.7    31.8
DATA> 25.9    10.5    15.9    27.5    18.1    17.9     9.4    24.1    20.1    28.5
MTB > MEAN C1
   MEAN    =        18.896
MTB > STAN C1
   ST.DEV. =         5.6565
```

FIGURE 3.4 *Computer printout for the mean of the sulfur oxides emission data.*

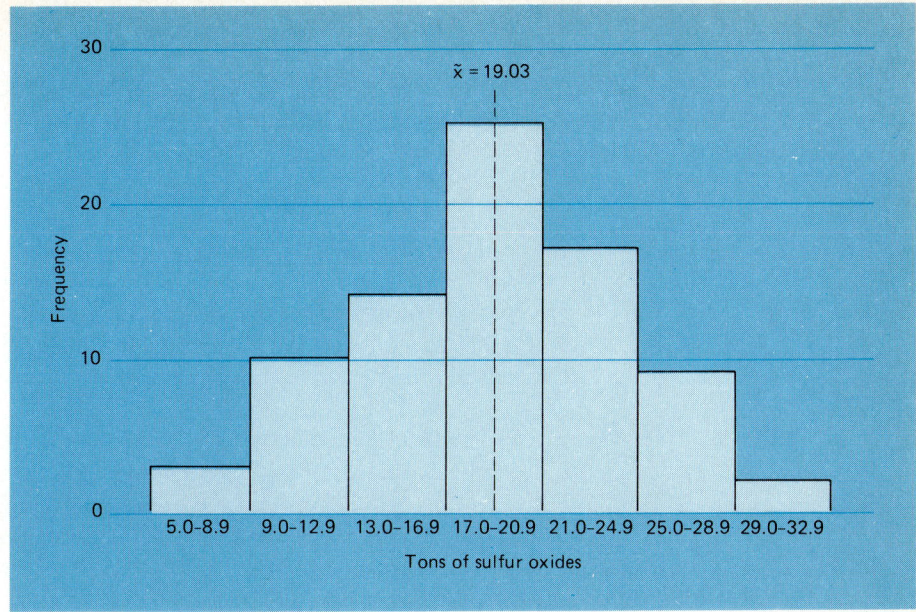

FIGURE 3.5 *The median of the distribution of the sulfur oxides emission data.*

that $3 + 10 + 14 = 27$ of the values fall into the first three classes, and that $3 + 10 + 14 + 25 = 52$ of the values fall into the first four classes. Therefore, we must count $40 - 27 = 13$ more values beyond the 27 which fall into the first three classes. On the assumption that the 25 values in the fourth class are spread evenly throughout that class, we can do this by adding $\frac{13}{25}$ of the class interval of 4 to 16.95, the lower boundary of the fourth class. This gives us

$$\tilde{x} = 16.95 + \frac{13}{25} \cdot 4 = 19.03$$

for the median of the distribution.

In general, if L is the lower boundary of the class into which the median must fall, f is its frequency, c is its class interval, and j is the number of items we still lack when we reach L, then the median of the distribution is given by

Median of grouped data

$$\tilde{x} = L + \frac{j}{f} \cdot c$$

If we prefer, we can find the median of a distribution by starting to count at the other end (beginning with the largest values) and subtracting an appropriate fraction of the class interval from the upper boundary of the class into which the median must fall.

EXAMPLE Use this alternate approach to find the median of the distribution of the sulfur oxides emission data.

Solution Since $2 + 9 + 17 = 28$ of the values fall above 20.95, we need $40 - 28 = 12$ of the 25 values which fall into the next class to reach the median, and we write

$$\tilde{x} = 20.95 - \frac{12}{25} \cdot 4 = 19.03$$

The result is, of course, the same.

Note that the median of a distribution can be found regardless of whether the class intervals are all equal; in fact, it can usually be found even when either or both of the classes at the top and at the bottom of a distribution are open (see Exercise 3.52).

The method by which we found the median of a distribution can also be used to determine other fractiles. For instance, the three quartiles (which are intended to divide a set of data into four more or less equal parts) are defined for grouped data so that 25 percent of the total area of the rectangles of the histogram lies to the left of Q_1, 25 percent lies between Q_1 and Q_2, 25 percent lies between Q_2 and Q_3, and 25 percent lies to the right of Q_3. Similarly, the nine deciles (which are intended to divide a set of data into ten more or less equal parts) are defined for grouped data so that 10 percent of the total area of the rectangles of the histogram lies to the left of D_1, 10 percent lies between D_1 and $D_2, \ldots$, and 10 percent lies to the right of D_9. And finally, the ninety-nine percentiles (which are intended to divide a set of data into a hundred more or less equal parts) are defined for grouped data so that 1 percent of the total area of the rectangles of the histogram lies to the left of P_1, 1 percent lies between P_1 and $P_2, \ldots$, and 1 percent lies to the right of P_{99}. Note that Q_2, D_5, and P_{50} are all equal to the median, and that P_{25} equals Q_1 and P_{75} equals Q_3. See also Exercise 3.43 on page 59.

EXAMPLE Find Q_1 and Q_3 for the distribution of the sulfur oxides emission data.

Solution To find Q_1 we must count $\frac{80}{4} = 20$ of the items starting at the bottom of the distribution. Since there are $3 + 10 = 13$ values in the first two classes, we must count $20 - 13 = 7$ of the 14 values in the third class to reach Q_1, and we get

$$Q_1 = 12.95 + \frac{7}{14} \cdot 4 = 14.95$$

To find Q_3 we must count 20 of the items starting at the other end of the distribution. Since $2 + 9 = 11$ of the values fall into the two classes at the top of the distribution, we must count $20 - 11 = 9$ of the 17 values in the next class to reach Q_3, and we get

$$Q_3 = 24.95 - \frac{9}{17} \cdot 4 = 22.83$$

EXAMPLE Find D_8 and P_2 for the distribution of the sulfur oxides emission data.

Solution Counting $80(0.20) = 16$ of the items starting with the largest values, we get

$$D_8 = 24.95 - \frac{5}{17} \cdot 4 = 23.77$$

and counting $80(0.02) = 1.6$ of the items starting with the smallest values, we get

$$P_2 = 4.95 + \frac{1.6}{3} \cdot 4 = 7.08$$

Note that when we determine a fractile of a distribution, the number of items we have to count and the quantity j in the formula on page 63 need not be a whole number.

EXERCISES

★ 3.52 For each of the following distributions determine whether it is possible to find the mean and/or the median:

(a)

Grade	Frequency
40–49	5
50–59	18
60–69	27
70–79	15
80–89	6

(b)

IQ	Frequency
Less than 90	3
90– 99	14
100–109	22
110–119	19
More than 119	7

(c)

Weight	Frequency
100 or less	41
101–110	13
111–120	8
121–130	3
131–140	1

★ 3.53 Find the mean of the following distribution of the grades which 500 students obtained in a geography test:

Grade	Number of students
10–24	44
25–39	70
40–54	92
55–69	147
70–84	115
85–99	32

★ 3.54 With reference to the distribution of the preceding exercise, find
(a) the median;
(b) the quartiles Q_1 and Q_3.

★ 3.55 With reference to the distribution of Exercise 3.53, find
(a) the deciles D_1 and D_9;
(b) the percentiles P_5 and P_{95}.

★ 3.56 Find the mean of the following distribution of the percentages of the students belonging to a certain ethnic group in a sample of 50 elementary schools:

Percentage	Number of schools
0– 4	18
5– 9	15
10–14	9
15–19	7
20–24	1

★ 3.57 With reference to the distribution of the preceding exercise, find
  (a) the median;
  (b) the quartiles Q_1 and Q_3.

★ 3.58 With reference to the distribution of Exercise 3.56, find
  (a) the deciles D_3 and D_7;
  (b) the percentiles P_5 and P_{98}.

★ 3.59 Find the mean of the following distribution of the ages of the members of a labor union:

Age (years)	Frequency
15–19	16
20–24	35
25–29	44
30–34	27
35–39	17
40–44	8
45–49	2
50–54	1

★ 3.60 With reference to the distribution of the preceding exercise, find
  (a) the median;
  (b) the quartiles Q_1 and Q_3.

★ 3.61 With reference to the distribution of Exercise 3.59, find the two fractiles which divide the distribution into three equal parts.

★ 3.62 Find the mean of the distribution obtained in Exercise 2.22 on page 28 for the weights of the 50 rats. Also, use a computer package or a calculator to find the mean of the original (ungrouped) data, and thus determine the size of the grouping error.

★ 3.63 Find the mean of the distribution obtained in Exercise 2.27 on page 28 for the mileages obtained with 40 tanksful of gas. Also, use a computer package or a calculator to find the mean of the original (ungrouped) data, and thus determine the size of the grouping error.

★ 3.64 Suppose that data values have been grouped into a distribution with class interval c. It is guaranteed that the grouping error in computing the mean cannot exceed c. For each of the two preceding exercises, compare the grouping error to the class interval.

3.8

TECHNICAL NOTE (SUMMATIONS)

In the notation introduced on page 42, $\sum x$ does not tell us which, or how many, values of x we must add. This is taken care of by the more explicit notation

$$\sum_{i=1}^{n} x_i = x_1 + x_2 + \cdots + x_n$$

where it is made clear that we are adding the x's whose subscripts i are $1, 2, \ldots,$ and n. We are not using the more explicit notation in this text to simplify the overall appearance of the formulas, assuming that it is clear in each case what x's we are referring to and how many there are.

Using the $\sum$ notation, we shall also have occasion to write such expressions as $\sum x^2, \sum xy, \sum x^2 f, \ldots,$ which (more explicitly) represent the sums

$$\sum_{i=1}^{n} x_i^2 = x_1^2 + x_2^2 + x_3^2 + \cdots + x_n^2$$

$$\sum_{j=1}^{m} x_j y_j = x_1 y_1 + x_2 y_2 + \cdots + x_m y_m$$

$$\sum_{i=1}^{n} x_i^2 f_i = x_1^2 f_1 + x_2^2 f_2 + \cdots + x_n^2 f_n$$

Working with two subscripts, we shall also have the occasion to evaluate **double summations** such as

$$\sum_{j=1}^{3} \sum_{i=1}^{4} x_{ij} = \sum_{j=1}^{3} (x_{1j} + x_{2j} + x_{3j} + x_{4j})$$

$$= x_{11} + x_{21} + x_{31} + x_{41} + x_{12} + x_{22} + x_{32} + x_{42}$$

$$+ x_{13} + x_{23} + x_{33} + x_{43}$$

To verify some of the formulas involving summations that are stated but not proved in the text, the reader will need the following rules:

Rules for summations

Rule A: $\displaystyle\sum_{i=1}^{n} (x_i \pm y_i) = \sum_{i=1}^{n} x_i \pm \sum_{i=1}^{n} y_i$

Rule B: $\displaystyle\sum_{i=1}^{n} k \cdot x_i = k \cdot \sum_{i=1}^{n} x_i$

Rule C: $\displaystyle\sum_{i=1}^{n} k = k \cdot n$

The first of these rules states that the summation of the sum (or difference) of two terms equals the sum (or difference) of the individual summations, and it can be extended to the sum or difference of more than two terms. The second rule states that we can, so to speak, factor a constant out of a summation, and the third rule states that the summation of a constant is simply n times that constant. All of these rules can be proved by actually writing out in full what each of the summation represents.

EXERCISES

3.65 Write each of the following in full; that is, without summation signs:

(a) $\displaystyle\sum_{i=1}^{6} x_i$;

(d) $\displaystyle\sum_{j=1}^{8} x_j f_j$;

(b) $\displaystyle\sum_{i=1}^{5} y_i$;

(e) $\displaystyle\sum_{i=3}^{7} x_i^2$;

(c) $\displaystyle\sum_{i=1}^{3} x_i y_i$;

(f) $\displaystyle\sum_{j=1}^{4} (x_j + y_j)$.

3.66 Write each of the following without summation signs and simplify if possible:

(a) $\displaystyle\sum_{i=1}^{5} (x_i + 1)$;

(b) $\displaystyle\sum_{j=1}^{4} (3y_j)$;

(c) $\displaystyle\sum_{i=1}^{4} x_i + \sum_{j=1}^{4} (2x_j)$.

3.67 Write each of the following as summations:

(a) $z_1 + z_2 + z_3 + z_4$;

(b) $x_6 + x_7 + x_8 + x_9 + x_{10} + x_{11}$;

(c) $x_1 f_1 + x_2 f_2 + x_3 f_3 + x_4 f_4 + x_5 f_5 + x_6 f_6$;

(d) $y_1^2 + y_2^2 + y_3^2 + y_4^2 + y_5^2$;

(e) $3x_1 + 3x_2 + 3x_3 + 3x_4 + 3x_5 + 3x_6$;

(f) $(x_1 - y_1) + (x_2 - y_2) + (x_3 - y_3) + (x_4 - y_4)$;

(g) $(w_1 - 5) + (w_2 - 5) + (w_3 - 5)$;

(h) $a_1 b_1 c_1 + a_2 b_2 c_2 + a_3 b_3 c_3 + a_4 b_4 c_4$.

3.68 Given $x_1 = 1$, $x_2 = 3$, $x_3 = 5$, $x_4 = 7$, $x_5 = 9$, $f_1 = 1$, $f_2 = 5$, $f_3 = 10$, $f_4 = 3$, and $f_5 = 2$, find

(a) $\displaystyle\sum_{i=1}^{5} x_i$;

(c) $\displaystyle\sum_{i=1}^{5} x_i \cdot f_i$;

(b) $\displaystyle\sum_{i=1}^{5} f_i$;

(d) $\displaystyle\sum_{i=1}^{5} x_i^2 \cdot f_i$.

3.69 Given $x_1 = -2$, $x_2 = 3$, $x_3 = 1$, and $x_4 = 4$ find

(a) $\displaystyle\sum_{i=1}^{4} x_i$;

(b) $\displaystyle\sum_{i=1}^{4} x_i^2$.

3.70 Given $x_{11} = 3$, $x_{12} = 1$, $x_{13} = -2$, $x_{14} = 2$, $x_{21} = 1$, $x_{22} = 4$, $x_{23} = -2$, $x_{24} = 5$, $x_{31} = 3$, $x_{32} = -1$, $x_{33} = 2$, and $x_{34} = 3$, find

(a) $\displaystyle\sum_{i=1}^{3} x_{ij}$ separately for $j = 1$, 2, 3, and 4;

(b) $\displaystyle\sum_{j=1}^{4} x_{ij}$ separately for $i = 1$, 2, and 3.

3.71 With reference to the preceding exercise, evaluate the double summation $\displaystyle\sum_{i=1}^{3} \sum_{j=1}^{4} x_{ij}$ using

(a) the results of part (a) of that exercise;

(b) the results of part (b) of that exercise.

3.72 Show that $\displaystyle\sum_{i=1}^{n} (x - \bar{x}) = 0$ for any set of x's whose mean is $\bar{x}$.

3.73 Is it true in general that $\left(\displaystyle\sum_{i=1}^{n} x_i\right)^2 = \displaystyle\sum_{i=1}^{n} x_i^2$? (*Hint:* Check whether the equation holds for $n = 2$.)

3.9

CHECKLIST OF KEY TERMS

(with page references to their definitions)

3.10
REVIEW EXERCISES

3.74 The following are the estimated television audience ratings for sixteen professional football games:

1.46 1.32 1.58 1.88 1.32 1.39 1.72 1.66
1.82 1.21 1.36 1.76 1.86 1.63 1.55 1.57

Find the median and the mode for these values.

★ **3.75** Find the hinges for the audience ratings given in the previous problem.

3.76 In a benefit sale, a service organization sold 120 books at a mean price of $2.10, 80 cakes at a mean price of $2.75, and 50 craft items at a mean price of $4.55. Find the total sales amount, and give the mean price per item sold.

★ **3.77** The following is the distribution of the number of days it rained in Seattle in 60 months:

Number of days	Frequency
5–7	5
8–10	9
11–13	12
14–16	18
17–19	13
20–22	3

Calculate the mean.

★ **3.78** With reference to Exercise 3.77, find
(a) the median of the distribution;
(b) the quartiles of the distribution.

★ **3.79** With reference to Exercise 3.77, find the 60th percentile of the distribution.

3.80 A producer of television commercials knows exactly how much money was spent on the production of each of ten one-minute commercials. Give one example each of a problem in which these data would be looked upon as
(a) a population;
(b) a sample.

3.81 Find the median position for
(a) $n = 31$;
(b) $n = 80$.

3.82 The following are the 1990 earnings, in thousands of dollars, of ten women working in industrial sales:

28.2 30.5 25.8 20.4 23.2
32.5 33.0 26.4 28.8 27.3

(a) Find the median of these earnings.
(b) Find the mean of these earnings.

★ **3.83** In the data of the previous problem, find the hinges of these women's 1990 earnings.

3.84 In a certain class, a student scored 78 on the first test, 83 on the second test, and 88 on the final. The instructor considers the second test to be twice as important as the first test and the final to be three times as important as the second test. What is the student's weighted mean score on the three tests?

3.85 During the three weeks before Christmas, twelve persons shopped on the average in 5.75 clothing stores. It is possible that at least seven of them shopped in ten or more stores?

3.86 Twenty registered voters were asked whether they consider themselves Democrats, Republicans, or Independents. Use the following results to determine their modal choice:

Democrat	Democrat	Democrat
Democrat	Independent	Independent
Independent	Independent	Republican
Democrat	Independent	Independent
Independent	Republican	Independent
Republican	Democrat	Democrat
Independent	Republican	

★ **3.87** Given a sample of 23 observations, find the positions of the median and the two hinges.

★ **3.88** Given a sample of 24 observations, find the positions of the median and the two hinges.

3.89 The following values are the percent sulfur retention in 42 watershed areas of the Northeastern United States, reported in 1989:

18.3	41.9	32.4	33.2	25.9
60.6	19.0	54.7	48.8	43.0
57.4	44.2	56.2	59.6	43.5
17.8	37.6	66.2	49.7	39.0
53.7	39.8	21.8	43.2	24.3
42.0	19.3	42.1	53.9	28.2
47.8	34.3	33.0	33.2	42.2
31.7	38.0	45.3	41.7	45.7
37.0	51.0			

Construct a stem-and-leaf display and use it to find the median of these data.

3.90 The following are the numbers of passengers on 46 sight-seeing buses in Boston, Massachusetts:

Number of persons	Frequency
15 or less	5
16–20	8
21–25	11
26–30	13
31–35	6
36–40	3

If possible, find the mean and the median.

⋆ 3.91 With reference to the distribution of the preceding exercise, find, if possible,
(a) the lower quartile;
(b) the upper quartile;
(c) the 40th percentile;
(d) the 80th percentile.

3.92 Given $x_1 = 3$, $x_2 = 5$, $x_3 = -2$, $x_4 = 1$, $x_5 = 3$, $x_6 = -4$, $x_7 = 2$, and $x_8 = 4$, find

(a) $\sum_{i=1}^{6} x_i$;   (b) $\sum_{i=3}^{8} x_i$.

3.93 If the mean salary of all male employees of company A exceeds that of all male employees of company B, and the mean salary of all female employees of company A exceeds that of all female employees of company B, does it follow that the mean salary of all employees of company A exceeds that of all employees of company B? Explain your answer.

3.94 Baseball batting averages are obtained by dividing the number of hits that a player has by his number of at-bats, expressing the result as a three-figure decimal. For example, a player with 30 hits and 100 at-bats has a batting average of 0.300. Here are some values from the 1990 baseball season:

Player	League	Hits	At-bats
George Brett	American	179	544
Willie McGee	American	31	113
Willie McGee	National	168	501
Eddie Murray	National	184	558

George Brett had the highest batting average in the American League, and Willie McGee had the highest batting average in the National League. McGee, however, played in both leagues. Show that Eddie Murray had the highest overall batting average, even though he was not best in his league.

3.11

REFERENCES

Informal discussions of the ethics involved in choosing among averages and other questions of ethics in statistics in general are given in

HOOKE, R., *How to Tell the Liars from the Statisticians.* New York: Marcel Dekker, Inc., 1983.
HUFF, D., *How to Lie with Statistics.* New York: W. W. Norton & Company, Inc., 1954.

For further information about the use and interpretation of hinges, see the books on exploratory data analysis referred to on page 38.

Simplification of calculations in grouped data by use of "coding" is discussed in Appendix C of

HAMBURG, M., *Basic Statistics: A Modern Approach.* New York: Harcourt Brace, Jovanovich, Inc., 1974.

A discussion of the mean of an infinite population is found on page 187 of

NETER, J., WASSERMAN, W., and WHITMORE, G., *Applied Statistics, 2nd ed.* Boston: Allyn and Bacon, Inc., 1982.

SUMMARIZING DATA: MEASURES OF VARIATION

In most sets of data the values are not all alike. The extent to which they vary is a serious concern of statistics. Consider the following examples:

In a hospital where each patient's pulse rate is taken three times a day, that of patient *A* is 72, 76, and 74, while that of patient *B* is 72, 91, and 59. The mean pulse rate of the two patients is the same, 74, but observe the difference in variability. Whereas patient *A*'s pulse rate is stable, that of patient *B* fluctuates widely.

A supermarket stocks certain 1-pound bags of mixed nuts, which on the average contain 12 almonds per bag. If all the bags contain anywhere from 10 to 14 almonds, the product is consistent and satisfactory, but the situation is quite different if some of the bags have no almonds while others have 20 or more.

Consider a basketball player who has scored 22, 26, and 24 points in his first three games. A teammate has scored 41, 13, and 18 in these games. Both players have the same average, 24, but the first player is much more consistent.

Measuring variability is of special importance in statistical inference. Suppose, for instance, that we have a coin which is slightly bent and we wonder whether there is still a fifty-fifty chance for heads. What if we toss the coin 100 times and get 28 heads and 72 tails? Does the shortage of heads—only 28 where we might have expected 50—imply that the count is not "fair"? To answer such questions we must have some idea about the magnitude of the fluctuations, or variations, that are brought about by chance when coins are tossed 100 times.

We need to assess the extent to which data disperse; the measures which will provide this information are called **measures of variation**. In Sections 4.1 through 4.3 we present the most widely used measures of variation and some of their special applications. Measuring the variation of grouped data and some statistical descriptions other than measures of location and variation are discussed in Sections 4.4 and 4.5, which are optional.

4.1

THE RANGE

To introduce a simple way of measuring variability, let us refer to the first of the three examples above, where the pulse rate of patient *A* varied from 72 to 76 while that of patient *B* varied from 59 to 91. These extreme (smallest and largest) values are indicative of the variability of the two sets of data, and just about the same information is conveyed if we take the differences between the respective extremes. So, let us make the following definition:

The range of a set of data is the largest value minus the smallest.

For patient *A* of the above example we get a range of $76 - 72 = 4$ and for patient *B* we get a range of $91 - 59 = 32$. Also, for the sulfur oxides emission data on page 23 the smallest value is 6.2, the largest value is 31.8, and the range is $31.8 - 6.2 = 25.6$; and for the lifetimes of the five light bulbs on page 43 the smallest value is 922, the largest value is 967, and the range is $967 - 922 = 45$.

The range is easy to calculate and easy to understand, and there is a natural curiosity about the minimum and maximum values. Nonetheless, it is not generally a useful measure of variation. Its main shortcoming is that is does not tell us anything about the dispersion of the values which fall between the two extremes. Each of the following sets of data

$$
\begin{array}{llllllllll}
Set\ 1: & 5, & 20, & 20, & 20, & 20, & 20, & 20, & 20 \\
Set\ 2: & 5, & 5, & 5, & 5, & 20, & 20, & 20, & 20 \\
Set\ 3: & 5, & 7, & 9, & 12, & 15, & 17, & 19, & 20
\end{array}
$$

has a range of $20 - 5 = 15$, but the dispersion is entirely different in each case. Thus, the range is used mainly as a "quick and easy" indication of variability, for instance, in industrial quality control to keep a close check on raw materials or products by observing, and charting, the range of small samples taken at regular intervals of time.

Whereas the range covers all the values in a sample, a similar measure of variation covers (more or less) the middle 50 percent. It is the **interquartile range** $Q_3 - Q_1$, where Q_1 and Q_3 can be defined in various ways for ungrouped or grouped data as in Sections 3.5 and 3.7. For instance, for the nine history test scores on page 54 we might use $Q_3 - Q_1 = 89 - 73.5 = 15.5$, and for the distribution of the sulfur oxides emission data we might use the values of Q_1 and Q_3 on page 64 and write $22.83 - 14.95 = 7.88$. Some statisticians also use the **semi-interquartile range** $\frac{1}{2}(Q_3 - Q_1)$, which is sometimes called the **quartile deviation**.

4.2

THE VARIANCE AND THE STANDARD DEVIATION

To define the **standard deviation**, by far the most generally useful measure of variation, let us observe that the dispersion of a set of data is small if the values are closely bunched about their mean, and that it is large if the values are scattered widely about their mean. It would seem reasonable, therefore, to measure the variation of a set of data in terms of the amounts by which the values deviate from their mean. If a set of numbers $x_1, x_2, x_3, \ldots,$ and x_n, constituting a sample, has the mean $\bar{x}$, the differences $x_1 - \bar{x}$, $x_2 - \bar{x}$, $x_3 - \bar{x}, \ldots,$ and $x_n - \bar{x}$ are called the **deviations from the mean**, and it suggests itself that we might use their average (namely, their mean) as a measure of the variation of the sample. Unfortunately, this will not do. Unless the x's are all equal, some of the deviations will be positive, some will be negative, and as the reader was asked to show in Exercise 3.72 on page 68, the sum of the deviations from the mean, $\sum (x - \bar{x})$, and hence also their mean, is always zero.

Since we are really interested in the magnitude of the deviations, and not in whether they are positive or negative, we might simply ignore the signs and define a measure of variation in terms of the absolute values of the deviations from the mean. Indeed, if we add the deviations from the mean as if they were all positive or zero and divide by n, we obtain the statistical measure which is called the **mean deviation**. This measure has intuitive appeal, but because of the absolute values it leads to serious theoretical difficulties in problems of inference, and it is rarely used.

An alternative approach is to work with the squares of the deviations from the mean, as this will also eliminate the effect of the signs. Squares of real numbers cannot be negative; in fact, squares of the deviations from a mean are all positive unless a value happens to coincide with the mean. Then, if we average the squared deviations from the mean and take the square root of the result (to compensate for the fact that the deviations were squared), we get

$$\sqrt{\frac{\sum (x - \bar{x})^2}{n}}$$

and this is how, traditionally, the standard deviation used to be defined. Expressing literally what we have done here mathematically, it is also called the **root-mean-square deviation**.

It is customary to modify this formula by dividing the sum of the squared deviations from the mean by $n - 1$ instead of n. Following this practice, which will be explained below, let us define the **sample standard deviation**, denoted by s, as

Sample standard deviation

$$s = \sqrt{\frac{\sum (x - \bar{x})^2}{n - 1}}$$

and its square, the **sample variance**, as

Sample variance

$$s^2 = \frac{\sum (x - \bar{x})^2}{n - 1}$$

These formulas for the standard deviation and the variance apply to samples, but if we substitute μ for $\bar{x}$ and N for n, we obtain analogous formulas for the standard deviation and the variance of a population. It has become fairly general practice to denote the **population standard deviation** by σ (*sigma*, the Greek letter for lowercase s) when dividing by N, and by S when dividing by $N - 1$. Thus, for σ we write

Population standard deviation

$$\sigma = \sqrt{\frac{\sum (x - \mu)^2}{N}}$$

Ordinarily, the purpose of calculating a sample statistic (such as the mean, the standard deviation, or the variance) is to estimate the corresponding population parameter. If we actually took many samples from a population which has the mean μ, calculated the sample means $\bar{x}$, and then averaged all these estimates of μ, we should find that their average is very close to μ. However, if we calculated the variance of each sample by means of the formula $\dfrac{\sum (x - \bar{x})^2}{n}$, and then averaged all these supposed estimates of σ^2, we would probably find that their average is less than σ^2. Theoretically, it can be shown that we can compensate for this by dividing by $n - 1$ instead of n in the formula for s^2. Estimators which have the desirable property that their values will on the average equal the quantity they are supposed to estimate are said to be **unbiased**; otherwise, they are said to be **biased**. So, we say that $\bar{x}$ is an unbiased estimator of the population mean μ, and that s^2 is an unbiased estimator of the population variance σ^2. It does not follow from this that s is also an unbiased estimator of σ, but when n is large the bias is small and can usually be ignored.

In calculating the sample standard deviation using the formula by which it is defined, we must (1) find $\bar{x}$, (2) determine the n deviations from the mean $x - \bar{x}$, (3) square these deviations, (4) add all the squared deviations, (5) divide by $n - 1$, and (6) take the square root of the result obtained in step 5. In actual practice, this formula is rarely used, but we shall illustrate it here to emphasize what is really measured by σ and s.

EXAMPLE On six consecutive Sundays, a tow-truck operator received 9, 7, 11, 10, 13, and 7 service calls. Calculate s.

Solution First calculating the mean, we get

$$\bar{x} = \frac{9 + 7 + 11 + 10 + 13 + 7}{6} = \frac{57}{6} = 9.5$$

and the work required to find $\sum (x - \bar{x})^2$ may be arranged as in the following table:

x	$x - \bar{x}$	$(x - \bar{x})^2$
9	-0.5	0.25
7	-2.5	6.25
11	1.5	2.25
10	0.5	0.25
13	3.5	12.25
7	-2.5	6.25
	0.0	27.50

Then, dividing by $6 - 1 = 5$ and taking the square root, we get

$$s = \sqrt{\frac{27.50}{5}} = \sqrt{5.5} = 2.3$$

rounded to one decimal. Note in the table above that the total for the middle column is zero; since this must always be the case, it provides a check on the calculations.

It was easy to calculate s in this example because the data were whole numbers and the mean was exact to one decimal. Otherwise, the calculations required by the formula defining s can be quite tedious, and, unless we can get s directly with a statistical calculator or a computer, it helps to use the computing formula

Computing formula for the sample standard deviation

$$s = \sqrt{\frac{S_{xx}}{n-1}} \qquad \text{where } S_{xx} = \sum x^2 - \frac{(\sum x)^2}{n}$$

EXAMPLE   Use this computing formula for s to rework the preceding example.

Solution   First we calculate $\sum x$ and $\sum x^2$, getting

x	x^2
9	81
7	49
11	121
10	100
13	169
7	49
57	569

Then, substituting $\sum x = 57$ and $\sum x^2 = 569$, together with $n = 6$, we find that

$$S_{xx} = 569 - \frac{(57)^2}{6} = 27.50$$

This gives us

$$s = \sqrt{\frac{27.50}{6-1}} = \sqrt{5.5} = 2.3,$$

agreeing with the result we obtained before.

The computing formula for s can also be used to find σ; the denominator $n - 1$ must be replaced by N.

4.3
APPLICATIONS OF THE STANDARD DEVIATION

In subsequent chapters, sample standard deviations will be used mainly to estimate population standard deviations in problems of inference. To get more of a feeling for what a standard deviation really measures, we shall devote this section to some applications.

In the argument which led to the definition of the standard deviation, we observed that the dispersion of a set of data is small if the values are bunched closely about their mean, and that it is large if the values are scattered widely about their mean. Correspondingly, we can now say that if the standard deviation of a set of data is small, the values are concentrated near the mean, and if the standard deviation is large, the values are scattered widely about the mean. This idea is expressed more formally by the following theorem, called **Chebyshev's theorem** after the Russian mathematician P. L. Chebyshev (1821–1894):

Chebyshev's theorem

For any set of data (population or sample) and any constant k greater than 1, the proportion of the data that must lie within k standard deviations on either side of the mean is at least

$$1 - \frac{1}{k^2}$$

Thus, we can be sure that at least $1 - \frac{1}{2^2} = \frac{3}{4}$, or 75 percent, of the values in any set of data must lie within two standard deviations on either side of the mean; at least $1 - \frac{1}{3^2} = \frac{8}{9}$, or 88.9 percent, must lie within three standard deviations on either side of the mean; and at least $1 - \frac{1}{5^2} = \frac{24}{25}$, or 96 percent, must lie within five standard deviations on either side of the mean. Here we arbitrarily let $k = 2$, 3, and 5.

EXAMPLE If all the 1-pound cans of coffee filled by a food processor have a mean weight of 16.00 ounces with a standard deviation of 0.02 ounce, at least what percentage of the cans must contain between 15.80 and 16.20 ounces of coffee?

Solution Since k standard deviations, or $k(0.02)$, equals

$$16.20 - 16.00 = 16.00 - 15.80 = 0.20$$

we have $k(0.02) = 0.20$ and $k = \frac{0.20}{0.02} = 10$. According to Chebyshev's theorem, it follows that at least $1 - \frac{1}{10^2} = 0.99$, or 99 percent, of the cans must contain between 15.80 and 16.20 ounces of coffee.

Chebyshev's theorem applies to any kind of data, but it tells us only "at least what percentage" must lie between certain limits. For nearly all sets of data, the

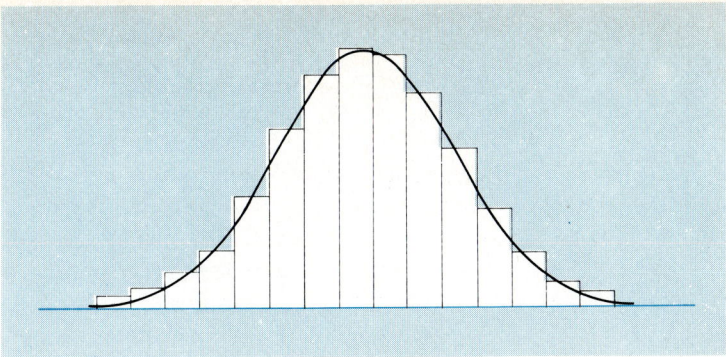

FIGURE 4.1 *Bell-shaped distribution.*

actual percentage of data lying between the limits is much greater than that specified by Chebyshev's theorem.

For distributions having the general shape of the cross section of a bell (see Figure 4.1), we can make the following much stronger statement:

(1) **about 68% of the values will lie within one standard deviation of the mean, that is, between $\bar{x} - s$ and $\bar{x} + s$;**

(2) **about 95% of the values will lie within two standard deviations of the mean, that is, between $\bar{x} - 2s$ and $\bar{x} + 2s$;**

(3) **about 99.7% of the values will lie between three standard deviations of the mean, that is, between $\bar{x} - 3s$ and $\bar{x} + 3s$.**

This result is sometimes referred to as the **empirical rule**, presumably because such percentages are observed in practice. Actually, it is a theoretical result based on the normal distribution, which we shall study in Chapter 9 (in particular, see Exercise 9.12 on page 231).

EXAMPLE   Use the values of $\bar{x}$ and s given in the computer printout of Figure 3.4 to determine what percentage of the sulfur oxides emission data on page 23 actually falls within one standard deviation of the mean, within two standard deviations of the mean, and within three standard deviations of the mean.

Solution   Since $\bar{x} = 18.896$ and $s = 5.6565$, we shall first have to determine what percentage of the data falls between

$$18.896 - 5.6565 = 13.2395 \quad \text{and} \quad 18.896 + 5.6565 = 24.5525.$$

Counting 14 values below 13.2395 and 14 values above 24.5525, we find that $80 - 28 = 52$ values, and hence $\frac{52}{80} \cdot 100\% = 65$ percent of the data, falls between the two limits. Similarly, we find that $\frac{78}{80} \cdot 100\% = 97.5$ percent of the data falls within two standard deviations of the mean, and that $\frac{80}{80} \cdot 100\% = 100$ percent of the data fall within three standard deviations of the mean.

The results which we have obtained here are not exactly 68, 95, and 99.7 percent, but it would seem reasonable to say that they are "about 68 percent," "about 95 percent," and "about 99.7 percent." Also, Figure 2.2 on page 29 shows that the distribution of the data has the shape of a somewhat lopsided bell, so we cannot really expect perfect results.

On page 73 we gave examples in which knowledge about the variability of the data was of importance. This is also the case when we want to compare numbers belonging to different sets of data. To illustrate, suppose that the final examination in a French course consists of two parts, vocabulary and grammar, and that a certain student scored 66 points in the vocabulary part and 80 points in the grammar part. At first glance it would seem that the student did much better in grammar than in vocabulary, but suppose that all the students in the class averaged 51 points in the vocabulary part with a standard deviation of 12, and 72 points in the grammar part with a standard deviation of 16. Thus, we can argue that the student's score in the vocabulary part is $\frac{66 - 51}{12} = 1.25$ standard deviations above the average for the class, while her score in the grammar part is only $\frac{80 - 72}{16} = 0.50$ standard deviation above the average for the class. Whereas the original scores cannot be meaningfully compared, these new scores, expressed in terms of standard deviations, can. Clearly, the given student rates much higher on her command of French vocabulary than on her knowledge of French grammar, compared to the rest of the class.

What we have done here consisted of converting the grades into **standard units** or **z-scores**. In general, if x is a measurement belonging to a set of data having the mean $\bar{x}$ (or μ) and the standard deviation s (or σ), then its value in standard units, denoted by z, is

Formula for converting to standard units

$$z = \frac{x - \bar{x}}{s} \quad or \quad z = \frac{x - \mu}{\sigma}$$

depending on whether the data constitute a sample or a population. In these units, z tells us how many standard deviations a value lies above or below the mean of the set of data to which it belongs. Standard units will be used frequently in later chapters.

EXAMPLE   Two-year-old models of a certain kind of car have been selling on the average for $7,860 with a standard deviation of $820, while three-year-old models of the same kind of car have been selling on the average for $6,400 with a standard deviation of

$960. Leaving all other considerations aside, is a two-year-old model priced at $6,960 a greater bargain than a three-year-old model priced at $5,400?

Solution Converting both prices into standard units, we get

$$\frac{6{,}960 - 7{,}860}{820} = -1.10$$

for the two-year-old car, and

$$\frac{5{,}400 - 6{,}400}{960} = -1.04$$

for the three-year-old car. Even though the two-year-old model is priced $900 below average and the three-year-old model is priced $1,000 below average, the two-year-old model is priced relatively lower for cars of the same kind and, hence, it is a greater bargain.

One disadvantage of the standard deviation as a measure of variation is that it depends on the units of measurement. For instance, the weights of certain objects may have a standard deviation of 0.1 ounce or 2,835 milligrams, which is the same, but neither value really tells us whether it reflects a great deal of variation or very little variation. If we are weighing the eggs of small birds, either figure would reflect a great deal of variation, but this would not be the case if we are weighing 100-pound bags of potatoes. What we need in a situation like this is a **measure of relative variation**, such as the **coefficient of variation**

Coefficient of variation

$$V = \frac{s}{\bar{x}} \cdot 100\% \qquad or \qquad V = \frac{\sigma}{\mu} \cdot 100\%$$

which expresses the standard deviation as a percentage of what is being measured, at least on the average.

EXAMPLE Several measurements of the diameter of a ball bearing made with one micrometer had a mean of 2.49 mm and a standard deviation of 0.012 mm, and several measurements of the unstretched length of a spring made with another micrometer had a mean of 0.75 in. with a standard deviation of 0.002 in. Which of the two micrometers is relatively more precise?

Solution Calculating the two coefficients of variation, we get

$$\frac{0.012}{2.49} \cdot 100\% = 0.48\% \qquad and \qquad \frac{0.002}{0.75} \cdot 100\% = 0.27\%$$

Thus, the measurements of the length of the spring are relatively less variable, and this means that the second micrometer is more precise. Both coefficients of variation are less than one percent, so that both micrometers qualify as highly precise.

EXERCISES

4.1 The following are the response times of a smoke alarm after the release of smoke from a fixed source: 12, 9, 11, 7, 9, 14, 6, and 10 seconds. Find the range.

4.2 The numbers that follow are the yields (in trays per acre) for sun-dried raisins over a ten-year period in California.

$$
\begin{array}{ccccc}
715 & 825 & 640 & 900 & 790 \\
965 & 895 & 700 & 915 & 945
\end{array}
$$

Find the range.

4.3 The following are the closing prices for two stocks on five consecutive Fridays:

Stock A: $15\frac{7}{8}$ $15\frac{1}{2}$ $16\frac{3}{8}$ $16\frac{5}{8}$ $15\frac{3}{4}$

Stock B: $22\frac{1}{8}$ 22 $21\frac{7}{8}$ $22\frac{1}{8}$ $22\frac{1}{4}$

Find the range for each stock.

4.4 Twenty-five employees of a motel chain, given a course in first aid, obtained these scores on a test administered after the course:

$$
\begin{array}{ccccc}
17 & 19 & 14 & 20 & 17 \\
17 & 12 & 15 & 15 & 16 \\
16 & 19 & 18 & 15 & 16 \\
16 & 17 & 18 & 17 & 13 \\
17 & 16 & 14 & 18 & 17
\end{array}
$$

Find the range.

★ 4.5 Find the hinges of the data above. Interpreting the hinges as quartiles, find the interquartile range. Should it be regarded as surprising that the interquartile range is less than half the range?

4.6 Find the range of the stopping distances in Exercise 3.38 on page 59.

★ 4.7 Find the interquartile range of the stopping distances in Exercise 3.38 on page 59.

4.8 The example on page 51 gives the numbers of passengers on 50 runs of a ferryboat. Find the range.

★ 4.9 In the ferryboat example on page 51, find the two hinges. Interpreting these hinges as quartiles, find the semi-interquartile range.

4.10 In five attempts, it took a person 12, 18, 14, 11, and 15 minutes to change the oil in a particular make of automobile. Calculate the standard deviation of this sample using
  (a) the formula which defines s;
  (b) the computing formula for s.

4.11 Four purchases of jelly beans in bags labelled "one pound" contained 16.2, 15.9, 15.8, and 16.1 ounces. Calculate s using
  (a) the formula which defines s;
  (b) the computing formula for s.

4.12 The numbers of misdirected luggage cases reported for six consecutive weeks at a small airport were 13, 8, 15, 11, 3, and 10. Find the variance of these figures using
  (a) the formula by which s^2 is defined;
  (b) the computing formula for s^2.

4.13 On four days it took a person 37, 32, 35, and 41 minutes to drive to work.
  (a) Use the computing formula which defines s to compute the standard deviation of these data.
  (b) Subtract 30 from each and then use the computing formula for s to calculate the standard deviation. What general rule does this suggest for simplifying the calculation of s?

4.14 Each of the following lists contains an even number of items. Further, each list contains just two different values. Find the standard deviation for each list. Can you relate this standard deviation to the difference between the values? For this exercise, the formula which defines s will be somewhat easier to use.
  (a) 16, 16, 20, 20.
  (b) 100, 100, 100, 200, 200, 200.
  (c) 60, 60, 60, 60, 80, 80, 80, 80.

4.15 Each of the following lists contains just one value which is different from all the others. Find the standard deviation for each list. Can you relate this standard deviation to the difference between the values?
  (a) 6, 6, 6, 10.
  (b) 6, 10, 10, 10.
  (c) 20, 20, 20, 20, 30.

4.16 It has been claimed that for samples of size $n = 4$, the range should be roughly twice as large as the standard deviation. Check this claim with reference to the following data, representing the numbers of emergency surgeries that were performed at a hospital on four days: 3, 6, 2, and 6.

4.17 It has been claimed that for samples of size $n = 10$, the range should be roughly three times as large as the standard deviation. Check this claim with reference to the following data, representing the emission scores of ten automobiles:

$$18 \quad 21 \quad 16 \quad 24 \quad 28$$
$$20 \quad 22 \quad 29 \quad 19 \quad 25$$

4.18 If we add the same constant c to each item in a set of data, the mean and the median of the new set equal the mean and the median of the original set plus the constant c, while the range and the standard deviation remain unchanged.
   (a) Verify that for a sample consisting of the values $-3, 4, 1, 5, 3, 4$, and 0 the mean is 2, the median is 3, the range is 8, and the standard deviation is $\sqrt{8}$, and that after we add 3 to each value the mean becomes 5, the median becomes 6, but the range is still 8 and the standard deviation is still $\sqrt{8}$.
   (b) Find the standard deviation of the data in Exercise 4.4 after subtracting 10 from each value.

4.19 If we multiply each item in a set of data by the same positive constant b, the mean, median, range, and standard deviation of the new set equal the mean, median, range, and standard deviation of the original set multiplied by b.
   (a) With reference to the seven sample values of part (a) of the preceding exercise, show that if each value is multiplied by 2 the mean becomes 4, the median becomes 6, the range becomes 16, and the standard deviation becomes $\sqrt{32} = 2\sqrt{8}$.
   (b) During four pit stops, the front-tire man changed a racing car's right front tire in 10.8, 12.0, 10.5, and 10.7 seconds. Calculate the standard deviation by first multiplying each figure by 10, subtracting 110, determining s for the resulting figures, and then dividing by 10.

4.20 According to Chebyshev's theorem, what can we assert about the proportion of any set of data that must lie within k standard deviations of the mean when
   (a) $k = 6$;
   (b) $k = 12$?

4.21 According to Chebyshev's theorem, what can we assert about the percentage of any set of data that must lie within k standard deviations of the mean when
   (a) $k = 4$;
   (b) $k = 9$?

4.22 According to Chebyshev's theorem, what can we assert about the percentage of any set of data that must lie within k standard deviations on either side of the mean when
   (a) $k = 5$;
   (b) $k = 8$;
   (c) $k = 10$;
   (d) $k = 20$?

4.23 An airline's records show that its flights between two cities arrive on the average 5.4 minutes late with a standard deviation of 1.4 minutes. At least what percentage of its flights between the two cities arrive anywhere between
   (a) 2.6 minutes late and 8.2 minutes late;
   (b) 1.6 minutes early and 12.4 minutes late?

4.24 A study of the nutritional value of a certain kind of bread shows that on the average one slice contains 0.260 milligram of thiamine (vitamin B_1) with a standard deviation of 0.005 milligram. According to Chebyshev's theorem, between what values must the thiamine content be of

   (a) at least $\dfrac{35}{36}$ of all slices of this bread;

   (b) at least $\dfrac{80}{81}$ of all slices of this bread?

4.25 With reference to the preceding exercise, at least what percentage of the slices of the given kind of bread must have a thiamine content between 0.245 and 0.275 milligram? What can we say about this percentage if it can be assumed that the distribution of the thiamine contents of the slices of bread is bell-shaped?

4.26 The data below are the numbers of customers a restaurant served for lunch on 120 weekdays. These were given in Exercise 2.25 on page 28.

50	64	55	51	60	41	71	53	63	64
46	59	66	45	61	57	65	62	58	65
55	61	50	55	53	57	58	66	53	56
64	46	59	49	64	60	58	64	42	47
59	62	56	63	61	68	57	51	61	51
60	59	67	52	52	58	64	43	60	62
48	62	56	63	55	73	60	69	53	66
54	52	56	59	65	60	61	59	63	56
62	56	62	57	57	52	63	48	58	64
59	43	67	52	58	47	63	53	54	67
57	61	76	78	60	66	63	58	60	55
61	59	74	62	49	63	65	55	61	54

(a) Use a computer package to determine $\bar{x}$ and s for these data.

(b) Use the results of (a) to find the percentages of the data values that lie within one, two, and three standard deviations of the mean. Compare these with the percentages expected according to the rule on page 79. (Exercise 2.25 indicates that the distribution of these values is reasonably bell-shaped.)

4.27 In a city in the Southeast, supermarkets charge on the average $3.67 per pound for sirloin steak (with a standard deviation of $0.40), $1.12 per pound for chicken drumsticks (with a standard deviation of $0.11), and $8.18 per pound for leg of veal (with a standard deviation of $0.92). If a supermarket in this city charges $3.59 per pound for sirloin steak, $1.09 per pound for chicken drumsticks, and $7.69 per pound for leg of veal, which of the three items is, relatively speaking, the best bargain?

4.28 Among two men on a weight-reducing diet, the first belongs to an age/body type group for which the mean weight is 145 pounds with a standard deviation of 15 pounds. The second belongs to an age/body type group for which the mean weight is 165 pounds with a standard deviation of 20 pounds. If their respective weights are 178 and 204 pounds, which is more seriously overweight for his age/body type group?

4.29 A laboratory technician studied recent measurements made with two different instruments. The first measured the diameter of a ball bearing and obtained a mean of 4.96 mm with a standard deviation of 0.022 mm. The second measured the diameter of a metal rod and obtained a mean of 6.48 mm with a standard deviation of 0.032 mm. Which of the two instruments was relatively more precise?

4.30 One patient's blood pressure was measured daily for several weeks. These measurements had mean 188 with standard deviation 14.2. A second patient was also measured daily, obtaining an average of 136 with standard deviation 8.6. Which patient's blood pressure is relatively more variable?

4.31 On sixteen days a restaurant had the following numbers of orders for chicken and steak:

Chicken:	46	55	43	48	54	65	36	40
	51	53	64	32	41	46	53	47
Steak:	39	41	25	30	46	36	37	23
	30	33	50	44	41	28	35	37

Calculate the two coefficients of variation to determine for which item the number of orders is relatively more variable.

★ 4.32 An alternative measure of relative variation is the **coefficient of quartile variation**, defined as $\dfrac{Q_3 - Q_1}{Q_3 + Q_1} \cdot 100\%$ where Q_1 and Q_3 are the lower and upper quartiles. Use the results of Exercise 4.7 on page 82 to find the coefficient of quartile variation for the stopping distance of Exercise 3.38 on page 59.

4.4

THE DESCRIPTION OF GROUPED DATA ★

To determine the standard deviation of a distribution, we make the same approximation as on page 61—**to each value falling into a class we assign the value of the corresponding class mark**. Again letting $x_1, x_2, \ldots,$ and x_k denote the class marks, and $f_1, f_2, \ldots,$ and f_k the corresponding class frequencies, then the sum of all the measurements or observations is represented by

$$x_1 f_1 + x_2 f_2 + x_3 f_3 + \cdots + x_k f_k = \sum x \cdot f$$

and the sum of their squares is represented by

$$x_1^2 f_1 + x_2^2 f_2 + x_3^2 f_3 + \cdots + x_k^2 f_k = \sum x^2 \cdot f$$

The formula for $\bar{x}$ and the computing formula for s can be written as

$$\bar{x} = \frac{\sum x \cdot f}{n} \quad \text{and} \quad S_{xx} = \sum x^2 \cdot f - \frac{(\sum x \cdot f)^2}{n}$$

Then $\quad\quad s = \sqrt{\frac{S_{xx}}{n-1}}$

To get the corresponding formulas for $\bar{x}$ and S_{xx} for a population, we replace n by N; then in the formula for s, we replace $n-1$ by N.

EXAMPLE Find the mean and the standard deviation of the following distribution, giving the amounts of time that 80 college students devoted to leisure activities during a typical school week:

Hours	Frequency
10–14	8
15–19	28
20–24	27
25–29	12
30–34	4
35–39	1

Solution To obtain $\sum x \cdot f$ and $\sum x^2 \cdot f$, we perform the calculations shown in the following table:

Class mark x	x^2	Frequency f	$x \cdot f$	$x^2 \cdot f$
12	144	8	96	1,152
17	289	28	476	8,092
22	484	27	594	13,068
27	729	12	324	8,748
32	1,024	4	128	4,096
37	1,369	1	37	1,369
		80	1,655	36,525

Then, substitution into the formulas yields

$$\bar{x} = \frac{1,655}{80} = 20.6875$$

or $\bar{x} = 20.69$ rounded to two decimals, and

$$S_{xx} = 36,525 - \frac{(1,655)^2}{80} = 2287.19 \quad \text{and then} \quad s = \sqrt{\frac{2287.19}{79}} = 5.38$$

4.5

SOME FURTHER DESCRIPTIONS ★

So far we have discussed only statistical descriptions which come under the general heading of measures of location or measures of variation. Actually, there is no limit to the number of ways in which statistical data can be described, and statisticians continually develop new methods of describing characteristics of numerical data that are of interest in particular problems. In this section we shall consider briefly the problem of describing the overall shape of a distribution.

Although frequency distributions can take on almost any shape or form, most of the distributions we meet in practice can be described fairly well by one or another of a few standard types. Among these, foremost in importance is the aptly described symmetrical **bell-shaped distribution** shown at the top of Figure 4.2; it is symmetrical about the dashed vertical line. The other two distributions of Figure 4.2 can still, by a stretch of the imagination, be called bell-shaped, but they are not symmetrical. Distributions like these, having a "tail" on one side or the other, are said to be **skewed**; if the tail is on the left we say that they are **negatively skewed** and if the tail is on the right we say that they are **positively skewed**. Distributions of

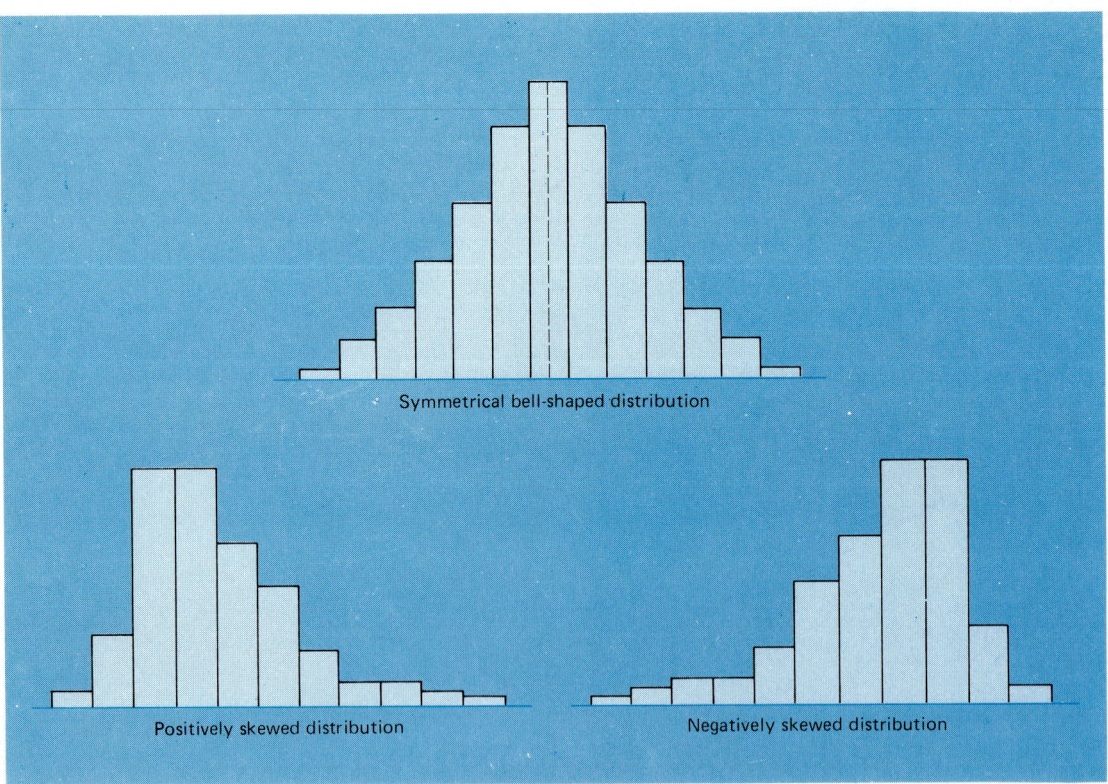

Symmetrical bell-shaped distribution

Positively skewed distribution

Negatively skewed distribution

FIGURE 4.2 *Bell-shaped distributions.*

incomes or wages are often positively skewed because of the presence of some relatively high values that are not offset by correspondingly low values.

The concepts of symmetry and skewness apply to any kind of data, not only distributions. Of course, for a large set of data we may just group the data and draw and study a histogram, but if that is not enough, we can use any one of several statistical **measures of skewness**. A relatively easy one is based on the fact that when there is perfect symmetry as in the distribution at the top of Figure 4.2, the mean and the median will coincide; when there is positive skewness and some of the high values are not offset by correspondingly low values, the mean will be greater than the median (see Figure 4.3); and when there is negative skewness and some of the low values are not offset by correspondingly high values, the mean will be smaller than the median. This relationship between the mean and the median can be used to define a relatively simple measure of skewness. It is called the **Pearsonian coefficient of skewness**, and it is given by

Pearsonian coefficient of skewness

$$SK = \frac{3(mean - median)}{standard\ deviation}$$

For a perfectly symmetrical distribution the value of SK is 0, and in general its values must fall between -3 and 3. (Division by the standard deviation makes SK independent of the scale of measurement.)

EXAMPLE Use the results obtained on pages 61–62 and 64, along with $S = 5.55$, to find the Pearsonian coefficient of skewness for the distribution of the sulfur oxides emission data.

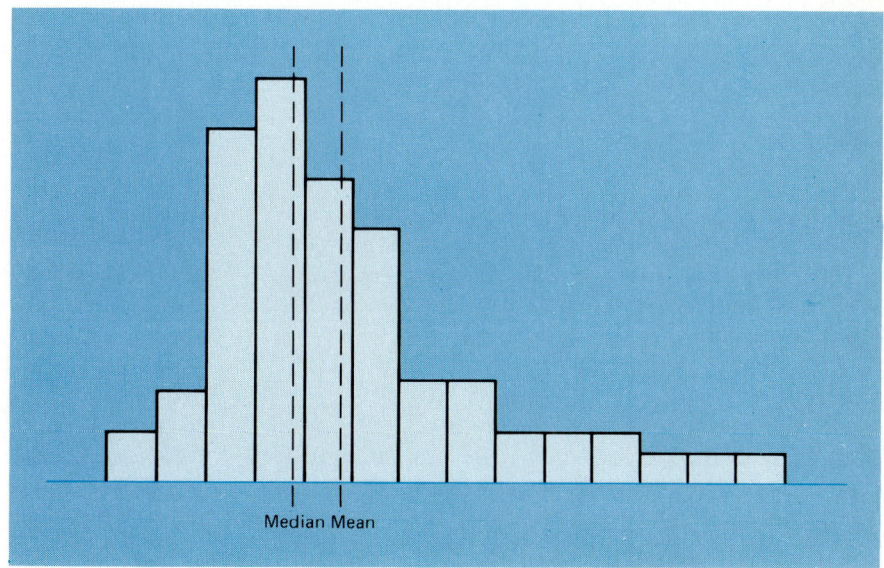

FIGURE 4.3 *The mean and the median of a positively skewed distribution.*

Solution Substituting $\bar{x} = 18.85$, $\tilde{x} = 19.03$, and $s = 5.55$ into the formula, we get

$$SK = \frac{3(18.85 - 19.03)}{5.55} = -0.01$$

Since this value is so close to zero, we can say that the distribution is nearly symmetrical. This is also apparent from the histogram of Figure 2.2 on page 29.

Besides the distributions we have discussed in this section, two others sometimes met in practice are the **reverse J-shaped** and **U-shaped** distributions shown in Figure 4.4. As can be seen from this figure, the names of these distributions literally describe their shape. Examples of such distributions may be found in Exercises 4.44 and 4.46 on page 89.

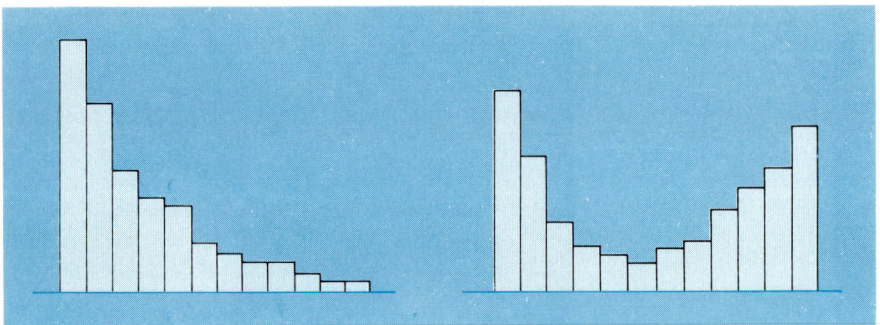

FIGURE 4.4 *Reverse J-shaped and U-shaped distributions.*

Note that the U-shaped distribution has two peaks, so that it is often described as **bimodal**; distributions with one peak are called **unimodal**.

Box plots, introduced in Section 3.5, can also be useful in describing distributions.

EXERCISES

★ **4.33** In a factory, the time during working hours in which a machine is not operating as a result of breakage or failure is called a downtime. The following distribution shows a sample of the length of the downtimes of a certain machine:

Downtime (minutes)	Frequency
0– 9	2
10–19	15
20–29	17
30–39	13
40–49	3

Find
   (a) the mean and the median;
   (b) the standard deviation.

4.34 Find $\bar{x}$ and s for the following distribution of the weekly earnings of 125 wage earners. Observe that the class marks are $124.995, $134.995, $144.995, and so on. The computations will be somewhat easier if you raise these by $0.005 to $125, $135, $145, and so on. This action will have the effect of raising each value in the list, on average, by one-half cent.

Weekly earnings (dollars)	Frequency
120.00–129.99	9
130.00–139.99	20
140.00–149.99	36
150.00–159.99	30
160.00–169.99	15
170.00–179.99	11
180.00–189.99	4

★ 4.35 Calculate the Pearsonian coefficient of skewness for the distribution of the preceding exercise and discuss the symmetry or skewness of the data.

★ 4.36 Find s for the grade distribution of Exercise 3.53 on page 65.

★ 4.37 Use the results of Exercise 3.53, part (a) of Exercise 3.54, and the preceding exercise, to calculate the Pearsonian coefficient of skewness for the grade distribution of Exercise 3.53. Discuss the symmetry or skewness of the data.

★ 4.38 In a sample of 30 three-minute intervals, a fast-food restaurant served 4, 5, 5, 8, 7, 3, 5, 6, 9, 5, 6, 5, 4, 7, 3, 5, 10, 6, 4, 5, 6, 9, 4, 5, 3, 8, 6, 7, 4, and 5 customers. Calcuate the Pearsonian coefficient of skewness and discuss the symmetry or skewness of these data.

★ 4.39 Construct a boxplot for the data of the preceding exercise and discuss the symmetry or skewness of the data.

4.40 Draw a boxplot for the data of Exercise 3.37 on page 59 and discuss their symmetry or skewness.

★ 4.41 Draw a boxplot for the data of Exercise 3.38 on page 59 and discuss the symmetry or skewness of the stopping distances.

★ 4.42 Draw a boxplot for the data of Exercise 3.28 on page 58 and discuss their symmetry or skewness.

★ 4.43 Draw a boxplot for the NBA data of Exercise 3.30 on page 58 and discuss their symmetry or skewness.

★ 4.44 The following are the numbers of 6's obtained in fifty rolls of four dice: 0, 0, 1, 0, 0, 0, 2, 0, 0, 1, 0, 0, 0, 0, 1, 1, 0, 1, 2, 0, 0, 1, 0, 0, 0, 1, 1, 0, 1, 0, 0, 1, 2, 1, 0, 0, 3, 1, 1, 0, 0, 0, 0, 1, 2, 1, 0, 0, 1, and 1. Construct a frequency distribution and a histogram, and discuss the overall shape of the data.

★ 4.45 Draw a boxplot for the data of the preceding exercise. What features of the boxplot suggest that the data have a somewhat unusual shape?

★ 4.46 If a coin is flipped five times in a row, the result may be represented by a sequence of H's and T's, where H stands for heads and T for tails. Having obtained such a sequence of H's and T's, we can then check after each successive flip whether the number of heads exceeds the number of tails. For instance, for the sequence HHTTH, heads is ahead after the first flip, after the second flip, after the third flip, not after the fourth flip, but again after the fifth flip; altogether, it is ahead four times. Actually repeating this "experiment" sixty times, we got TTHHH, THHTT, ..., and TTTHT, and found that heads was ahead 1, 1, 5, 0, 0, 5, 0, 1, 2, 0, 1, 0, 5, 1, 0, 0, 5, 0, 0, 0, 1, 0, 0, 5, 0, 2, 0, 1, 0, 5, 5, 0, 5, 4, 3, 5, 0, 5, 0, 1, 5, 0, 1, 5, 3, 1, 5, 5, 1, 2, 4, 2, 3, 0, 5, 5, 0, 0, and 0 times. Construct a frequency distribution and a histogram, and discuss the overall shape of the data.

★ 4.47 Draw a boxplot for the data of the preceding exercise. What feature of the boxplot suggests that the data have a very unusual shape?

4.6
CHECKLIST OF KEY TERMS
(with page references to their definitions)

Bell-shaped distribution, 86
Biased estimator, 76
Bimodal, 88
Chebyshev's theorem, 78

4.7

REVIEW EXERCISES

4.48 Find the standard deviation for a set of data for which $n = 15$, $\sum x = 202$, and $\sum x^2 = 3{,}452$.

4.49 The purchase amounts for the customers at a particular supermarket on a Saturday morning between 8 a.m. and noon had a mean of $56.42 and a standard deviation of $12.40. At least what proportion of the amounts must lie between $37.82 and $75.02?

★ **4.50** The following is the distribution of the sizes of a sample of 60 orders receiving by a mail-order seed business:

Size of order	Number of orders
$ 0.00–$ 19.99	6
$ 20.00–$ 39.99	16
$ 40.00–$ 59.99	12
$ 60.00–$ 79.99	11
$ 80.00–$ 99.99	8
$100.00–$119.99	7
Total	60

Calculate
(a) the quartiles Q_1 and Q_3;

(b) the interquartile range;
(c) the coefficient of quartile variation.

★ **4.51** The following are the numbers of false alarms which a fire department recorded during seventeen consecutive weeks: 8, 3, 12, 5, 6, 12, 6, 3, 4, 11, 8, 7, 5, 6, 8, 8, and 4. Construct a boxplot and discuss the symmetry or skewness of these data.

4.52 A survey of 218 families in Massachusetts indicated that the standard deviation of the number of empty deposit bottles that they had on hand was 14.6. If the coefficient of variation was 74%, what is the mean of the sample data?

4.53 Calculate σ^2 for the population which consists of the integers 1, 2, 3, 4, and 5. Does your calculation support the claim that for a population consisting of the first k positive integers the variance is $\dfrac{k^2 - 1}{12}$?

4.54 A quality control inspector examined 15 crates of ceramic tiles, each containing 144 tiles. The numbers of cracked tiles in these boxes were 2, 5, 3, 4, 2, 0, 1, 5, 7, 3, 0, 2, 2, 4, and 3. Calculate s for these data.

★ **4.55** The following is the distribution of the number of mistakes made by 200 students taking German in a multiple-choice quiz on vocabulary:

Number of mistakes	Number of students
6–10	12
11–15	73
16–20	52
21–25	39
26–30	24
Total	200

Find
(a) the mean;
(b) the standard deviation.

★ **4.56** Calculate the Pearsonian coefficient of skewness for the distribution of the preceding exercise and discuss the symmetry or skewness of the data.

4.57 For a certain variety of lightbulb, a bulb that lasts 1,020 hours has a standard score of $z = 2$. Given that the lifetimes of these bulbs have a coefficient of variation of 14%, find the mean and standard deviation of the lifetimes.

4.58 According to Chebyshev's theorem, what can we assert about the percentage of any set of data that must lie within k standard deviations on either side of the mean when
(a) $k = 6$;
(b) $k = 8$;
(c) $k = 15$?

4.59 The following are the number of accidents that occurred in July of 1990 in a certain town at eighteen intersections without left-turn arrows: 8, 29, 31, 14, 35, 28, 12, 18, 22, 13, 6, 32, 2, 10, 26, 22, 32, and 25. Find
(a) the median;
(b) Q_1 and Q_3.

4.60 Use the results of the preceding exercise to draw a boxplot for the accident data.

4.61 For a large group of students the mean score on a historical events quiz is 160 points, with a standard deviation of 22 points. At least what percentage of the scores must lie between
(a) 127 and 193 points;
(b) 116 and 204 points;
(c) 105 and 215 points?

4.62 If a set of measurements has the mean $\bar{x} = 48$ and the standard deviation $s = 12$, convert each of the following into standard units:
(a) 54;
(b) 72;
(c) 78.

4.63 Find the variance of the sample data of Exercise 4.59.

4.64 Explain why it is impossible to have $n = 10$, $\sum x = 40$, $\sum x^2 = 140$ for a given set of data.

4.8

REFERENCES

A proof that division by $n - 1$ makes the sample variance an unbiased estimator of the population variance may be found in most textbooks on mathematical statistics; for instance, in

Freund, J. E., *Mathematical Statistics*, 5th ed. Englewood Cliffs, N.J.: Prentice-Hall, Inc., 1987.

An unbiased estimator of the population standard deviation σ is given on pages 62–63 of

Johnson, N. L., and Kotz, S., *Distributions in Statistics: Continuous Univariate Distributions*. Boston: Houghton Mifflin, 1970.

For further information about the use and interpretation of boxplots, see the books on exploratory data analysis listed on page 38.

Some information about the effect of grouping on the calculation of various statistical descriptions may be found in some of the older textbooks on statistics; for instance, in

MILLS, F. C., *Introduction to Statistics.* New York: Holt, Rinehart and Winston, 1956.

POSSIBILITIES AND PROBABILITIES

The central problem of statistics is dealing with chance and uncertainty. Chance events have always been regarded as mysterious. The Book of Job pondered long ago the role of divine intent in chance occurrences, and it was many centuries later that the power of mathematics was used to explain randomness. The scientific advances of the centuries following the Renaissance, emphasizing careful observation and experimentation, brought forth **probability theory** to study the laws of nature and the problems of everyday life.

In this chapter we shall see how uncertainties can actually be measured, how they can be assigned numbers, and how these numbers are to be interpreted. In subsequent chapters we shall see how these numbers, called **probabilities**, can be used to live with uncertainties, enabling us to make decisions which are likely to be fruitful.

In Sections 5.1 through 5.3 we present mathematical preliminaries dealing with the question of "what is possible" in given situations. After all, we can hardly predict the outcome of a football game unless we know what teams are playing, and we cannot very well predict what will happen in an election unless we know what candidates are running for office. Then, in Section 5.4 we shall learn how to judge also "what is probable;" that is, we shall learn about several ways in which probabilities are defined, or interpreted, and their values are determined.

5.1
COUNTING

In the study of "what is possible," there are essentially two kinds of problems. There is the problem of listing everything that can happen in a given situation, and then there is the problem of determining how many different things can happen (without actually constructing a complete list). The second kind of problem is especially important, because there are many situations in which we do not need a complete list and, hence, can save ourselves a great deal of work. Although the first kind of problem may seem straightforward and easy, the following example illustrates that this is not always the case:

EXAMPLE   A governmental agency is committed to the purchase of three vehicles from a local automobile dealer. Each of these vehicles will be either a jeep, a pickup truck, a minivan, or some other vehicle. List the different ways in which the purchases can be made.

Solution   There are many possibilities. All three vehicles may be jeeps; two may be jeeps while the other is a minivan; two may be pickup trucks while the other is a minivan; and so forth. If we continue in this manner, it is likely that we will omit some of the possibilities.

The problem can be handled systematically by drawing a **tree diagram** such as Figure 5.1. This shows that there are four possibilities (four branches) corre-

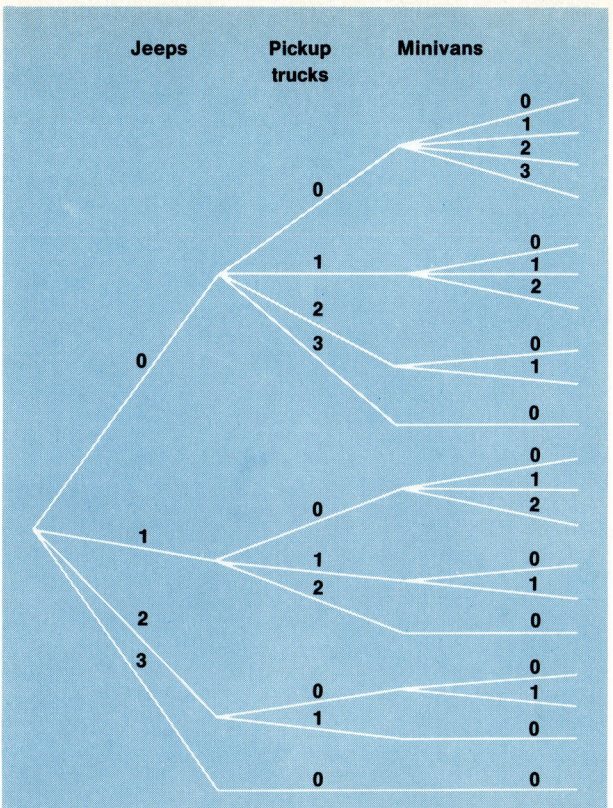

FIGURE 5.1 *Tree diagram for choice of vehicle example.*

sponding to 0, 1, 2, or 3 of the vehicles being jeeps. For pickup trucks there are four branches coming from the top branch (0 jeeps), three branches coming from the next branch (1 jeep), two branches coming from the following branch (2 jeeps), and just one branch coming from the bottom branch. For the number of minivans, the reasoning is similar. You can see that twenty branches end at the right side of Figure 5.1. In other words, there are altogether twenty possibilities.

EXAMPLE In a medical study, patients are classified according to whether they have blood type A, B, AB, or O, and also according to whether their blood pressure is low, normal, or high. In how many different ways can a patient thus be classified?

Solution As is apparent from the tree diagram of Figure 5.2, the answer is 12. Starting at the top, the first path along the "branches" corresponds to a patient having blood type A and low blood pressure, the second path corresponds to a patient having blood type A and normal blood pressure,..., and the twelfth path corresponds to a patient having blood type O and high blood pressure.

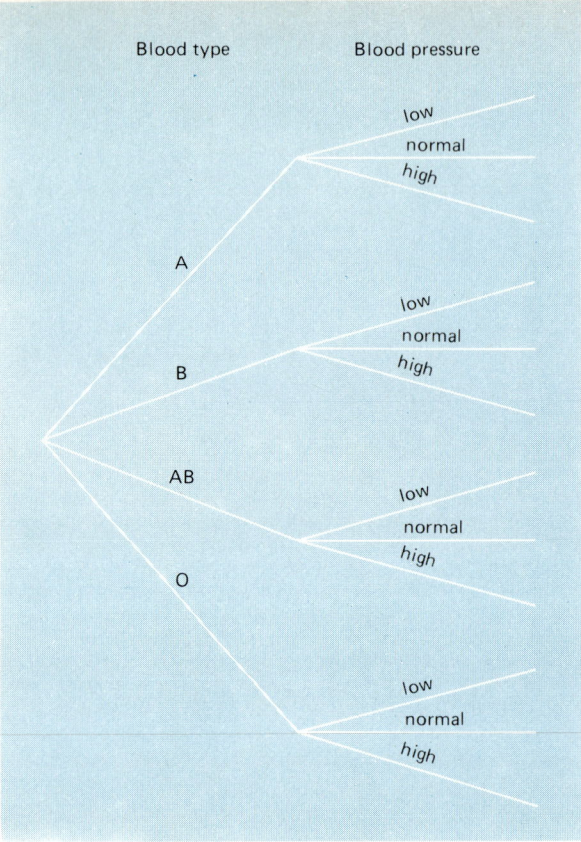

FIGURE 5.2 *Tree diagram for classification of patients.*

The answer we got in this example is $4 \cdot 3 = 12$, namely, the product of the number of blood types and the number of blood pressure levels. Generalizing from this example, let us state the following rule:

Multiplication of choices

If a choice consists of two steps, of which the first can be made in m ways and for each of these the second can be made in n ways, then the whole choice can be made in m · n ways.

To prove this we have only to draw a tree diagram like that of Figure 5.2. First there are *m* branches corresponding to the *m* possibilities in the first step, and then there are *n* branches emanating from each of these branches corresponding to the *n* possibilities in the second step. This leads to *m · n* paths along the branches of the tree diagram, and hence to *m · n* possibilities.

EXAMPLE   If a travel agency offers special weekend trips to 12 different cities, by air, rail, or bus, in how many different ways can such a trip be arranged?

Solution   Since $m = 12$ and $n = 3$, there are $12 \cdot 3 = 36$ ways in which such a trip can be arranged.

EXAMPLE   If an ice-cream shop offers one-scoop mini-sundaes with a choice of twenty ice-cream flavors along with a choice of eight toppings, in how many different ways can a customer order a mini-sundae?

Solution   Since $m = 20$ and $n = 8$, there are $20 \cdot 8 = 160$ ways in which a customer can order a mini-sundae.

By using appropriate tree diagrams, we can easily generalize the above rule so that it will apply to choices involving more than two steps. For k steps, where k is a positive integer, we get the following rule:

Multiplication of choice

> *If a choice consists of k steps, of which the first can be made in n_1 ways, for each of these the second can be made in n_2 ways, ..., and for each of these the kth can be made in n_k ways, then the whole choice can be made in $n_1 \cdot n_2 \cdot \cdots \cdot n_k$ ways.*

We simply keep multiplying the numbers of ways in which the different steps can be made.

EXAMPLE   A luncheonette offers a special meal consisting of a sandwich (using one of eight different meats and one of four different types of bread), one of four different types of soup, and one of three different beverages. In how many ways can a person choose one of these special meals?

Solution   Since $n_1 = 8$, $n_2 = 4$, $n_3 = 4$, and $n_4 = 3$, there are $8 \cdot 4 \cdot 4 \cdot 3 = 384$ ways in which the special meal may be chosen.

EXAMPLE   A test consists of fifteen multiple-choice questions, with each question having four possible answers. In how many different ways can a student check off one answer to each question?

Solution   Since $n_1 = n_2 = n_3 = \cdots = n_{15} = 4$, there are altogether

$$4 \cdot 4 \cdot 4 \cdot 4 \cdot 4 \cdot 4 \cdot 4 \cdot 4 \cdot 4 \cdot 4 \cdot 4 \cdot 4 \cdot 4 \cdot 4 \cdot 4 = 1{,}073{,}741{,}824$$

different ways in which a student can check off one answer to each question. Note that in only one of the 1,073,741,824 possibilities all the answers are correct and in

$$3 \cdot 3 \cdot 3 \cdot 3 \cdot 3 \cdot 3 \cdot 3 \cdot 3 \cdot 3 \cdot 3 \cdot 3 \cdot 3 \cdot 3 \cdot 3 \cdot 3 = 14{,}348{,}907$$

of them all the answers are wrong.

5.2

PERMUTATIONS

The rule for the multiplication of choices and its generalization are often used when several choices are made from one and the same set and we are concerned with the order in which they are made.

EXAMPLE If twenty paintings are entered in an art show, in how many different ways can the judges award a first prize and a second prize?

Solution Since the first prize can be awarded in $m = 20$ ways and the second prize must be awarded to one of the other $n = 19$ paintings, there are altogether $20 \cdot 19 = 380$ ways in which the judges can make the two awards.

EXAMPLE In how many different ways can the 52 members of a labor union choose a president, a vice-president, a secretary, and a treasurer?

Solution Since $n_1 = 52$, $n_2 = 51$, $n_3 = 50$, and $n_4 = 49$ (regardless of which officer is chosen first, second, third, and fourth), there are altogether $52 \cdot 51 \cdot 50 \cdot 49 = 6,497,400$ different possibilities.

In general, if r objects are selected from a set of n distinct objects, any particular arrangement (order) of these objects is called a **permutation**. For instance, 4 1 2 3 is a permutation of the first four positive integers; Maine, Vermont, and Connecticut is a permutation (a particular ordered arrangement) of three of the six New England states; and

Red Sox, Tigers, Indians, Orioles

Blue Jays, Red Sox, Yankees, Tigers

are two different permutations (ordered arrangements) of four of the seven baseball teams in the Eastern Division of the American League.

EXAMPLE Determine the number of different permutations of two of the five vowels a, e, i, o, and u, and list them all.

Solution Since $m = 5$ and $n = 4$, there are $5 \cdot 4 = 20$ different permutations, and they are

ae	*ai*	*ao*	*au*	*ei*	*eo*	*eu*	*io*	*iu*	*ou*
ea	*ia*	*oa*	*ua*	*ie*	*oe*	*ue*	*oi*	*ui*	*uo*

A formula for the total number of permutations of r objects selected from a set of n distinct objects, such as the seven baseball teams, would be useful. Observe that the first selection is made from the whole set of n objects, the second selection is made from the $n - 1$ objects which remain after the first selection has been made, the third selection is made from the $n - 2$ objects which remain after the first two selections have been made, . . . , and the rth and final selection is made from the

$n - (r - 1) = n - r + 1$ objects which remain after the first $r - 1$ selections have been made. Therefore, direct application of the generalized rule for the multiplication of choices yields the result that the total number of permutations of r objects selected from a set of n distinct objects, which we shall denote by $_nP_r$, is

$$n(n - 1)(n - 2) \cdots \cdot (n - r + 1)$$

Since products of consecutive integers arise in many problems relating to permutations and other kinds of special arrangements or selections, it is convenient to introduce here the **factorial notation**. In this notation, the product of all positive integers less than or equal to the positive integer n is called "n factorial" and denoted by $n!$. Thus,

$$1! = 1$$
$$2! = 2 \cdot 1 = 2$$
$$3! = 3 \cdot 2 \cdot 1 = 6$$
$$4! = 4 \cdot 3 \cdot 2 \cdot 1 = 24$$
$$5! = 5 \cdot 4 \cdot 3 \cdot 2 \cdot 1 = 120$$
$$6! = 6 \cdot 5 \cdot 4 \cdot 3 \cdot 2 \cdot 1 = 720$$

$$. \quad . \quad . \quad . \quad . \quad .$$

and in general $n! = n(n - 1)(n - 2) \cdots \cdot 3 \cdot 2 \cdot 1$. Also, to make various formulas more generally applicable, we let $0! = 1$ by definition.

The factorials grow so quickly that some have claimed that the exclamation point indicates surprise. The value of $10!$ exceeds three million, and $70!$ exceeds the memory limit of most hand-held calculators.

To express the formula for $_nP_r$ in terms of factorials, we note, for example, that $12 \cdot 11 \cdot 10! = 12!$, $9 \cdot 8 \cdot 7 \cdot 6! = 9!$, and $37 \cdot 36 \cdot 35 \cdot 34 \cdot 33! = 37!$. Similarly,

$$_nP_r \cdot (n - r)! = n(n - 1)(n - 2) \cdots \cdot (n - r + 1) \cdot (n - r)!$$
$$= n!$$

so that $_nP_r = \dfrac{n!}{(n - r)!}$. To summarize:

Number of permutations of n objects taken r at a time

> *The number of permutations of r objects selected from a set of n distinct objects is*
>
> $$_nP_r = n(n - 1)(n - 2) \cdots \cdot (n - r + 1)$$
>
> *or, in factorial notation,*
>
> $$_nP_r = \frac{n!}{(n - r)!}$$

where either formula can be used for $r = 1, 2, \ldots$, or n. (The second formula, but not the first, could be used also for $r = 0$, for which we would get the trivial result that there is

$$_nP_0 = \frac{n!}{(n-0)!} = 1$$

way of selecting none of n objects.) The first formula is generally easier to use because it requires fewer steps, but many students find the one in factorial notation easier to remember.

EXAMPLE   Find the number of permutations of four objects selected from a set of 12 distinct objects (say, the number of ways in which four of 12 basketball teams can be ranked first, second, third, and fourth by a panel of coaches).

Solution   For $n = 12$ and $r = 4$, the first formula yields

$$_{12}P_4 = 12 \cdot 11 \cdot 10 \cdot 9 = 11{,}880$$

and the second formula yields

$$_{12}P_4 = \frac{12!}{(12-4)!} = \frac{12!}{8!} = \frac{12 \cdot 11 \cdot 10 \cdot 9 \cdot \cancel{8!}}{\cancel{8!}} = 11{,}880$$

Essentially, the work is the same, but the second formula requires a few extra steps.

To find the formula for the number of permutations of n distinct objects taken all together, we substitute $r = n$ into either formula for $_nP_r$ and get

Number of permutations of n objects taken n at a time

$$_nP_n = n!$$

EXAMPLE   In how many ways can ten instructors be assigned to ten sections of a course in economics?

Solution   Substituting $n = 10$, we get

$$_{10}P_{10} = 10! = 3{,}628{,}800$$

Throughout this discussion it has been assumed that the n objects are all distinct. When this is not the case, the formula for $_nP_n$ can be easily modified, but we shall not go into that in this book. (The modification of $_nP_r$ with $r < n$ is complicated.) Exercises 5.26 and 5.27 deal with non-distinct objects in the case $r = n$.

5.3
COMBINATIONS

There are many problems in which we want to know the number of ways in which r objects can be selected from a set of n objects, but we do not care about the order in which the selection is made. For instance, we may want to know in how many ways a committee of four can be selected from among the 45 members of a college fraternity, or the number of ways in which the IRS can choose five of 36 tax returns for a special audit. To derive a formula which applies to problems like these, let us first examine the following 24 permutations of three of the first four letters of the alphabet:

abc	acb	bac	bca	cab	cba
abd	adb	bad	bda	dab	dba
acd	adc	cad	cda	dac	dca
bcd	bdc	cbd	cdb	dbc	dcb

If we do not care about the order in which the three letters are chosen from among the four letters a, b, c, and d, there are only four ways in which the selection can be made: abc, abd, acd, and bcd. Note that these are the groups of letters shown in the first column of the table, and that each row contains the $_3P_3 = 3! = 6$ permutations of the three letters in the first column.

In general, there are $_rP_r = r!$ permutations of r distinct objects, so that the $_nP_r$ permutations of r objects selected from among n distinct objects contain each group of r objects $r!$ times. (In our example, the $_4P_3 = 4 \cdot 3 \cdot 2 = 24$ permutations of three letters selected from among the first four letters of the alphabet contain each group of three letters $_3P_3 = 3! = 6$ times.) Therefore, to get a formula for the number of ways in which r objects can be selected from a set of n distinct objects *without regard to their order*, we divide $_nP_r$ by $r!$. Referring to such a selection as a **combination** of n objects taken r at a time, we denote the number of combinations of n objects taken r at a time by $_nC_r$ or $\binom{n}{r}$, and write

Number of combinations of n objects taken r at a time

The number of ways in which r objects can be selected from a set of n distinct objects is

$$\binom{n}{r} = \frac{n(n-1)(n-2) \cdot \cdots \cdot (n-r+1)}{r!}$$

or, in factorial notation,

$$\binom{n}{r} = \frac{n!}{r!(n-r)!}$$

Like the two formulas for $_nP_r$, either formula can be used for $r = 1, 2, \ldots,$ or n, but the second one only for $r = 0$. Again, the first formula is generally easier to use because it requires fewer steps, but many students find the one in factorial notation easier to remember.

For $n = 0$ to $n = 20$, the values of $\binom{n}{r}$ may be read from Table X, where these quantities are referred to as **binomial coefficients**. The reason for this is explained in Exercise 5.41 on page 107.

EXAMPLE   In how many ways can a person choose three books from a list of eight best-sellers?

Solution   As it is assumed here that the order in which the three books are selected does not matter, we substitute $n = 8$ and $r = 3$ into the first formula and get

$$\binom{8}{3} = \frac{8 \cdot 7 \cdot 6}{3!} = 8 \cdot 7 = 56$$

Similarly, substitution into the second formula yields

$$\binom{8}{3} = \frac{8!}{3!5!} = \frac{8 \cdot 7 \cdot 6 \cdot 5!}{3! 5!} = \frac{8 \cdot 7 \cdot 6}{3 \cdot 2 \cdot 1} = 56$$

Essentially, the work is the same, but the first formula required fewer steps.

EXAMPLE   In how many different ways can a committee of five be selected from the 62 members of the clerical staff at a large legal firm?

Solution   Since the order in which the committee members are selected does not matter, we substitute $n = 62$ and $r = 5$ into the first formula to get

$$\binom{62}{5} = \frac{62 \cdot 61 \cdot 60 \cdot 59 \cdot 58}{5!} = 6,471,002$$

Note that the result of the first example, but not that of the second, can be verified in Table X.

EXAMPLE   In how many different ways can the director of a research laboratory choose two chemists from among seven applicants and three physicists from among nine applicants?

Solution   The two chemists can be selected in $\binom{7}{2}$ ways, the three physicists can be selected in $\binom{9}{3}$ ways, so that, by the multiplication of choices, all five of them can be selected in

$$\binom{7}{2} \cdot \binom{9}{3} = 21 \cdot 84 = 1,764$$

ways. The values of the two binomial coefficients were obtained from Table X.

Using $r = n$ in either formula for $\binom{n}{r}$ yields $\binom{n}{n} = 1$. In other words, there is one and only one way in which we can select all n of the elements which constitute a set.

When we take 7 objects from a set of 10 distinct objects, then $10 - 7 = 3$ of the objects are left. Thus, there are as many ways of leaving (or selecting) 3 objects from a set of 10 distinct objects as there are ways of selecting 7 objects, and we can write $\binom{10}{7} = \binom{10}{3}$. In general, when r objects are selected from a set of n distinct objects, $n - r$ of the objects are left, and consequently, there are as many ways of leaving (or selecting) $n - r$ objects from a set of n distinct objects as there are ways of selecting r objects. Symbolically, we write

Rule for binomial coefficients

$$\binom{n}{r} = \binom{n}{n-r} \qquad \text{for } r = 0, 1, 2, \dots, n$$

Sometimes this rule serves to simplify calculations and sometimes it is needed in connection with the use of Table X.

EXAMPLE Determine the value of $\binom{75}{72}$.

Solution To avoid having to write down the product $75 \cdot 74 \cdot 73 \cdot \dots \cdot 4$ and cancel $72 \cdot 71 \cdot 70 \cdot \dots \cdot 4$, we write directly

$$\binom{75}{72} = \binom{75}{3} = \frac{75 \cdot 74 \cdot 73}{3!} = 67{,}525$$

EXAMPLE Find the value of $\binom{18}{13}$.

Solution $\binom{18}{13}$ cannot be looked up directly in Table X, but making use of the fact that $\binom{18}{13} = \binom{18}{18 - 13} = \binom{18}{5}$, we look up $\binom{18}{5}$ and get 8,568.

If we substitute $r = 0$ into the second of the two formulas for $\binom{n}{r}$, or if we write $\binom{n}{0} = \binom{n}{n-0} = \binom{n}{n}$, we get $\binom{n}{0} = 1$. Evidently, there are as many ways of selecting none of the objects in a set as there are ways of choosing the n objects which are left.

EXERCISES

5.1 In a baseball World Series the winner is the first team to win four games. Suppose that the National League champion leads the American League champion three games to two. Construct a tree diagram to show the number of ways in which these teams may continue to the completion of the series.

5.2 A person with $3 in his pocket bets $1, even money, on the flip of a coin. He continues to bet $1 so long as he has any money. Draw a tree diagram to show the various things that can happen in the first four flips of the coin. In how many of the cases will he be
  (a) exactly $2 ahead;
  (b) exactly $1 ahead;
  (c) exactly even;
  (d) exactly $1 behind;
  (e) exactly $2 behind?

5.3 A student can study 0, 1, or 2 hours for a statistics test on any given night. Draw tree diagrams to find the number of ways in which the student can study
  (a) a total of exactly five hours on three consecutive nights;
  (b) a total of at least five hours on three consecutive nights.

5.4 A food specialty store receives two cherry cheesecakes each morning. Cheesecakes not sold by closing time are discarded. Construct a tree diagram to show the number of ways in which the store can sell a total of five cherry cheesecakes on four consecutive days.

5.5 The store of the previous exercise also receives three cream cakes each morning, and cream cakes which are not sold by closing time must be discarded. Construct a tree diagram to show the number of ways in which the store can sell a total of five cream cakes on three consecutive days.

5.6 In a union election, Mr. Brown, Ms. Green, and Ms. Jones are running for president. Mr. Adams, Ms. Roberts, and Mr. Smith are running for vice-president. Construct a tree diagram showing the nine possible outcomes, and use it to determine the number of ways in which the two unions officials will not be of the same sex.

5.7 In a political science survey, voters are classified into six income categories and five education categories. In how many different ways can a voter be classified?

5.8 A chain of furniture stores has three warehouses and twenty retail outlets. In how many different ways can they ship an item from one of the warehouses to one of the retail outlets?

5.9 A purchasing agent places his orders by phone, by fax, by mail, or by express carrier. He requests that the order be confirmed by phone or by fax. In how many different ways can one of his orders by placed and confirmed?

5.10 There are five routes between an executive's home and her place of work.
  (a) In how many different ways can she go to and from work?
  (b) In how many different ways can she go to and from work if she does not want to use the same route both ways?
  (c) If one of her five routes has been made into a one-way street, then in how many different ways can she go to and from work (assuming that she is willing to use the same route both ways)?
  (d) If one of her five routes has been made into a one-way street, then in how many different ways can she go to and from work (assuming that she does not want to use the same route both ways)?

5.11 In an optics kit there are six concave lenses, four convex lenses, two prisms, and two mirrors. In how many different ways can one choose a concave lens, a convex lens, a prism, and a mirror from this kit?

5.12 A psychologist is preparing three-letter nonsense words for use in a memory test. He chooses the first letter from the consonants k, m, w, and z. He chooses the middle letter from the vowels a, i, and u. He chooses the final letter from the consonants b, d, f, k, m, and t.
  (a) How many different three-letter nonsense words can he construct?
  (b) How many of these nonsense words will begin with the letter z?
  (c) How many of these nonsense words will end with either k or m?
  (d) How many of these nonsense words will begin and end with the same letter?

5.13 In a doctor's office, there are eight recent issues of *Newsweek*, six issues of the *New Yorker*, and five issues of the *Reader's Digest*. In how many different ways can a patient waiting to see the doctor glance at one of each kind of magazine if the order does not matter?

5.14 A true–false test consists of ten questions. In how many ways can a student mark one answer to each question?

5.15 A pizzeria offers ten toppings for its pizzas. The customer can order any combinations of these toppings, including all of them and none of them. How many different types of pizza are possible? (*Hint*: For each topping, the diner must decide whether or not to select it.)

5.16 Determine whether each of the following is true or false:

(a) $18! = 18 \cdot 17 \cdot 16!$;   (c) $6! = 30 \cdot 4!$;

(b) $4! \cdot 5! = 20!$;   (d) $15! = \dfrac{16!}{16}$.

5.17 Determine whether each of the following is true or false:

(a) $\dfrac{1}{2!} + \dfrac{1}{3!} = \dfrac{5}{6!}$;

(b) $5! + 2! = 7!$;

(c) $\dfrac{6!}{2! \cdot 4!} + \dfrac{6!}{3! \cdot 3!} = \dfrac{7!}{3! \cdot 4!}$;

(d) $\dfrac{1}{3!} + \dfrac{1}{3!} = \dfrac{1}{3}$.

5.18 On a vacation, a person would like to visit three of ten historical sights in Philadelphia. If the order of the visits matters, in how many different ways can the trip be planned?

5.19 A person spending seven nights in Cleveland has obtained a list containing the city's eight best Italian restaurants and nine best Chinese restaurants. In how many ways can this person eat seven evening meals at these restaurants, assuming that he wishes to try a different restaurant each night and also that he wishes to alternate between Italian and Chinese food?

5.20 In how many ways can four new corporate clients be assigned to eleven service representatives, assuming that each service representative can be given at most one of the corporate clients?

5.21 An amusement park has 28 different rides. In how many different ways can a person try four of these rides, assuming that order matters and that she does not want to try any ride more than once?

5.22 If there are nine horses in a race, in how many different ways can they finish first, second, and third?

5.23 In how many different ways can eight books be placed on a shelf?

5.24 In how many different ways can the manager of a baseball team arrange the batting order of the nine players in the starting lineup? In how many different ways can this be done if he insists that the pitcher should bat ninth?

5.25 Four married couples have purchased eight seats in a row for a football game. In how many different ways can they be seated if
(a) each couple is to sit together;
(b) all the men are to sit together and all the women are to sit together;
(c) all the men are to sit together;
(d) the men and women are to sit in alternate seats;
(e) no man is to sit next to another man?

★ **5.26** If among n objects r are alike, and the others are all distinct, the number of permutations of these n objects taken all together is $\dfrac{n!}{r!}$.
(a) How many permutations are there of the letters in the word "silly"?
(b) How many permutations are there of the letters in the word "bubble"?
(c) In how many ways can a radio director select six consecutive commercials, using one commercial for product A, one commercial for product B, one commercial for product C, and three identical commercials for product D?
(d) Present an argument to justify the formula given in this exercise.

★ **5.27** If among n objects r_1 are identical, another r_2 are identical, and the remaining (if any) are all distinct, the number of permutations of these n objects taken all together is $\dfrac{n!}{r_1! \cdot r_2!}$.
(a) How many permutations are there of the letters in the word "better"?
(b) How many permutations are there of the letters in the word "hubbub"?
(c) In how many ways can the radio director of the previous problem select six consecutive commercials, using three identical commercials for product E and three identical commercials for product F?
(d) Generalize the formula so that it applies to the permutations of n objects taken all together if there are r_1 identical, another r_2 identical, another r_3 identical, and so on.
(e) How many permutations are there of the letters in Mississippi?

★ **5.28** There is no simple formula for the number of permutations of n non-identical objects when fewer than n are selected.

  (a) List all possible permutations of three letters taken from the letters of the word "fever." Note that "e" may appear 0, 1, or 2 times among the selected letters.

  (b) List all possible permutations of four letters taken from the letters of the word "lullaby".

★ **5.29** The number of ways in which n distinct objects can be arranged in a circle is $(n - 1)!$.

  (a) Present an argument to justify this formula.

  (b) In how many ways can six persons be seated at a round table, if we care only who sits on whose left or right side?

  (c) In how many ways can eight persons form a circle for a folk dance?

  (d) Four men and four women are forming a circle for a folk dance. In how many ways can this be done if we require that the men and women alternate positions?

★ **5.30** The number of ways in which n distinct keys can be placed on a key ring is $(n - 1)!/2$.

  (a) Show that there is only one way in which three keys can be placed on a ring.

  (b) In how many ways can four keys be placed on a ring?

  (c) In how many ways can six keys be placed on a ring?

5.31 The giftshop of a tourist resort has fifteen different scenic pictures. In how many ways can a person select four of these as souvenirs?

5.32 A pizza shop offers ten different toppings for its pizza. In how many ways can a customer select three toppings for his pizza?

5.33 A bookstore has a sale in which a customer gets a special price if buying four of the current ten best-sellers. In how many ways can a customer make such a selection?

5.34 A true–false test contains twelve questions. Calculate the numbers of ways in which a student can mark each question either true or false and get

  (a) eight right and four wrong;

  (b) ten right and two wrong.

5.35 A high-school student doing a report on ancient Greece has found fifteen books on this subject in the school library. The library rules permit him to check out only five books at a time. Find the number of ways in which the student can select five books, and verify your answer in Table X.

5.36 A carton of twelve eggs contains one egg that is spoiled. In how many ways can a person choose three of these eggs and

  (a) get the one egg that is spoiled;

  (b) not get the one egg that is spoiled?

5.37 A ten-pack of batteries has two defective batteries. In how many ways can one select three of these batteries and get

  (a) neither of the defective batteries;

  (b) one of the defective batteries;

  (c) both of the defective batteries?

5.38 Among the eight nominees for two vacancies on a school board are four men and four women. In how many ways can these vacancies be filled

  (a) with any two of the eight nominees;

  (b) with any two of the female nominees;

  (c) with any two of the male nominees;

  (d) with one of the male nominees and one of the female nominees?

5.39 A men's clothing story carries eight kinds of sweaters, six kinds of slacks, and ten kinds of shirts. In how many ways can two of each kind be chosen for a special sale?

★ **5.40** Counting the number of outcomes in games of chance has been popular for many centuries. Not only was gambling involved, but the outcomes were also taken as indications of divine intent. It was just about one thousand years ago that a bishop in what is now Belgium determined that there are 56 different ways in which three dice can fall, provided one is interested only in the overall result and not in the outcomes of the individual dice. He assigned a virtue to each of these possibilities and each sinner had to concentrate for some time on the virtue which corresponded to his cast of the dice.

  (a) Find the number of ways in which three dice can all come up with the same number of points.

  (b) Find the number of ways in which two of the three dice can come up with the same number of points while the third die comes up with a different number of points.

  (c) Find the number of ways in which all three of the dice can come up with a different number of points.

  (d) Use (a), (b), and (c) to verify the bishop's calculation that there are altogether 56 possibilities.

5.41 The quantity $\binom{n}{r}$ is called a **binomial coefficient** because it is the coefficient of $a^{n-r}b^r$ in the binomial expansion of $(a+b)^n$.

    (a) Verify this fact for $n = 1$.
    (b) Verify this fact for $n = 2$ by expanding $(a+b)^2$.
    (c) Verify this fact for $n = 3$ by expanding $(a+b)^3$.
    (d) Verify this fact for $n = 4$ by expanding $(a+b)^4$.
    (e) Compare your findings to the corresponding entries of Table X.

5.42 Use Table X to determine the values of the following binomial coefficients:

    (a) $\binom{18}{6}$;   (c) $\binom{18}{15}$;

    (b) $\binom{14}{6}$;   (d) $\binom{16}{11}$.

5.43 Verify that

    (a) $\binom{12}{8} = 3 \cdot \binom{11}{8}$;

    (b) $10 \cdot \binom{14}{4} = 14 \cdot \binom{13}{4}$.

    (c) $\dfrac{\binom{10}{4}}{\binom{9}{3}} = \dfrac{5}{2}$.

★ 5.44 A table of binomial coefficients is easy to construct by following the pattern below.

```
                    1
                 1     1
              1     2     1
           1     3     3     1
        1     4     6     4     1
     1     5    10    10     5     1
   . . . . . . . . . . . . . . . . . . . . . . . . . .
```

This pattern is called **Pascal's triangle**. In this arrangement, each row begins with a 1, ends with a 1, and each entry is the sum of the nearest two values from the row above. Construct the next three rows of Pascal's triangle and verify from Table X that they are, respectively, the binomial coefficients corresponding to $n = 6$, $n = 7$, and $n = 8$.

★ 5.45 In Pascal's triangle, denote the row 1 2 1 as row 2, the row 1 3 3 1 as row 3, the row 1 4 6 4 1 as row 4, and so on. The row beginning 1 n ... will be row n. Within row n, label the $n + 1$ positions as 0, 1, 2, ..., n. Observe that $\binom{n}{r}$ appears in position r of row n.

    (a) What is row 7?

    (b) Give the value of $\binom{7}{3}$.

    (c) The triangle is constructed by the relationship

$$\binom{n}{r} = \binom{n-1}{r-1} + \binom{n-1}{r}$$

Verify this relationship by expressing the binomial coefficients in factorial form.

★ 5.46 Susan is one of seven office workers in a small business. Three of these workers will be selected to serve on a committee.

    (a) In how many different ways can three of these persons be chosen to be on the committee?

    (b) In how many different ways can three of these persons be chosen so that Susan is *not* on the committee?

    (c) In how many different ways can three of the persons be chosen so that Susan is one of the chosen persons?

    (d) Verify that your solutions to (b) and (c) add up to the solution to (a). This is an example of the result given by the previous problem.

5.4
PROBABILITY

So far in this chapter we have studied only what is possible in a given situation. In some instances we listed all the possibilities and in others we merely determined how many different possibilities there are. Now we shall go one step further and judge also what is probable and what is improbable.

The most common way of measuring the uncertainties connected with events (say, the outcome of a presidential election, the side effects of a new medication, the durability of an exterior paint, or the total number of points we may roll with a pair of dice) is to assign them **probabilities** or to specify the **odds** at which it would be fair to bet that the events will occur. In this section we shall learn how probabilities are interpreted and how their numerical values are determined; odds will be discussed in Section 6.3.

Historically, the oldest way of measuring uncertainties is the **classical probability concept**. It was developed originally in connection with games of chance, and it lends itself most readily to bridging the gap between possibilities and probabilities. The classical probability concept applies only when all possible outcomes are equally likely, in which case we say that

The classical probability concept

> *If there are n equally likely possibilities, of which one must occur and s are regarded as favorable, or as a "success," then the probability of a "success" is $\frac{s}{n}$.*

In the application of this rule, the terms "favorable" and "success" are used rather loosely—what is favorable to one player is unfavorable to his opponent, and what is a success from one point of view is a failure from another. Thus, the terms "favorable" and "success" can be applied to any particular kind of outcome, even if "favorable" means that a television set does not work, or "success" means that someone catches the flu. This usage dates back to the days when probabilities were quoted only in connection with games of chance.

EXAMPLE   What is the probability of drawing an ace from a well-shuffled deck of 52 playing cards?

Solution   By "well-shuffled" we mean that each card has the same chance of being drawn, so that the classical probability concept can be applied. Since there are $s = 4$ aces among the $n = 52$ cards, we find that the probability of drawing an ace is

$$\frac{s}{n} = \frac{4}{52} = \frac{1}{13}$$

EXAMPLE   What is the probability of rolling a 3 or a 4 with a balanced die?

Solution   By "balanced" we mean that each face of the die has the same chance, so that the classical probability concept can be applied. Since $n = 6$ and $s = 2$, we find that the probability of rolling a 3 or a 4 is

$$\frac{s}{n} = \frac{2}{6} = \frac{1}{3}$$

EXAMPLE If H stands for heads and T for tails, the four possible outcomes for two flips of a coin are HH, HT, TH, and TT. If it can be assumed that these four possibilities are equally likely, what are the probabilities of getting zero, one, or two heads?

Solution Since $n = 4$ and $s = 1$ for zero heads, $s = 2$ for one head, and $s = 1$ for two heads, the probability of getting zero heads is $\frac{1}{4}$, the probability of getting one head is $\frac{2}{4} = \frac{1}{2}$, and the probability of getting two heads is $\frac{1}{4}$.

Although equally likely possibilities are found mostly in games of chance, the classical probability concept applies also in a great variety of situations where gambling devices are used to make **random selections**—say, when offices are assigned to research assistants by lot, when laboratory animals are chosen for an experiment so that each one has the same chance of being selected (perhaps, by the method which is described in Section 10.1), when each family in a township has the same chance of being included in a survey, or when machine parts are chosen for inspection so that each part produced has the same chance of being selected.

EXAMPLE If three of twenty tires are defective and four of them are randomly chosen for inspection, what is the probability that one of the defective tires will be included?

Solution There are $n = \binom{20}{4} = 4{,}845$ ways of choosing four of the twenty tires, which may be regarded as equally likely by virtue of the random selection. The number of favorable outcomes is the number of ways in which one of the three defective tires and three of the seventeen nondefective tires can be selected, namely,

$$s = \binom{3}{1}\binom{17}{3} = 3 \cdot 680 = 2{,}040$$

It follows that the probability of getting one defective tire and three nondefective tires is

$$\frac{s}{n} = \frac{2{,}040}{4{,}845} = \frac{8}{19}$$

or approximately 0.42. (The values of the various binomial coefficients were read directly from Table X.)

A major shortcoming of the classical probability concept is its limited applicability, for there are many situations in which the various possibilities cannot all be regarded as equally likely. This would be the case, for instance, if we are concerned with the question of whether it will rain on a certain day; when we wonder whether a person will get a raise; when we want to predict the outcome of an election or the score of a baseball game; or when we want to judge whether a stock market index will go up or down.

Among the various probability concepts, most widely held is the **frequency interpretation**, according to which probabilities are interpreted as follows:

The frequency interpretation of probability

> *The probability of an event (happening or outcome) is the proportion of the time that events of the same kind will occur in the long run.*

If we say that that the probability is 0.78 that a jet from San Francisco to Phoenix will arrive on time, we mean that such flights arrive on time 78 percent of the time. Also, if the Weather Service predicts that there is a 40 percent chance for rain (that the probability is 0.40 that it will rain), they mean that under the same weather conditions it will rain 40 percent of the time. More generally, we say that an event has a probability of, say, 0.90, in the same sense in which we might say that our car will start in cold weather 90 percent of the time. We cannot guarantee what will happen on any particular occasion—the car may start and then it may not—but if we kept records over a long period of time, we should find that the proportion of "successes" is very close to 0.90.

In accordance with the frequency interpretation of probability, we estimate the probability of an event by observing what fraction of the time similar events have occurred in the past.

EXAMPLE　If an airline's records show that (over a period of time) 468 of 600 of its jets from San Francisco to Phoenix arrived on time, what is the probability that any one of the airline's jets from San Francisco to Phoenix will arrive on time?

Solution　Since in the past $\frac{468}{600} = 0.78$ of the flights arrived on time, we use this figure as an estimate of the desired probability.

EXAMPLE　If records show that 504 of 813 automatic dishwashers sold by a large retailer required repairs within the warranty year, what is the probability that an automatic dishwasher sold by the retailer will not require repairs within the warranty year?

Solution　Since $813 - 504 = 309$ of the dishwashers did not require repairs, we estimate the probability as $\frac{309}{813} = 0.38$.

When probabilities are estimated in this way, it is only reasonable to ask whether the estimates are any good. In Chapter 13 we shall answer this question in some detail, but for now let us refer to an important theorem called the **Law of Large Numbers**. Informally, this theorem may be stated as follows:

The Law of Large Numbers

> *If a situation, trial, or experiment is repeated again and again, the proportion of successes will tend to approach the probability that any one outcome will be a success.*

```
MTB > BRANDOM 1ØØ N=1 P=.5 C1
  1ØØ BINOMIAL EXPERIMENTS WITH N =   1  AND P =  .5ØØØ
   Ø.     Ø.     1.     1.     1.     1.     1.     Ø.     Ø.     1.
   1.     Ø.     Ø.     1.     Ø.     1.     1.     1.     Ø.     1.
   Ø.     Ø.     1.     Ø.     1.     1.     Ø.     1.     Ø.     Ø.
   1.     1.     Ø.     1.     Ø.     Ø.     1.     1.     1.     Ø.
   1.     Ø.     1.     Ø.     Ø.     Ø.     Ø.     1.     Ø.     Ø.
   1.     1.     Ø.     Ø.     Ø.     Ø.     Ø.     1.     Ø.     Ø.
   1.     1.     Ø.     Ø.     1.     1.     1.     Ø.     1.     1.
   1.     Ø.     1.     1.     Ø.     1.     1.     Ø.     Ø.     Ø.
   Ø.     Ø.     Ø.     1.     Ø.     Ø.     1.     Ø.     1.     1.
   1.     Ø.     1.     1.     1.     1.     Ø.     1.     Ø.     1.

SUMMARY

VALUE       FREQUENCY
  Ø           49
  1           51
```

FIGURE 5.3 *Computer simulation of 100 flips of a balanced coin.*

This theorem is known informally as the "law of averages." It is a statement about the long-run proportion of successes, and it has little to say about any single trial.

An easy illustration of the Law of Large Numbers can be obtained through a **computer simulation** of the repeated flipping of a balanced coin. This is shown in Figure 5.3, where the 1's and 0's denote heads and tails.

Reading across successive rows, we find that among the first five simulated flips there are 3 heads, among the first ten there are 6 heads, among the first fifteen there are 8 heads, among the first twenty there are 12 heads, among the first twenty-five there are 14 heads,..., and among all hundred there are 51 heads. The corresponding proportions, plotted in Figure 5.4, are $\frac{3}{5} = 0.60$, $\frac{6}{10} = 0.60$, $\frac{8}{15} = 0.53$, $\frac{12}{20} = 0.60$, $\frac{14}{25} = 0.56$,..., and $\frac{51}{100} = 0.51$. Observe that the proportion of heads fluctuates but comes closer and closer to 0.50, the probability of heads for each flip of the coin.

In the frequency interpretation, the probability of an event is defined in terms of what happens to similar events in the long run, so let us examine briefly whether it is at all meaningful to talk about the probability of an event which can occur only once. For instance, can we assign a probability to the event that Ms. Bertha Jones will be able to leave the hospital within four days after having an appendectomy, or to the event that a certain major-party candidate will win an upcoming gubernatorial election? If we put ourselves in the position of Ms. Jones' doctor, we might check medical records, discover that patients left the hospital within four days after an appendectomy in, say 34 percent of hundreds of cases, and apply this figure to Ms. Jones. This may not be of much comfort to Ms. Jones, but it does provide a meaning for a probability statement about her leaving the hospital within four days—the probability is 0.34.

This illustrates that when we make a probability statement about a specific (nonrepeatable) event, the frequency interpretation of probability leaves us no

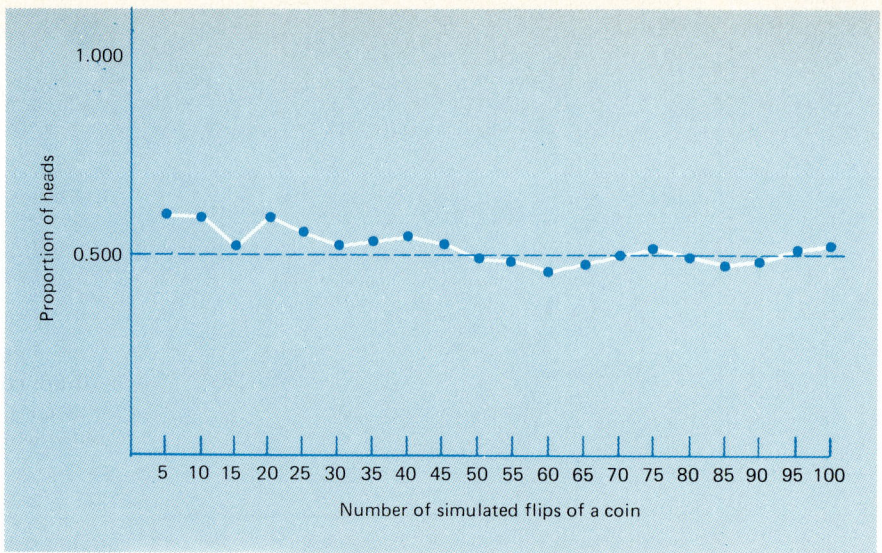

FIGURE 5.4 *Graph illustrating Law of Large Numbers.*

choice but to refer to a set of similar events. As can well be imagined, however, this can easily lead to complications, since the choice of "similar" events is generally neither obvious nor straightforward. With reference to Ms. Jones' appendectomy, we might consider as "similar" only cases in which the patients were of the same sex, only cases in which the patients were also of the same age as Ms. Jones, or only cases in which the patients were also of the same height and weight as Ms. Jones. Ultimately, the choice of "similar" events is a matter of personal judgment, and it is by no means contradictory that we can arrive at different probability estimates, all valid, concerning the same event.

With regard to the question whether a certain major-party candidate will win an upcoming gubernatorial election, suppose that we ask the persons who have conducted a poll "how sure" they are that the candidate will win. If they say they are "95 percent sure" (that is, if they assign a probability of 0.95 to the candidate's winning the election), this is not meant to imply that he would win 95 percent of the time if he ran for office a great number of times. Rather, it means that the pollsters' prediction is based on a method which "works" 95 percent of the time. It is in this way that we must interpret many of the probabilities attached to statistical results.

Finally, let us mention a third probability concept which is currently gaining in favor. According to this point of view, probabilities are interpreted as **personal** or **subjective** evaluations. They reflect one's belief with regard to the uncertainties that are involved, and they apply especially when there is little or no direct evidence, so that there really is no choice but to consider collateral (indirect) information, "educated guesses," and perhaps intuition and other subjective factors. Subjective probabilities are sometimes determined by putting the issues in question on a "put up or shut up" basis, as will be explained in Sections 6.3 and 7.1.

EXERCISES

5.47 When one card is drawn from a well-shuffled deck of 52 standard playing cards, what are the probabilities of getting
  (a) a black queen;
  (b) a jack, queen, or king of any suit;
  (c) a black card;
  (d) any one of a 4, 5, 6, or 7;
  (e) a heart?

5.48 Two cards are dealt from a well-shuffled deck. What are the probabilities of getting
  (a) two red cards;
  (b) two kings?

5.49 If three cards are dealt from a well-shuffled deck, find the probabilities of getting
  (a) three spades;
  (b) two kings and one queen;
  (c) two diamonds and one heart.

5.50 If we roll a balanced die, what are the probabilities of getting
  (a) a 6;
  (b) an even number?

5.51 If we roll a pair of balanced dice, what are the probabilities of getting
  (a) a 5;
  (b) a 7;
  (c) an 11;
  (d) either a 7 or an 11?

5.52 If H stands for heads and T for tails, the sixteen possible outcomes for four flips of a coin are

HHHH	HTHH	THHH	TTHH
HHHT	HTHT	THHT	TTHT
HHTH	HTTH	THTH	TTTH
HHTT	HTTT	THTT	TTTT

If these sixteen outcomes are equally likely, what are the probabilities of getting 0, 1, 2, 3, or 4 heads?

5.53 A bowl contains 15 red beads, 30 white beads, 20 blue beads, and 7 black beads. If one of the beads is drawn at random, what are the probabilities that it will be
  (a) red;
  (b) white or blue;
  (c) black;
  (d) neither white nor black?

5.54 The balls used in selecting numbers for BINGO carry the numbers 1, 2, 3, . . . , 75. If one of the balls is selected at random, what are the probabilities that it will be
  (a) an even number;
  (b) a number which is 15 or below;
  (c) a number which is 60 or above?

5.55 If a game has n equally likely outcomes, what is the probability of each individual outcome?

5.56 Among the 15 applicants for three positions at a newspaper, 10 are college graduates. If the selections are random, what are the probabilities that the positions will be filled with
  (a) three applicants with college degrees;
  (b) two applicants with college degrees and one without;
  (c) three applicants without college degrees?

5.57 A carton of 24 light bulbs includes two that are defective. If two of the bulbs are chosen at random, what are the probabilities that
  (a) neither bulb will be defective;
  (b) one of the bulbs will be defective;
  (c) both bulbs will be defective?

5.58 A hoard of medieval coins discovered in Spain includes 24 struck in Seville and 16 struck in Toledo. If a person chooses four of these coins at random, what is the probability that he or she will get two coins from each city?

5.59 The seven cities in the United States with the most murders in 1990 were New York, Los Angeles, Chicago, Detroit, Houston, Philadelphia, and Washington. If a television news program randomly selects two of these cities as a subject for a special report, what is the probability that the selection
  (a) will include Chicago;
  (b) will consist of Detroit and Houston?

5.60 On a tray there are six pieces of chocolate cake and five pieces of walnut cake. If a waiter randomly picks two pieces of cake from the tray and gives them to diners who ordered chocolate cake, what is the probability that he is making a mistake?

5.61 If 226 of 300 randomly selected newspaper subscribers indicated that they read the comics section daily, estimate the probability that any one subscriber selected randomly would also read the comics section daily.

5.62 If 103 of 150 randomly selected riders of a bus line felt that the busses were too dirty, estimate the probability that any one rider selected randomly would feel that the busses were too dirty.

5.63 Among the 842 armed robberies that occurred in a certain city during the last five years, 143 were never solved. Assuming that conditions have not changed, estimate the probability that an armed robbery in this city will not be solved.

5.64 With reference to the preceding exercise, give an example of a change in conditions which would make the estimate of the probability unreasonable.

5.65 In a sample of 278 cars randomly stopped at a roadblock at various times of the day, 126 of the drivers had their seatbelts fastened. Estimate the probability that a driver on that road will have his or her seatbelt fastened.

5.66 If 1,558 of 2,050 persons visiting a national park said that they would like to return, estimate the probability that any randomly chosen visitor to the park would like to return.

5.67 You can get a feeling for the Law of Large Numbers by flipping coins. Flip a coin 100 times, and plot the accumulated proportion of heads after each five flips, as in Figure 5.4 on page 112. By how much do these proportions differ from 0.50 after the first 50 flips?

5.68 Record the last digit on the license plates of 300 cars. After each 10 cars plot the accumulated proportions of 5's, as in Figure 5.4 on page 112. By how much do these proportions differ from 0.10 after the first 200 cars?

5.69 Record the first 200 numbers encountered in a newspaper, beginning with page one and proceeding in any convenient systematic fashion. Include also numbers appearing in advertisements. For each of these numbers note the leftmost digit and record the proportions of 1's, 2's, 3's, . . . , and 9's. Note that 0 cannot be a leftmost digit. In the decimal number 0.0056, the leftmost digit is the 5. The results will be quite surprising, but the Law of Large Numbers tells you that you must be estimating correctly.

5.5
CHECKLIST OF KEY TERMS
(with page references to their definitions)

Binomial coefficients, 102
Classical probability concept, 108
Combinations, 101
Computer simulation, 111
Factorial notation, 99
Frequency interpretation of probability, 110
Law of Large Numbers, 110
Multiplication of choices, 96
Odds, 108

★ *Pascal's triangle*, 107
Permutations, 98
Personal probability, 112
Probability, 108
Random selection, 109
Subjective probability, 112
Tree diagram, 94

5.6
REVIEW EXERCISES

5.70 Certain government employees are classified into six categories according to age and four categories according to marital status. In how many ways can one of these employees be classified?

5.71 In the meat department of a supermarket there are ten rib steaks and twelve sirloin steaks. If the first customer buys one of the steaks and the second customer buys one of each kind, how many choices does the

second customer have if
   (a) the first customer buys a rib steak;
   (b) the first customer buys a sirloin steak?

5.72 Determine whether each of the following is true or false:

   (a) $\dfrac{1}{3!} + \dfrac{1}{4!} = \dfrac{5}{4!}$;

   (b) $4! + 5! = 6 \cdot 4!$;

   (c) $\dfrac{16!}{13!} = 16 \cdot 15 \cdot 14 \cdot 13$;

   (d) $2! + 2! + 1! + 0! = 3!$.

5.73 A market research organization surveyed 625 people and found that 102 of them would like more baseball games shown on network television. Estimate the probability that any one person would express this opinion.

5.74 The five finalists in a college art competition are identified as BG, CW, RM, JE, and MK. Draw a tree diagram showing the different ways in which the judges can choose the winner and the first runner-up.

5.75 What is the probability of rolling a total of five with a pair of balanced dice?

5.76 In how many different ways can a person arrange eight books on a shelf?

5.77 A personality inventory consists of ten questions, each with four different answers. In how many different ways can a person choose one answer for each question?

5.78 In how many ways can three newspaper reporters, A, B, and C, divide up twelve assignments so that A will cover three, B will cover five, and C will cover four?

5.79 A business employs three persons named Jones: Harry Jones, Norma Jones, and Richard Jones. Draw a tree diagram to show the different ways in which the payroll department can distribute their paychecks so that each of them receives a check made out to a Jones. In how many of the possibilities will
   (a) none of them get the right check;
   (b) only one of them get the right check;
   (c) only two of them get the right check;
   (d) all three of them get the right check?

5.80 In how many ways can a person buy a pound each of three of the sixteen kinds of cheese carried by a gourmet food shop?

5.81 A small women's clothing shop has three mannequins for the left, center, and right positions in its small display window. The store manager has seven different spring outfits, and she wants to select three of these for display. In how many different ways can she design the store window
   (a) if the mannequins are identical;
   (b) if the mannequins are distinguishable and can be moved among the three positions in the window?

5.82 Suppose that someone flips a coin 100 times and gets 32 heads, which is far short of the number of heads she might expect. Then she flips the coin another 100 times and gets 44 heads, which is again short of the number of heads she might expect. Can she accuse the Law of Large Numbers of "letting her down"? Explain.

5.83 How many different gin rummy hands, consisting of ten cards, can be dealt from an ordinary deck of 52 playing cards?

5.84 If H stands for heads and T for tails, the 32 possible outcomes for five flips of a coin are

HHHHH	HTHHH	THHHH	TTHHH
HHHHT	HTHHT	THHHT	TTHHT
HHHTH	HTHTH	THHTH	TTHTH
HHHTT	HTHTT	THHTT	TTHTT
HHTHH	HTTHH	THTHH	TTTHH
HHTHT	HTटHT	THTHT	TTTHT
HHTTH	HTTTH	THTTH	TTTTH
HHTTT	HTTTT	THTTT	TTTTT

If these 32 outcomes are equally likely, what are the probabilities of getting 0, 1, 2, 3, 4, or 5 heads?

5.85 If one letter is chosen at random from the word "anemone," what is the probability that it will be a consonant?

5.86 Among 800 married women interviewed, 648 said that they prefer to be called "Mrs." rather than "Ms." Estimate the probability that a married woman prefers to be called "Mrs." rather than "Ms."

5.87 A restaurant has on its menu eight fish dishes, four potato dishes, and three green vegetables. In how many ways can a diner choose
   (a) one of the fish dishes and one of the potato dishes;
   (b) one of the fish dishes, one of the potato dishes, and one of the green vegetables?

5.88 A carton of 16 hair dryers contains one with serious damage, two which have minor blemishes, and the others in perfect condition. If three of the dryers are randomly selected from the carton, find the probability that

(a) all three will be in perfect condition;
(b) two will be in perfect condition and the third will have a minor blemish;
(c) two will be in perfect condition and the third will have serious damage;
(d) two will have minor blemishes and the third will have serious damage.

5.89 If 756 of 1,200 letters mailed by a government agency were delivered within 48 hours, estimate the probability that any one letter mailed by the agency will be delivered within 48 hours.

5.90 In an ice cream store, a customer can order an ice cream dish in one of 24 flavors, and also with or without chocolate syrup, with or without whipped cream, with or without chopped nuts, with or without crushed pineapple, and with or without marshmallow syrup. In how many different ways can a person order an ice cream dish?

5.91 To cut down on smoking, a person is allowed to smoke at most three cigarettes per day. Construct a tree diagram to determine the number of ways in which the person can smoke at least four cigarettes on two days.

5.92 The food editor and three friends are trying out a new restaurant. If the menu lists 19 different entrees, in how many ways can they each order a different entree?

5.7

REFERENCES

Informal introductions to probability, written primarily for the layman, may be found in

GARVIN, A. D., *Probability in Your Life.* Portland, Maine: J. Weston Walch Publisher, 1978.

HUFF, D., and GEIS, I., *How to Take a Chance.* New York: W. W. Norton & Company, Inc., 1959.

KOTZ, S., and STROUP, D. E., *Educated Guessing*: *How to Cope in an Uncertain World.* New York: Marcel Dekker, Inc., 1983.

LEVINSON, H. C., *Chance, Luck, and Statistics.* New York: Dover Publications, Inc., 1963.

MOSTELLER, F., KRUSKAL, W. H., LINK, R. F., PIETERS, R. S., and RISING, G. R., *Statistics by Example*: *Weighing Chances.* Reading, Mass.: Addison-Wesley Publishing Company, Inc., 1973.

WEAVER, W., *LADY LUCK*: *The Theory of Probability.* New York: Dover Publications, Inc., 1982.

For fascinating reading on the history of probability, see

DAVID, F. N., *Games, Gods and Gambling.* New York: Hafner Press, 1962.

and the first three chapters of

STIGLER, S. M., *The History of Statistics.* Cambridge, Mass.: Harvard University Press, 1986.

SOME RULES
OF
PROBABILITY

In the study of probability there are three fundamental kinds of questions:

1. **What do we mean when we say that the probability of an event is, say, 0.50, 0.78, or 0.44?**
2. **How are the numbers we call probabilities determined, or measured in practice?**
3. **What are the mathematical rules which probabilities must obey?**

For the most part, we have already studied the first two kinds of questions in Chapter 5. In the classical probability concept we are concerned with equally likely possibilities, count the ones that are favorable, and use the formula *s/n*. In the frequency interpretation we are concerned with proportions of "successes" in the long run and base our estimates on what has happened in the past. When it comes to subjective probabilities we are concerned with a measure of a person's belief. In Section 6.3, and also in Section 7.1, we shall see how such subjective probabilities can actually be determined.

In this chapter, after some preliminaries in Section 6.1, we shall concentrate on the rules which probabilities must obey, namely, on the **theory of probability**. This includes the basic postulates, the relationship between probabilities and odds, the addition rules, the definition of conditional probability, the multiplication rules, and finally Bayes' theorem.

6.1

SAMPLE SPACES AND EVENTS

In statistics, the word "experiment" is used in a very wide, and unconventional, sense. For lack of a better term, "experiment" refers to any process of observation or measurement. Thus, an **experiment** may consist of counting how many times a student has been absent; it may consist of the simple process of noting whether a light is on or off, or whether a person is single or married; or it may consist of the very complicated process of obtaining and evaluating data to predict trends in the economy, to find the source of social unrest, or to study the cause of a disease. The results one obtains from an experiment, whether they are instrument readings, counts, "yes or no" answers, or values obtained through extensive calculations, are called the **outcomes** of the experiment.

For each experiment, the set of all possible outcomes is called the **sample space** and it is usually denoted by the letter *S*. For instance, if a zoologist must choose three of 24 guinea pigs for an experiment, the sample space consists of the $\binom{24}{3} = 2{,}024$ ways in which the selection can be made, if the dean of a college must assign two of 84 faculty members as advisors to a political science club, the sample space consists of the $\binom{84}{2} = 3{,}486$ ways in which this can be done. Also, if we are concerned with the number of days it rains in Chicago during the month of

January, the sample space is the set

$$S = \{0, 1, 2, 3, 4, \ldots, 30, 31\}$$

When we study the outcomes of an experiment, we usually identify the various possibilities with numbers, points, or some other kinds of symbols, so that we can treat all questions about them mathematically, without having to go through long verbal descriptions of what has taken place, is taking place, or will take place. For instance, if there are eight candidates for a scholarship and we let a, b, c, d, e, f, g, and h denote that it is awarded to Ms. Adam, Mr. Bean, Miss Clark, and so on, then the sample space for this experiment is the set

$$S = \{a, b, c, d, e, f, g, h\}$$

The use of points rather than letters or numbers has the advantage that it makes it easier to visualize the various possibilities, and perhaps discover special features which several of the outcomes may have in common.

EXAMPLE A used-car dealer has two 1984 Chevrolet Camaros on his lot and we are interested in how many of them each of two salespersons will sell in a given week.
(a) Using two coordinates so that (0, 1), for example, represents the outcome that the first salesperson will sell neither of the Camaros and the second salesperson will sell one, (1, 1) represents the outcome that each of the two salespersons will sell one of the Camaros, and (2, 0) represents the outcome that the first salesperson will sell them both, list all possible outcomes of this experiment.
(b) Draw a figure showing the corresponding points of the sample space.

Solution (a) The six possible outcomes are (0, 0), (1, 0), (0, 1), (2, 0), (1, 1), and (0, 2).
(b) The corresponding points are shown in Figure 6.1, from which it is apparent, for instance, that they sell equally many 1984 Camaros in two of the six possibilities, and that they sell both cars in three of the six possibilities.

Usually, we classify sample spaces according to the number of elements, or points, which they contain. The ones we have mentioned so far in this section contained 2,024, 3,486, 32, 8, and 6 elements, and we refer to them all as **finite**. In this chapter we shall consider only sample spaces that are finite, but in later chapters we shall consider also samples spaces that are **infinite**. An infinite sample space arises, for example, when we throw a dart at a target and there is a continuum of points we may hit.

In statistics, any subset of a sample space is called an **event**. By subset we mean any part of a set, including the set as a whole and the **empty set**, denoted by $\emptyset$, which has no elements at all. For instance, for the example dealing with the number of days that it rains in Chicago during the month of January,

$$F = \{18, 19, 20, 21, 22, 23, 24\}$$

is the event that there will be from 18 to 24 rainy days, and

$$G = \{20, 21, 22, \ldots, 30, 31\}$$

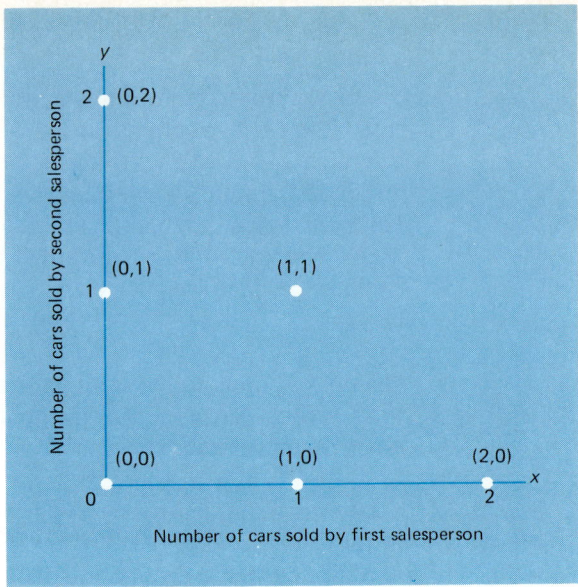

FIGURE 6.1 *Sample space for two-salespersons example.*

is the event that there will be at least 20 rainy days. As is the custom, we denoted these events by capital letters.

EXAMPLE With reference to Figure 6.1, express in words what events are represented by
  (a) $C = \{(0, 0), (1, 1)\}$;
  (b) $D = \{(1, 0), (1, 1)\}$;
  (c) $E = \{(0, 2)\}$.

Solution **(a)** C is the event that the two salespersons will sell equally many of the 1984 Chevrolet Camaros.
  (b) D is the event that the first salesperson will sell one and only one of the two cars.
  (c) E is the event that the second salesperson will sell both cars.

In this example, the events C and E have no elements in common. Such events are called **mutually exclusive**, meaning that they cannot occur at the same time. If the two salespersons sell equally many of these cars (event C), then it is impossible for the second salesperson to sell them both (event E). Observe also that events C and D are not mutually exclusive since they both contain the outcome (1, 1), in which each of the salespersons sells one of the cars.

In many probability problems we are interested in events that can be expressed in terms of two or more events by forming **unions**, **intersections**, and **complements**. In general, the union of two events A and B, denoted by $A \cup B$, is the event which consists of all the elements (outcomes) contained in A, in B, or in both;

EXAMPLE   If X is the event that hamburgers will be served at the company picnic, Y is the event that beer will be served, and Z is the event that watermelon will be served, express in words the events which are represented by the following regions of the Venn diagram of Figure 6.3:

(a) region 3;

(b) regions 1 and 2 together;

(c) regions 4, 6, 7, and 8 together.

Solution   (a) Since this region is part of Y and part of Z, but not part of X, it represents the event that beer and watermelon will be served at the picnic, but hamburgers will not be served.

(b) Since this region is part of both X and Y, it represents the event that hamburgers and beer will be served.

(c) Since these are the entire region outside Y, the event is that beer will not be served.

EXERCISES

6.1 With reference to the illustration on page 119 suppose that a, b, c, d, e, f, g, and h denote the events that Ms. Adam, Mr. Bean, Miss Clark, Mrs. Daly, Mr. Earl, Ms. Fuentes, Ms. Gardner, and Mr. Hall is awarded the scholarship, and that $U = \{b, e, h\}$ and $V = \{a, c, e, f, g\}$. List the outcomes which comprise each of the following events and also express the events in words:

(a) U';

(b) $U \cup V$;

(c) $U \cap V$.

6.2 With reference to the preceding exercise, are the two events U and V mutually exclusive?

6.3 In an experiment, persons are asked to pick a number from 1 to 10, so that for each person the sample space is the set $S = \{1, 2, \ldots, 9, 10\}$. If $C = \{1, 2, 3, 4, 5, 6\}$ and $D = \{5, 6, 7, 8, 9\}$, express in terms of these events the event that a person will

(a) pick a number greater than 6;

(b) pick a 5 or a 6.

6.4 With reference to the preceding exercise, express in terms of C and D the event that a person will pick the number 10.

6.5 To construct sample spaces for experiments in which we deal with categorical data, we often code the various alternatives by assigning them numbers. For instance, if persons are asked whether their favorite color is red, yellow, blue, green, brown, white, purple, or some other color, we might assign these alternatives the codes 1, 2, 3, 4, 5, 6, 7, and 8. If $A = \{3, 4\}$, $B = \{1, 2, 3, 4, 5, 6, 7\}$, and $C = \{6, 7, 8\}$, list the outcomes which comprise each of the following events and also express the events in words:

(a) B';   (c) $B \cap C'$;

(b) $A \cap C$;   (d) $A \cup B'$.

6.6 With reference to the preceding exercise, which of the pairs of events, A and B, A and C, and B and C, are mutually exclusive?

6.7 With reference to the two automobile salesmen and Figure 6.1, describe each of the following events in words:

(a) $N = \{(1, 0), (1, 1)\}$;

(b) $O = \{(0, 1), (0, 2)\}$;

(c) $P = \{(0, 0), (1, 1)\}$;

(d) $Q = \{(1, 0), (2, 0)\}$.

6.8 With reference to the previous exercise, which of the six pairs of events, N and O, N and P, N and Q, O and P, O and Q, and P and Q, are mutually exclusive?

6.9 With reference to the two automobile salesmen and Figure 6.1, list the points of the sample space which comprise the following events:

(a) Between them, the two salesmen will sell both cars.
(b) The first salesman will sell more cars than the second salesman.
(c) The first salesman will sell at least as many cars as the second salesman.

6.10 A movie critic has two days in which to view some of the pictures that have recently been released. She wants to see at least three of the movies but not more than three on either day.

(a) Using two coordinates so that (3, 1), for example, represents the event that she will see three of the movies on the first day and one on the second day, draw a diagram similar to that of Figure 6.1 showing the ten points of the corresponding sample space.
(b) If T is the event that altogether she will see three of the movies, U is the event that she will see more of the movies on the second day than on the first, V is the event that she will see three of the movies on the first day, and W is the event that she will see equally many movies on both days, list the outcomes comprising each of these events.

6.11 With reference to the preceding exercise, which of the six pairs of events, T and U, T and V, T and W, U and V, U and W, and V and W, are mutually exclusive?

6.12 With reference to Exercise 6.10, List the outcomes which comprise each of the following events and also express the events in words:

(a) V';
(b) $T \cap U$;
(c) $V' \cap W$.

6.13 A small marina has three fishing boats which are sometimes in dry dock for repairs.

(a) Using two coordinates so that (2, 1), for example, represents the event that two of the fishing boats are in dry dock and one is rented out for the day, and (0, 2) represents the event that none of the boats is in dry dock and two are rented out for the day, draw a diagram similar to that of Figure 6.1 showing the ten points of the corresponding sample space .
(b) If K is the event that at least two of the boats are rented out for the day, L is the event that more boats are in dry dock than are rented out for the

day, and M is the event that all the boats that are not in dry dock are rented out for the day, list the outcomes which comprise each of these events.
(c) Which of the three pairs of events, K and L, K and M, and L and M, are mutually exclusive?

6.14 With reference to the preceding exercise, list the outcomes which comprise each of the following events and also express the events in words:

(a) K';   (b) $L \cap M$.

6.15 There are six applicants for an executive job, and the table below gives some facts about their backgrounds:

Applicant	Native born?	College graduate?	Marital status
A	No	Yes	Single
B	No	No	Married
C	Yes	Yes	Married
D	Yes	No	Single
E	Yes	Yes	Married
F	Yes	No	Married

One of these applicants is to get the job, and the event that the job is given to a college graduate, for example, is denoted $\{A, C, E\}$. State in a similar manner the event that the job is given to

(a) a single person;
(b) a native-born college graduate;
(c) a married person who is foreign born;
(d) a college graduate.

6.16 Which of the following pairs of events are mutually exclusive? Explain your answers.

(a) A driver getting a ticket for speeding and a ticket for going through a red light.
(b) Being foreign-born and being President of the United States.
(c) A baseball player getting a walk and hitting a home run in the same at bat.
(d) A baseball player getting a walk and hitting a home run in the same game.
(e) Having rain and sunshine on July 4, 1990.

6.17 In Figure 6.4, L is the event that a driver has liability insurance and C is the event that he has collision insurance. Explain in words what events are represented by regions 1, 2, 3, and 4.

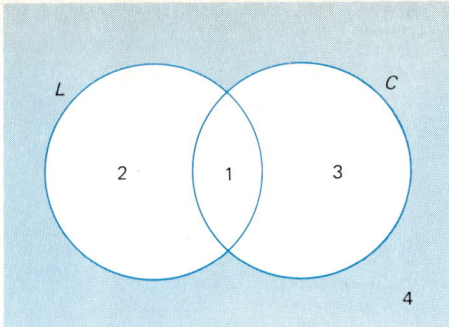

FIGURE 6.4 *Venn diagram for Exercise* 6.17.

6.18 With reference to the preceding exercise, what events are represented by
(a) regions 1 and 2 together;
(b) regions 2 and 4 together;
(c) regions 1, 2, and 3 together?

6.19 In Figure 6.5, D is the event that a flight leaves Denver on time and H is the event that it arrives in Houston on time. Explain in words what events are represented by regions 1, 2, 3, and 4.

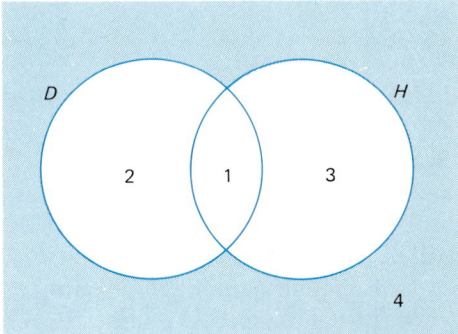

FIGURE 6.5 *Venn diagram for Exercise* 6.19.

6.20 With reference to the preceding exercise, what events are represented by
(a) regions 1 and 3 together;
(b) regions 3 and 4 together;
(c) regions 2, 3, and 4 together?

★ **6.21** Venn diagrams are also useful in determining the numbers of outcomes associated with various events. One of the 240 members of a tennis club is to be named Player of the Year. If 145 of the members are

women, 85 use a two-handed backhand, and 50 are women who use a two-handed backhand, how many of the outcomes correspond to the choice of a man who does not use a two-handed backhand?

★ **6.22** One of the 200 business majors at a college is to be chosen for the student senate. If 77 of these students are enrolled in a course in accounting, 64 are enrolled in a course in business law, and 92 are not enrolled in either course, how many of the outcomes correspond to the choice of a business major who is enrolled in both courses?

6.23 With reference to the example of page 123 and Figure 6.3, what regions or combinations of regions represent the events that
(a) hamburgers will not be served;
(b) hamburgers will be served, but watermelon will not;
(c) beer and watermelon will be both be served;
(d) neither beer nor watermelon will be served?

6.24 In Figure 6.6, E, T, and N are the events that a car brought to a garage needs an engine overhaul, transmission repairs, or new tires. Express in words what events are represented by
(a) region 1;
(b) region 3;
(c) region 7;
(d) regions 1 and 4 together;
(e) regions 2 and 5 together;
(f) regions 3, 5, 6, and 8 together;

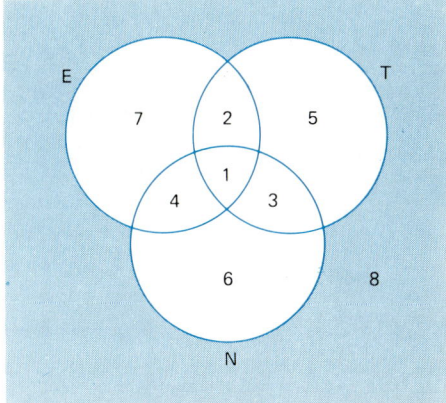

FIGURE 6.6 *Venn diagram for Exercise* 6.24.

6.25 With reference to the preceding exercise and the Venn diagram of Figure 6.6, list the regions or combinations of regions which represent the events that a car brought to the garage needs

  (a) transmission repairs, but neither an engine overhaul nor new tires;
  (b) an engine overhaul and transmission repairs;
  (c) transmission repairs or new tires, but not an engine overhaul;
  (d) new tires.

★ 6.26 As we pointed out in Exercise 6.21, Venn diagrams are also useful in determining the numbers of outcomes associated with various events. Among 60 houses advertised for sale there are 8 with swimming pools, three or more bedrooms, and wall-to-wall carpeting, 5 with swimming pools and three or more bedrooms, but no wall-to-wall carpeting, 3 with swimming pools and wall-to-wall carpeting, but fewer than three bedrooms, 8 with swimming pools but neither wall-to-wall carpeting nor three or more bedrooms, 24 with three or more bedrooms, but neither a swimming pool nor wall-to-wall carpeting, 2 with three or more bedrooms and wall-to-wall carpeting, but no swimming pool, 3 with wall-to-wall carpeting, but neither a swimming pool nor three or more bedrooms, and 7 without any of these features. If one of these houses is to be chosen for a television commercial, how many outcomes correspond to the choice of

  (a) a house with a swimming pool;
  (b) a house with wall-to-wall carpeting?

6.27 Venn diagrams are often used to verify relationships among sets, subsets, or events, without requiring formal proofs based on the algebra of sets. We simply check whether the expressions which are supposed to be equal are represented by the same region of a Venn diagram. Use Venn diagrams to show that

  (a) $A \cup (A \cap B) = A$;
  (b) $(A \cap B) \cup (A \cap B') = A$;
  (c) $(A \cap B)' = A' \cup B'$ and also $(A \cup B)' = A' \cap B'$;
  (d) $A \cup B = (A \cap B) \cup (A \cap B') \cup (A' \cap B)$;
  (e) $A \cap (B \cup C) = (A \cap B) \cup (A \cap C)$.

6.2
THE POSTULATES OF PROBABILITY

Probabilities always pertain to the occurrence of events, and now that we have learned how to deal mathematically with events, let us turn to the rules which probabilities must obey. To formulate these rules, we shall continue the practice of denoting events by capital letters and write the probability of event A as $P(A)$, the probability of event B as $P(B)$, and so forth. As before, we shall denote the set of all possible outcomes, the sample space, by the letter S.

Most basic among all the rules of probability are the three **postulates**, which, as we shall state them here, apply when the sample space S is finite. Beginning with the first two, we write

First two postulates of probability

> 1. *Probabilities are positive real numbers or zero; symbolically, $P(A) \geq 0$ for any event A.*
> 2. *Every sample space has probability 1; symbolically, $P(S) = 1$ for any sample space S.*

To justify these two postulates, as well as the third one which follows, let us show that they are in agreement with the classical probability concept as well as the frequency interpretation. In Section 6.3 we shall see to what extent the postulates are compatible also with subjective probabilities.

The first two postulates are in agreement with the classical probability concept because the fraction $\frac{s}{n}$ is always positive or zero, and for the entire sample space (which includes all n outcomes) the probability is $\frac{s}{n} = \frac{n}{n} = 1$. When it comes to the frequency interpretation, the proportion of the time that an event will occur cannot be a negative number, and one of the outcomes in the sample space must occur 100 percent of the time, that is, with probability 1.

Although a probability of 1 is thus identified with certainty, in actual practice we also assign a probability of 1 to events that are "practically certain" to occur. For instance, we would assign a probability of 1 to the event that at least one person will vote in the next presidential election, even though it is not logically impossible. Similarly, we would assign a probability of 1 to the event that not every student entering college in the Fall of 1990 will apply for admission to Princeton University.

The third postulate of probability is especially important, but it is not quite as obvious as the other two.

Third
postulate of
probability

3. *If two events are mutually exclusive, the probability that one or the other will occur equals the sum of their probabilities. Symbolically,*

$$P(A \cup B) = P(A) + P(B)$$

for any two mutually exclusive events A and B.

For instance, if the probability that weather conditions will improve during a certain week is 0.62 and the probability that they will remain unchanged is 0.23, then the probability that they will either improve or remain unchanged is $0.62 + 0.23 = 0.85$. Similarly, if the probabilities that a student will get an A or a B in a course are 0.13 and 0.29, then the probability that he or she will get either an A or a B is $0.13 + 0.29 = 0.42$.

To show that the third postulate is also compatible with the classical probability concept, let s_1 and s_2 denote the number of equally likely possibilities which comprise events A and B. Since A and B are mutually exclusive, no two of these possibilities are alike and all $s_1 + s_2$ of them comprise event $A \cup B$. Thus,

$$P(A) = \frac{s_1}{n} \qquad P(B) = \frac{s_2}{n} \qquad P(A \cup B) = \frac{s_1 + s_2}{n}$$

and $P(A) + P(B) = P(A \cup B)$.

So far as the frequency interpretation is concerned, if one event occurs, say, 36 percent of the time, another event occurs 41 percent of the time, and they cannot both occur at the same time (that is, they are mutually exclusive), then one or the other will occur $36 + 41 = 77$ percent of the time. This is in agreement with the third postulate.

By using the three postulates of probability, we can derive many further rules according to which probabilites must "behave"—some of them are easy to prove and some are not, but they all have important applications. Among the immediate consequences of the three postulates we find that probabilities can never be greater than 1, that an event which cannot occur has probability 0, and that the probabilities that an event will occur and that it will not occur always add up to 1. Symbolically,

Further rules of probability

$$P(A) \leq 1 \qquad \text{for any event } A$$

$$P(\emptyset) = 0$$

$$P(A) + P(A') = 1 \qquad \text{for any event } A$$

The first of these results simply expresses the fact that there cannot be more favorable outcomes than there are outcomes, or that an event cannot occur more than 100 percent of the time. The second result expresses the fact that when an event cannot occur there are $s = 0$ favorable outcomes, or that such an event occurs zero percent of the time. In actual practice, we also assign 0 probability to events which are so unlikely that we are "practically certain" they will not occur. For instance, in flipping a coin we will assign 0 probability to the event that the coin lands on its edge.

The third result can also be derived from the postulates of probability, and it can easily be seen that it is compatible with the classical probability concept and the frequency interpretation. In the classical concept, if there are s "successes" there are $n - s$ "failures," the corresponding probabilities are $\dfrac{s}{n}$ and $\dfrac{n-s}{n}$, and their sum is

$$\frac{s}{n} + \frac{n-s}{n} = \frac{n}{n} = 1$$

In accordance with the frequency interpretation, we can say that if some given investments are successful 22 percent of the time, then they are not successful 78 percent of the time, the corresponding probabilities are 0.22 and 0.78, and their sum is 1.

The examples which follow show how the postulates and the rules we gave above are put to use in actual practice.

EXAMPLE If A is the event that a student will stay home to study and B is the event that she will instead go to a movie, $P(A) = 0.64$, and $P(B) = 0.21$, find
(a) $P(A')$;   (b) $P(A \cup B)$;   (c) $P(A \cap B)$.

Solution (a) Using the final rule, we find that the probability of A', the event that the student will not stay home to study, is $1 - P(A) = 1 - 0.64 = 0.36$.
(b) Since A and B are mutually exclusive, we can use the third postulate and write $P(A \cup B) = P(A) + P(B) = 0.64 + 0.21 = 0.85$ for the probability that the student will either stay home to study or go to a movie.

(c) Since A and B are mutually exclusive, they cannot possibly both occur and, hence, $P(A \cap B) = P(\emptyset) = 0$.

In problems like this, it often helps to draw a Venn diagram, fill in the probabilities associated with the various regions, and then read the answers directly off the diagram.

EXAMPLE If C is the event that at 9:30 A.M. a certain doctor is in his office and D is the event that he is in the hospital, $P(C) = 0.48$, and $P(D) = 0.27$, find $P(C' \cap D')$, which is the probability that he is neither in his office nor in the hospital.

Solution Drawing the Venn diagram as in Figure 6.7, we first put a 0 probability into region 1 because the events C and D are mutually exclusive. It follows that the 0.48 probability of event C must go into region 2, the 0.27 probability of event D must go into region 3, and since the probability for the entire sample space must total 1, we put $1 - (0.48 + 0.27) = 0.25$ into region 4. Since the event $C' \cap D'$ is represented by the region outside both circles, namely, region 4, we find that the answer is $P(C' \cap D') = 0.25$.

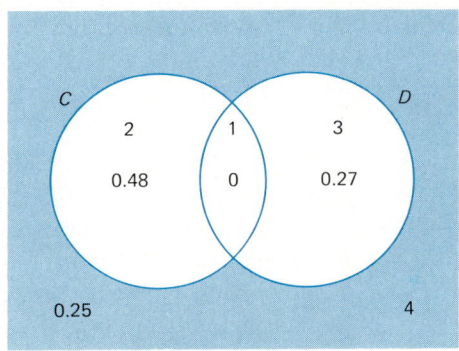

FIGURE 6.7 *Venn diagram.*

6.3
PROBABILITIES AND ODDS

If an event is twice as likely to occur than not to occur, we say that the **odds** are 2 to 1 that it will occur; if an event is three times as likely to occur than not to occur, we say that the odds are 3 to 1; if an event is ten times as likely to occur than not to occur, we say that the odds are 10 to 1; and so forth. In general,

> **The odds that an event will occur are given by the ratio of the probability that it will occur to the probability that it will not occur.**

Symbolically, if the probability of an event is p, the odds for its occurrence are a to b, where a and b are positive values such that

$$\frac{a}{b} = \frac{p}{1-p}$$

It is customary to express odds in terms of positive integers having no common factors.

EXAMPLE What are the odds for the occurrence of an event if its probability is
(a) $\frac{5}{9}$;
(b) 0.85?

Solution (a) By definition, the odds are $\frac{5}{9}$ to $1 - \frac{5}{9} = \frac{4}{9}$, or 5 to 4.
(b) By definition, the odds are 0.85 to $1 - 0.85 = 0.15$, 85 to 15, or better, 17 to 3.

If an event is more likely not to occur than to occur, it is customary to quote the odds that it will not occur rather than the odds that it will occur.

EXAMPLE What are the odds, if the probability of an event is 0.20?

Solution The odds for the occurrence of the event are 0.20 to $1 - 0.20 = 0.80$, or 1 to 4, but it is customary to say instead that the odds against the occurrence of the event are 4 to 1.

In betting, the word "odds" is also used to denote the ratio of the wager of one party to that of another. For instance, if a gambler says that he will give 3 to 1 odds on the occurrence of an event, he means that he is willing to bet $3 against $1 (or perhaps $30 against $10 or $1,500 against $500) that the event will occur. If such **betting odds** actually equal the odds that the event will occur, we say that the betting odds are **fair**. If a gambler really believes that a bet is fair, then he is, at least in principle, willing to bet on either side. The gambler in this situation would also be willing to bet $1 against $3 (or $10 against $30 or $500 against $1,500) that the event will not occur.

EXAMPLE Records show that $\frac{1}{12}$ of the trucks weighed at a certain check point in Nevada carry too heavy a load. If someone offers to bet $40 against $4 that the next truck weighed at this check point will not carry too heavy a load, are these betting odds fair?

Solution Since the probability is $1 - \frac{1}{12} = \frac{11}{12}$ that the truck will not carry too heavy a load, the odds are 11 to 1, and the bet would be fair if the person offered to bet $44 against $4 that the next truck weighed at the check point will not carry too heavy a load. Thus, the $40 against $4 bet is not fair; it favors the person offering the bet.

In most betting situations outside of casinos the actual odds that an event will occur are unknown and opinions will be different. Suppose, for example, that a football game between the Wildcats and the Magpies is about to take place.

Suppose also that Al believes that the odds that the Wildcats will win are 2 to 3. It happens that Bob believes that the odds that the Wildcats will win are 1 to 1.

Observe that Al is willing to bet $20 on the Wildcats to win $30. He is also willing to bet $30 on the Magpies to win $20.

Since Bob believes that the game is even, he is willing to bet $25 on either team to win $25.

If Al and Bob actually discuss the game with each other, they will probably make a bet somewhere around $27 to $23. In this example, Al will bet $27 on the Magpies and Bob will bet $23 on the Wildcats. Each believes that he has a good bet. (The actual dollar amounts depend on negotiating skill. It could well be that the amounts bet will be $29 to $21.)

This discussion of odds and betting odds provides the groundwork for a way of measuring **subjective probabilities**. If a businessman "feels" that the odds for the success of a new clothing store are 3 to 2, this means that he is willing to bet (or considers it fair to bet) $300 against $200, or perhaps $3,000 against $2,000, that the new store will be a success. In this way he is expressing his belief regarding the uncertainties connected with the success of the store, and to convert it into a probability we take the equation $\dfrac{a}{b} = \dfrac{p}{1 - p}$ and solve it for p. Leaving the details to the reader in Exercise 6.57, let us merely state the result that

Formula relating probabilities to odds

> *If the odds are a to b that an event will occur, the probability of its occurrence is*
>
> $$p = \frac{a}{a + b}$$

EXAMPLE Convert the businessman's 3 to 2 odds for the success of the new clothing store into a probability.

Solution Substituting $a = 3$ and $b = 2$ into the formula for p, we get

$$p = \frac{3}{3 + 2} = \frac{3}{5}$$

EXAMPLE If an applicant for a managerial position feels that the odds are 7 to 4 that she will get the job, what probability is she thus assigning to her getting the job?

Solution Substituting $a = 7$ and $b = 4$ into the formula for p, we get

$$p = \frac{7}{7 + 4} = \frac{7}{11}$$

or approximately 0.64.

Let us now see whether subjective probabilities, determined in this way, behave in accordance with the postulates of probability on pages 126 and 127. Insofar as the first postulate is concerned, this is easy to see. Since a and b are positive quantities, $\frac{a}{a+b}$ is certainly greater than or equal to zero. For the second postulate, observe that the surer we are that an event will occur, the "better" odds we should be willing to give—say 100 to 1, 1,000 to 1, or perhaps even one million to 1. The corresponding probabilities are

$$\frac{100}{100+1} = 0.99, \quad \frac{1,000}{1,000+1} = 0.999, \quad \text{and} \quad \frac{1,000,000}{1,000,000+1} = 0.999999,$$

and it can be seen that the surer we are that an event will occur, the closer its probability will be to 1.

This leaves only the third postulate, $-P(A \cup B) = P(A) + P(B)$, for any two mutually exclusive events A and B—and this rule is not necessarily satisfied when it comes to subjective probabilities. Indeed, proponents of the subjectivist point of view impose it as a **consistency criterion**, and this provides a means of "policing" a person's subjective probabilities.

EXAMPLE   An economist feels that the odds are 2 to 1 that the price of beef will go up during the next month, 1 to 5 that it will remain unchanged, and 8 to 3 that it will go up or remain unchanged. Are the corresponding probabilities consistent?

Solution   The corresponding probabilities that the price of beef will go up during the next month, that it will remain unchanged, and that it will go up or remain unchanged are, respectively, $\frac{2}{2+1} = \frac{2}{3}, \frac{1}{1+5} = \frac{1}{6}$, and $\frac{8}{8+3} = \frac{8}{11}$, and since $\frac{2}{3} + \frac{1}{6} = \frac{5}{6}$ and not $\frac{8}{11}$, the probabilities are not consistent. Hence, the economist's judgment must be questioned.

EXERCISES

6.28 In a study of the future needs of a community, C is the event that there will be enough capital for expansion and T is the event that there will be adequate transportation. State in words what probabilities are expressed by
  (a) $P(C')$;
  (b) $P(T')$;
  (c) $P(C \cup T)$;
  (d) $P(C \cap T)$.

6.29 With reference to the preceding exercise, express symbolically the probabilities that there will be

  (a) adequate transportation but not enough capital for expansion;
  (b) neither enough capital for expansion nor adequate transportation.

6.30 In a student evaluation of faculty, V is the event that a professor is well versed in his field, D is the event that he gives difficult tests, and R is the event that he is a rough grader. State in words what probabilities are expressed by

(a) $P(D')$;
(b) $P(D \cup R)$;
(c) $P(V' \cap D)$;
(d) $P(V \cap R')$.

6.31 With reference to the preceding exercise, express symbolically the probabilities that a professor

(a) is not a rough grader;
(b) does not give difficult tests but is a rough grader;
(c) is not well versed in his field and/or does not give difficult tests.

6.32 Which of the postulates of probability are violated by the following statements:

(a) Since there is no cloud in the sky, the probability that it will rain later on in the day is -0.90.
(b) The probability that a mineral specimen will contain copper is 0.28 and the probability that it will not contain copper is 0.55.
(c) The probability that a lawyer will win his case is 0.30, and the probability that he will lose it is five times as large.
(d) The probabilities that a person will spend an evening at the movies or at home watching television are 0.27 and 0.35, and the probability that it will be one or the other is 0.52.

Explain your answers.

6.33 Which of the rules on pages 126 and 127 are violated by the following statements:

(a) The probability that a chemistry experiment will succeed is 0.44 and the probability that it will fail is 0.53.
(b) According to the doctor, the probability that a certain patient has the flu is 1.12.
(c) The probability that two mutually exclusive events will both occur is always equal to 1.
(d) At a certain time of the year, the probability of seeing a whale off the San Diego coast is only 0.18, but the probability of seeing a seal is six times as large.

Explain your answers.

6.34 Express symbolically what general rule is violated by each of the following assertions:

(a) The probability that a person will buy a cake at a certain bakery is 0.48, and the probability that the person will buy a cake or a pie is 0.35.
(b) The probability that a student will get a passing grade in English is 0.77, and the probability that he or she will get a passing grade in English as well as psychology is 0.85.

6.35 On Friday and Saturday, John will go to the theater and the opera. The probability that he will enjoy the theater is 0.38, the probability that he will enjoy them both is 0.23, and the probability that he will enjoy the theater but not the opera is 0.17. Use part (b) of Exercise 6.27 on page 126 to show that these probabilities cannot all be correct.

★ **6.36** Since $A \cup A' = S$ in accordance with the definition of A', we can write $P(A \cup A') = P(S)$. Then, making use of the fact that A and A' are mutually exclusive, we get

$$P(A) + P(A') = P(S) \qquad \text{step 1}$$
$$P(A) + P(A') = 1 \qquad \text{step 2}$$

which proves the third of the three rules on page 128. State which of the postulates of probability justify steps 1 and 2.

★ **6.37** Since $S \cup \varnothing = S$ for any sample space S, we can write $P(S \cup \varnothing) = P(S)$. Then, making use of the fact that S and $\varnothing$ are mutually exclusive (by default), we get

$$P(S) + P(\varnothing) = P(S) \qquad \text{step 1}$$
$$1 + P(\varnothing) = 1 \qquad \text{step 2}$$

and $P(\varnothing) = 0$, which proves the second of the three rules on page 128. State which of the postulates of probability justify steps 1 and 2.

★ **6.38** Using the result of Exercise 6.36 we can write $P(A) = 1 - P(A')$, and hence $P(A) \leq 1$, which proves the first of the three rules on page 128. State which of the postulates of probability justifies the last step.

6.39 Given $P(M) = 0.31$ and $P(N) = 0.62$, where M and N are mutually exclusive, use the postulates and/or the rules on page 128 to find

(a) $P(M')$;
(b) $P(N')$;
(c) $P(M \cup N)$;
(d) $P(M' \cap N')$.

6.40 Use a Venn diagram to rework the preceding exercise.

6.41 The events A and B are mutually exclusive. Provide examples in which also

(a) A' and B' are not mutually exclusive;
(b) A' and B' are also mutually exclusive.

(*Hint*: Use Venn diagrams.)

6.42 The probability that a typist will make at most three mistakes when typing a long letter, or make from four to eight, are 0.57 and 0.33. Use the postulates and/or the rules on page 128 to find the probabilities that the typist will make

(a) at least four mistakes;
(b) at most eight mistakes;
(c) more than eight mistakes.

6.43 Use a Venn diagram to rework the preceding example.

6.44 The probabilities that a missile will explode during liftoff or have its guidance system fail in flight are 0.002 and 0.005. Use the postulates and/or the rules on page 128 to find the probabilities that the missile will
(a) not explode during lift-off;
(b) explode during lift-off or have its guidance system fail in flight;
(c) not explode during lift-off nor have its guidance system fail in flight.

6.45 Use a Venn diagram to rework the preceding exercise.

6.46 Convert each of the following probabilities to odds:
(a) The probability that the last digit of a car's license plate is a 2, 3, 4, 5, 6 or 7 is $\frac{6}{10}$.
(b) The probability of getting at least two heads in four flips of a balanced coin is $\frac{11}{16}$.
(c) The probability of rolling "7 or 11" with a pair of balanced dice is $\frac{2}{9}$.

6.47 If the probability is 0.65 that next year's inflation rate will exceed this year's, what are the corresponding odds?

6.48 If the probability is $\frac{4}{13}$ that a shipment of laboratory supplies will arrive on time, what are the odds that it will not arrive on time?

6.49 Convert each of the following odds to probabilities:
(a) If three eggs are randomly chosen from a carton of twelve eggs of which three are cracked, the odds are 34 to 21 that at least one of them will be cracked.
(b) If a person has eight $1 bills, five $5 bills, and one $20, and randomly selects three of them, the odds are 11 to 2 that they will not all be $1 bills.
(c) If we arbitrarily arrange the letters in the word "nest," the odds are 5 to 1 that we will not get a meaningful word in the English language.

6.50 A government economist claims that the odds are 3 to 1 that interest rates will go down and 5 to 1 that they will go up. Can these odds be right? Explain your answer.

6.51 A football fan is offered a bet of $12 against her $4 that her team will lose an upcoming game. What does it tell us about the subjective probability she assigns to her team winning the game, if
(a) she considers this bet to be fair;
(b) she is unwilling to bet?

6.52 A television producer is willing to bet $1,500 against $1,000, but not $2,000 against $1,000, that a new game show will be a success. What does this tell us about the probability which the producer assigns to the show's success?

6.53 Asked about his political future, a party official replies that the odds are 2 to 1 that he will not run for the House of Representatives and 4 to 1 that he will not run for the Senate. Furthermore, he feels that the odds are 7 to 5 that he will run for one or the other. Are the corresponding probabilities consistent?

6.54 A high school principal feels that the odds are 7 to 5 against her getting a $1,000 raise and 11 to 1 against her getting a $2,000 raise. Furthermore, she feels that it is an even-money bet that she will get one of these raises or the other. Discuss the consistency of the corresponding subjective probabilities.

6.55 There are two Porsches in a race, and a reporter feels that the odds against their winning are, respectively, 4 to 1 and 5 to 1. To be consistent, what odds should he assign to the event that neither car will win?

6.56 Some events are so unlikely that we choose to assign them probabilities of zero. Would you assign probability zero to the event that
(a) a coin flip will result in the coin's landing on its edge;
(b) ten successive flips of a coin will all be heads;
(c) a monkey sitting at a typewriter and striking randomly at the keys will correctly type Lincoln's Gettysburg Address;
(d) three United States commercial airline crashes will occur on the same day?

★ 6.57 Verify algebraically that the equation $\dfrac{a}{b} = \dfrac{P}{1-p}$, solved for p, yields $p = \dfrac{a}{a+b}$.

6.58 A gambler believes that the true odds favoring team Y in a coming game are 5 to 3. Which of the following bets would be attractive to him?
(a) Bet $5 on team Y to win $5.
(b) Bet $6 on team Y to win $4.
(c) Bet $4 against team Y to win $6.
(d) Bet $40 against team Y to win $60.

6.59 Alan and Ben are debating the coming football game between the Rockets and the Pythons. Alan believes that the odds are 14 to 11 in favor of the Rockets. Ben believes that the odds are 3 to 2 in favor of the Rockets. Will the two men be able to agree on a wager? If so, what will it be?

6.4
ADDITION RULES

The third postulate of probability applies only to two mutually exclusive events, but it can easily be generalized so that it applies to more than two mutually exclusive events. We say that k events are mutually exclusive if no two of them have any elements in common. Repeatedly using the third postulate, we can show that

Generalization of Postulate 3

> *If k events are mutually exclusive, the probability that one of them will occur equals the sum of their individual probabilities; symbolically,*
>
> $$P(A_1 \cup A_2 \cup \cdots \cup A_k) = P(A_1) + P(A_2) + \cdots + P(A_k)$$
>
> *for any mutually exclusive events $A_1, A_2, \ldots,$ and A_k.*

Here again, we read $\cup$ as "or."

EXAMPLE A person is looking for a new car. The probabilities that she will buy a Chrysler, a Ford, or a Toyota are 0.17, 0.22, and 0.08, respectively. Assuming that she will buy just one car, what is the probability that she will buy one of these three?

Solution Since the three possibilities are mutually exclusive, direct substitution into the formula yields
$$0.17 + 0.22 + 0.08 = 0.47$$

EXAMPLE The probabilities that a consumer testing service will rate a new washing machine very poor, poor, fair, good, very good, or excellent are 0.06, 0.13, 0.17, 0.32, 0.22, and 0.10. What are the probabilities that it will rate the new washing machine
(a) very poor, poor, fair, or good;
(b) good, very good, or excellent?

Solution Since the six possibilities are mutually exclusive, direct substitution into the formula yields
(a) $0.06 + 0.13 + 0.17 + 0.32 = 0.68$;
(b) $0.32 + 0.22 + 0.10 = 0.64$.

The job of assigning probabilities to all possible events connected with a given situation can be very tedious. Indeed, it can be shown that if a sample space has ten elements (points or outcomes) we can form more than a thousand different events, and if a sample space has twenty elements we can form more than a million.[†] Fortunately, it is seldom necessary to assign probabilities to all possible events (that is,

---

[†] In general, if a sample space has n elements, we can form 2^n different events. Each element is either included or excluded for a given event, so by the multiplication of choices there are $2 \cdot 2 \cdot 2 \cdots \cdot 2 = 2^n$ possibilities. Note that $2^{10} = 1,024$ and $2^{20} = 1,048,576$.

to all possible subsets of a sample space). The following rule, which is a direct application of the above generalization of the third postulate, makes it easy to determine the probability of any event on the basis of the probabilities assigned to the individual elements of a sample space:

Rule for calculating the probability of an event

> *The probability of any event A is given by the sum of the probabilities of the individual outcomes comprising A.*

In the special case where the outcomes are all equiprobable, this rule leads to the formula $P(A) = \dfrac{s}{n}$, which we used earlier in connection with the classical probability concept. Here, n is the total number of outcomes in the sample space and s is the number of "successes," namely, the number of outcomes in event A.

EXAMPLE Referring again to the example on page 119, which dealt with the two salespersons trying to sell two 1984 Chevrolet Camaros, suppose that the six points of the sample space have the probabilities shown in Figure 6.8. Find the probabilities that
(a) the first salesperson will not sell either of the two cars;
(b) both cars will be sold;
(c) the second salesperson will sell at least one of the two cars.

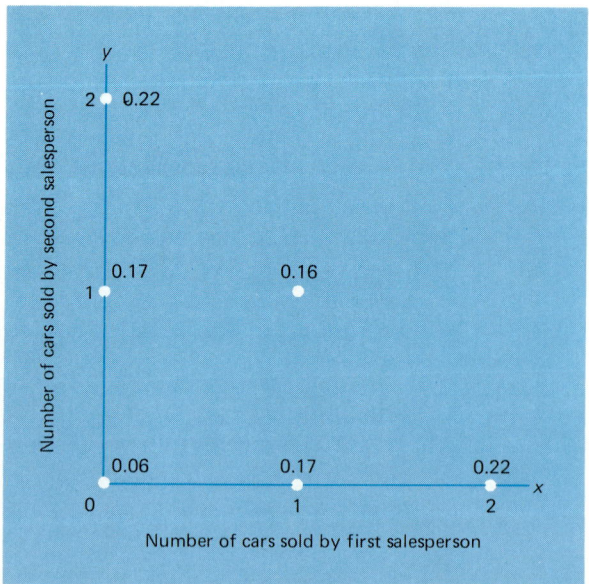

FIGURE 6.8 *Sample space for two-salespersons example with probabilities assigned to the outcomes.*

Solution (a) Adding the probabilities associated with the points (0, 0), (0, 1), and (0, 2) we get $0.06 + 0.17 + 0.22 = 0.45$.

(b) Adding the probabilities associated with the points (2, 0), (1, 1), and (0, 2), we get $0.22 + 0.16 + 0.22 = 0.60$.

(c) Adding the probabilities associated with the points (0, 1), (1, 1), and (0, 2), we get $0.17 + 0.16 + 0.22 = 0.55$.

EXAMPLE Assuming that the 44 points (outcomes) of the sample space of Figure 6.9 are all equiprobable, find $P(A)$.

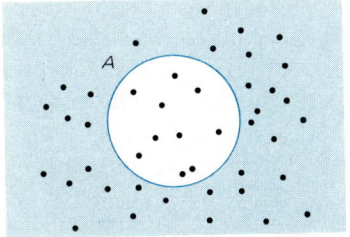

FIGURE 6.9 *Sample space with* 44 *outcomes.*

Solution Since there are $s = 10$ outcomes in A and the $n = 44$ outcomes of the sample space are all equiprobable, it follows that $P(A) = \frac{10}{44} = \frac{5}{22}$ or approximately 0.23.

Since the third postulate and its generalization apply only to mutually exclusive events, they cannot be used, for example, to find the probability that at least one of two friends will pass a language examination, the probability that a bird-watcher will spot a roadrunner, a cactus wren, or a Gila woodpecker, or the probability that a customer will buy a shirt, a sweater, a belt, or a tie at a department store. Both friends can pass the examination, a bird-watcher can spot more than one of these birds, and the customer of the department store can buy any number of these items.

To find a formula for $P(A \cup B)$ which holds regardless of whether A and B are mutually exclusive, let us consider the Venn diagram of Figure 6.10. It concerns

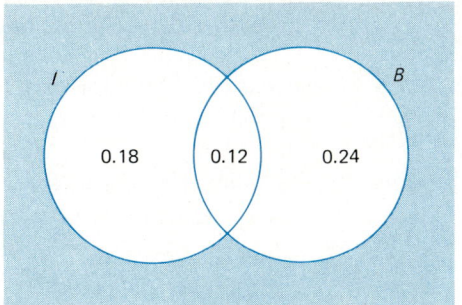

FIGURE 6.10 *Venn diagram.*

the job applications of a recent business school graduate, and the letters I and B denote her getting a job offer from an investment broker or a bank. It follows from the figures in the Venn diagram that

$$P(I) = 0.18 + 0.12 = 0.30$$

$$P(B) = 0.12 + 0.24 = 0.36$$

and

$$P(I \cup B) = 0.18 + 0.12 + 0.24 = 0.54$$

where we could add the respective probabilities because they pertain to mutually exclusive events (nonoverlapping regions of the Venn diagram).

Had we erroneously used the third postulate to calculate $P(I \cup B)$, we would have obtained $P(I) + P(B) = 0.30 + 0.36 = 0.66$, which exceeds the correct value by 0.12. This error results from adding $P(I \cap B) = 0.12$ in twice, once in $P(I) = 0.30$ and once in $P(B) = 0.36$, and we could correct for this by subtracting 0.12 from 0.66. Thus, we could write

$$P(I \cup B) = P(I) + P(B) - P(I \cap B)$$

$$= 0.30 + 0.36 - 0.12$$

$$= 0.54$$

and this agrees, as it should, with the result obtained before.

Since the argument used in this example holds for any two events A and B, we can now state the following **general addition rule**, which applies regardless of whether A and B are mutually exclusive events:

General addition rule

$$P(A \cup B) = P(A) + P(B) - P(A \cap B)$$

When A and B are mutually exclusive, $P(A \cap B) = 0$ and the above formula reduces to that of the third postulate of probability. In this connection, the third postulate is also referred to as the **special addition rule**. To add to this terminology, the generalization of the third postulate on page 136 is sometimes referred to as the **generalized** (special) **addition rule**.

EXAMPLE   If one card is drawn from an ordinary deck of 52 playing cards, what is the probability that it will be either a club or a face card (king, queen, or jack)?

Solution   If C denotes drawing a club and F denotes drawing a face card, then $P(C) = \frac{13}{52}$, $P(F) = \frac{12}{52}$, and $P(C \cap F) = \frac{3}{52}$, so that

$$P(C \cup F) = \frac{13}{52} + \frac{12}{52} - \frac{3}{52} = \frac{22}{52}$$

EXAMPLE If the probabilities are, respectively, 0.92, 0.33, and 0.29 that a person vacationing in Washington, D.C., will visit the Capitol building, the Smithsonian Institution, or both, what is the probability that a person vacationing there will visit at least one of these buildings?

Solution Substituting into the formula, we get

$$0.92 + 0.33 - 0.29 = 0.96$$

Note that if we had incorrectly used the special addition rule (for mutually exclusive events), we would have obtained the impossible answer $0.92 + 0.33 = 1.25$.

The general addition rule can be generalized further so that it will apply to more than two events, but we shall not go into that in this book.

EXERCISES

6.60 A police department needs new tires for its patrol cars and the probabilities that it will buy Uniroyal, Goodyear, Michelin, General, or Goodrich tires are 0.18, 0.23, 0.20, 0.25, and 0.14. Find the probabilities that it will buy
  (a) Goodyear or Goodrich tires;
  (b) Uniroyal, Michelin, General, or Goodrich tires;
  (c) Uniroyal, Michelin, or General tires.

6.61 The probabilities that a student will get an A, a B, or a C in a history course are 0.09, 0.15, and 0.53. What is the probability that the student will get a grade lower than C?

6.62 The probabilities that a television station will receive at most 5 complaints, 6 complaints, 7 complaints, . . . , 13 complaints, or 14 or more complaints after showing a controversial program are 0.01, 0.03, 0.07, 0.15, 0.19, 0.18, 0.14, 0.12, 0.09, and 0.02. What are the probabilities that after showing such a program the station will receive
  (a) at most 10 complaints;
  (b) at least 12 complaints;
  (c) anywhere from 6 to 10 complaints?

6.63 With reference to Figure 6.9 on page 137, suppose that each outcome in A is three times as likely as each outcome in A'. Find $P(A)$.

6.64 Figure 6.11 pertains to the number of persons who are invited to a conference and the number of persons who attend. If each of the 35 points of the sample space has

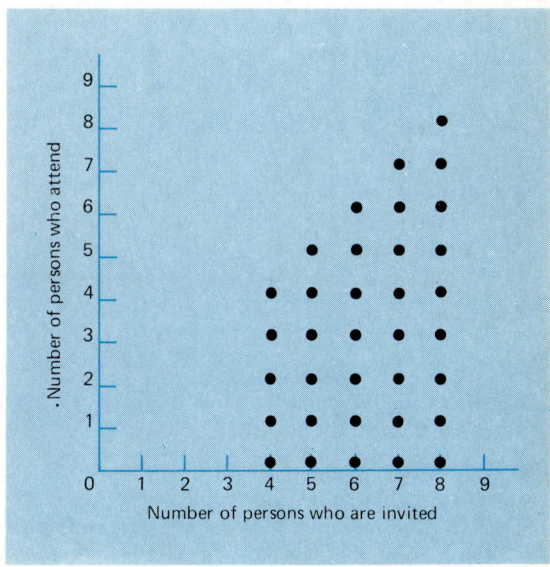

FIGURE 6.11 *Sample space for Exercise 6.64.*

the probability $\frac{1}{35}$, what are the probabilities that
  (a) at most three persons will attend;
  (b) at least six persons will be invited;
  (c) one invited person will not attend?

6.65 With reference to Exercise 6.10, on page 124, suppose that each of the ten points of the sample space has the probability 0.10. Find the probabilities of events T, U, V, and W.

6.66 With reference to Figure 6.3 on page 122, suppose that probabilities are assigned to the regions as follows:

Region	Probability
1	0.15
2	0.13
3	0.10
4	0.17
5	0.15
6	0.12
7	0.10
8	0.08

Find the probabilities that
  (a) hamburgers will be served;
  (b) beer will be served;
  (c) neither hamburgers nor watermelon will be served;
  (d) beer and/or hamburgers will be served.

6.67 If H stands for heads and T for tails, the 32 possible outcomes for five flips of a coin are

```
HHHHH   HTHHH   THHHH   TTHHH
HHHHT   HTHHT   THHHT   TTHHT
HHHTH   HTHTH   THHTH   TTHTH
HHHTT   HTHTT   THHTT   TTHTT
HHTHH   HTTHH   THTHH   TTTHH
HHTHT   HTTHT   THTHT   TTTHT
HHTTH   HTTTH   THTTH   TTTTH
HHTTT   HTTTT   THTTT   TTTTT
```

If these 32 outcomes are equally likely, what are the probabilities of getting 0, 1, 2, 3, 4, or 5 heads?

6.68 The probabilities that a person convicted of reckless driving will be fined, have his license revoked, or both are 0.88, 0.62, and 0.55. What is the probability that a person convicted of reckless driving will be fined and/or have his license revoked.

6.69 A geology professor has two graduate assistants helping her with her research. The probability that the older of the two will be absent on any given day is 0.08, the probability that the younger of the two will be absent on any given day is 0.06, and the probability that they will both be absent on any given day is 0.02. Find the probability that either or both of the graduate assistants will be absent on any given day.

6.70 The probabilities that a person stopping at a gas station will ask to have his tires checked is 0.14, the probability that he will ask to have his oil checked is 0.27, and the probability that he will ask to have them both checked is 0.09. What are the probabilities that a person stopping at this gas station will have
  (a) his tires, his oil, or both checked;
  (b) neither his tires nor his oil checked?

6.71 Among the 64 doctors on the staff of a hospital, 58 carry malpractice insurance, 33 are surgeons, and 31 of the surgeons carry malpractice insurance. If one of these doctors is chosen by lot to represent the hospital staff at an AMA convention, so that each doctor has the probability $\frac{1}{64}$ of being selected, what is the probability that he or she will not be a surgeon nor carry malpractice insurance?

6.72 The probabilities that in any one year a teenage girl will attend a professional football game, a professional tennis match, or both are 0.37, 0.13, and 0.10. Draw a Venn diagram, fill in the probabilities associated with the various regions, and thus determine the probabilities that in any one year a teenage girl will attend
  (a) a professional football game but no professional tennis match;
  (b) a professional football game and/or a professional tennis match;
  (c) neither a professional football game nor a professional tennis match.

6.73 A teacher feels that the odds are 3 to 2 against his getting a promotion, it is an even bet that he will get a raise, and the odds are 4 to 1 against him getting both. What are the odds that he will get a promotion and/or a raise?

6.74 Explain why there must be a mistake in each of the following statements:
  (a) The probabilities that the University of Georgia football team will win the first game of the season, the second game of the season, or the first two games of the season are 0.84, 0.71, and 0.53.

(b) At a certain medical clinic, the probabilities that a patient with a fever will get a shot, some medicine but no shot, or no medicine and no shot are 0.48, 0.36, and 0.12.

6.75 It can be shown that for any three events A, B, and C, the probability that at least one of them will occur is given by

$$P(A \cup B \cup C) = P(A) + P(B) + P(C)$$
$$- P(A \cap B) - P(A \cap C) - P(B \cap C)$$
$$+ P(A \cap B \cap C)$$

Use this formula in the following problems:

(a) The probability that a person visiting his dentist will have

his teeth cleaned is 0.47;

a cavity filled is 0.29;

a tooth extracted is 0.09;

his teeth cleaned and a cavity filled is 0.13;

his teeth cleaned and a tooth extracted is 0.03;

a cavity filled and a tooth extracted is 0.02;

his teeth cleaned, a cavity filled, and a tooth extracted is 0.01.

Find the probability that the person will have at least one of these things done.

(b) For a visitor to Disneyland, the probability that he or she will go on

the Jungle Cruise is 0.74;

the Monorail is 0.70;

the Matterhorn ride is 0.62;

the Jungle Cruise and the Monorail is 0.52;

the Jungle Cruise and the Matterhorn ride is 0.46;

the Monorail and the Matterhorn ride is 0.44;

the Jungle Cruise, the Monorail, and the Matterhorn ride is 0.34.

What is the probability that a person visiting Disneyland will go on at least one of these three rides?

6.5

CONDITIONAL PROBABILITY

If we ask for the probability of an event without specifying the sample space, we can easily get different answers and they may all be correct. For instance, if we ask for the probability that a lawyer will make more than $120,000 a year within ten years after passing the bar, we may get one answer which applies to all persons practicing law in the United States, another one which applies to corporation lawyers, another one which applies to lawyers employed by the federal government, another one which applies to lawyers who specialize in divorce cases, and so forth. Since the choice of the sample space is by no means always self-evident, it helps to use the symbol $P(A|S)$ to denote the **conditional probability** of event A relative to the sample space S, or as we often call it "the probability of A given S." The symbol $P(A|S)$ makes it explicit that we are referring to a particular sample space S, and it is generally preferable to the abbreviated notation $P(A)$ unless the tacit choice of S is clearly understood. It is also preferable when we have to refer to different sample spaces in the same problem.

To elaborate on the idea of a conditional probability, suppose that a consumer research organization has studied the service under warranty provided by the 200 tire dealers in a large city, and that their findings are summarized in the following table:

	Good service under warranty	Poor service under warranty	Total
Name-brand tire dealers	64	16	80
Off-brand tire dealers	42	78	120
Total	106	94	200

If one of these tire dealers is randomly selected (that is, each one has the probability $\frac{1}{200}$ of being selected), we find that the probabilities of choosing a name-brand dealer, a dealer who provides good service under warranty, or a name-brand dealer who provides good service under warranty are

$$P(N) = \frac{80}{200} = 0.40$$

$$P(G) = \frac{106}{200} = 0.53$$

and

$$P(N \cap G) = \frac{64}{200} = 0.32$$

All these probabilities were calculated by means of the formula $\frac{s}{n}$ for equally likely possibilities.

Since the second of these possibilities is particularly disconcerting—there is almost a fifty–fifty chance of choosing a dealer who does not provide good service under warranty—let us see what will happen if we limit the choice to name-brand dealers. This reduces the sample space to the 80 choices corresponding to the first row of the table, and we find that the probability of choosing a name-brand dealer who will provide good service under warranty is

$$P(G|N) = \frac{64}{80} = 0.80$$

This is quite an improvement over $P(G) = 0.53$, as might have been expected. Note that the conditional probability which we have obtained here, $P(G|N) = 0.80$, can also be written as

$$P(G|N) = \frac{\frac{64}{200}}{\frac{80}{200}} = \frac{P(N \cap G)}{P(N)}$$

namely, as the ratio of the probability of choosing a name-brand dealer who provides good service under warranty to the probability of choosing a name-brand dealer.

Generalizing from this example, let us now make the following definition of conditional probability, which applies to any two events A and B belonging to a given sample space S:

Definition of conditional probability

> If $P(B)$ is not equal to zero, then the conditional probability of A relative to B, namely, the probability of A given B, is
>
> $$P(A|B) = \frac{P(A \cap B)}{P(B)}$$

When $P(B)$ is equal to zero, the conditional probability of A relative to B is undefined.

EXAMPLE With reference to the tire dealers of the example above, what is the probability that a dealer will provide good service under warranty given that he is not a name-brand dealer.

Solution As can be seen from the table,

$$P(G \cap N') = \frac{42}{200} = 0.21 \quad \text{and} \quad P(N') = \frac{120}{200} = 0.60,$$

so that substitution into the formula yields

$$P(G|N') = \frac{P(G \cap N')}{P(N')} = \frac{0.21}{0.60} = 0.35$$

Of course, by writing $\frac{42}{120} = 0.35$, we could have obtained this result directly from the second row of the table on page 142.

Although we introduced the formula for $P(A|B)$ by means of an example in which the possibilities were all equally likely, this is not a requirement for its use.

EXAMPLE At a certain elementary school, the probability that a student who is randomly selected will come from a two-parent home is 0.75, and the probability that he or she will come from a two-parent home and be a low achiever (get mostly D's and F's) is 0.18. What is the probability that such a randomly selected student will be a low achiever given that he or she comes from a two-parent home?

Solution Letting L denote a low achiever and T a student from a two-parent home, we have $P(T) = 0.75$ and $P(L \cap T) = 0.18$, and we get

$$P(L|T) = \frac{P(L \cap T)}{P(T)} = \frac{0.18}{0.75} = 0.24$$

To introduce another concept which is important in the study of probability, let us consider the following example:

EXAMPLE   If H stands for heads and T for tails, the four equally likely outcomes for two flips of a balanced coin are HH, HT, TH, and TT. If A and B denote the respective events of getting heads on the first and second flips of the coin, find
  (a) $P(A)$ and $P(B)$;
  (b) $P(B \cap A)$;
  (c) $P(B|A)$.

Solution   Using the formula $\dfrac{s}{n}$ for equally likely outcomes, we get

  (a) $P(A) = \frac{2}{4} = 0.50$ and $P(B) = \frac{2}{4} = 0.50$;
  (b) $P(B \cap A) = \frac{1}{4} = 0.25$;

and, hence,

  (c) $P(B|A) = \dfrac{P(B \cap A)}{P(A)} = \dfrac{0.25}{0.50} = 0.50.$

What is special, and interesting, about this result is that $P(B|A) = P(B) = 0.50$, so that the probability of event B is the same regardless of whether event A has occurred. Actually, this should not come as a surprise, since the coin has no memory and what happens in the second flip is in no way affected by what happened in the first.

If $P(B)$ is not equal to zero and if $P(A|B) = P(A)$, we say that event A is **independent** of event B; that is,

> **Event A is independent of event B if the probability of A is not affected by the occurrence or nonoccurrence of B.**

Since it can be shown that event B is independent of event A whenever event A is independent of event B, it is customary to say simply that **A and B are independent** whenever one is independent of the other (see Exercise 6.97 on page 150). If two events A and B are not independent, we say that they are **dependent**.

This notion of independence is satisfied by events of probability zero, even though certain conditional probabilities are not defined. Some people use the special multiplication rule on page 146 as the definition of independence.

EXAMPLE   The probabilities that a student will get passing grades in Mathematics, in English, or in both are $P(M) = 0.70$, $P(E) = 0.80$, and $P(M \cap E) = 0.56$. Check whether events M and E are independent.

Solution   Substituting into the formula for a conditional probability, we get

$$P(M|E) = \dfrac{P(M \cap E)}{P(E)} = \dfrac{0.56}{0.80} = 0.70$$

Since $P(M|E) = 0.70 = P(M)$, we find that events M and E are independent.

EXAMPLE The probabilities that it will rain or snow in a given city on Christmas Day, on New Year's Day, or on both days are $P(C) = 0.60$, $P(N) = 0.60$, and $P(C \cap N) = 0.42$. Check whether events N and C are independent.

Solution Substituting into the formula for a conditional probability, we get

$$P(N|C) = \frac{P(C \cap N)}{P(C)} = \frac{0.42}{0.60} = 0.70$$

Since $P(N|C) = 0.70$ does not equal $P(N) = 0.60$, we find that events N and C are dependent.

In the next section we shall see that there is another way of working the two preceding examples.

Note that we did not use Venn diagrams in discussing independence. Independence is not easily displayed graphically.

6.6

MULTIPLICATION RULES

In Section 6.5 we used the formula $P(A|B) = \dfrac{P(A \cap B)}{P(B)}$ only to calculate conditional probabilities, but if we multiply on both sides of the equation by $P(B)$, we get the following formula, called the **general multiplication rule**, which enables us to calculate the probability that two events will both occur:

General multiplication rule

$$P(A \cap B) = P(B) \cdot P(A|B)$$

In words, this formula states that the probability that two events will both occur is the product of the probability that one of the events will occur and the conditional probability that the other event will occur given that the first event has occurred (occurs, or will occur). As it does not matter which event is referred to as A and which is referred to as B, the above formula can also be written as

General multiplication rule

$$P(A \cap B) = P(A) \cdot P(B|A)$$

EXAMPLE A jury consists of nine persons who are native-born and three persons who are foreign-born. If two of the jurors are randomly picked for an interview, what is the probability that they will both be foreign-born?

Solution Let A be the event that the first juror is foreign-born and B the event that the second juror is foreign-born. If we assume equal probabilities for each choice (which is, in fact, what we mean by the selection being random), the probability that the first juror picked will be foreign-born is $P(A) = \frac{3}{12}$. Then, if the first juror picked is foreign-born, the probability that the second juror will also be foreign-born is $P(B|A) = \frac{2}{11}$. Hence, the probability of getting two jurors who are both foreign-born is

$$P(A \cap B) = P(A) \cdot P(B|A) = \frac{3}{12} \cdot \frac{2}{11} = \frac{1}{22}$$

When A and B are independent, we can substitute $P(A)$ for $P(A|B)$ in the first of the two formulas for $P(A \cap B)$, or $P(B)$ for $P(B|A)$ in the second, and we obtain

Special multiplication rule (independent events)

> *If A and B are independent events, then*
>
> $$P(A \cap B) = P(A) \cdot P(B)$$

In words, the probability that two independent events will both occur is simply the product of their respective probabilities. This rule is sometimes used as the definition of independence; in any case, it may be used to check whether two given events are independent.

EXAMPLE   What is the probability of getting two heads in two flips of a balanced coin?

Solution   Since the probability of heads is $\frac{1}{2}$ for each flip of the coin, the answer is $\frac{1}{2} \cdot \frac{1}{2} = \frac{1}{4}$.

EXAMPLE   If $P(C) = 0.65$, $P(D) = 0.40$, and $P(C \cap D) = 0.26$, are the events C and D independent?

Solution   Since $P(C) \cdot P(D) = (0.65)(0.40) = 0.26 = P(C \cap D)$, the two events are independent.

EXAMPLE   With reference to the example on page 144, use the special multiplication rule to verify that events M and E are independent.

Solution   Since $P(M) \cdot P(E) = (0.70)(0.80) = 0.56 = P(M \cap E)$, the special multiplication rule is satisfied and the two events are independent.

EXAMPLE   With reference to the example on page 145, use the special multiplication rule to verify that events N and C are independent.

Solution   Since $P(C) \cdot P(N) = (0.60)(0.60) = 0.36$ does not equal $P(C \cap N) = 0.42$, the special multiplication rule is not satisfied and the two events are not independent; that is, they are dependent.

EXAMPLE   What is the probability of getting two aces in a row when two cards are drawn from an ordinary deck of 52 playing cards, if
    (a) the first card is replaced before the second card is drawn;
    (b) the first card is not replaced before the second card is drawn?

Solution   **(a)** Since there are four aces among the 52 cards, we get

$$\frac{4}{52} \cdot \frac{4}{52} = \frac{1}{169}.$$

(b) Since there are only three aces among the 51 cards which remain after one ace has been removed from the deck, we get

$$\frac{4}{52} \cdot \frac{3}{51} = \frac{1}{221}.$$

The distinction between the two parts of this example is important in statistics. What we did in part (a) is called **sampling with replacement** and what we did in part (b) is called **sampling without replacement**.

The special multiplication rule can easily be generalized so that it applies to the occurrence of three or more independent events—again, we multiply together all the individual probabilities.

EXAMPLE   What is the probability of getting three heads in three flips of a balanced coin?

Solution   Since the flips of the coin are independent, we get

$$\frac{1}{2} \cdot \frac{1}{2} \cdot \frac{1}{2} = \frac{1}{8}.$$

EXAMPLE   If the probability is 0.70 that any one person interviewed at a shopping mall will be against rezoning a certain piece of property for industrial development, what is the probability that among four persons interviewed at the mall the first three will be against the rezoning but the fourth one will not be against it?

Solution   Assuming independence, we multiply the respective probabilities and get

$$(0.70)(0.70)(0.70)(0.30) = 0.1029$$

or approximately 0.10.

When three or more events are not independent, the multiplication rule becomes more complicated—we form the product of the probability that one of the events will occur, the probability that a second event will occur given that the first event has occurred, the probability that a third event will occur given that the first two events have occurred, and so forth.

EXAMPLE   With reference to the example on page 145, what is the probability that if three of the twelve jurors are randomly picked for an interview, they will all be foreign born?

Solution   Since the probabilities are $\frac{3}{12}$, $\frac{2}{11}$, and $\frac{1}{10}$ that the first juror picked will be foreign born, that the second juror picked will be foreign born given that the first juror picked is foreign born, and that the third juror picked will be foreign born given that the first two jurors picked are foreign born, the probability is

$$\frac{3}{12} \cdot \frac{2}{11} \cdot \frac{1}{10} = \frac{1}{220}.$$

EXERCISES

6.76 If W is the event that a worker is well trained and Q is the event that he or she meets the production quota, express symbolically the probabilities that
   (a) a worker who is well trained will meet the production quota;
   (b) a worker who meets the production quota is not well trained;
   (c) a worker who is not well trained will not meet the production quota.

6.77 With reference to the preceding exercise, state in words what probabilities are expressed by
   (a) $P(W|Q)$;
   (b) $P(Q'|W)$;
   (c) $P(W'|Q')$.

6.78 A guidance department gives students various kinds of tests. If I is the event that a student scores high in intelligence, A is the event that a student rates high on a social adjustment scale, and N is the event that a student displays neurotic tendencies, express symbolically the probabilities that
   (a) a student who scores high in intelligence will display neurotic tendencies;
   (b) a student who does not rate high on the social adjustment scale will not score high in intelligence;
   (c) a student who displays neurotic tendencies will neither score high in intelligence nor rate high on the social adjustment scale.

6.79 With reference to the preceding exercise, state in words what probabilities are expressed by
   (a) $P(I|A)$;
   (b) $P(A'|N')$;
   (c) $P(N'|I \cap A)$.

6.80 If E is the event that an applicant for a home mortgage is employed, G is the event that she has a good credit rating, and A is the event that the application is approved, express symbolically the probabilities that
   (a) an applicant who has a good credit rating will be employed;
   (b) an applicant whose application is not approved will not have a good credit rating;
   (c) an applicant who is employed and has a good credit rating will have the application approved.

6.81 With reference to the preceding exercise, state in words what probabilities are expressed by
   (a) $P(A|E)$;
   (b) $P(A'|E')$;
   (c) $P(E' \cup G'|A')$.

6.82 There are 80 applicants seeking to obtain a fast food franchise. Some of these persons are college graduates and some are not. Some have prior experience in the food service industry and some do not. The exact breakdown is

	College graduates	Not college graduates
Prior food service experience	24	36
No prior food service experience	12	8

If the order in which the applicants are processed is random, G is the event that the first applicant processed is a college graduate, and E is the event that the first applicant processed has prior food service experience, determine each of the following probabilities directly from the entries and the row and column

totals of the table:
  (a) $P(G)$;
  (b) $P(E')$;
  (c) $P(G \cap E)$;
  (d) $P(G' \cap E')$;
  (e) $P(E|G)$;
  (f) $P(G'|E')$;
  (g) $P(E'|G')$;
  (h) $P(E \cup G)$.

6.83 Use the results of the preceding exercise to verify that

(a) $P(E|G) = \dfrac{P(E \cap G)}{P(G)}$;

(b) $P(G'|E') = \dfrac{P(E' \cap G')}{P(E')}$.

6.84 Among the 400 inmates of a prison, some are first offenders, some are hardened criminals, some serve terms of less than five years, and some serve longer terms, with the exact breakdown being

	Terms of less than five years	Longer terms
First offenders	120	40
Hardened criminals	80	160

If one of the inmates is to be selected at random to be interviewed about prison conditions, H is the event that he is a hardened criminal, and L is the event that he is serving a longer term, determine each of the following probabilities directly from the entries and the row and column totals of the table:
  (a) $P(H)$;
  (b) $P(L)$;
  (c) $P(L \cap H)$;
  (d) $P(H' \cap L)$;
  (e) $P(L|H)$;
  (f) $P(H'|L)$.

6.85 Use the results of the preceding exercise to verify that

(a) $P(L|H) = \dfrac{P(L \cap H)}{P(H)}$;

(b) $P(H'|L) = \dfrac{P(H' \cap L)}{P(L)}$.

6.86 The probabilities are 0.45, 0.36, and 0.18 that a person traveling in the Northeastern United States will visit Boston, Providence, or both. Find the probabilities that
  (a) such a traveler who is visiting Boston will also visit Providence;

(b) such a traveler who is visiting Providence will also visit Boston.

6.87 A survey of women in executive positions showed that the probability is 0.60 that such a woman will enjoy making financial decisions, and 0.42 that such a woman will enjoy making financial decisions and be willing to assume substantial risks. What is the probability that a woman in an executive position who enjoys making financial decisions will be willing to assume substantial risks?

6.88 The probability that a bus from Cleveland to Chicago will leave on time is 0.80, and the probability that it will leave on time and also arrive on time is 0.72.
  (a) What is the conditional probability that if such a bus leaves on time it will also arrive on time?
  (b) If the probability is 0.75 that such a bus will arrive on time, what is the conditional probability that if such a bus does not leave on time it will nevertheless arrive on time?

6.89 Given $P(A) = 0.50$, $P(B) = 0.30$, and $P(A \cap B) = 0.15$, verify that
  (a) $P(A|B) = P(A)$;
  (b) $P(A|B') = P(A)$;
  (c) $P(B|A) = P(B)$;
  (d) $P(B|A') = P(B)$.

6.90 The probability that a certain person will go out for breakfast is 0.40, and the probability that if she goes out for breakfast she will spend more than $5.00 is 0.75. What is the probability that she will go out for breakfast and spend more than $5.00?

6.91 The probability that a student attending a Western college will buy a personal computer is 0.50, and the probability that if he buys a personal computer his grades will go up is 0.72. What is the probability that a student attending this college will buy a personal computer and have his grades go up?

6.92 If two cards are drawn from an ordinary deck of 52 playing cards, what are the probabilities that they will both be diamonds if the drawing is
  (a) with replacement;
  (b) without replacement?

6.93 Among 60 pieces of luggage loaded on a plane in San Francisco, 45 are destined for Seattle and 15 for Vancouver. If two of the pieces of luggage are sent to Portland by mistake and the "selection" is random, what are the probabilities that
  (a) both should have gone to Seattle;
  (b) both should have gone to Vancouver?

Use the results of parts (a) and (b) to find the probability that

    (c) one should have gone to Seattle and one to Vancouver.

6.94 In a fifth-grade class of 18 boys and 12 girls, one pupil is chosen each week by lot to act as an assistant to the teacher. What is the probability that a girl will be chosen two weeks in a row if

    (a) the same pupil cannot serve two weeks in a row;

    (b) the restriction of part (a) is removed?

6.95 If A and B are independent events and $P(A) = 0.20$ and $P(B) = 0.45$, find

    (a) $P(A \cap B)$;   (e) $P(A \cap B')$;

    (b) $P(A|B)$;   (f) $P(B'|A')$;

    (c) $P(B|A)$;   (g) $P(A' \cap B')$;

    (d) $P(A' \cap B)$;   (h) $P(A \cup B)$.

6.96 If $P(A) = 0$ and $P(B) > 0$, show that A is independent of B, in the sense that $P(A|B) = P(A)$.

6.97 If $P(A) > 0$ and $P(B) > 0$, and if A is independent of B, show that B is independent of A.

6.98 If $P(A) = 0.80$, $P(C) = 0.95$, and $P(A \cap C) = 0.76$, are events A and C independent?

6.99 If $P(K) = 0.15$, $P(L) = 0.82$, and $P(K \cap L) = 0.12$, are events K and L independent?

6.100 If the odds are 5 to 3 that event M will not occur, 2 to 1 that event N will occur, and 4 to 1 that they will not both occur, are the two events M and N independent?

6.101 What is the probability of getting eight heads in a row with a balanced coin?

6.102 What is the probability of getting no 3's in four rolls of a balanced die?

6.103 Some special cards are prepared for an unusual game. There are four varieties: cards with red letter X, cards with green letter X, cards with red letter Y, and cards with green letter Y. Deck A has forty of these cards, and the table below gives the contents of this deck.

	Red	Green
X	5	14
Y	5	16

If a card is selected at random from this deck, it follows that

$$P(X|\text{Red, Deck } A) = \frac{5}{10}$$

since 5 of the 10 red cards contain the letter X. Also,

$$P(X|\text{Green, Deck } A) = \frac{14}{30} \approx 0.47.$$

Deck B also has forty of these cards, and the table below gives the makeup of Deck B.

	Red	Green
X	5	2
Y	20	13

    (a) Show that

        $P(X|\text{Red, Deck } A) > P(X|\text{Green, Deck } A)$.

    (b) Show that

        $P(X|\text{Red, Deck } B) > P(X|\text{Green, Deck } B)$.

    (c) Suppose that the two decks are combined to make a single deck of eighty cards. How many cards of each of the four varieties are in this deck?

    (d) Suppose that a single card is to be selected randomly from this deck. Is it true that

      $P(X|\text{Red, Combined Deck})$

            $> P(X|\text{Green, Combined Deck})$?

Note: Conditional probabilities behave in peculiar ways. This particular problem in an example of **Simpson's paradox**.

6.104 Abel and Baker are quarterbacks for a football team. The coach uses both players during each game. In the first game of the season, Abel got to throw 20 passes, and 11 of them were complete. Also in the first game, Baker threw five passes, completing three of them.

    In the second game, Abel threw five passes, completing only one. Baker threw 16 passes, completing six.

    (a) Show that Baker's proportion of pass completions was better than Abel's proportion of pass completions in the first game.

    (b) Show that Baker's proportion of pass completions was better than Abel's proportion of pass completions in the second game.

(c) A newspaper combined the data from the two games, to get the "season-to-date" data. According to the combined information, which quarterback has the higher proportion of pass completions?

6.105 If five of a company's ten delivery trucks do not meet emission standards and three of them are chosen for inspection, what is the probability that none of the trucks chosen will meet emission standards?

6.106 If a person randomly picks four of the fifteen gold coins a dealer has in stock, and six of the coins are counterfeits, what is the probability that the coins picked will all be counterfeits?

6.107 In a certain city, the probability that it will rain on a November day is 0.60, the probability that a rainy November day will be followed by another rainy day is 0.80, and the probability that a sunny November day will be followed by a rainy day is 0.30. What is the probability that it will rain, rain, not rain, and rain in this city on four consecutive November days?

6.108 A department store which bills its charge-account customers once a month has found that if a customer pays promptly one month, the probability is 0.90 that he will also pay promptly the next month; however, if a customer does not pay promptly one month, the probability that he will pay promptly the next month is only 0.40.

(a) What is the probability that a customer who pays promptly one month will also pay promptly the next three months?

(b) What is the probability that a customer who does not pay promptly one month will also not pay promptly the next two months and then make a prompt payment the month after that?

6.7

BAYES' THEOREM ★

Although the symbols $P(A|B)$ and $P(B|A)$ may look alike, there is a great difference between the probabilities which they represent. For instance, on page 142 we calculated the probability $P(G|N)$ that a name-brand tire dealer will provide good service under warranty, but what do we mean when we write $P(N|G)$? This is the probability that a tire dealer who provides good service under warranty is a name-brand dealer. To give another example, suppose that B represents the event that a person committed a burglary and G represents the event that he or she is found guilty of the crime. Then, $P(G|B)$ is the probability that the person who committed the burglary will be found guilty of the crime, and $P(B|G)$ is the probability that the person who is found guilty of the burglary actually committed it. Thus, in both of these examples we turned things around—cause, so to speak, became effect and effect became cause.

Since there are many problems in statistics which involve such pairs of conditional probabilities, let us find a formula which expresses $P(B|A)$ in terms of $P(A|B)$ for any two events A and B. To this end we equate the expressions for $P(A \cap B)$ in the two forms of the general multiplication rule on page 145 and we get

$$P(A) \cdot P(B|A) = P(B) \cdot P(A|B)$$

and, hence,

$$P(B|A) = \frac{P(B) \cdot P(A|B)}{P(A)}$$

after we divide by $P(A)$.

EXAMPLE   In a state where cars have to be tested for the emission of pollutants, 25 percent of all cars emit excessive amounts of pollutants. When tested, 99 percent of all cars that emit excessive amounts of pollutants will fail, but 17 percent of the cars that do not emit excessive amounts of pollutants will also fail. What is the probability that a car which fails the test actually emits excessive amounts of pollutants?

Solution   Letting A denote the event that a car fails the test and B the event that it emits excessive amounts of pollutants, we can translate the given percentages into probabilities and write

$$P(B) = 0.25, \qquad P(A|B) = 0.99, \quad \text{and} \quad P(A|B') = 0.17.$$

Before we can calculate $P(B|A)$ by means of the formula given on the previous page, we will first have to determine $P(A)$, and to this end let us look at the tree diagram of Figure 6.12. Here A is reached either along the branch which passes through B or along the branch which passes through B', and the probabilities of this happening are

$$(0.25)(0.99) = 0.2475 \text{ and } (0.75)(0.17) = 0.1275.$$

Since the alternatives represented by the two branches are mutually exclusive, we find that $P(A) = 0.2475 + 0.1275 = 0.3750$, and substitution into the formula for $P(B|A)$ yields

$$P(B|A) = \frac{P(B) \cdot P(A|B)}{P(A)} = \frac{(0.25)(0.99)}{0.3750} = 0.66$$

This is the probability that a car which fails the test actually emits excessive amounts of pollutants.

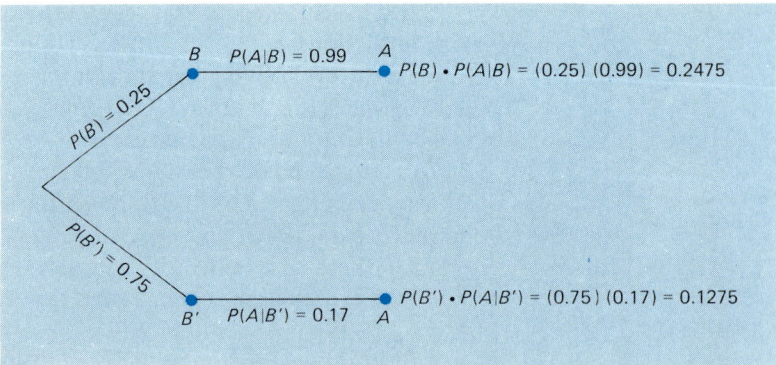

FIGURE 6.12   *Tree diagram for emission-testing example.*

With reference to the tree diagram of Figure 6.12 we can say that $P(B|A)$ is the probability that event A was reached via the upper branch of the tree, and we showed that its value is given by the ratio of the probability associated with that branch of the tree to the sum of the probabilities associated with both branches.

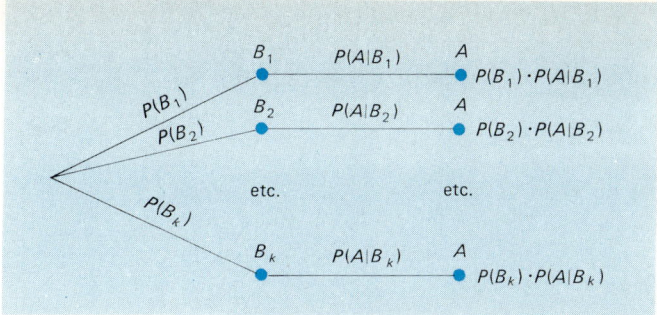

FIGURE 6.13 *Tree diagram for Bayes' theorem.*

This argument can be generalized to the case where there are more than two possible "causes," namely, more than two branches leading to an event A. With reference to Figure 6.13 we can say that $P(B_i|A)$ is the probability that event A was reached via the ith branch of the tree (for $i = 1, 2, \ldots$, or k), and it can be shown that its value is given by the ratio of the probability associated with the ith branch to the sum of the probabilities associated with all the branches leading to A. Formally, we write

Bayes' theorem

> *If B_1, B_2, \ldots, and B_k are mutually exclusive events of which one must occur, then*
>
> $$P(B_i|A) = \frac{P(B_i) \cdot P(A|B_i)}{P(B_1) \cdot P(A|B_1) + P(B_2) \cdot P(A|B_2) + \cdots + P(B_k) \cdot P(A|B_k)}$$
>
> *for $i = 1, 2, \ldots$, or k.*

Note that the expression in the denominator actually equals $P(A)$. This formula for calculating $P(A)$ when A is reached via one of several intermediate steps, is called the **Rule of Elimination** or the **Rule of Total Probability**.

EXAMPLE   In a cannery, assembly lines I, II, and III account for 50, 30, and 20 percent of the total output. If 0.4 percent of the cans from assembly line I are improperly sealed, and the corresponding percentages for assembly lines II and III are 0.6 and 1.2 percent, what is the probability that
  (a) a can produced by this cannery will be improperly sealed;
  (b) an improperly sealed can (discovered at the final inspection of outgoing products) will have come from assembly line I?

Solution   (a) Letting A denote the event that a can is improperly sealed, and B_1, B_2, and B_3 denote the events that a can comes from assembly lines I, II, or III, we can translate the given percentages into probabilities and write $P(B_1) = 0.50$, $P(B_2) = 0.30$, $P(B_3) = 0.20$, $P(A|B_1) = 0.004$, $P(A|B_2) = 0.006$, and

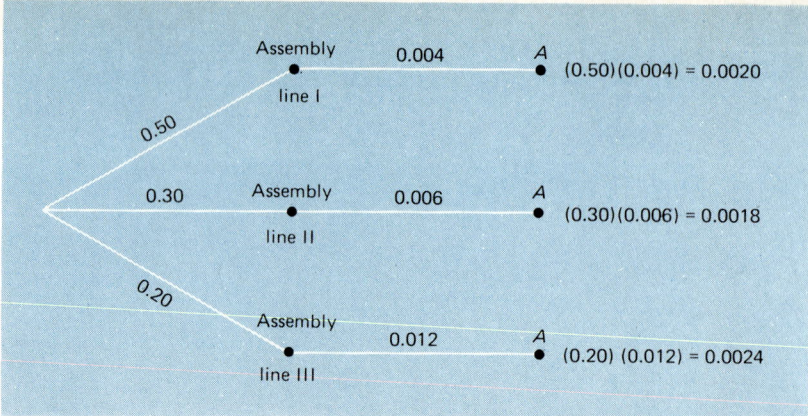

FIGURE 6.14 *Tree diagram for cannery example.*

$P(A|B_3) = 0.012$. Thus, the probabilities associated with the three branches of the tree diagram of Figure 6.14 are $(0.50)(0.004) = 0.0020$, $(0.30)(0.006) = 0.0018$, and $(0.20)(0.012) = 0.0024$, and the rule of elimination yields

$$P(A) = 0.0020 + 0.0018 + 0.0024 = 0.062$$

(b) Substituting this result together with the probability associated with the first branch of the tree diagram into the formula for Bayes' theorem, we get

$$P(B_1|A) = \frac{0.0020}{0.0062} = 0.32$$

rounded to two decimals.

Problems like these may be solved by creating a fictitious production run. Suppose that we imagine a run of, say, 10,000 cans. Of these cans, 5,000 will come from line I, 3,000 will come from line II, and 2,000 will come from line III. Of the 5,000 cans from line I, there will be $0.004 \cdot 5{,}000 = 20$ cans with improper seals. Similar logic applied to the other two lines lets us fill in the chart below:

Production line	Number of cans	Number with improper seals
I	5,000	20
II	3,000	18
III	2,000	24
Total	10,000	62

This fictitious run produces, on average, 62 improperly sealed cans. Of these, 20 came from line I, so the conditional probability $P(B_1|A)$ is $\frac{20}{62} = 0.32$, exactly as before. The method does not allow for chance variation in the numbers of cans produced, so it is a bit unrealistic. It does, however, lead easily to the solution.

As can be seen from the two examples of this section, Bayes' formula is a relatively simple mathematical rule. There can be no question about its validity, but criticism has frequently been raised about its applicability. This is because it involves a "backward" or "inverse" sort of reasoning, namely, reasoning from effect to cause. For instance, in the example on page 152 we wondered whether a car's failing the test was brought about, or caused, by its emitting excessive amounts of pollutants. Similarly, in the example immediately above we wondered whether an improperly sealed can was produced, or caused, by assembly line I. It is precisely this aspect of Bayes' theorem which makes it play an important role in statistical inference, where our reasoning goes from sample data that are observed to the populations from which they came. Brief discussions of such inferences, appropriately called **Bayesian inferences**, may be found in Sections 11.3 and 13.2.

EXERCISES

★ **6.109** The probability that a one-car accident is due to faulty brakes is 0.04, the probability that a one-car accident is correctly attributed to faulty brakes is 0.82, and the probability that a one-car accident is incorrectly attributed to faulty brakes is 0.03. What is the probability that
  (a) a one-car accident will be attributed to faulty brakes;
  (b) a one-car accident attributed to faulty brakes was actually due to faulty brakes?

★ **6.110** In a T-maze, a rat is given food if it turns left and an electric shock if it turns right. On the first trial there is a fifty-fifty chance that a rat will turn either way; then, if it receives food on the first trial, the probability that it will turn left on the second trial is 0.68, and if it receives a shock on the first trial, the probability that it will turn left on the second trial is 0.84. What is the probability that
  (a) a rat will turn left on the second trial;
  (b) a rat which turns left on the second trial will have turned left also on the first trial?

★ **6.111** At an electronics plant, it is known from past experience that the probability is 0.86 that a new worker who has attended the company's training program will meet his production quota, and that the corresponding probability is 0.35 for a new worker who has not attended the company's training program. If 80 percent of all new workers attend the training program, what is the probability that

  (a) a new worker will not meet his production quota;
  (b) a new worker who does not meet his production quota will not have attended the company's training program?

★ **6.112** Two firms V and W consider bidding on a road-building job which may or may not be awarded depending on the amounts of the bids. Firm V submits a bid and the probability is $\frac{3}{4}$ that it will get the job provided firm W does not bid. The odds are 3 to 1 that W will bid, and if it does, the probability that V will get the job is only $\frac{1}{3}$.
  (a) What is the probability that V will get the job?
  (b) If V gets the job, what is the probability that W did not bid?

★ **6.113** In a certain community, 8 percent of all adults over 50 have diabetes. If a doctor in this community correctly diagnoses 95 percent of all persons with diabetes as having the disease and incorrectly diagnoses 2 percent of all persons without diabetes as having the disease, what is the probability that an adult over 50 diagnosed by this doctor as having diabetes actually has the disease?

★ **6.114** With reference to the example on page 152, suppose that the state agency doing the testing can reduce to 0.03 the probability that a car which does not emit excessive amounts of pollutants will fail the test. How will this affect the probability that a car which fails the test actually emits excessive amounts of pollutants?

★ **6.115** A blood disease is found in 2 percent of the persons in a certain population. A new blood test will correctly identify 96 percent of the persons with the disease and 94 percent of the persons without the disease.

(a) What is the probability that a person who is called positive by the blood test actually has the disease?

(b) What is the probability that a person who is called negative by the blood test actually does not have the disease?

★ **6.116** It is known from experience that in a certain industry 60 percent of all labor-management disputes are over wages, 15 percent are over working conditions, and 25 percent are over fringe issues. Also, 45 percent of the disputes over wages are resolved without strikes, 70 percent of the disputes over working conditions are resolved without strikes, and 40 percent of the disputes over fringe issues are resolved without strikes. What is the probability that a labor–management dispute in this industry will be resolved without a strike?

★ **6.117** With reference to the preceding exercise, what is the probability that if a labor–management dispute in this industry is resolved without a strike, it was over wages?

★ **6.118** A retailer of automobile parts has four employees, K, L, M, and N, who make mistakes in filling orders one time in 100, four times in 100, six times in 100, and two times in 100. Of all the orders filled, K, L, M, and N fill 20, 30, 10, and 40 percent. What is the probability that

(a) a mistake will be made in an order;

(b) if a mistake is made in an order, the order was filled by K;

(c) if a mistake is made in an order, the order was filled by N?

★ **6.119** (From I. Miller and J. E. Freund, *Probability and Statistics for Engineers*, 3rd edition, Englewood Cliffs, New Jersey: Prentice Hall, 1985.) An explosion in a liquefied natural gas tank undergoing repair could have occurred as the result of static electricity, malfunctioning electrical equipment, an open flame in contact with the liner, or purposeful action (industrial sabotage). Interviews with engineers who were analyzing the risks led to estimates that such an explosion would occur with probability 0.25 as a result of static electricity, 0.20 as a result of malfunctioning electric equipment, 0.40 as a result of an open flame, and 0.75 as a result of purposeful action. These interviews also yielded subjective estimates of the probabilities of the four causes of 0.30, 0.40, 0.15, and 0.15, respectively. Based on all this information, what is the most likely cause of the explosion?

★ **6.120** To get answers to sensitive questions, we sometimes use a method called the **randomized response technique**. Suppose, for instance, that we want to determine what percentage of the students at a large university smoke marijuana. We construct 20 flash cards, write "I smoke marijuana at least once a week" on 12 of the cards, where 12 is an arbitrary choice, and "I do not smoke marijuana at least once a week" on the others. Then, we let each student (in the sample interviewed) select one of the cards at random, and respond "yes" or "no" without divulging the question.

(a) Establish a relationship between $P(Y)$, the probability that a student will give a "yes" response, and $P(M)$, the probability that a student randomly selected at that university smokes marijuana at least once a week.

(b) If 106 of 250 students answered "yes" under these conditions, use the result of part (a) and $\frac{106}{250}$ as an estimate of $P(Y)$ to estimate $P(M)$.

★ **6.121** The following story illustrates what can happen when we use "common sense," or intuition, in connection with probabilities:

"Among three indistinguishable boxes one contains two pennies, one contains a penny and a dime, and one contains two dimes. We randomly choose one of the three boxes, and randomly pick one of the two coins in that box without looking at the other. If the coin we pick is a penny, what is the probability that the other coin in the box is also a penny? Without giving the matter too much thought, we might argue that there is a fifty-fifty chance that the other coin is also a penny. After all, the coin we picked must have come from the box with the penny and the dime or from the box with the two pennies. In the first case the other coin is a dime, in the second case the other coin is a penny, and it would seem reasonable to say that these two possibilities are equally likely."

Use Bayes' theorem to show that the probability is actually $\frac{2}{3}$ that the other coin is also a penny.

6.8

CHECKLIST OF KEY TERMS
(with page references to their definitions)

Addition rules, 135
★ Bayes' theorem, 151
Betting odds, 130
Complement, 120
Conditional probability, 141
Consistency criterion, 132
Dependent events, 144
Empty set, 119
Event, 119
Experiment, 118
Fair odds, 130
Finite sample space, 119
General addition rule, 138
General multiplication rule, 145
Generalized addition rule, 138
Independent events, 144
Infinite sample space, 119
Intersection, 120

Multiplication rules, 145
Mutually exclusive events, 120
Odds, 129
Outcome, 118
Postulates of probability, 126
★ *Randomized response technique*, 156
★ Rule of elimination, 153
★ Rule of total probability, 153
Sample space, 118
Sampling with replacement, 147
Sampling without replacement, 147
Simpson's paradox, 150
Special addition rule, 138
Special multiplication rule, 146
Subjective probabilities, 131
Theory of probability, 118
Union, 120
Venn diagram, 121

6.9

REVIEW EXERCISES

6.122 Convert each of the following probabilities to odds:
  (a) The probability of rolling eight or less with pair of balanced dice is $\frac{26}{36}$.
  (b) If a researcher randomly selected 8 of 80 households to be included in a study, the probability is $\frac{1}{10}$ that any particular household will be included.
  (c) The probability of getting at least three heads in six flips of a balanced coin is $\frac{42}{64}$.

6.123 A small real estate office has five part-time salespersons. Using two coordinates so that (3, 1), for example, represents the event that three of the salespersons are at work and one of them is busy with a customer, and (2, 0) represents the event that two of the salespersons are at work but neither of them is busy with a customer, draw a diagram similar to that of Figure 6.1, showing the 21 points of the corresponding sample space.

6.124 With reference to the preceding exercise, assume that each of the 21 points of the sample space has the probability $\frac{1}{21}$, find the probabilities that
  (a) there are no salespersons at work;
  (b) there is at least one salesperson at work, and all the salespersons that are at work are busy with customers;
  (c) at least three salespersons are at work;
  (d) at least three salespersons are busy with customers;
  (e) none of the salespersons are busy with customers;
  (f) there is at least one salesperson who is at work but not busy with customers.

★ **6.125** A hotel gets cars for its guests from three rental agencies, 20 percent from agency X, 40 percent from agency Y, and 40 percent from agency Z. If 14 percent of the cars from X, 4 percent from Y, and 8 percent from Z need tune-ups, what is the probability that

    (a) a car needing a tune-up will be delivered to one of the guests;

    (b) if a car needing a tune-up is delivered to one of the guests, it came from agency Z?

6.126 if R and D are the events that a person will have regular coffee or decaffeinated coffee after dinner at a certain restaurant, $P(R) = 0.63$ and $P(D) = 0.28$, find the probabilities that after dinner at this restaurant a person will

    (a) not have decaffeinated coffee;

    (b) have regular or decaffeinated coffee;

    (c) have neither kind of coffee.

6.127 If the probability is 0.26 that any one woman will name yellow or orange as her favorite color, what is the probability that four women, selected at random, will all name yellow or orange as their favorite color?

6.128 If Q is the event that a person is qualified for a job and G is the event that he or she will get the job, express in words what probabilities are represented by

    (a) $P(G|Q)$;   (c) $P(Q|G)$;

    (b) $P(G'|Q')$;   (d) $P(Q|G')$.

6.129 If $P(M) = 0.55$, $P(N) = 0.18$, and $P(M \cap N) = 0.099$, are the events M and N independent or dependent?

6.130 Discuss the following assertion: Since probabilities are measures of uncertainty, the probability we assign to a future event will always increase when we get more information.

6.131 Convert each of the following odds to probabilities:

    (a) The odds are 19 to 5 that a given horse will not win the Kentucky Derby.

    (b) If five cards are drawn with replacement from an ordinary deck of 52 playing cards, the odds are 13 to 3 that at most three of them will be red.

    (c) If two persons are chosen at random from a group of ten men and twelve women, the odds are 40 to 37 that one man and one woman will be selected.

6.132 Suppose that the numbers 1, 2, 3, 4, 5, and 6 are used to denote that a committee of parents and teachers decides that a certain education program is terrible, poor, fair, good, very good, or excellent. If $L = \{2, 3, 4, 5\}$ and $R = \{4, 5, 6\}$, list the elements of the sample space comprising each of the following events,

and also express the events in words:

    (a) L';   (c) $L \cap R$;

    (b) $L \cup R$;   (d) $L \cap R'$.

6.133 If someone feels that 17 to 8 are fair odds that a paint job will be finished on time, what subjective probability does he assign to this event?

6.134 The probabilities are 0.15, 0.26, and 0.08 that a family driving through a Western city will spend the night at one of its hotels, at one of its motels, or at its campground. What is the probability that a family driving through this city will spend the night at one of these kinds of facilities?

6.135 The probabilities that a newspaper will receive 0, 1, 2, ..., 7, or at least 8 letters to the editor about an unpopular decision of the school board are 0.01, 0.02, 0.05, 0.14, 0.16, 0.20, 0.18, 0.15, and 0.09. What are the probabilities that the newspaper will receive

    (a) at most 4 letters to the editor about the school board decision;

    (b) at least 6;

    (c) from 3 to 5?

6.136 Mr. Sokol is willing to bet Mrs. Sokol $24 to her $8, but not $24 to her $6, that they will be late for the opera. What does this tell us about the probability which he assigns to their being late for the opera?

6.137 Joe and Walter both enjoy sports betting, and the two are currently discussing the next Rams–49ers football game. Joe feels that the odds should be 7 to 4 in favor of the 49ers, while Walter feels that the odds should be 5 to 2 in favor of the 49ers. Describe a bet that would be agreeable to both men.

6.138 If each point of the sample space of Figure 6.15 represents an outcome having the probability $\frac{1}{32}$, find

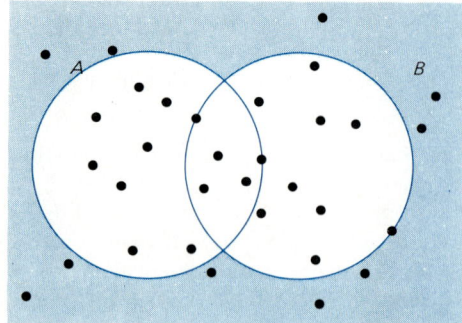

FIGURE 6.15 *Sample space for Exercise* 6.138.

(a) $P(A)$;   (d) $P(A \cup B)$;
(b) $P(B)$;   (e) $P(A' \cap B)$;
(c) $P(A \cap B)$;   (f) $P(A' \cap B')$.

6.139 With reference to the preceding exercise, find
(a) $P(A|B)$;
(b) $P(B|A')$.

6.140 As part of a promotional scheme in Arizona and New Mexico, a company distributing frozen foods will award a grand prize of $100,000 to some person sending in his or her name on an entry blank, with the option of including a label from one of the company's products. A breakdown of the 225,000 entries received is shown in the following table:

	With label	Without label
Arizona	120,000	42,000
New Mexico	30,000	33,000

If the winner of the grand prize is chosen by lot, A represents the event that it will be won by an entry from Arizona, and L represents the event that it will be won by an entry which included a label, find each of the following probabilities:
(a) $P(A)$;   (d) $P(L|A)$;
(b) $P(L)$;   (e) $P(A'|L')$;
(c) $P(A|L)$;   (f) $P(L|A')$.

6.141 Suppose that in the preceding exercise the drawing is rigged so that by including a label each entry's probability of winning the grand prize is doubled. Recalculate the probabilities of parts (a) through (f).

★ **6.142** A box contains 100 beads, some red and some white. One bead will be drawn at random, and you are asked to call beforehand whether it is going to be red or white. What odds would constitute a fair bet on this game if
(a) you have no idea how many of the beads are red and how many are white;
(b) you are told that 50 of the beads are red and 50 are white?

6.143 A library received 40 new books including 12 historical novels. If four of these books are selected at random, what is the probability that not one of them is a historical novel?

6.144 If A is the event that a university's football team is rated among the top twenty by AP and U is the event that it is rated among the top twenty by UPI, what events are represented by the four regions of the Venn diagram of Figure 6.16?

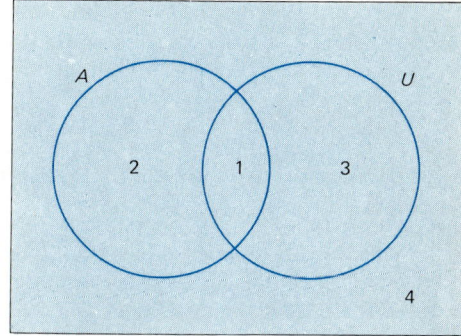

FIGURE 6.16 *Venn diagram for Exercise* 6.144.

6.145 A movie producer feels that the odds are 8 to 1 that his new movie will not be rated G, 15 to 3 that it will not be rated PG, and 13 to 5 that it will not get either of these two ratings. Are the corresponding probabilities consistent?

6.146 The probability that George will get an M.A. degree is 0.40, and the probability that with an M.A. degree he will get a well-paying job is 0.85. What is the probability that he will get an M.A. degree and a well-paying job?

★ **6.147** Lie detectors have been used during wartime to uncover security risks. As is well known, lie detectors are not infallible. Let us suppose that the probability is 0.10 that the lie detector will fail to detect a person who is a security risk and that the probability is 0.08 that the lie detector will incorrectly label a person who is not a security risk. If 2% of the persons who are given the test are actually security risks, what is the probability that
(a) a person labeled a security risk by a lie detector test is in fact a security risk;
(b) a person who is cleared by a lie detector test is in fact not a security risk?

6.148 Explain why there must be a mistake in each of the following statements:
(a) The probability that a new safety feature in cars will be able to prevent injuries is -0.02.

(b) The probability that a student will get a B in a course is 0.11, but she is ten times as likely to get a C.

(c) The probability that a teachers' conference will be well attended is 0.59, and the probability that it will not be well attended is 0.31.

6.149 If a student answers the 12 questions on a true–false test by flipping a balanced coin, what is the probability that he will answer all questions
(a) correctly;
(b) incorrectly?

6.150 Provide an example in which $P(A|B)$ is positive while $P(B|A') = 0$.

6.151 In Figure 6.17, B is the event that a person traveling in Europe will visit Belgium, P is the event that he or she will visit Portugal, and E is the event that he or she will visit England. Explain in words what events are represented by the following regions or combinations of regions of the Venn diagram:
(a) region 5;
(b) regions 1 and 4 together;
(c) regions 3 and 6 together;
(d) regions 3, 5, and 6 together;
(e) regions 5 and 8 together.

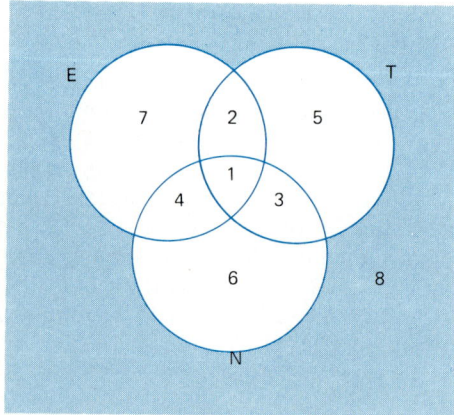

FIGURE 6.17 *Venn diagram for Exercise 6.151.*

★ **6.152** The four checkout clerks of a supermarket are supposed to ask for two identifications when they cash a check. George, who cashes 30 percent of all the checks, fails to do so 1 time in 50; Tracy, who also cashes 30 percent of all the checks, fails to do so 1 time in 40; Charles, who cashes 20 percent of all the checks, fails to do so 1 time in 10; and Susan, who also cashes 20 percent of all the checks, fails to do so 1 time in 20.
(a) What is the probability that a check will be cashed at this supermarket without two identifications?
(b) If a check is found to have been cashed without two identifications, what is the probability that it was cashed by Tracy or Charles?

6.153 Among a company's replacement parts for a given assembly, 20 percent are defective and the rest are good, 60 percent were bought from external sources and the rest were made by the company itself, and of those bought from external sources 80 percent are good and the rest are defective. What are the probabilities that a replacement part, randomly selected from this stock, is
(a) bought from an external source and good;
(b) bought from an external source and defective;
(c) company made and defective;
(d) company made given that it is defective?

★ **6.154** Surgical operations can be classified as to level of risk. The data below give the numbers of operations and the numbers of deaths at each of two hospitals:

Risk	Hospital A Operations	Hospital A Deaths	Hospital B Operations	Hospital B Deaths
Low	120	2	40	0
Medium	100	6	90	4
High	60	20	160	36

(a) Show that the death rate is higher in hospital A for low-risk operations, for medium-risk operations, and for higher-risk operations.
(b) Show that the overall death rate is higher in hospital B.
(c) Explain the apparent inconsistency between the answers to (a) and (b).

6.155 A horse breeder has entered one of his horses in a race in Florida and a race in Kentucky. If the odds are 3 to 1 that it will not win in Florida and 7 to 1 that it will not win in Kentucky, what are the odds that it will not win either race? Assume independence.

6.10
REFERENCES

More detailed, though still elementary, treatments of probability may be found in

BARR, D, R., and ZEHNA, P. W., *Probability*: *Modeling Uncertainty*. Reading, Mass.: Addison-Wesley Publishing Company, Inc., 1983.

DRAPER, N. R., and LAWRENCE, W. E., *Probability*: *An Introductory Course*. Chicago: Markham Publishing Co., 1970.

FREUND, J. E., *Introduction to Probability*. Encino, Calif.: Dickenson Publishing Company, Inc., 1973.

GOLDBERG, S., *Probability—An Introduction*. Englewood Cliffs, N.J.: Prentice-Hall, Inc., 1960.

HODGES, J. L., and LEHMANN, E. L., *Elements of Finite Probability*. San Francisco: Holden-Day, Inc., 1965.

MOSTELLER, F., ROURKE, R. E. K., and THOMAS, G. B., *Probability with Statistical Applications, 2nd ed.* Reading, Mass.: Addison-Wesley Publishing Company, Inc., 1970.

SCHEAFFER, R. L. and MENDENHALL, W., *Introduction to Probability*: *Theory and Applications*. Boston: Duxbury Press, 1975.

and interesting data on probabilities and odds in

NEFT, D. S., COHEN, R. M., and DEUTCH, J. A., *The World Book of Odds*. New York: Grosset & Dunlap, Inc., 1978.

The following is an introduction to the mathematics of gambling and various games of chance:

PACKEL, E. W., *The Mathematics of Games and Gambling*. Washington, D. C.: Mathematics Association of America, 1981.

The following is a collection of entertaining, and sometimes difficult, problems in probability:

MOSTELLER, F., *Fifty Challenging Problems in Probability with Solutions*. Reading, Mass.: Addison-Wesley Publishing Company, Inc., 1965.

7

EXPECTATIONS AND DECISIONS

When decisions are made in the face of uncertainty, they are rarely based on probabilities alone; in most cases we must also know something about the potential consequences (profits, losses, penalties, or rewards). If we must decide whether to buy a new car, knowing that our old car will soon require repairs is not enough—to make an intelligent decision we must know, among other things, the cost of the repairs and the trade-in value of the old car. Also, suppose that a building contractor has to decide whether to bid on a construction job which promises a profit of $120,000 with probability 0.20 or a loss of $27,000 (perhaps, due to a strike) with probability 0.80. The probability that the contractor will make a profit is not very high, but on the other hand, the amount he stands to gain is much greater than the amount he stands to lose. Both of these examples demonstrate the need for a method of combining probabilities and consequences, and this is why we introduce the concept of a **mathematical expection** in Section 7.1.

In Chapters 11 through 17 we deal with many different problems of inference. We estimate unknown quantities, test hypotheses (assumptions or claims), and make predictions, and in problems like these it is essential that, directly or indirectly, we pay attention to the consequences of what we do. After all, if there is nothing at stake, no penalties or rewards, and nobody cares, why not estimate the average weight of gray squirrels as 315.2 pounds, why not accept the claim that by adding water to gasoline we can get 150 miles per gallon with the old family car, and why not predict that by the year 2500 the average person will live to be 200 years old? Why not—if there is nothing at stake, no penalties or rewards, and no one cares? In section 7.2, we give some examples which show how mathematical expectations based on penalties, rewards, and other kinds of payoffs can be used in making decisions, and in Section 7.3 we show how such factors may have to be considered in choosing appropriate statistical techniques.

7.1
MATHEMATICAL EXPECTATION

If a mortality table tells us that in the United States a 50-year-old woman can expect to live 31 more years, this does not mean that anyone really expects a 50-year-old woman to live until her 81st birthday and then die the next day. Similarly, if we read that in the United States a person can expect to eat 104.4 pounds of beef and drink 39.6 gallons of soft drinks per year, or that a child in the age group from 6 to 16 can expect to visit a dentist 2.2 times a year, it must be apparent that the word "expect" is not being used in its colloquial sense. A child cannot go to the dentist 2.2 times, and it would be surprising, indeed, if we found somebody who actually ate 104.4 pounds of beef and drank 39.6 gallons of soft drinks in any given year. So far as 50-year-old women are concerned, some will live another 12 years, some will live another 20 years, some will live another 33 years, ..., and the life expectancy of "31 more years" will have to be interpreted as an average, or as we call it here, a **mathematical expectation**.

Originally, the concept of a mathematical expectation arose in connection with games of chance, and in its simplest form it is the product of the amount a player stands to win and the probability that he or she will win.

EXAMPLE What is our mathematical expectation when we will receive $10 if and only if a balanced coin comes up heads?

Solution If we assume that the coin is balanced and randomly tossed, namely, that the probability of heads is $\frac{1}{2}$, our mathematical expectation is $10 \cdot \frac{1}{2} = \$5$.

EXAMPLE What is our mathematical expectation if we buy one of 2,000 raffle tickets for a television set worth $640?

Solution Since the probability that we will win the television set is $\frac{1}{2,000} = 0.0005$, our mathematical expectation is $640(0.0005) = \$0.32$. This means that it would be foolish to spend more than 32 cents for the ticket, unless the proceeds of the raffle go to a worthy cause or the difference can be credited to whatever pleasure we may get from placing a bet.

In each of these two examples there was a single prize, but two possible payoffs—$0 or $10 in the first example and $0 or the television set worth $640 in the other. Indeed, in the second example we can argue that one of the tickets will pay $640 (in merchandise), each of the other 1,999 tickets will pay $0, so that, altogether, the 2,000 tickets will pay $640 (in merchandise) or on the average $0.32 per ticket. This is the mathematical expectation.

To generalize the concept of a mathematical expectation, let us consider the following change in the raffle of the preceding example.

EXAMPLE What is our mathematical expectation if we buy one of 2,000 raffle tickets for a first prize of a television set worth $640, a second prize of a tape recorder worth $120, and a third prize of a radio worth $40?

Solution Now we can argue that one of the tickets will pay $640 (in merchandise), another will pay $120 (in merchandise), a third will pay $40 (in merchandise), and each of the other 1,997 tickets will pay $0. Altogether, the 2,000 tickets will thus pay $640 + 120 + 40 = \$800$ (in merchandise), or on the average $\frac{800}{2,000} = \$0.40$ per ticket. Again, this is the mathematical expectation.

Looking at the preceding example in a different way, we could argue that if the raffle were repeated many times, we would lose $\frac{1,997}{2,000} \cdot 100\% = 99.85\%$ of the time (or with probability 0.9985) and win each of the prizes $\frac{1}{2,000} \cdot 100\% = 0.05\%$ of the time (or with probability 0.0005). On the average we would thus win

$$0(0.9985) + 640(0.0005) + 120(0.0005) + 40(0.0005) = \$0.40$$

which is the sum of the products obtained by multiplying each payoff by the corresponding proportion or probability. Generalizing from this example, let us now give the following definition:

Mathematical expectation

> *If the probabilities of obtaining the amounts $a_1, a_2, \ldots,$ or a_k are $p_1, p_2, \ldots,$ and p_k, where $p_1 + p_2 + \cdots + p_k = 1$, then the mathematical expectation is*
>
> $$E = a_1 p_1 + a_2 p_2 + \cdots + a_k p_k$$

Each amount is multiplied by the corresponding probability, and the mathematical expectation, E, is given by the sum of all these products. In the $\sum$ notation, $E = \sum a \cdot p$.

So far as the a's are concerned, it is important to keep in mind that they are positive when they represent profits, winning, or gains (namely, amounts which we receive), and that they are negative when they represent losses, penalties, or deficits (namely, amounts which we have to pay).

EXAMPLE What is our mathematical expectation if we win \$10 when a die comes up 1 or 6, and lose \$5 when it comes up 2, 3, 4, or 5?

Solution The amounts are $a_1 = 10$ and $a_2 = -5$, and the probabilities are $p_1 = \frac{2}{6} = \frac{1}{3}$ and $p_2 = \frac{4}{6} = \frac{2}{3}$ (if we can assume that the die is balanced and randomly tossed). Thus, the mathematical expectation is

$$E = 10 \cdot \tfrac{1}{3} + (-5) \cdot \tfrac{2}{3} = 0$$

EXAMPLE The probabilities are 0.22, 0.36, 0.28, and 0.14 that an investor will be able to sell a piece of property at a profit of \$2,500, at a profit of \$1,500, at a profit of \$500, or at a loss of \$500. What is the investor's expected profit?

Solution Substituting $a_1 = 2{,}500$, $a_2 = 1{,}500$, $a_3 = 500$, $a_4 = -500$, $p_1 = 0.22$, $p_2 = 0.36$, $p_3 = 0.28$, and $p_4 = 0.14$ into the formula for E, we get

$$E = 2{,}500(0.22) + 1{,}500(0.36) + 500(0.28) - 500(0.14)$$
$$= \$1{,}160$$

The first of these two examples illustrates what we mean by an **equitable** or **fair game**. It is a game which does not favor either player; namely, a game in which each player's mathematical expectation is zero.

Although we referred to the quantities $a_1, a_2, \ldots,$ and a_k as "amounts," they need not be cash winnings, penalties, or rewards. When we said on page 163 that a child in the age group from 6 to 16 can expect to go to a dentist 2.2 times a year, we referred to a result which is the sum of the products obtained by multiplying $0, 1, 2, 3, 4, \ldots,$ by the corresponding probabilities that a child in that age group will go to a dentist that many times a year.

EXAMPLE   If the probabilities are 0.06, 0.21, 0.24, 0.18, 0.14, 0.10, 0.04, 0.02, and 0.01 that an airline office at a certain airport will receive 0, 1, 2, 3, 4, 5, 6, 7, or 8 complaints per day about its luggage handling, how many such complaints can it expect per day?

Solution   Substituting into the formula for a mathematical expectation, we get

$$E = 0(0.06) + 1(0.21) + 2(0.24) + 3(0.18) + 4(0.14)$$
$$+ 5(0.10) + 6(0.04) + 7(0.02) + 8(0.01)$$
$$= 2.75$$

In all of the examples of this section we were given the values of a and p (or the values of the a's and p's) and calculated E. Now let us consider an example in which we are given values of a and E to arrive at some result about p, and also an example in which we are given values of p and E to arrive at some result about a.

EXAMPLE   To defend a client in a liability suit resulting from a car accident, a lawyer must decide whether to charge a straight fee of $1,500 or a contingent fee which she will get only if her client wins. How does she feel about her client's chances if
   (a) she prefers the straight fee of $1,500 to a contingent fee of $5,000;
   (b) she prefers a contingent fee of $12,000 to the straight fee of $1,500?

Solution   **(a)** If she feels that the probability is p that her client will win, the lawyer associates a mathematical expectation of $5,000p$ with the contingent fee of $5,000. Since she feels that $1,500 is preferable to this expectation, we can write

$$1,500 > 5,000p$$

and, hence, $p < \dfrac{1,500}{5,000} = 0.30$.

   (b) Now the mathematical expectation associated with the contingent fee is $12,000p$, and since she feels that this is preferable to $1,500, we can write

$$12,000p > 1,500$$

and, hence, $p > \dfrac{1,500}{12,000} = 0.125$.

Combining the results of parts (a) and (b) of the preceding example, we have shown here that $0.125 < p < 0.30$, where p is the lawyer's personal probability about her client's success. To narrow it down further, we might vary the contingent fees as in Exercises 7.13 and 7.14 on page 168.

EXAMPLE   A friend says that he would "give his right arm" for our two tickets to an NBA play-off game. To put this on a cash basis, we propose that he pay us $60 (the actual price of the two tickets), but he will get the tickets only if he draws a jack, queen, king, or ace from an ordinary deck of 52 playing cards; otherwise, we keep the tickets and his $60. What are the two tickets worth to our friend, if he feels that this arrangement is fair?

Solution Since there are four jacks, four queens, four kings, and four aces, the probability that our friend will get the tickets is $\frac{16}{52}$. Hence, the probability that he will not get the tickets is $1 - \frac{16}{52} = \frac{36}{52}$, and the mathematical expectation associated with the gamble is

$$E = a \cdot \frac{16}{52} + 0 \cdot \frac{36}{52} = a \cdot \frac{16}{52}$$

where a is the value he places on these tickets. Putting this mathematical expectation equal to $60, which he considers a fair price to pay for taking the risk. we get

$$a \cdot \frac{16}{52} = 60 \quad \text{and} \quad a = \frac{52 \cdot 60}{16} = \$195$$

This is what the two tickets are worth to our friend.

EXERCISES

7.1 If a service club sells 600 raffle tickets for a cash prize of $120, what is the mathematical expectation of a person who buys one of the tickets?

7.2 At a bazaar held to raise money for a charity, it costs $1.00 to try one's luck at drawing an ace from an ordinary deck of 52 playing cards. What is the bazaar's organizers' expected profit per customer, if they pay $10 if and only if a customer draws an ace?

7.3 If someone gives us $15 if we draw a spade from an ordinary deck of 52 playing cards, how much should we pay him if we draw a heart, diamond, or club, so as to make the game fair?

7.4 The winnder of a tennis tournament gets $40,000 and the runner-up gets $15,000. What are the two finalists' mathematical expectations if
(a) they are evenly matched;
(b) their probabilities of winning are 0.60 and 0.40;
(c) their probabilities of winning are 0.70 and 0.30?

7.5 Box 1 contains twenty slips of paper of which 19 are marked $0 and the other is marked $5; box 2 contains fifty slips of paper of which 49 are marked $0 and the other is marked $14. If a person gets whatever is on the slip he or she draws, would it be smarter to draw a slip of paper from box 1 or from box 2?

7.6 As part of a promotional scheme, the manufacturer of a new breakfast food offers a prize of $50,000 to someone willing to try the new product (distributed without charge) and send in his or her name on the label. The winner is to be drawn at random from all the entries received.
(a) What is each entrant's mathematical expectation if 200,000 persons send in their names?
(b) Was sending in the label worth the cost of postage?

7.7 If the two league champions are evenly matched, the probabilities that a "best of seven" basketball play-off will take 4, 5, 6, or 7 games are $\frac{1}{8}$, $\frac{1}{4}$, $\frac{5}{16}$, and $\frac{5}{16}$. Under these conditions, how many games can we expect such a play-off to last?

7.8 A student's parents promise her a gift of $100 if she gets an A in statistics, $50 if she gets a B, and otherwise no reward. What is her mathematical expectation if the probabilities of her getting an A or B are 0.32 and 0.40?

7.9 The wage negotiator of a labor union feels that the odds are 3 to 1 that the members of the union will get a $1.00 raise in their hourly wage, 17 to 3 that they will not get a $1.40 raise in their hourly wage, and 9 to 1 that they will not get a $2.00 raise in their hourly wage. What is the corresponding expected raise in the hourly wage?

7.10 An importer is offered a shipment of Finnish cheeses for $22,000. The probability that he will be able to sell them for $18,000 is 0.32, and the probability that he will be able to sell them for $30,000 is 0.68. What is the importer's expected gross profit?

7.11 A police chief knows that the probabilities of 0, 1, 2, 3, or 4 burglaries on any given day are 0.12, 0.25, 0.39, 0.18, and 0.06. How many burglaries can the police chief expect per day? It is assumed here that the probability of more than 4 burglaries is negligible.

7.12 The probabilities that a person who enters "The Department Store" will make 0, 1, 2, 3, 4, or 5 purchases are 0.11, 0.33, 0.31, 0.12, 0.09, and 0.04. How many purchases can a person entering this store be expected to make?

7.13 With reference to the example on page 166, how does the lawyer feel about her clients chances if
  (a) she prefers the straight fee of $1,500 to a contingent fee of $6,000;
  (b) she prefers a contingent fee of $10,000 to the straight fee of $1,500?

7.14 With reference to the example on page 166, how does the lawyer feel about her clients chances if she cannot decide whether to accept the straight fee of $1,500 or a contingent fee of $7,500?

7.15 A baseball pitcher has to choose between a straight salary of $2,000,000 and a salary of $1,800,000 with a bonus of $350,000 if he wins twenty or more games.

How does this pitcher feel about his chances of winning twenty games if he decides to take the straight salary of $2,000,000?

7.16 One contractor offers to do a road repair job for $45,000, while another contractor offers to do the job for $50,000 with a penalty of $12,500 if the job is not finished on time. If the person who lets out the contract for the job prefers the second offer, what does this tell us about her assessment of the probability that the second contractor will not finish the job on time?

7.17 Mr. Smith feels that it is just about a toss-up whether to accept a cash prize of $26 or to gamble on two flips of a coin, where he is to receive an electric drill if the coin comes up heads both times, while otherwise he is to receive $5. What cash value does he attach to owning the drill?

7.18 Mr. Jones would like to beat Mr. Brown in an upcoming golf tournament, but this chances are nil unless he takes $400 worth of lessons, which (according to the pro at his club) will give him a fifty–fifty chance. If Mr. Jones can expect to break even if he takes the lessons and bets Mr. Brown $1,000 against x dollars that he will win, find x.

7.2

DECISION MAKING ★

In the face of uncertainty, mathematical expectations can often be used to great advantage in making decisions. In general, if we must choose between two or more alternatives, it is considered rational to select the one with the "most promising" mathematical expectation—the one which maximizes expected profits, minimizes expected costs, maximizes expected tax advantages, minimizes expected losses, and so on.

EXAMPLE A furniture manufacturer must decide whether to expand his plant capacity now or wait at least another year. His advisors tell him that if he expands now and economic conditions remain good, there will be a profit of $328,000 during the next fiscal year; if he expands now and there is a recession, there will be a loss (negative profit) of $80,000; if he waits at least another year and economic conditions remain good, there will be a profit of $160,000; and if he waits at least another year and there is a recession, there will be a small profit of $16,000. If the furniture manufacturer feels that the probabilities for economic conditions remaining good or there being a recession are $\frac{1}{3}$ and $\frac{2}{3}$, will expanding his plant capacity now maximize his expected profit?

Solution In problems like this it often helps to present the given information in a table such as the following:

	Expand now	Delay expansion
Economic conditions remain good	$328,000	$160,000
There is a recession	−$80,000	$16,000

As can be seen from this table, it will be advantageous to expand the plant capacity right away only if economic conditions remain good, and the furniture manufacturer's decision will, therefore, have to depend on the chances that this will be the case. Using the manufacturer's probabilities of $\frac{1}{3}$ and $\frac{2}{3}$ for economic conditions remaining good or there being a recession, we find that the expected profit is

$$328,000 \cdot \frac{1}{3} + (-80,000) \cdot \frac{2}{3} = \$56,000$$

if he expands his plant capacity right away, and

$$160,000 \cdot \frac{1}{3} + 16,000 \cdot \frac{2}{3} = \$64,000$$

if the expansion is delayed. Since the second of these figures exceeds the first, it follows that delaying the expansion maximizes the furniture manufacturer's expected profit.

The way in which we have studied this problem is called a **Bayesian analysis**. In this kind of analysis, probabilities are assigned to the alternatives about which uncertainties exist (the **states of nature**, which in our example were economic conditions remaining good and a recession); then we choose whichever alternative promises the greatest expected profit or the smallest expected loss.

This approach to decision making has great intuitive appeal, but it is not without complications. If mathematical expectations are to be used for making decisions, it is essential that out appraisals of all relevant probabilities and payoffs be very close to correct. An additional difficulty is that two persons dealing with the same problem may have different opinions about the probabilities or the payoffs and come to different conclusions.

EXAMPLE Suppose that an expert forecaster feels that the probabilities for economic conditions remaining good or there being a recession are 0.40 and 0.60. Based on these probabilities, what should the furniture manufacturer do so as to maximize his expected profit?

Solution Now the expected profit is

$$328{,}000(0.40) + (-80{,}000)(0.60) = \$83{,}200$$

if the furniture manufacturer expands his plant capacity right away, and

$$160{,}000(0.40) + 16{,}000(0.60) = \$73{,}600$$

if the expansion is delayed. Since the first of these two figures exceeds the second, the furniture manufacturer will maximize his expected profit if he expands his plant capacity right away. Note that by changing the probabilities from $\frac{1}{3}$ and $\frac{2}{3}$ to 0.40 and 0.60, the decision has been reversed.

EXAMPLE Now suppose that the furniture manufacturer is told by his accountant that the \$328,000 figure is incorrect and that it should be \$352,000. If the furniture manufacturer uses his own appraisal of the probabilities for economic conditions remaining good or there being a recession, will this change in the payoffs affect his decision?

Solution With this change, the expected profit is

$$352{,}000 \cdot \frac{1}{3} + (-80{,}000) \cdot \frac{2}{3} = \$64{,}000$$

if the manufacturer expands his plant capacity right away, and

$$160{,}000 \cdot \frac{1}{3} + 16{,}000 \cdot \frac{2}{3} = \$64{,}000$$

if the expansion is delayed. Since the two figures are the same, it does not matter now whether the furniture manufacturer expands his plant capacity right away or delays the expansion. Again, a minor change in the figures has affected the result.

7.3
STATISTICAL DECISION PROBLEMS ★

Modern statistics, with its emphasis on inference, may be looked upon as the art, or science, of decision making under uncertainty. This approach to statistics, called **decision theory**, dates back only to the middle of this century and the publication of John von Neumann and Oscar Morgenstern's *Theory of Games and Economic Behavior* in 1944 and Abraham Wald's *Statistical Decision Functions* in 1950. Since the study of decision theory is quite complicated mathematically, we shall limit our discussion here to an example in which the method of the preceding section is applied to a problem that is of a statistical nature.

EXAMPLE   On the five teams appointed by a government agency to study racial discrimination, there are 1, 2, 5, 1, and 6 members who favor liberal causes. The teams are randomly assigned to various cities, and the mayor of one city hires a consultant to predict how many of the members of the team sent to her city will favor liberal causes. If the consultant is paid $100 plus a bonus of $200 which he receives only if his prediction is correct, what prediction maximizes the amount of money he can expect to get?

Solution   If the consultant's prediction is 1, which is the mode of the five numbers, he will make $100 with probability $\frac{3}{5}$ and $300 with probability $\frac{2}{5}$. So, he can expect to make

$$100 \cdot \frac{3}{5} + 300 \cdot \frac{2}{5} = \$180$$

and it can easily be verified that this is the best he can do. If his prediction is 2, 5, or 6 he can expect to make

$$100 \cdot \frac{4}{5} + 300 \cdot \frac{1}{5} = \$140$$

and for any other prediction his expectation is only $100.

This illustrates the (perhaps obvious) fact that if one has to pick an exact value on the nose and there is no reward for being close, the best prediction is the mode.

To illustrate further how the consequences of one's decisions may dictate the choice of a statistical method of decision or prediction, let us consider the following variation of our example:

EXAMPLE   Suppose that the consultant is paid $300 minus an amount of money equal in dollars to 40 times the magnitude of his error. What prediction will maximize the amount of money he can expect to get?

Solution   Now it is the median which yields the best predictions. If the consultant's prediction is 2, which is the median of the five numbers, the magnitude of the error will be 1, 0, 3, or 4, depending on whether 1, 2, 5, or 6 of the members of the team sent to the city will favor liberal causes. Consequently, he will get $260, $300, $180, or $140 with probabilities $\frac{2}{5}$, $\frac{1}{5}$, $\frac{1}{5}$, and $\frac{1}{5}$, and he can expect to make

$$260 \cdot \frac{2}{5} + 300 \cdot \frac{1}{5} + 180 \cdot \frac{1}{5} + 140 \cdot \frac{1}{5} = \$228$$

It can be shown that the consultant's expectation is less than $228 for any value other than 2, but let us verify it here merely for 3, which is the mean of the five numbers. In that case the magnitude of the error will be 2, 1, 2, or 3 depending on whether 1, 2, 5, or 6 of the members of the team sent to the city will favor liberal causes, the consultant will get \$220, \$260, \$220, or \$180 with probabilities $\frac{2}{5}$, $\frac{1}{5}$, $\frac{1}{5}$, and $\frac{1}{5}$, and he can expect to make

$$220 \cdot \frac{2}{5} + 260 \cdot \frac{1}{5} + 220 \cdot \frac{1}{5} + 180 \cdot \frac{1}{5} = \$220$$

The mean comes into its own right when the penalty, the amount subtracted, increases more rapidly with the size of the error, namely, when it is proportional to its square.

EXAMPLE Suppose that the consultant is paid \$300 minus an amount of money equal in dollars to 20 times the square of the error. What prediction will maximize the amount of money he can expect to get?

Solution If the consultant's prediction is 3, which is the mean of the five numbers, the squares of the errors will be 4, 1, 4, or 9 depending on whether 1, 2, 5, or 6 of the members of the team sent to the city will favor liberal causes. Correspondingly, the consultant will get \$220, \$280, \$220, or \$120 with probabilities $\frac{2}{5}$, $\frac{1}{5}$, $\frac{1}{5}$, and $\frac{1}{5}$, and he can expect to make

$$220 \cdot \frac{2}{5} + 280 \cdot \frac{1}{5} + 220 \cdot \frac{1}{5} + 120 \cdot \frac{1}{5} = \$212$$

As can be verified, the consultant's expectation is less than \$212 for any other prediction (see Exercise 7.33).

This third case is of special importance in statistics, as it ties in closely with the **method of least squares**. We shall study this method in Chapter 15, where it is used in fitting curves to observed data, but besides this it has other important applications in the theory of statistics. The idea of working with the squares of the errors is justified on the grounds that in actual practice the seriousness of an error often increases rapidly with the size of the error, more rapidly than the magnitude of the error itself.

The greatest difficulty in applying the methods of this chapter to realistic problems in statistics is that we seldom know the exact values of all the risks that are involved; that is, we seldom know the exact values of the "payoffs" corresponding to the various eventualities. For instance, if the FDA must decide whether or not to release a new drug for general use, how can it put a cash value on the damage that might be done by not waiting for a more thorough analysis of possible side effects, or on the lives that might be lost by not making the drug available to the public right away? Similarly, if a faculty committee must decide which of several applicants should be admitted to a medical school or, perhaps, receive a scholarship, how can they possibly foresee all the consequences that might be involved?

The fact that we seldom have adequate information about relevant probabilities also provides obstacles to finding suitable decision criteria; without them, is it "reasonable" to base decisions, say, on pessimism or optimism as in Exercises 7.28 and 7.29 on page 174? Questions like these are difficult to answer, but their analysis serves the important purpose of revealing the logic that underlies statistical thinking.

EXERCISES

★ **7.19** A grab-bag contains 5 packages worth $1 apiece, 5 packages worth $3 apiece, and 10 packages worth $5 apiece. Is it rational to pay $4 for the privilege of selecting one of these packages at random?

★ **7.20** A contractor must choose between two jobs. The first promises a profit of $120,000 with a probability of $\frac{3}{4}$ or a loss of $30,000 (due to strikes and other delays) with a probability of $\frac{1}{4}$; the second job promises a profit of $180,000 with a probability of $\frac{1}{2}$ or a loss of $45,000 with a probability of $\frac{1}{2}$. Which job should the contractor choose so as to maximize his expected profit?

★ **7.21** A landscape architect must decide whether to bid on the landscaping of a public building. What should she do if she figures that the job promises a profit of $10,800 with probability of 0.40 or a loss of $7,000 (due to a lack of rain or perhaps an early frost) with probability 0.60, and it is not worth her time unless the expected profit is at least $1,000?

★ **7.22** A truck driver has to deliver a load of building materials to one of two construction sites, which are 18 and 22 miles from the lumberyard. He has misplaced the order telling him where the load should go. He must return to the lumberyard after the delivery. The construction sites are 8 miles apart, and to complicate matters, the telephone at the lumberyard is broken. If the driver feels that the probability is $\frac{1}{6}$ that the load should go to the site which is 18 miles from the lumberyard and $\frac{5}{6}$ that it should go to the other site, where should he go first so as to minimize the expected distance that he will have to drive?

★ **7.23** With reference to the preceding exercise, where should the driver go first so as to minimize the expected distance he will have to drive, if instead of $\frac{1}{6}$ and $\frac{5}{6}$ the

two probabilities are
(a) $\frac{1}{3}$ and $\frac{2}{3}$;
(b) $\frac{1}{4}$ and $\frac{3}{4}$?

★ **7.24** The management of a mining company must decide whether to continue an operation at a certain location. If they continue and are successful, they will make a profit of $4,500,000; if they continue and are not successful, they will lose $2,700,000; if they do not continue but would have been successful if they had continued, they will lose $1,800,000 (for competitive reasons); and if they do not continue and would not have been successful if they had continued, they will make a profit of $450,000 (because funds allocated to the operation remain unspent). What decision would maximize the company's expected profit if it is felt that there is a fifty-fifty chance for success?

★ **7.25** With reference to the preceding exercise, show that it does not matter what they decide to do if it is felt that the probabilities for and against success are $\frac{1}{3}$ and $\frac{2}{3}$.

★ **7.26** A small delicatessen has shelf space for four highly perishable cream pies which are destroyed at the end of the day if they are not sold. The cream pies cost $2 each and sell for $6 each, so that the profit is $4 for each cream pie sold. How many cream pies should the retailer stock so as to maximize his expected profit, if he knows that the probabilities of a demand for 0, 1, 2, 3, or 4 cream pies are, respectively, 0.10, 0.30, 0.35, 0.15, and 0.10?

★ **7.27** With reference to the furniture manufacturer on page 168, suppose that the $80,000 loss is in error and should be a $120,000 loss. What decision would maximize his expected profit, if the probabilities are 0.40 and 0.60 that the economic conditions will remain good or that there will be a recession?

★ **7.28** In the absence of any information about relevant probabilities, a pessimist may well try to minimize the maximum loss or maximize the minimum profit. This person would be using the *minimax* or *maximin criterion*.

(a) With reference to the furniture manufacturer on page 168, suppose that he has no idea about the probabilities that economic conditions will remain good or that there will be a recession. What decision would maximize his maximum profit?

(b) With regard to Exercise 7.26, suppose that the delicatessen owner has no idea about the probabilities for the demand for the cream pies. How many cream pies should he stock if he wishes to minimize the maximum loss to which he could be exposed? Discuss the reasonableness of using the minimax criterion in a problem of this kind.

★ **7.29** In the absence of any information about relevant probabilities, an optimist may well try to minimize the minimun loss or maximize the maximum profit. This person would be using the *minimin* or *maximax criterion*.

(a) With reference to the furniture manufacturer on page 168, suppose that he has no idea about the probabilities that economic conditions will remain good or that there will be a recession. What decision would maximize his maximum profit?

(b) With regard to Exercise 7.26, suppose that the delicatessen owner has no idea about the probabilities for the demand for the cream pies. How many cream pies should he stock if he wishes to maximize his maximum profit?

★ **7.30** With reference to Exercise 7.24, suppose that the management of the mining company has no idea about the chances for success. What decision will

(a) maximize their minimum profit;
(b) maximize their maximum profit?

★ **7.31** With reference to the furniture manufacturer on page 168, suppose that before he must decide whether or not to expand the capacity of his plant, he has the option of paying an infallible consultant $25,000 to find out for sure whether economic conditions will remain good or whether there will be a recession. This raises the question whether it is worthwhile for him to spend the $25,000. To answer this question, let us use his $\frac{1}{3}$ and $\frac{2}{3}$ probabilities that economic conditions will remain good or that there will be a recession. If he knew for sure what was going to happen, the right decision would yield a profit of $328,000 or a profit of

$16,000. Since the corresponding probabilities are $\frac{1}{3}$ and $\frac{2}{3}$, we find that (with the help of the infallible expert) the furniture manufacturer's expected profit is

$$328,000 \cdot \frac{1}{3} + 16,000 \cdot \frac{2}{3} = \$120,000$$

This is called the **expected profit with perfect information**. On page 166 we showed that without the help of the infallible expert, the expected profit was either $56,000 or $64,000, so that there is an improvement of at least $120,000 − $64,000 = $56,000, and this makes the $25,000 fee well worthwhile. It is customary to refer to the amount by which perfect information improves one's expectation, $56,000 in our example, as the **expected value of perfect information**.

(a) With reference to Exercise 7.22, find the expected distance with perfect information and the expected value (in miles) of perfect information.

(b) With reference to Exercise 7.24, find the expected profit with perfect information and the expected value of perfect information. Would it be worthwhile to spend $500,000 beforehand to find out for sure whether the operation will be a success?

★ **7.32** There are situations where the various criteria we have discussed are all outweighed by special considerations. For instance, with reference to the furniture manufacturer on page 168, what may well be his decision if he knows that

(a) he will be bankrupt unless he makes a profit of at least $200,000 during the next fiscal year;

(b) he will be bankrupt unless he shows a profit, no matter how small, during the next fiscal year?

★ **7.33** With reference to the example on page 172, where the consultant is paid $300 minus an amount of money equal to 20 times the square of the error, what can the consultant expect to make if

(a) his prediction is 1;
(b) his prediction is 2?

In general, it can be shown that for any set of numbers, $x_1, x_2, \ldots,$ and x_n, the quantity $\sum_{i=1}^{n} (x_i - k)^2$ is smallest when $k = \bar{x}$. In this case, the amount subtracted from $300 is smallest when the prediction is $\bar{x} = 3$.

★ **7.34** The ages of the seven entries in an essay contest are 17, 17, 17, 18, 20, 21, and 23, and their chances of winning are equal. We want to predict the age of the winter. What prediction maximizes the expected reward if

    (a) there is a reward for being right, but none for being close;

    (b) there is a penalty proportional to the size of the error;

    (c) there is a penalty proportional to the square of the error?

★ 7.35 Some of the used cars on a lot are priced at $895, some are priced at $1,395, some are priced at $1,795, and some are priced at $2,495. If we want to predict the price of the car which will be sold first, what prediction minimizes the maximum size of the error? What name did we give to this statistic in one of the exercises of Chapter 3?

★ 7.36 There are sometimes occasions in which we select an option which has an inferior mathematical expectation. Suppose that you have a choice between two investments for $10,000. One option is a bank certifi-cate of deposit (CD) which will certainly return to you the amount $10,800 after one year. The second option consists of shares in a silver-prospecting company; in one year these shares will be worth nothing with prob-ability 0.80 and will be worth $60,000 with probability 0.20. Which investment do you prefer? Why?

★ 7.37 The situation of Exercise 7.26 is sometimes described as the **newsboy problem**. Suppose that a newspaper seller buys papers at the beginning of the day for 15 cents each and sells them for 40 cents each. If he buys more papers than he can sell, the leftovers must be dis-carded as a complete loss. Suppose that the probability is 0.01 that he will have a demand for one paper, 0.01 that he will have a demand for two papers, 0.01 that he will have a demand for three papers, . . . , 0.01 that he will have a demand for 100 papers. How many papers should he purchase at the beginning of the day so as to maximize his expected profit?

7.4
CHECKLIST OF KEY TERMS
(with page references to their definitions)

7.5
REVIEW EXERCISES

7.38 The probabilities that a person shopping at "The Bookstore" will buy 0, 1, 2, 3, or 4 books are 0.22, 0.54, 0.17, 0.06, and 0.01. How many books can a person shopping at this bookstore be expected to buy?

7.39 A playwright feels that she stands to make $250,000 if her new play is a success, but nothing if it is a flop. How does she feel about the chances that the play will be a success, if prior to its opening she sells the play outright for $150,000?

★ **7.40** The mortgage manager of a bank figures that if an applicant for a $150,000 home mortgage is a good risk and the bank accepts him, the bank's profit will be $8,000. If he is a bad risk and the bank accepts him, the bank will lose $14,000. If the mortgage manager turns down the applicant, there will be no profit or loss either way. Which decision maximizes the bank's expected profit if the mortgage manager feels that the probabilities are 0.10 and 0.90, respectively, that a particular applicant is a bad risk or a good risk?

★ **7.41** With reference to the preceding exercise, would the credit manager's decision be the same if he feels that the probabilities are 0.20 and 0.80 that the loan applicant is a bad risk or a good risk?

★ **7.42** With reference to Exercise 7.40, what would the credit manager do if he had no idea about the probabilities that the loan applicant is a good risk or a bad risk and he wants to minimize the maximum loss?

7.43 If the probabilities are 0.09, 0.25, 0.29, 0.18, 0.14, 0.03, and 0.02 that a child in the age group from 6 to 16 will visit a dentist 0, 1, 2, 3, 4, 5, or 6 times a year, how many times can a child in this age group expect to visit a dentist in any given year?

7.44 Mrs. Black feels that it is about a toss-up whether to accept $20 cash or to gamble on drawing a bead from an urn containing 15 white beads and 45 red beads, where she is to receive $2 if she draws a white bead or a bottle of fancy perfume if she draws a red bead. What value, or utility, does she attach to the bottle of perfume?

★ **7.45** The five kinds of cars considered by a police department average 41, 40, 40, 43, and 40 miles per gallon, and their chances of being chosen are all equal. If we want to predict the average miles per gallon of the kind of car they will choose and there is a reward for being right, but none for being close, what prediction maximizes the expected reward?

★ **7.46** With reference to the preceding exercise, what prediction maximizes the expected reward if a penalty proportional to the square of the error is subtracted from the reward?

7.47 An athlete is willing to give odds of 2 to 1, but not odds of 3 to 1, that he will be able to beat his roommate at arm wrestling. What does this tell us about the probability he assigns to his being able to win?

7.48 A person gets $40 if he or she gets three heads in three random flips of a balanced coin; otherwise, there is no reward. How much should the person pay to play this game so as to make it fair?

7.49 The manufacturer of a new battery additive has to decide whether to sell his product for $1.00 a can, or for $1.25 with a "double-your-money-back-if-not-satisfied guarantee." How does he feel about the chances that a person will actually ask for double his or her money back if
  (a) he decides to sell the product for $1.00;
  (b) he decides to sell the product for $1.25 with the guarantee;
  (c) he cannot make up his mind?

★ **7.50** Ms. Cooper is planning to attend a convention in San Diego, and she must send in her room reservations immediately. The convention is so large that the activities are held partly in hotel A and partly in hotel B, and Ms. Cooper does not know whether the particular session she wants to attend will be held in hotel A or hotel B. She is planning to stay only one day, which would cost her $80.00 at hotel A and $72.80 at hotel B, but it will cost her an extra $12.00 for cab fare if she stays at the wrong hotel. Where should she make her reservation if she feels that the probability is 0.75 that the session she wants to attend will be held at hotel A and she wants to minimize her expected cost?

★ **7.51** With reference to the preceding exercise, would it be worthwhile to spend $2.40 on a long-distance call to find out where the session will be held?

★ **7.52** With reference to Exercise 7.50, where should Ms. Cooper make her reservation if she feels that the probability is 0.80 that the session she wants to attend will be held at hotel A?

★ **7.53** With reference to Exercise 7.50, where should Ms. Cooper make her reservation if she is a confirmed optimist but has no idea about the probability that the session she wants to attend will be held at hotel A?

7.54 An insurance company agrees to pay the promoter of a drag race $8,000 in case the event has to be canceled because of rain. If the company's actuary feels that a fair net premium for this risk is $1,280, what probability does she assign to the possibility that the race will have to be canceled because of rain?

★ **7.55** With reference to Exercise 7.45, suppose that there is a penalty of 100 times the size of the error. What prediction will minimize the maximum penalty?

7.56 Two friends are betting on repeated flips of a balanced coin. One has $5 at the start and the other has $3, and after each flip the loser pays the winner $1. If p is the probability that the one who starts with $5 will win his friend's $3 before he loses his own $5, expxlain why $3p - 5(1 - p)$ should equal 0, and then solve the equation $3p - 5(1 - p) = 0$ for p. Generalize this result to the case where two players start with a dollars and b dollars, respectively.

7.6

REFERENCES

More detailed treatments of the subject matter of this chapter may be found in

BROSS, I. D. J., *Design for Decision*. New York: Macmillan Publishing Co., Inc., 1953.
JEFFREY, R. C., *The Logic of Decision*. New York: McGraw-Hill Book Company, 1965.

and in many textbooks on business statistics; for instance, in Chapter 7 of

FREUND, J. E., WILLIAMS, F. J., and PERLES, B. M., *Elementary Business Statistics*: *The Modern Approach*, 5th ed. Englewood Cliffs, N.J.: Prentice-Hall, Inc., 1988.

Decision theory is discussed in Chapter 15 of

EPPEN, G. D., GOULD, F. J., and SCHMIDT, C. P., *Introductory Management Science*, 3rd ed. Englewood Cliffs, N.J.: Prentice-Hall, Inc., 1991.

"The Dowry Problem" and "A Tie Is Like Kissing Your Sister" are two amusing examples of decision making given in

HOLLANDER, M., and PROSCHAN, F., *The Statistical Exorcist*: *Dispelling Statistics Anxiety*. New York: Marcel Dekker, Inc., 1984.

PROBABILITY DISTRIBUTIONS

In most problems of statistics we are interested only in one aspect, or at most in a few aspects, of the outcomes of experiments. For instance, a student taking a true–false test may be interested only in the number of questions he answers correctly and not which ones; a geologist may be interested only in the age of a rock sample and not in its hardness; and a sociologist may be interested only in the socioeconomic status of a person interviewed in a survey and not in her age or weight. Also, an agronomist may be interested in determining not only the yield per acre of a new variety of corn but also the temperature at which it will germinate; and an automotive engineer may be interested in the brightness and the durability of the headlights proposed for a new model car and also in their projected cost.

In these five examples, the student, the geologist, the sociologist, the agronomist, and the automotive engineer are all interested in numbers that are associated with the outcomes of situations involving an element of chance, or more specifically, in values of **random variables**. Since random variables are neither random nor variables, why do they go by this name? This is hard to say, but a mathematics professor with a good sense of humor likened them to alligator pears, or avocados, which are neither alligators nor pears.

In the study of random variables we are usually interested in the probabilities with which they take on the various values within their range, namely, in their **probability distributions**. The general introduction of random variables and probability distributions in Sections 8.1 and 8.2 will be followed by the discussion of some of the most important probability distributions in Sections 8.3 through 8.6. Then, we discuss some ways of describing the most relevant features of probability distributions in Sections 8.7 and 8.8. Section 8.9 contains optional material on the simulation of values of random variables.

8.1
RANDOM VARIABLES

To be more explicit about the concept of a random variable, let us consider Figure 8.1, which shows the sample space for a situation in which an Allstate agent and a State Farm agent are trying to sell three insurance policies, home, automobile, and life to a family that has just moved. The point (2, 1), for example, labels the outcome in which the Allstate agent sells two of the policies and the State Farm agent sells one. In Figure 8.1 we place another number (in white) next to each point—0 to the point (0, 0); 1 to the points (1, 0) and (0, 1); 2 to the points (2, 0), (1, 1), and (0, 2); and 3 to the points (3, 0), (2, 1), (1, 2), and (0, 3). In this way we have associated with each point of the sample space the number of policies which, between them, the two agents sell to the family.

Since associating numbers with the points of a sample space is just a way of defining a function over the points of the sample space, random variables are really functions and not variables. Conceptually, though, most beginners find it easier to think of random variables simply as quantities which can take on different values depending on chance. For instance, the number of speeding tickets issued each day on the freeway between Indio and Blythe in California is a

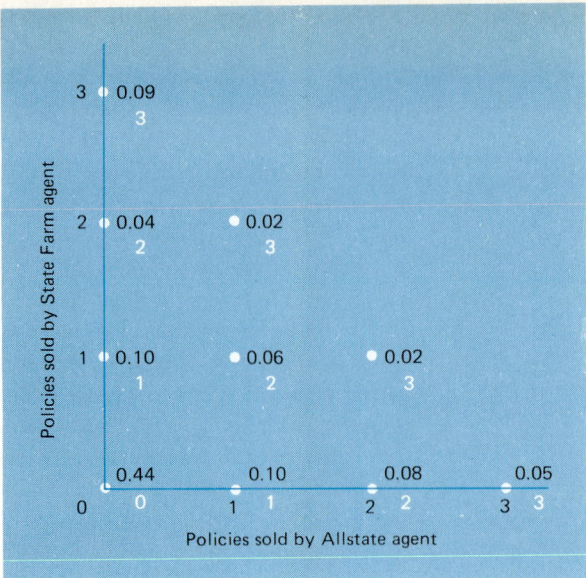

FIGURE 8.1 *Sample space with values of random variable.*

random variable, and so is the annual production of coffee in Brazil, the number of persons visiting Disneyland each week, the wind velocity at Kennedy airport, the size of the audience at a baseball game, and the number of mistakes a person makes when typing a report.

It is customary to classify random variables according to the number of values which they can assume, and in this chapter we shall consider only random variables which are **discrete**; namely, random variables which can take on a finite number of values or a countable infinity (as many values as there are whole numbers). For nearly all discrete random variables, the possible values form a subset of the integers. For instance, the number of policies which, between them, the two insurance agents sell to the given family is a discrete random variable which can take on a finite number of values, the four values 0, 1, 2, and 3. In contrast, the number of the roll on which a die comes up 6 for the first time is a discrete random variable which can take on the countable infinity of values 1, 2, 3, 4, ... It is possible, though highly unlikely, that it will take a thousand rolls of the die, a million rolls or even more, until we finally get a 6. There are also **continuous random variables**, which arise when we deal with quantities measured on a continuous scale, say, time, weight, or distance. These will be taken up in Chapter 9.

8.2
PROBABILITY DISTRIBUTIONS

The points of the sample space illustrated in Figure 8-1 are also labelled with their probabilities (in black). For instance, the value 0.06 on the point (1, 1) means that the probability is 0.06 that each of the agents will sell one policy. We find from this figure that the random variable "the number of policies which, between them, they sell to the given family" takes on the value 0 with probability 0.44, the value 1 with probability $0.10 + 0.10 = 0.20$, the value 2 with probability $0.08 + 0.06 + 0.04 = 0.18$, and the value 3 with probability $0.05 + 0.02 + 0.02 + 0.09 = 0.18$. All this is summarized in the following table:

Number of policies sold	Probability
0	0.44
1	0.20
2	0.18
3	0.18

This table, and the two that follow, serve to illustrate what we mean by a **probability distribution**. As this table shows, **a probability distribution is a correspondence which assigns probabilities to the values of a random variable**. Another example of such a correspondence is given by the following table, which pertains to the number of points we roll with a balanced die:

Number of points we roll with a die	Probability
1	$\frac{1}{6}$
2	$\frac{1}{6}$
3	$\frac{1}{6}$
4	$\frac{1}{6}$
5	$\frac{1}{6}$
6	$\frac{1}{6}$

Finally, for four flips of a balanced coin there are the sixteen equally likely possibilities HHHH, HHHT, HHTH, HTHH, THHH, HHTT, HTHT, HTTH, THHT, THTH, TTHH, HTTT, THTT, TTHT, TTTH, and TTTT, where H stands for heads and T for tails. Counting the number of heads in each case and using the formula $\frac{s}{n}$ for equiprobable outcomes, we get the following probability distribution for the total number of heads:

Number of heads	Probability
0	$\frac{1}{16}$
1	$\frac{4}{16}$
2	$\frac{6}{16}$
3	$\frac{4}{16}$
4	$\frac{1}{16}$

When possible, we try to express probability distributions by means of formulas which enable us to calculate the probabilities associated with the various values of a random variable. For instance, for the number of points we roll with a balanced die we can write

$$f(x) = \frac{1}{6} \qquad \text{for } x = 1, 2, 3, 4, 5, \text{ and } 6$$

where $f(1)$ denotes the probability of rolling a 1, $f(2)$ denotes the probability of rolling a 2, and so on, in the usual functional notation. Here we wrote the probability that the random variable will take on the value x as $f(x)$, but we could just as well write it as $g(x)$, $h(x)$, $m(x)$, etc.

EXAMPLE Verify that for the number of heads obtained in four flips of a balanced coin the probability distribution is given by

$$f(x) = \frac{\binom{4}{x}}{16} \qquad \text{for } x = 0, 1, 2, 3, \text{ and } 4$$

Solution By direct calculation, or by using Table X at the end of the book, we find that $\binom{4}{0} = 1$, $\binom{4}{1} = 4$, $\binom{4}{2} = 6$, $\binom{4}{3} = 4$, and $\binom{4}{4} = 1$. Thus, the probabilities for $x = 0, 1, 2, 3,$ and 4 are $\frac{1}{16}$, $\frac{4}{16}$, $\frac{6}{16}$, $\frac{4}{16}$, and $\frac{1}{16}$, which agrees with the results obtained above.

Since the values of probability distributions are probabilities, and since random variables have to take on one of their values, we have the following two rules which apply to any probability distribution:

The values of a probability distribution must be numbers on the interval from 0 to 1.

The sum of all the values of a probability distribution must be equal to 1.

These rules enable us to determine whether or not a function (given by an equation or by a table) can serve as the probability distribution of some random variable.

EXAMPLE Check whether the correspondence given by

$$f(x) = \frac{x + 3}{15} \qquad \text{for } x = 1, 2, \text{ and } 3$$

can serve as the probability distribution of some random variable.

Solution Substituting $x = 1, 2,$ and 3 into $\frac{x + 3}{15}$, we get $f(1) = \frac{4}{15}$, $f(2) = \frac{5}{15}$, and $f(3) = \frac{6}{15}$. Since none of these values is negative or greater than 1, and since their sum is

$\frac{4}{15} + \frac{5}{15} + \frac{6}{15} = 1$, the given function can serve as the probability distribution of some random variable.

8.3
THE BINOMIAL DISTRIBUTION

There are many applied problems in which we are interested in the probability that an event will occur x times out of n. For instance, we may be interested in the probability of getting 45 responses to 400 questionnaires sent out as part of a sociological study, the probability that 5 of 12 mice will survive for a given length of time after the injection of a cancer-inducing substance, the probability that 45 of 300 drivers stopped at a road block will be wearing their seat belts, or the probability that 66 of 200 television viewers (interviewed by a rating service) will recall what products were advertised on a given program. To borrow from the language of games of chance, we could say that in each of these examples we are interested in the probability of getting "x **successes** in n **trials**," or in other words, "x successes and $n - x$ failures in n attempts."

In the problems we shall study in this section, we shall always make the following assumptions:

There is a fixed number of trials.

The probability of a success is the same for each trial.

The trials are all independent.

This means that the theory we shall develop will not apply, for example, if we are interested in the number of dresses a woman may try on before she buys one (where the number of trials is not fixed), if we check every hour whether traffic is congested at a certain intersection (where the probability of "success" is not constant), or if we are interested in the number of times that a person voted for the Republican candidate in the last five presidential elections (where the trials are not independent).

To solve problems which do meet the conditions listed in the preceding paragraph, we use a formula obtained in the following way: If p and $1 - p$ are the probabilities of a success and a failure on any given trial, then the probability of getting x successes and $n - x$ failures *in some specific order* is $p^x(1 - p)^{n-x}$; clearly, in this product of p's and $(1 - p)$'s there is one factor p for each success, one factor $1 - p$ for each failure, and the x factors p and $n - x$ factors $1 - p$ are all multiplied together by virtue of the generalization of the special multiplication rule for more than two independent events. Since this probability applies to any point of the sample space which represents x successes and $n - x$ failures (in some specific order), we have only to count how many points of this kind there are, and then multiply $p^x(1 - p)^{n-x}$ by this number. Clearly, the number of ways in which we can choose the x trials on which the successes are to occur is $\binom{n}{x}$, and we have thus arrived at the following result:

Binomial
distribution

> *The probability of getting x successes in n independent trials is*
>
> $$f(x) = \binom{n}{x}p^x(1 - p)^{n-x} \qquad \text{for } x = 0, 1, 2, \ldots, \text{ or } n$$
>
> *where p is the constant probability of a success for each trial.*

$$\binom{n}{x} = \frac{n!}{x!(n-x)!}$$

It is customary to say here that the number of successes in n trials is a random variable having the **binomial probability distribution**, or simply the **binomial distribution**. The binomial distribution is called by this name because for $x = 0, 1, 2, \ldots,$ and n, the values of the probabilities are the successive terms of the binomial expansion of $[(1 - p) + p]^n$.

EXAMPLE   Verify that the formula which we gave on page 182 for the probability of getting x heads in four flips of a balanced coin is, in fact, the one for the binomial distribution with $n = 4$ and $p = \frac{1}{2}$.

Solution   Substituting $n = 4$ and $p = \frac{1}{2}$ into the formula for the binomial distribution, we get

$$f(x) = \binom{4}{x}\left(\frac{1}{2}\right)^x\left(1 - \frac{1}{2}\right)^{4-x} = \binom{4}{x}\left(\frac{1}{2}\right)^4 = \frac{\binom{4}{x}}{16}$$

for $x = 0, 1, 2, 3,$ and 4. This is exactly the formula which we gave on page 182.

EXAMPLE   If the probability is 0.70 that any one registered voter (randomly selected from official rolls) will vote in a given election, what is the probability that two of five registered voters will vote in the election?

Solution   Substituting $x = 2, n = 5, p = 0.70,$ and $\binom{5}{2} = 10$ into the formula for the binomial distribution, we get

$$f(2) = \binom{5}{2}(0.70)^2(1 - 0.70)^{5-2}$$

$$= 10(0.70)^2(0.30)^3$$

$$= 0.132$$

rounded to three decimals.

The following is an example in which we calculate all the probabilities of a binomial distribution:

EXAMPLE The probability is 0.30 that a person shopping at a certain supermarket will take advantage of its special promotion of ice cream. Find the probabilities that among six persons shopping at this market there will be 0, 1, 2, 3, 4, 5, or 6 who will take advantage of the promotion. Also draw a histogram of this probability distribution.

Solution Assuming that the selection is random, we substitute $n = 6$, $p = 0.30$, and, respectively, $x = 0, 1, 2, 3, 4, 5$, and 6 into the formula for the binomial distribution, and we get

$$f(0) = \binom{6}{0}(0.30)^0(0.70)^6 = 0.118$$

$$f(1) = \binom{6}{1}(0.30)^1(0.70)^5 = 0.303$$

$$f(2) = \binom{6}{2}(0.30)^2(0.70)^4 = 0.324$$

$$f(3) = \binom{6}{3}(0.30)^3(0.70)^3 = 0.185$$

$$f(4) = \binom{6}{4}(0.30)^4(0.70)^2 = 0.060$$

$$f(5) = \binom{6}{5}(0.30)^5(0.70)^1 = 0.010$$

$$f(6) = \binom{6}{6}(0.30)^6(0.70)^0 = 0.001$$

where all the probabilities are rounded to three decimals. A histogram of this distribution is shown in Figure 8.2.

In case the reader does not care much for ice cream and is not particularly interested in personal buying habits, let us stress the importance of the binomial distribution as a **statistical model**. The results of the preceding example apply also if the probability is 0.30 that the energy cell of a watch will last two years under normal usage, and we want to know the probabilities that, among six of these cells, 0, 1, 2, 3, 4, 5, or 6 will last two years under normal usage; if the probability is 0.30 that an embezzler will be caught and brought to trial, and we want to know the probabilities that, among six embezzlers, 0, 1, 2, 3, 4, 5, or 6 will be caught and brought to trial; if the probability is 0.30 that the head of a household owns at least one life insurance policy, and we want to know the probabilities that, among six heads of households, 0, 1, 2, 3, 4, 5, or 6 will own at least one life insurance policy; or if the probability that a person having a certain disease will live for another ten years is 0.30, and we want to know the probabilities that, among six persons having the disease, 0, 1, 2, 3, 4, 5, or 6 will live another ten years. The argument we have

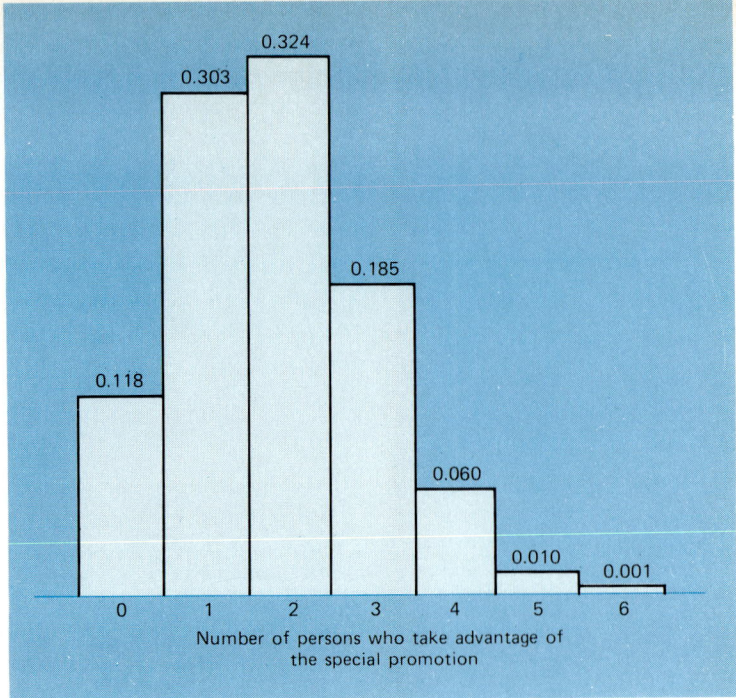

FIGURE 8.2 *Histogram of binomial distribution with n = 6 and p = 0.30.*

presented here is precisely like the one we used in Section 1.2, where we tried to impress upon the reader the generality of statistical techniques.

In actual practice, binomial probabilities are seldom found by direct substitution into the formula. Sometimes we use approximations such as those discussed later in this chapter and in Chapter 9, but more often we refer to special tables such as Table V at the end of this book, the more detailed tables listed among the references at the end of this chapter, or computers. Table V is limited to the binomial probabilities for $n = 2$ to $n = 20$ and $p = 0.05, 0.1, 0.2, 0.3, 0.4, 0.5, 0.6, 0.7, 0.8, 0.9$, and 0.95, all rounded to three decimals. Where values are omitted in this table, they are 0.0005 or less and, hence, 0.000 rounded to three decimals.

EXAMPLE   Suppose that the probability is 0.60 that a car stolen in a certain Southern city will be recovered. Use Table V to find the probabilities that
(a) at most three of ten cars stolen in this city will be recovered;
(b) at least seven of ten cars stolen in this city will be recovered.

Solution   (a) For $n = 10$ and $p = 0.60$, the entries in Table V corresponding to $x = 0, 1, 2$, and 3 are 0.000, 0.002, 0.011, and 0.042. Thus, the probability that at most three of ten cars will be recovered is

$$0.000 + 0.002 + 0.011 + 0.042 = 0.055$$

(b) For $n = 10$ and $p = 0.60$, the entries in Table V corresponding to $x = 7, 8, 9$, and 10 are 0.215, 0.121, 0.040, and 0.006. Thus, the probability that at least seven of ten cars will be recovered is

$$0.215 + 0.121 + 0.040 + 0.006 = 0.382$$

EXAMPLE If the probability is 0.05 that there will be a serious accident at a certain intersection on a weekday, what is the probability that there will be a serious accident at that intersection on at least three of twenty weekdays?

Solution For $n = 20$ and $p = 0.05$, the entries in Table V corresponding to $x = 3, 4$, and 5 are 0.060, 0.013, and 0.002, and those corresponding to $x = 6, 7, 8, \ldots$, and 20 are all 0.000. Thus, the probability that there will be a serious accident at the intersection on at least three of twenty weekdays is

$$0.060 + 0.013 + 0.002 = 0.075$$

In the first of the two examples immediately above, we could not have used Table V if the probability that a stolen car will be recovered in the given city had been 0.63 instead of 0.60. In general, if n is greater than 20 or p takes on a value other than 0.05, 0.1, 0.2, ..., 0.9, or 0.95, we will have to use one of the more detailed tables referred to on pages 217 and 218, use a computer printout as in the example which follows, or, as a last resort, refer to the formula for the binomial distribution.

EXAMPLE Rework the example dealing with the stolen cars, using 0.63 instead of 0.60 as the probability that a car stolen in the given city will be recovered.

Solution **(a)** Using the computer printout of Figure 8.3, we find that the probabilities corresponding to $x = 0, 1, 2$, and 3 are 0.0000, 0.0008, 0.0063, and

```
MTB > BINOMIAL N=10 P=0.63

  BINOMIAL PROBABILITIES FOR N =   10   AND P =   .630000

      K            P( X = K)          P(X LESS OR = K)
      0              .0000                .0000
      1              .0008                .0009
      2              .0063                .0071
      3              .0285                .0356
      4              .0849                .1205
      5              .1734                .2939
      6              .2461                .5400
      7              .2394                .7794
      8              .1529                .9323
      9              .0578                .9902
     10              .0098               1.0000
```

FIGURE 8.3 *Computer printout of the binomial distribution with $n = 10$ and $p = 0.63$.*

0.0285. Thus, the probability that at most three of ten cars will be recovered is

$$0.0000 + 0.0008 + 0.0063 + 0.0285 = 0.0356$$

Note that this answer is actually shown on the printout in the column on the right, which contains the cumulative "or less" probabilities. There, corresponding to $K = 3$, we find that the desired probability is 0.0356.

(b) Also using the computer printout of Figure 8.3, we find that the probabilities corresponding to $x = 7$, 8, 9, and 10 are 0.2394, 0.1529, 0.0578, and 0.0098. Thus, the probability that at least seven of ten cars will be recovered is

$$0.2394 + 0.1529 + 0.0578 + 0.0098 = 0.4599$$

Since the probability of "at least 7" is 1 minus the probability of "6 or less," we also get $1 - 0.5400 = 0.4600$, where 0.5400 is the value in the right-hand column of the printout corresponding to $K = 6$. The difference between the two results is due to rounding.

When we observe a value of a random variable having the binomial distribution—for instance, when we observe the number of heads in 25 flips of a coin, the number of seeds (in a package of 24 seeds) that germinate, the number of students (among 200 interviewed) who are opposed to a change in student activity fees, or the number of automobile accidents (among 20 investigated) that are due to drunk driving—we say that we are **sampling a binomial population**. This terminology is widely used in statistics.

EXERCISES

8.1 In each case determine whether the given values can serve as the values of a probability distribution of some random variable which can take on only the values 1, 2, and 3, and explain your answers:
  (a) $f(1) = 0.42$, $f(2) = 0.31$, and $f(3) = 0.37$;
  (b) $f(1) = 0.08$, $f(2) = 0.12$, and $f(3) = 1.03$;
  (c) $f(1) = \frac{10}{33}$, $f(2) = \frac{1}{3}$, and $f(3) = \frac{12}{33}$.

8.2 In each case determine whether the given values can serve as the values of a probability distribution of some random variable which can take on only the values 1, 2, 3, and 4, and explain your answers:
  (a) $f(1) = 0.25$, $f(2) = 0.75$, $f(3) = 0.25$, and $f(4) = -0.25$;
  (b) $f(1) = 0.15$, $f(2) = 0.27$, $f(3) = 0.29$, and $f(4) = 0.29$;
  (c) $f(1) = \frac{1}{19}$, $f(2) = \frac{10}{19}$; $f(3) = \frac{2}{19}$, and $f(4) = \frac{5}{19}$.

8.3 For each of the following, determine whether it can serve as the probability distribution of some random variable:

  (a) $f(x) = \dfrac{1}{5}$ for $x = 0$, 1, 2, 3, 4, 5;

  (b) $f(x) = \dfrac{x + 1}{14}$ for $x = 1$, 2, 3, 4.

8.4 For each of the following, determine whether it can serve as the probability distribution of some random variable:

  (a) $f(x) = \dfrac{x - 2}{5}$ for $x = 1$, 2, 3, 4, 5;

(b) $f(x) = \dfrac{x^2}{30}$ for $x = 0, 1, 2, 3, 4$.

8.5 In a given city, medical expenses are given as the rea- for 60 percent of all personal bankruptcies. What is the probability that medical expenses will be given as the reason for four of the next six personal bankruptcies filed in that city?

8.6 If the probability is 0.15 that a set of tennis between two given professional players will go into a tie breaker, what is the probability that two of three sets between these two players will go into tie breakers?

8.7 Incompatibility is given as the legal reason for 55% of all divorce cases filed in a given county. Find the probability that incompatability will be given as the reason in four of the next six divorce cases filed in that county.

8.8 A multiple-choice test consists of ten questions with four answers to each question, of which only one is correct. A student answers each question by flipping a dime and a quarter and checking the first answer if he gets heads on both coins, the second answer if he gets heads on the dime and tails on the quarter, the third answer if he gets tails on the dime and heads on the quarter, and the fourth answer if he gets tails on both coins. Find the probabilities that he will get
  (a) exactly three correct answers;
  (b) no correct answers;
  (c) at most four correct answers.

8.9 If the probability is 0.65 that a person traveling on a certain airline will pay extra to see a movie, what is the probability that only three of six persons traveling on this airline will pay extra to see a movie?

8.10 If 40 percent of the mice used in an experiment will become very aggressive within one minute after having been administered an experimental drug, find the probability that exactly four of 10 mice which have been administered the drug become very aggressive within one minute, using
  (a) the formula for the binomial distribution;
  (b) Table V.

8.11 If it is true that 80 percent of all industrial accidents can be prevented by paying strict attention to safety regulations, find the probability that four of seven industrial accidents can thus be prevented, using
  (a) the formula for the binomial distribution;
  (b) Table V.

8.12 Suppose that a civil service examination is designed so that 70 percent of all persons with an IQ of 90 can pass it. Use Table V to find the probabilities that among 15 persons with an IQ of 90 who take the test
  (a) at most six will pass;
  (b) at least 12 will pass;
  (c) from eight through 12 will pass.

8.13 A study shows that 50 percent of the families in a certain large metropolitan area have at least two cars. Use Table V to find the probabilities that among 16 families randomly selected in this metropolitan area
  (a) exactly nine have at least two cars;
  (b) at most six have at least two cars;
  (c) anywhere from eight to twelve have at least two cars.

8.14 An agricultural cooperative claims that 95 percent of the watermelons shipped out are ripe and ready to eat. Find the probabilities that among eighteen watermelons that are shipped out
  (a) all eighteen are ripe and ready to eat;
  (b) at least sixteen are ripe and ready to eat;
  (c) at most fourteen are ripe and ready to eat.

8.15 It is known that 20 percent of all persons given a certain medication get very drowsy within two minutes. Find the probabilities that among fourteen persons given the medication
  (a) at most two will get very drowsy within two minutes;
  (b) at least five will get very drowsy within two minutes;
  (c) two, three, or four will get very drowsy within two minutes.

8.16 A food distributor claims that 80 percent of her 6-ounce cans of mixed nuts contain at least three pecans. To check on this, a consumer testing service decides to examine eight of these 6-ounce cans of mixed nuts from a very large production lot, and reject the claim if fewer than six of them contain at least three pecans. Find the probabilities that the testing service will commit the error of
  (a) rejecting the claim even though it is true;
  (b) not rejecting the claim when in reality only 60 percent of the cans of mixed nuts contain at least three pecans;
  (c) not rejecting the claim when in reality only 40 percent of the cans of mixed nuts contain at least three pecans.

8.17 A study shows that 70 percent of all patients coming to a certain medical clinic have to wait at least fifteen minutes to see their doctor. Find the probabilities that among ten patients coming to this clinic 0, 1, 2, 3,..., or 10 have to wait at least fifteen minutes to see their doctor, and draw a histogram of this probability distribution.

8.18 A quality control engineer wants to check whether, in accordance with specifications, 95% of the products shipped are in perfect condition. To this end, he randomly selects 10 items from each large lot ready to be shipped and passes the lot only if all 10 are in perfect condition. If all 10 are not in perfect condition, then he holds the lot for a complete inspection. Use Table V to find the probabilities that he will commit the error of
   (a) holding a lot for further inspection even though 95% of the items are in perfect condition;
   (b) letting a lot pass through even though only 90% of the items are in perfect condition;
   (c) letting a lot pass through even though only 80% of the items are in perfect condition;
   (d) letting a lot pass through even though only 70% of the items are in perfect condition;

★ **8.19** With reference to Exercise 8.15, suppose that 24 percent of all persons given the medication get very drowsy within two minutes. Use a suitable table or a computer printout of the binomial distribution with $n = 14$ and $p = 0.24$ to rework the three parts of that exercise.

★ **8.20** If we have to determine all the values of a binomial distribution, it is sometimes helpful to calculate $f(0)$ using the formula for the binomial distribution, and then calculate the other values, one after the other, using the formula

$$\frac{f(x+1)}{f(x)} = \frac{n-x}{x+1} \cdot \frac{p}{1-p}$$

   (a) Verify this formula by substituting for $f(x)$ and $f(x+1)$ the corresponding expressions given by the formula for the binomial distribution.

   (b) Use this method to find all the values of the binomial distribution with $n = 6$ and $p = \frac{1}{4}$, writing them as common fractions with the denominator 4,096.

★ **8.21** In some situations where otherwise the binomial distribution applies, we are interested in the probability that the first success will occur on a given trial. For this to happen on the xth trial, it must be preceded by $x - 1$ failures for which the probability is $(1 - p)^{x-1}$, and it follows that the probability that the first success will occur on the xth trial is

Geometric distribution

$$f(x) = p(1-p)^{x-1}$$
$$\text{for } x = 1, 2, 3, 4,...$$

This distribution is called the **geometric distribution** (because its successive values constitute a geometric progression) and it should be observed that there is a countable infinity of possibilities.[†] Using the formula, we find, for example, that for repeated rolls of a balanced die the probability that the first 6 will occur on the fifth roll is

$$\frac{1}{6}\left(\frac{5}{6}\right)^{5-1} = \frac{625}{7,776} \text{ or approximately } 0.080.$$

   (a) When taping a television commercial, the probability that a certain actor will get his lines straight on any one take is 0.40. What is the probability that this actor will get his lines straight for the first time on the fourth take?
   (b) Suppose the probability is 0.25 that any given person will believe a rumor about the private life of a certain politician. What is the probability that the fifth person to hear the rumor will be the first one to believe it?
   (c) The probability is 0.70 that a child exposed to a certain contagious disease will catch it. What is the probability that the third child exposed to the disease will be the first one to catch it?

[†] As formulated in Chapter 6, the postulates of probability apply only when the sample space is finite. When the sample space is countably infinite, as is the case here, the third postulate must be modified accordingly. This will be explained on pages 195 and 196.

8.22 A florist needs 12 exotic potted ferns for a floral exhibition to take place in six months. He knows that
the probability is 0.70 that a fern planted now will be
suitable for the exhibition in six months. His concern
is the minimum number of ferns he should plant if he
wants the probability of having 12 suitable ferns in six
months to be at least 0.90.

   (a) Will 15 be enough? Find the probability that,
      if he plants 15 ferns, he will have at least 12
      suitable.

   (b) Will 20 be enough? Find the probability that,
      that, if he plants 20 ferns, he will have at least
      12 suitable.

   (c) Will 30 be enough? Find the probability that,
      if he plants 30 ferns, he will have at least 12
      suitable.

   (d) Find the smallest number of ferns for which
      the probability of 12 or more suitable is 0.90
      (or more).

8.23 It is known in a certain seaside town that 43% of
lobsters caught during spawning season will have edi-
ble roe. A restaurant owner would like to have 20
lobsters with roe in order to prepare a special banquet.
What is the minimum number of lobsters she should
order if she wants the probability of having at least 20
with roe to be 95% or more?

8.4
THE HYPERGEOMETRIC DISTRIBUTION

In Chapter 6 we spoke of sampling with and without replacement to illustrate the
multiplication rules for independent and dependent events. The binomial distribution applies when we sample with replacement and the trials are all independent,
but not when we sample without replacement. To introduce a probability distribution which applies when we sample without replacement, let us consider the
following example: A factory ships tape recorders in lots of 24, and when they
arrive at their destination, an inspector randomly selects three from each lot. If
they are all in good working condition, the whole lot is accepted; otherwise, the
whole lot is inspected. Since a lot can be accepted without further inspection even
though quite a few of the tape recorders are defective, there is a considerable
risk. So, let us find the probability that a lot will be accepted without further
inspection even though, say, six of the 24 tape recorders are defective. This means
that we must find the probability of three successes (three nondefective tape
recorders) in three trials (among the three tape recorders inspected), and we might
be tempted to argue that since 18 of the 24 tape recorders are not defective, the
probability of a success is $\frac{18}{24} = \frac{3}{4}$ and, hence, the desired probability is

$$f(3) = \binom{3}{3}\left(\frac{3}{4}\right)^3\left(1 - \frac{3}{4}\right)^{3-3} = 0.42$$

rounded to two decimals.

   This result, obtained with the formula for the binomial distribution, would be
correct if sampling is with replacement, but that is not what we do in realistic
problems of sampling inspection. To get the correct answer for our problem when
sampling is without replacement, we might argue as follows: There are altogether
$\binom{24}{3} = 2{,}024$ ways of choosing three of the tape recorders, and they are all

equiprobable by virtue of the assumption that the selection is random. Among these, there are $\binom{18}{3} = 816$ ways of selecting three of the nondefective tape recorders, and it follows by the special formula $\dfrac{s}{n}$ for equiprobable outcomes that the desired probability is $\dfrac{816}{2{,}024} = 0.40$ rounded to two decimals.

To generalize the method we used here, suppose that n objects are to be chosen from a set of a objects of one kind (successes) and b objects of another kind (failures), the selection is without replacement, and we are interested in the probability of getting x successes and $n - x$ failures. Arguing as before, we find that the n objects can be chosen from the whole set of $a + b$ objects in $\binom{a + b}{n}$ ways, and that x of the a successes and $n - x$ of the b failures can be chosen in $\binom{a}{x} \cdot \binom{b}{n - x}$ ways. It follows that for sampling without replacement the probability of "x successes in n trials" is

Hypergeometric distribution

$$f(x) = \frac{\binom{a}{x} \cdot \binom{b}{n - x}}{\binom{a + b}{n}} \qquad \textit{for } x = 0, 1, 2, \ldots, \textit{or } n$$

where x cannot exceed a and $n - x$ cannot exceed b. This is the formula for the **hypergeometric distribution**.

The following are two examples where we sample without replacement and, hence, can use the formula for the hypergeometric distribution:

EXAMPLE   A mailroom clerk is supposed to send six of 15 packages to Europe by airmail, but he gets them all mixed up and randomly puts airmail postage on six of the packages. What is the probability that only three of the packages which are supposed to go by airmail will go by airmail?

Solution   Substituting $a = 6$, $b = 9$, $n = 6$, and $x = 3$ into the formula for the hypergeometric distribution, we get

$$f(3) = \frac{\binom{6}{3} \cdot \binom{9}{6 - 3}}{\binom{15}{6}} = \frac{20 \cdot 84}{5{,}005} = 0.336$$

rounded to three decimals.

EXAMPLE   Among a department store's 16 delivery trucks, five emit excessive amounts of pollutants. If eight of the trucks are randomly picked for inspection, what is the probability that this sample will include at least three of the trucks which emit excessive amounts of pollutants?

Solution   The probability we must find is $f(3) + f(4) + f(5)$, where each term in this sum is a value of the hypergeometric distribution with $a = 5$, $b = 11$, and $n = 8$. Substituting these quantities together with $x = 3, 4$, and 5 into the formula for the hypergeometric distribution, we get

$$f(3) = \frac{\binom{5}{3} \cdot \binom{11}{5}}{\binom{16}{8}} = \frac{10 \cdot 462}{12,870} = 0.359$$

$$f(4) = \frac{\binom{5}{4} \cdot \binom{11}{4}}{\binom{16}{8}} = \frac{5 \cdot 330}{12,870} = 0.128$$

$$f(5) = \frac{\binom{5}{5} \cdot \binom{11}{3}}{\binom{16}{8}} = \frac{1 \cdot 165}{12,870} = 0.013$$

and the probability that the sample will include at least three of the trucks which emit excessive amounts of pollutants is

$$0.359 + 0.128 + 0.013 = 0.500$$

In the beginning of this section we gave an example where we erroneously used the binomial distribution instead of the hypergeometric distribution. The error was quite small, however—we got 0.42 instead of 0.40—and in actual practice the binomial distribution is often used to approximate the hypergeometric distribution. It is generally agreed that this approximation is satisfactory if n does not exceed 5 percent of $a + b$, namely, if

$$n \leq (0.05)(a + b)$$

The main advantages of the approximation are that the binomial distribution has been tabulated much more extensively than the hypergeometric distribution, and that, between the two formulas, the one for the binomial distribution is easier to use. The calculations are much less involved.

EXAMPLE   In a federal prison, 120 of the 300 inmates are serving time for drug-related offenses. If eight of them are to be chosen at random to appear before a legislative committee, what is the probability that three of the eight will be serving time for drug-related offenses?

Solution Since $n = 8$, $a + b = 300$, and 8 is less than $0.05(300) = 15$, we can use the binomial approximation. From Table V we find that for $n = 8$, $p = \frac{120}{300} = 0.40$, and $x = 3$, the probability asked for is 0.279. Fairly extensive calculations will show that the error of this approximation is 0.003.

Also, the binomial calculations are generally less complicated. Observe that the binomial distribution is described by two parameters (n and p) while the hypergeometric distribution requires three (a, b, and n).

EXERCISES

8.24 Among the 12 houses for sale in a development, nine have air conditioning. If four of the houses are randomly chosen for a full-page newspaper ad, what is the probability that three of them will have air conditioning?

8.25 Among the 20 solar collectors on display at a trade show, 12 are flat-plate collectors and the others are concentrating collectors. If a person visiting the show randomly selects six of the solar collectors to check out, what is the probability that three of them will be flat-plate collectors?

8.26 Among the 12 male applicants for a job with the postal service, nine have working wives. If two of the applicants are randomly chosen for further consideration, what are the probabilities that
  (a) neither has a working wife;
  (b) only one has a working wife;
  (c) both have working wives?

8.27 A customs inspector decides to inspect three of 16 shipments that arrived from Madrid by plane. If the selection is random and five of the shipments contain contraband, find the probabilities that the customs inspector will catch
  (a) none of the shipments with contraband;
  (b) one of the shipments with contraband;
  (c) two of the shipments with contraband;
  (d) three of the shipments with contraband.

8.28 To pass a quality control inspection, two batteries are chosen from each lot of 12 car batteries, and the lot is passed only if neither battery has any defects; otherwise, each of the batteries in the lot is checked. If the selection of the batteries is random, find the probabilities that a lot will
  (a) pass the inspection when one of the 12 batteries is defective;
  (b) fail the inspection when three of the batteries are defective;
  (c) fail the inspection when six of the batteries are defective.

8.29 Check in each case whether the condition for the binomial approximation to the hypergeometric distribution is satisfied:
  (a) $a = 140$, $b = 60$, and $n = 12$;
  (b) $a = 220$, $b = 280$, and $n = 20$;
  (c) $a = 250$, $b = 390$, and $n = 30$;
  (d) $a = 220$, $b = 220$, and $n = 25$.

8.30 A shipment of 200 burglar alarms contains 10 that are defective. Five of the burglar alarms are randomly selected to be shipped to a customer.
  (a) Use the hypergeometric distribution to find the probability that the customer will receive exactly one defective burglar alarm.
  (b) Use the binomial approximation to the hypergeometric distribution to find the probability that the customer will receive exactly one defective burglar alarm.
  (c) Find the error of the approximation.

8.31 Among the 180 employees of a company, 144 are union members the others are not. Five of these are to be chosen by lot to serve on the pension fund advisory committee.
  (a) Use the hypergeometric distribution to find the probability that three of the five chosen for the committee will be union members.

(b) Use the binomial approximation to the hypergeometric distribution to find the probability that three of the five chosen for the committee will be union members.

(c) Find the error of the approximation.

★ **8.32** In certain sampling inspection schemes, the sampling continues until a defective item is found. We will assume that the lot contains a defective and b nondefective items. We are interested in the trial number at which the first defective item is encountered. For this to happen on the xth trial, it must be preceded by $x - 1$ nondefectives. Accordingly, the probability that the first defective item is encountered on the xth trial is

Negative hypergeometric distribution

$$f(x) = \frac{\binom{b}{x-1}}{\binom{a+b}{x-1}} \cdot \frac{a}{a + b - (x-1)}$$

$$\text{for } x = 1, 2, 3, \ldots, b, b + 1$$

This distribution is called the **negative hypergeometric distribution**. Using this formula for a lot with $a = 10$ defectives and $b = 30$ nondefectives, we find that the probability that the first defective will be encountered on the fourth trial is

$$\frac{\binom{30}{3}}{\binom{40}{3}} \cdot \frac{10}{37} = \frac{30 \cdot 29 \cdot 28 \cdot 10}{40 \cdot 39 \cdot 38 \cdot 37}$$

or approximately 0.111.

(a) Find the probability that in a lot with 20 defectives and 180 nondefectives, the first defective will be encountered at the fifth trial.

(b) Find the probability that in a lot with 8 defectives and 42 nondefectives, the first defective will be encountered at the first trial.

(c) Simplify the negative hypergeometric formula when $x = 1$.

★ **8.33** The negative hypergeometric distribution can also be used for defectives beyond the first. Give the formula for the probability that the second defective will be encountered at the xth trial in selecting from a lot with a defectives and b nondefectives.

8.5
THE POISSON DISTRIBUTION

When n is large and p is small, binomial probabilities are often approximated by means of the formula

Poisson approximation to binomial distribution

$$f(x) = \frac{(np)^x \cdot e^{-np}}{x!} \qquad \text{for } x = 0, 1, 2, 3, \ldots$$

which is a special form of the **Poisson distribution**, named after the French mathematician and physicist S. D. Poisson (1781–1840); the more general form is given on page 199. In this formula, the irrational number $e = 2.71828\ldots$ is the base of the system of natural logarithms, and the necessary values of e^{-np} may be obtained from Table XIII at the end of the book. Note also that, as in Exercise 8.21 on page 190, we are faced here with a random variable which can take on a countable infinity of values (namely, as many values as there are whole numbers). Correspondingly, the third postulate of probability must be modified so that for any sequence of mutually exclusive events $A_1, A_2, A_3, \ldots$, the probability that one of them will occur is

$$P(A_1 \cup A_2 \cup A_3 \cup \cdots) = P(A_1) + P(A_2) + P(A_3) + \cdots$$

It is difficult to give precise conditions under which the Poisson approximation to the binomial distribution may be used; that is, explain precisely what we mean here by "when n is large and p is small." Although other books may give less stringent rules of thumb, we shall play it relatively safe and use the Poisson approximation to the binomial distribution only when

$$n \geq 100 \quad \text{and} \quad np < 10$$

To get some idea about the closeness of the Poisson approximation to the binomial distribution, consider the computer printouts of Figure 8.4, which show, one above the other, the binomial distribution with $n = 150$ and $p = 0.05$, and the Poisson distribution with $np = 150(0.05) = 7.5$. Comparing the probabilities in the columns headed $P(X = K)$, we find that the greatest difference, corresponding to $K = 8$, is $0.1410 - 0.1373 = 0.0037$.

EXAMPLE It is known that 2 percent of the books bound at a certain bindery have defective bindings. Use the Poisson approximation to the binomial distribution to find the probability that 5 of 400 books bound by this bindery will have defective bindings.

Solution Substituting $x = 5$ and $np = 400(0.02) = 8$ into the formula for the Poisson distribution and getting $e^{-8} = 0.00034$ from Table XIII, we find that

$$f(5) = \frac{8^5 \cdot e^{-8}}{5!} = \frac{(32{,}768)(0.00034)}{120} = 0.093$$

EXAMPLE Records show that the probability is 0.00006 that a car will have a flat tire while driving through a certain tunnel. Use the Poisson approximation to the binomial distribution to find the probability that at least 2 of 10,000 cars passing through this tunnel will have flat tires.

Solution Rather than add the probabilities for $x = 2, 3, 4, \ldots$, we shall subtract from 1 the sum of the probabilities for $x = 0$ and $x = 1$. Thus, substituting $np = 10{,}000(0.00006) = 0.6$, $e^{-0.6} = 0.549$ from Table XIII, and, respectively, $x = 0$ and $x = 1$ into the formula for the Poisson distribution, we get

$$f(0) = \frac{(0.6)^0 (0.549)}{0!} = 0.549$$

$$f(1) = \frac{(0.6)^1 (0.549)}{1!} = 0.329$$

and, hence, find that the answer is $1 - (0.549 + 0.329) = 0.122$.

In actual practice, Poisson probabilities are seldom obtained by direct substitution into the formula. Sometimes we refer to tables of Poisson probabilities, which may be found in more advanced texts or in handbooks of statistical

```
MTB > BINOMIAL N=150 P=0.05

   BINOMIAL PROBABILITIES FOR N = 150   AND P =   .050000

        K              P( X = K)         P(X LESS OR = K)
        0               .0005                .0005
        1               .0036                .0041
        2               .0141                .0182
        3               .0366                .0548
        4               .0708                .1256
        5               .1088                .2344
        6               .1384                .3729
        7               .1499                .5228
        8               .1410                .6638
        9               .1171                .7809
       10               .0869                .8678
       11               .0582                .9260
       12               .0355                .9615
       13               .0198                .9813
       14               .0102                .9915
       15               .0049                .9964
       16               .0022                .9986
       17               .0009                .9995
       18               .0003                .9998
       19               .0001                .9999


MTB > POISSON MU=7.5

   POISSON PROBABILITIES FOR MEAN =   7.500

        K              P(X = K)          P(X LESS OR = K)
        0               .0006                .0006
        1               .0041                .0047
        2               .0156                .0203
        3               .0389                .0591
        4               .0729                .1321
        5               .1094                .2414
        6               .1367                .3782
        7               .1465                .5246
        8               .1373                .6620
        9               .1144                .7764
       10               .0858                .8622
       11               .0585                .9208
       12               .0366                .9573
       13               .0211                .9784
       14               .0113                .9897
       15               .0057                .9954
       16               .0026                .9980
       17               .0012                .9992
       18               .0005                .9997
       19               .0002                .9999
       20               .0001               1.0000
```

FIGURE 8.4 *Computer printouts of the binomial distribution with n = 150 and p = 0.05 and the Poisson distribution with np = 7.5.*

```
MTB > POISSON MU=0.6

   POISSON PROBABILITIES FOR MEAN =     .600

   K              P(X = K)          P(X LESS OR = K)
   0               .5488                .5488
   1               .3293                .8781
   2               .0988                .9769
   3               .0198                .9966
   4               .0030                .9996
   5               .0004               1.0000
```

FIGURE 8.5 *Computer printout of the Poisson distribution with* $np = 0.60$.

tables, and sometimes we use a computer. For instance, had we used the computer printout of Figure 8.5 in the preceding example, the answer would have been $1 - 0.8781 = 0.1219$, which is 1 minus the value in the right-hand column corresponding to $K = 1$.

Since in some cases the hypergeometric distribution may be approximated by a binomial distribution, and the binomial distribution may in some cases be approximated by a Poisson distribution, there will exist situations in which the hypergeometric distribution may be approximated by a Poisson distribution. Consider the following example.

EXAMPLE   An auditor has been asked to investigate a collection of 4,000 sales invoices, of which 28 contain errors. A sample of 150 invoices is selected. What is the probability that this set of 150 invoices will contain exactly 2 with errors?

Solution   This is a hypergeometric problem with $a = 28$, $b = 3,972$, $n = 150$, and $x = 2$. Since $150 \le 0.05(4,000) = 200$, it is reasonable to use the binomial approximation. This binomial distribution has

$$n = 150 \qquad p = \frac{a}{a + b} = \frac{28}{4,000} = 0.007 \quad \text{and} \quad x = 2$$

Since $n \ge 100$ and $n \cdot p = 150 \cdot 0.007 = 1.05 < 10$, it would also be reasonable to use the Poisson approximation with $np = 1.05$. This yields

$$f(2) = \frac{1.02^2 \cdot e^{-1.05}}{2!}$$

$$= \frac{(1.0404) \cdot 0.349938}{2}$$

$$= 0.1820$$

Since auditors deal frequently with situations with large values of n and even larger values of $a + b$, they find the Poisson distribution to be quite useful. Observe that, aside from x, the hypergeometric distribution involves three parameters (*a*, *b*, and *n*), while the binomial distribution involves two (*n* and *p*), and the Poisson distribution involves only one (*np*).

The Poisson distribution has many important applications which have no direct connection with the binomial distribution. In that case *np* is replaced by the parameter λ (Greek lowercase *lambda*) and we calculate the probability of getting *x* successes by means of the formula

Poisson distribution

$$f(x) = \frac{\lambda^x \cdot e^{-\lambda}}{x!} \qquad for\ x = 0, 1, 2, 3, \ldots$$

where λ is interpreted as the expected, or average, number of successes (see discussion on page 204).

This formula applies to many situations where we can expect a fixed number of "successes" per unit time (or for some other kind of unit), say, when a bank can expect to receive six bad checks per day, when 1.6 accidents can be expected per day at a busy intersection, when eight small pieces of meat can be expected in a frozen meat pie, when 5.6 imperfections can be expected per roll of cloth, when 0.03 complaint per passenger can be expected by an airline, and so on.

EXAMPLE If a bank receives on the average $\lambda = 6$ bad checks per day, what is the probability that it will receive four bad checks on any given day?

Solution Substituting $\lambda = 6$ and $x = 4$ into the formula, we get

$$f(4) = \frac{6^4 \cdot e^{-6}}{4!} = \frac{(1,296)(0.0025)}{24} = 0.135$$

where the value of e^{-6} was obtained from Table XIII.

EXAMPLE If $\lambda = 5.6$ imperfections can be expected per roll of a certain kind of cloth, what is the probability that a roll will have three imperfections?

Solution Substituting $\lambda = 5.6$ and $x = 3$ into the formula, we get

$$f(3) = \frac{(5.6)^3 \cdot e^{-5.6}}{3!} = \frac{(175.616)(0.0037)}{6} = 0.108$$

8.6

THE MULTINOMIAL DISTRIBUTION ★

An important generalization of the binomial distribution arises when there are more than two possible outcomes for each trial, the probabilities of the various outcomes remain the same for each trial, and the trials are all independent. This is the case, for example, when we repeatedly roll a die, where each trial has six possible outcomes; when students are asked whether they like a certain new recording, dislike it, or don't care; or when a U.S. Department of Agriculture inspector grades beef as prime, choice, good, commercial, or utility.

If there are k possible outcomes for each trial and their probabilities are p_1, $p_2, \ldots$, and p_k, it can be shown that the probability of x_1 outcomes of the first kind, x_2 outcomes of the second kind, $\ldots$, and x_k outcomes of the kth kind in n trials is given by

Multinomial distribution

$$\frac{n!}{x_1! x_2! \cdot \cdots \cdot x_k!} \, p_1^{x_1} \cdot p_2^{x_2} \cdot \cdots \cdot p_k^{x_k}$$

This distribution is called the **multinomial distribution**.

EXAMPLE   In a very large city, network TV has 40 percent of the viewing audience on Friday nights, a local channel has 20 percent, cable TV has 30 percent, and 10 percent are viewing video cassettes. What is the probability that among seven television viewers randomly selected in that city on a Friday night, two will be viewing network TV, one will be watching the local channel, three will be watching cable TV, and one will be watching a video cassette?

Solution   Substituting $n = 7$, $x_1 = 2$, $x_2 = 1$, $x_3 = 3$, $x_4 = 1$, $p_1 = 0.40$, $p_2 = 0.20$, $p_3 = 0.30$, and $p_4 = 0.10$ into the formula we get

$$\frac{7!}{2! \cdot 1! \cdot 3! \cdot 1!} (0.40)^2 (0.20)^1 (0.30)^3 (0.10)^1 = 0.036$$

Strictly speaking, the multinomial distribution does not apply when we sample without replacement, but in the example above this does not matter since the sample is very small and the city is very large. See Exercise 8.50.

EXERCISES

8.34 Check in each case whether the values of n and p satisfy the rules of thumb which we gave on page 196 for using the Poisson approximation to the binomial distribution:
  (a) $n = 200$ and $p = \frac{1}{25}$;
  (b) $n = 400$ and $p = \frac{1}{32}$;
  (c) $n = 80$ and $p = \frac{1}{10}$.

8.35 It is known from experience that 2 percent of the calls received by a switchboard are wrong numbers. Use the Poisson approximation to the binomial distribution to

determine the probability that 3 of 200 calls received by the switchboard will be wrong numbers.

8.36 If 0.6 percent of the fuses delivered to an arsenal are defective, use the Poisson approximation to the binomial distribution to determine the probability that in a random sample of 500 fuses, four will be defective.

8.37 Records show that the probability is 0.0012 that a person will get food poisoning spending a day at a certain state fair. Use the Poisson approximation to the binomial distribution to find the probability that among 1,000 persons attending the fair, at most two will get food poisoning.

8.38 In a given city, 4 percent of all licensed drivers will be involved in at least one car accident in any given year. Use the Poisson approximation to the binomial distribution to determine the probability that among 150 licensed drivers randomly chosen in this city
    (a) only five will be involved in at least one accident in any given year;
    (b) at most three will be involved in at least one accident in any given year.

8.39 The number of monthly breakdowns of a computer is a random variable having the Poisson distribution with $\lambda = 1.8$. Find the probabilities that this computer will function for a month
    (a) without a breakdown;
    (b) with only one breakdown.

8.40 The number of complaints that a railroad ticket office receives per day is a random variable having the Poisson distribution with $\lambda = 3.6$. What is the probablity that it will receive only two complaints on any given day?

8.41 The number of emergency calls which an ambulance service gets per day is a random variable having the Poisson distribution with $\lambda = 5.5$. What is the probability that on any given day it will receive only four emergency calls?

8.42 The number of inquiries a person gets in response to a newspaper ad listing a piano for sale is a random variable having the Poisson distribution with $\lambda = 4.4$. What are the probabilities that in response to such an ad a person will receive
    (a) only two inquiries;
    (b) only three inquiries;
    (c) at most three inquiries?

8.43 The number of loan applications that a bank gets per day is a random variable having the Poisson distribution with $\lambda = 7.5$. Use the computer printout of Figure 8.4 on page 197 to find the probabilities that on any one day the bank will get
    (a) exactly six loan applications;
    (b) at most four loan applications;
    (c) at least eight loan applications;
    (d) anywhere from five to ten loan applications.

8.44 With reference to the example on page 196 and the computer printout of Figure 8.5 on page 198, find the probabilities that among 10,000 cars passing through the tunnel
    (a) two will have a flat tire;
    (b) four or more will have a flat tire.

★ 8.45 The probabilities are 0.30, 0.60, and 0.10 that in city driving a certain two-door sedan will average less than 20 miles per gallon, anywhere from 20 to 25 miles per gallon, or more than 25 miles per gallon. Find the probability that among six such cars tested, two will average less than 20 miles per gallon, three will average anywhere from 20 to 25 miles per gallon, and one will average more than 25 miles per gallon.

★ 8.46 As we have seen, the probabilities of getting 0, 1, 2, or 3 heads when flipping three balanced coins are $\frac{1}{8}$, $\frac{3}{8}$, $\frac{3}{8}$, and $\frac{1}{8}$. What are the probabilities of getting
    (a) three tails once, one head and two tails twice, two heads and one tail once, and three heads once in five flips of three balanced coins;
    (b) three tails once, one head and two tails three times, and two heads and one tail twice in six flips of three balanced coins?

★ 8.47 The probabilities are 0.60, 0.20, 0.10, and 0.10 that a state income tax form will be filled out correctly, that it will contain only errors favoring the tax payer, that it will contain only errors favoring the government, and that it will contain both kinds of errors. What is the probability that among ten such tax forms (randomly selected for audit) seven will be filled out correctly, one will contain only errors favoring the tax payer, one will contain only errors favoring the goverment, and one will contain both kinds of errors?

★ 8.48 According to the Mendelian theory of heredity, if plants with round yellow seeds are crossbred with plants with wrinkled green seeds, the probabilities of getting a plant that produces round yellow seeds, wrinkled yellow seeds, round green seeds, or wrinkled green seeds are, respectively, $\frac{9}{16}$, $\frac{3}{16}$, $\frac{3}{16}$, and $\frac{1}{16}$. What is the probability that among nine plants thus obtained there will be four that produce round yellow seeds, two that produce wrinkled yellow seeds, three that produce round green seeds, and none that produce wrinkled green seeds?

★ 8.49 Suppose that n objects are to be chosen without replacement from a group consisting of a_1 of type 1, a_2 of type 2, . . . , a_k of type k. The probability of obtaining in the sample x_1 of type 1, x_2 of type 2, . . . , x_k of type k is given by

Multivariate hypergeometric distribution

$$f(x) = \frac{\binom{a_1}{x_1}\binom{a_2}{x_2}\cdots\binom{a_k}{x_k}}{\binom{a_1 + a_2 + \cdots + a_k}{n}}$$

This distribution is called the **multivariate hypergeometric distribution**. In using this formula, it is assumed that there are exactly k different types of objects, and it is also assumed that $x_1 + x_2 + \cdots + x_k = n$. Use this formula to determine the probability that

(a) Among five cards drawn without replacement from a standard deck of 52, there will be 1 spade, 2 hearts, 1 diamond, and 1 club;

(b) among 10 balls drawn without replacement from a jar containing 20 red, 10 green, and 12 yellow balls, there will be 6 red, 1 green, and 3 yellow balls;

(c) among 5 cards drawn without replacement from a standard deck of 52, there will be 2 aces, 1 king, 1 queen, and 1 jack.

★ **8.50** If the drawings in the previous exercise were done with replacement, then the multinomial distribution would be utilized. The multinomial may be a reasonable approximation to the multivariate hypergeometric when n, the number of objects to be chosen, is small relative to $a = a_1 + a_2 + \cdots + a_k$. The approximation can be regarded as adequate when $n \leq 0.05a$, and it may still give reasonably good answers even when $n > 0.05a$. Recalculate (a), (b), and (c) of the previous exercise assuming sampling with replacement, and compare the solutions.

8.7

THE MEAN OF A PROBABILITY DISTRIBUTION

When we showed on page 166 that an airline office at a certain airport can expect 2.75 complaints per day about its luggage handling, we arrived at this result by using the formula for a mathematical expectation, namely, by adding the products obtained by multiplying $0, 1, 2, 3, \ldots$, by the corresponding probabilities that it will receive $0, 1, 2, 3, \ldots$, complaints about its luggage handling on any given day. Here, the number of complaints is a random variable and 2.75 is its **expected value**.

If we apply the same argument to the first example of Section 8.2, we find that, between them, the two insurance agents can expect to sell

$$0(0.44) + 1(0.20) + 2(0.18) + 3(0.18) = 1.10$$

insurance policies to the given family. In this case, the number of policies is a random variable and 1.10 is its expected value.

As we explained in Chapter 7, mathematical expectations must be interpreted as averages, or means, and it is customary to refer to the expected value of a random variable as its **mean**, or as the **mean of its probability distribution**. In general, if a random variable takes on the values $x_1, x_2, x_3, \ldots$, or x_k, with the probabilities $f(x_1), f(x_2), f(x_3), \ldots$, and $f(x_k)$, its expected value is

$$x_1 \cdot f(x_1) + x_2 \cdot f(x_2) + x_3 \cdot f(x_3) + \cdots + x_k \cdot f(x_k)$$

and in the $\sum$ notation we write

Mean of a probability distribution

$$\mu = \sum x \cdot f(x)$$

Like the mean of a population, it is denoted by the Greek letter μ (*mu*). The notation is the same, for as we pointed out in connection with the binomial distribution, when we observe a value of a random variable, we refer to its distribution as the population we are sampling. For instance, the histogram of Figure 8.2 on page 186 may be looked upon as the population we are sampling when we observe a value of a random variable having the binomial distribution with $n = 6$ and $p = 0.30$.

EXAMPLE   Find the mean of the second probability distribution of Section 8.2, which pertains to the number of points we roll with a balanced die.

Solution   Since the probabilities of rolling a 1, 2, 3, 4, 5, or 6 are all equal to $\frac{1}{6}$, we get

$$1 \cdot \tfrac{1}{6} + 2 \cdot \tfrac{1}{6} + 3 \cdot \tfrac{1}{6} + 4 \cdot \tfrac{1}{6} + 5 \cdot \tfrac{1}{6} + 6 \cdot \tfrac{1}{6} = 3\tfrac{1}{2}$$

EXAMPLE   With reference to the example on page 185, find the mean number of persons, among six shopping at the market, who will take advantage of the special promotion.

Solution   Substituting $x = 0, 1, 2, 3, 4, 5,$ and 6, and the probabilities on page 185 into the formula for μ, we get

$$\mu = 0(0.118) + 1(0.303) + 2(0.324) + 3(0.185) + 4(0.060)$$

$$+ \ 5(0.010) + 6(0.001)$$

$$= 1.802$$

When a random variable can take on many different values, the calculation of μ may become very laborious. For instance, if we want to know how many persons can be expected to contribute to a charity, when 2,000 are solicited for funds and the probability is 0.40 that any one of them will make a contribution, we might consider calculating the 2,001 probabilities corresponding to 0, 1, 2, 3,..., 1,999, or 2,000 of them making a contribution. Not seriously, though, and we might argue instead that in the long run 40 percent of the persons will make a contribution, 40 percent of 2,000 is 800, and hence we can expect that 800 of the 2,000 persons will make a contribution. Similarly, if a balanced coin is flipped 1,000 times, we might argue that in the long run heads will come up 50 percent of the time, and hence that we can expect $1,000(0.50) = 500$ heads. These two results are correct; both problems deal with random variables having binomial distributions, and it can be shown that in general

Mean of a binomial distribution

$$\mu = n \cdot p$$

In words, the mean of a binomial distribution is simply the product of the number of trials and the probability of success on an individual trial.

EXAMPLE   With reference to the example on page 185, use this formula to find the mean number of persons, among six shopping at the market, who will take advantage of the special promotion.

Solution   Since we are dealing with a binomial distribution with $n = 6$ and $p = 0.30$, we get

$$\mu = 6(0.30) = 1.80$$

The small difference of 0.002 between the values obtained here and on page 203 is due to rounding the probabilities to three decimals.

It is important to remember that the formula $\mu = n \cdot p$ applies only to binomial distributions. There are other formulas for other distributions; for instance, for the hypergeometric distribution the formula for the mean is

Mean of a hypergeometric distribution

$$\mu = \frac{n \cdot a}{a + b}$$

EXAMPLE   Among 12 school buses, five have worn brakes. If six of the school buses are randomly picked for inspection, how many of them can be expected to have worn brakes?

Solution   Since we are sampling without replacement, we have here a hypergeometric distribution with $a = 5$, $b = 7$, and $n = 6$. Substituting these values into the above formula, we get

$$\mu = \frac{6 \cdot 5}{5 + 7} = 2.5$$

This should not come as a surprise—half of the school buses are selected and half of the ones with worn brakes are expected to be included in the sample.

Also, the mean of the Poisson distribution with the parameter λ is $\mu = \lambda$, and this agrees with what we suggested earlier, namely, that λ is to be interpreted as an average. Derivations of all these special formulas may be found in textbooks on mathematical statistics.

8.8
THE STANDARD DEVIATION OF A PROBABILITY DISTRIBUTION

In Chapter 4 we saw that the most widely used measures of variation are variance and its square root, the standard deviation, which both measure variability by averaging the squared deviations from the mean. For probability distributions we measure variability in almost the same way, but instead of averaging the squared deviations from the mean, we find their expected value. In general, if a random variable takes on the values $x_1, x_2, x_3, \ldots,$ or x_k, with the probabilities $f(x_1), f(x_2), f(x_3), \ldots,$ and $f(x_k)$, and the mean of this probability distribution is μ, then the deviations from the mean are $x_1 - \mu, x_2 - \mu, x_3 - \mu, \ldots,$ and $x_k - \mu$, and the expected value of their squares is

$$(x_1 - \mu)^2 \cdot f(x_1) + (x_2 - \mu)^2 \cdot f(x_2) + \cdots + (x_k - \mu)^2 \cdot f(x_k)$$

Thus, in the $\sum$ notation we write

Variance of a probability distribution

$$\sigma^2 = \sum (x - \mu)^2 \cdot f(x)$$

which we refer to as the **variance of the random variable** or **the variance of its probability distribution**. As in the preceding section, and for the same reason, we denote this description of a probability distribution with the same symbol as the corresponding description of a population.

The square root of the variance defines the **standard deviation of a probability distribution** and we write

Standard deviation of a probability distribution

$$\sigma = \sqrt{\sum (x - \mu)^2 \cdot f(x)}$$

EXAMPLE With reference to the example on page 185, find the standard deviation of the number of persons, among six shopping at the market, who will take advantage of the special promotion.

Solution As we have seen, $\mu = 6(0.30) = 1.80$, so that we can arrange the calculations as follows:

Number of persons	Probability	Deviation from mean	Squared deviation from mean	$(x - \mu)^2 f(x)$
0	0.118	−1.8	3.24	0.38232
1	0.303	−0.8	0.64	0.19392
2	0.324	0.2	0.04	0.01296
3	0.185	1.2	1.44	0.26640
4	0.060	2.2	4.84	0.29040
5	0.010	3.2	10.24	0.10240
6	0.001	4.2	17.64	0.01764

$$\sigma^2 = 1.26604$$

The values in the column on the right were obtained by multiplying the squared deviations from the mean by their probabilities, and the total of this column is the variance of the distribution. Thus, the standard deviation is $\sigma = \sqrt{1.26604} = 1.13$.

The calculations were easy in this example because the deviations from the mean were small numbers given to one decimal. If the deviations from the mean are large numbers, or if they are given to several decimals, it is usually worthwhile to simplify the calculations by using the following computing formula, which does not require that we work with the deviations from the mean:

Computing formula for the variance of a probability distribution

$$\sigma^2 = \sum x^2 \cdot f(x) - \mu^2$$

EXAMPLE   On page 166 we showed that if the probabilities are 0.06, 0.21, 0.24, 0.18, 0.14, 0.10, 0.04, 0.02, and 0.01 that an airline office at a certain airport will receive 0, 1, 2, 3, 4, 5, 6, 7, or 8 complaints per day about its luggage handling, the mean of this probability distribution is $\mu = 2.75$. Find its standard deviation.

Solution   First calculating

$$\sum x^2 \cdot f(x) = 0^2(0.06) + 1^2(0.21) + 2^2(0.24) + 3^2(0.18)$$
$$+ 4^2(0.14) + 5^2(0.10) + 6^2(0.04)$$
$$+ 7^2(0.02) + 8^2(0.01)$$
$$= 10.59$$

we get

$$\sigma^2 = 10.59 - (2.75)^2 = 3.03$$

and, hence, $\sigma = \sqrt{3.03} = 1.74$. The calculations would have been much more tedious in this case if we had not used the computing formula.

As in the case of the mean, the calculation of the variance or the standard deviation can generally be simplified when we deal with special kinds of probability distributions. For instance, for the binomial distribution we have the formula

Standard deviation of a binomial distribution

$$\sigma = \sqrt{np(1 - p)}$$

EXAMPLE   Use this formula to verify the result obtained just above for the example dealing with the number of persons, among six shopping at the market, who will take advantage of the special promotion.

Solution   Since we are dealing with a binomial distribution with $n = 6$ and $p = 0.30$, the formula yields

$$\sigma = \sqrt{6(0.30)(0.70)} = 1.12$$

The difference, due to rounding, is only $1.13 - 1.12 = 0.01$.

There also exist formulas for the standard deviations of other special probability distributions (see, for example, Exercises 8.69 and 8.73).

In Chapter 4 we mentioned Chebyshev's theorem because it gave us some idea how the standard deviation "controls" the spread or variation of a set of data. If we replace "the proportion of the data that must lie within k standard deviations on either side of the mean" by "the probability that a random variable will take on a value within k standard deviations on either side of the mean," the theorem applies also to random variables. From page 78, this probability is "at least $1 - \dfrac{1}{k^2}$."

EXAMPLE   The number of customers which a restaurant serves on a Friday night is a random variable with $\mu = 160$ and $\sigma = 7.5$. According to Chebyshev's theorem, what is the probability that it will serve between 145 and 175 customers on a Friday night?

Solution   Since 175 is $\dfrac{175 - 160}{7.5} = 2$ standard deviations above the mean and 145 is

$\dfrac{160 - 145}{7.5} = 2$ standard deviations below the mean, $k = 2$ and the probability is at

least $1 - \dfrac{1}{2^2} = 0.75$.

EXAMPLE   What does Chebyshev's theorem with $k = 5$ tell us about the number of heads, and hence the proportion of heads, we will get in 400 flips of a balanced coin?

Solution   Since we are dealing with a random variable having the binomial distribution with $n = 400$ and $p = 0.50$,

$$\mu = 400(0.50) = 200 \quad \text{and} \quad \sigma = \sqrt{400(0.50)(0.50)} = 10.$$

Thus, the probability is at least $1 - \dfrac{1}{5^2} = 0.96$ that we will get between $200 - 5 \cdot 10 = 150$ and $200 + 5 \cdot 10 = 250$ heads, or that the proportion of heads will be between $\dfrac{150}{400} = 0.375$ and $\dfrac{250}{400} = 0.625$.

To continue with this example, the reader will be asked to show in Exercise 8.76 on page 210 that for 10,000 flips of a balanced coin the probability is at least

0.96 that the proportion of heads will be between 0.475 and 0.525, and that for 1,000,000 flips of a balanced coin the probability is at least 0.96 that the proportion of heads will be between 0.4975 and 0.5025. All this provides support for the Law of Large Numbers, which we introduced in Section 5.4 in connection with the frequency interpretation of probability.

EXERCISES

8.51 Suppose that the probabilities are 0.4, 0.3, 0.2, and 0.1 that among three recently married couples 0, 1, 2, or 3 will be divorced within two years. Use the formulas which define μ and σ^2 to find
   (a) the mean of this probability distribution;
   (b) the variance of this probability distribution.

8.52 Use the computing formula for σ^2 to rework part (b) of the preceding exercise.

8.53 The following table gives the probabilities that a probation officer will receive 0, 1, 2, 3, 4, or 5 reports of probation violations on any given day:

Number of violations	0	1	2	3	4	5
Probability	0.20	0.32	0.23	0.15	0.08	0.02

Use the formulas which define μ and σ to find
   (a) the mean of this probability distribution;
   (b) the standard deviation of this probability distribution.

8.54 Use the computing formula for σ^2 to rework part (b) of the preceding exercise.

8.55 The probabilities that a building inspector will observe 0, 1, 2, 3, 4, or 5 violations of the building code in a home built in a large development are, respectively, 0.41, 0.22, 0.17, 0.13, 0.05, and 0.02. Find the mean of this probability distribution.

8.56 With reference to the preceding exercise, use the computing formula for σ^2 to find the standard deviation of the probability distribution.

8.57 As we saw in the beginning of Section 8.2, the probabilities of getting 0, 1, 2, 3, or 4 heads in four flips of a balanced coin are $\frac{1}{16}$, $\frac{4}{16}$, $\frac{6}{16}$, $\frac{4}{16}$, and $\frac{1}{16}$. Find the mean of this random variable using
   (a) the formula which defines μ;

   (b) the special formula for the mean of a binomial distribution.

8.58 With reference to the preceding exercise, find the variance of the random variable using
   (a) the formula which defines σ^2;
   (b) the computing formula for σ^2;
   (c) the special formula for the standard deviation of a binomial distribution.

8.59 A study shows that 60 percent of all first-class letters between two cities are delivered within 48 hours. Find the mean and the variance of the number of first-class letters between the two cities, among eight randomly selected, which will be delivered within 48 hours, using
   (a) Table V, the formula which defines μ, and the computing formula for σ^2;
   (b) the special formulas for the mean and the standard deviation of a binomial distribution.

8.60 If 80 percent of certain videocasette recorders will function successfully through the 90-day warranty period, find the mean and standard deviation of the number of these videocasette recorders, among 10 randomly selected, which will function successfully through the 90-day warranty period, using
   (a) Table V, the formula which defines μ, and the computing formula for σ^2;
   (b) the special formulas for the mean and standard deviation of the binomial distribution.

8.61 A study shows that 80 percent of all patients coming to a certain medical building have to wait at least 30 minutes to see their doctor. Use the special formulas for the mean and the standard deviation of a binomial distribution to find μ and σ for the number of patients, among 20 coming to this medical building, who have to wait at least 30 minutes to see their doctor.

8.62 Find the mean and the standard deviation of each of the following binomial random variables:

(a) The number of heads obtained in 484 flips of a balanced coin.

(b) The number of 3's obtained in 720 rolls of a balanced die.

(c) The number of persons, among 600 invited, who will attend the opening of a new branch bank, when the probability is 0.30 that any one of them will attend.

(d) The number of defectives in a sample of 600 parts made by a machine, when the probability is 0.04 that any one of the parts is defective.

(e) The number of students, among 800 interviewed, who do not like the food served at the university cafeteria, when the probability is 0.65 that any one of them does not like the food.

8.63 Use the results of Exercise 8.26 on page 194 to find the mean number of applicants, among the two randomly selected, who will have working wives.

8.64 Use the special formula for the mean of a hypergeometric distribution to verify the result of the preceding exercise.

8.65 Use the results of Exercise 8.27 on page 194 to find the mean number of shipments, among the three randomly selected, which will contain contraband.

8.66 Use the special formula for the mean of a hypergeometric distribution to verify the result of the preceding exercise.

8.67 Among eight faculty members considered for promotions, four have Ph.D.'s and four do not.

(a) If four of them are chosen at random, find the probabilities that 0, 1, 2, 3, or all 4 of them have Ph.D.'s.

(b) Use the results of part (a) to determine the mean of this probability distribution.

(c) Use the special formula for the mean of a hypergeometric distribution to verify the result of part (b).

8.68 Use parts (a) and (c) of the preceding exercise to find the standard deviation of the number of faculty members, among the four chosen at random, who have Ph.D.'s.

★ **8.69** For the variance of a hypergeometric distribution with the parameters a, b, and n, there is the special formula

$$\sigma^2 = n \cdot \frac{ab}{(a+b)^2} \cdot \frac{a+b-n}{a+b-1}$$

(a) Use this formula to verify the results of the previous exercise.

(b) Use this formula to find the standard deviation of the random variable of Exercise 8.63, where $a = 9$, $b = 3$, and $n = 2$.

(c) Use this formula to find the standard deviation of the random variable of Exercise 8.65, where $a = 5$, $b = 11$, and $n = 3$.

★ **8.70** If sampling is done with replacement from a group consisting of a of one type and b of another type, then the binomial probability distribution applies, and the variance of the number of the first type, out of n drawings, is

$$\sigma^2 = n \cdot \frac{ab}{(a+b)^2}$$

Only the factor $\dfrac{a+b-n}{a+b-1}$ distinguishes the binomial and hypergeometric variances. Compute this factor for the following cases:

(a) $a + b = \quad 20, \qquad n = \quad 5$;

(b) $a + b = \quad 100, \qquad n = \quad 4$;

(c) $a + b = 1{,}000, \qquad n = 40$;

(d) $a + b = \quad 200, \qquad n = 60$;

8.71 The probabilities that there will be 0, 1, 2, 3, 4, or 5 fires caused by lightning during a summer storm are, respectively, 0.449, 0.360, 0.144, 0.038, 0.008, and 0.001. Calculate the mean of this Poisson distribution with $\lambda = 0.8$, and use the result to verify the special formula $\mu = \lambda$ mentioned on page 204.

8.72 If the number of gamma rays emitted per second by a certain radioactive substance is a random variable having the Poisson distribution with $\lambda = 2.4$, then the probabilities that it will emit certain numbers of gamma rays in 1 second are given by

Number of gamma rays	Probability
0	0.091
1	0.218
2	0.261
3	0.209
4	0.125
5	0.060
6	0.024
7	0.008
8	0.003
9	0.001

The probability of 10 or more gamma rays is less than 0.001. Calculate the mean and use the result to verify the special formula $\mu = \lambda$ for a random variable having the Poisson distribution with the parameter λ.

★ 8.73 For the standard deviation of a Poisson distribution there is the special formula $\sigma = \sqrt{\lambda}$. Calculate the standard deviation of the probability distribution of Exercise 8.71 and use the result to verify this special formula.

8.74 If a student answers the 144 questions of a true–false test by flipping a balanced coin—heads is "true" and tails is "false"—what does Chebyshev's theorem with $k = 4$ tell us about the number of correct answers he will get?

8.75 The annual number of rainy days in a certain city is a random variable with $\mu = 126$ and $\sigma = 9$.
   (a) What does Chebyshev's theorem with $k = 12$ tell us about the number of days it will rain in the given city in any one year?
   (b) According to Chebyshev's theorem, with what probability can we assert that it will rain in the given city between 96 and 156 days in any one year?

8.76 Use Chebyshev's theorem to show that the probability is at least 0.96 that
   (a) for 10,000 flips of a balanced coin the proportion of heads will be between 0.475 and 0.525;
   (b) for 1,000,000 flips of a balanced coin the proportion of heads will be between 0.4975 and 0.5025.

8.9

TECHNICAL NOTE (Simulation) ★

In the last few decades, simulation techniques have been applied to many problems in business and scientific research, and if the processes being simulated involve an element of chance, we refer to these techniques as **Monte Carlo methods**. Such methods have been used, for example, to study traffic flow on proposed freeways, to conduct war "games," to study the spread of epidemics or human behavior during times of natural disaster (say, a flood or an earthquake), and to study the scattering of neutrons or the collisions of photons with electrons. Also, in business research, such methods are used to solve inventory problems, production scheduling, or the effects of advertising campaigns, and to study many other situations involving overall planning and organization. In most cases, this will eliminate the cost of building and operating expensive equipment, and in some instances it can be used when direct experimentation is impossible.

Although Monte Carlo methods are sometimes executed with actual gambling devices, usually they are based on **random numbers** (also called **random digits**) found in published tables or generated by means of computers. Such tables consist of many pages on which the digits 0, 1, 2,..., and 9 are set down in a supposedly random fashion, much as they would appear if they were generated one at a time by a chance or gambling device giving each digit the same probability of $\frac{1}{10}$. Some early tables of random numbers were copied from pages of census data or from tables of 20-place logarithms, but they were found to be deficient in various ways. Nowadays, such tables are generated by means of computers (see references on page 218).

To illustrate the use of random numbers, let us show how we might play "heads or tails" without actually flipping a coin. Letting the even numbers represent heads and the odd numbers tails, we might arbitrarily choose the fourth column of the first page of Table XI, start at the top and go down the page. We would thus get, 3, 9, 8, 1, 5, 1, 6, 3, 2, 4, ..., and interpret this as tail, tail, head, tail, tail, tail, head, tail, head, head, ... Note that in the computer printout of Figure 5.3 we actually simulated 100 flips of a balanced coin. The command BRANDOM 100 N=1 P=.5 tells the computer to generate 100 values of a random variable having the binomial distribution with $n = 1$ and $p = 0.50$; here 1 and 0 denote heads and tails.

Repeated flips of any number of coins, say, three, could be simulated in the same way. If we arbitrarily choose the first three columns of the second page of Table XI, start with the 16th row and go down the page, we get the three-digit random numbers, 659, 900, 972, 219, 411, 237, 599, 826, 838, 619, ..., and, counting the number of even digits in each case, we interpret this as 1, 2, 1, 1, 1, 1, 0, 3, 2, 1, ..., heads.

To save ourselves the job of having to count the even digits, we might use instead the following scheme, where we made use of the fact that for three balanced coins the probabilities of 0, 1, 2, or 3 heads are $\frac{1}{8}, \frac{3}{8}, \frac{3}{8}$, and $\frac{1}{8}$, or 0.125, 0.375, 0.375, and 0.125 in decimal notation:

Number of heads x	Probability of x heads	Probability of x or less heads	Random numbers
0	0.125	0.125	000–124
1	0.375	0.500	125–499
2	0.375	0.875	500–874
3	0.125	1.000	875–999

Here we used three-digit random numbers because the probabilities are given to three decimals, and we allocated 125 of the 1,000 random numbers from 000 to 999 to 0 heads, 375 to 1 head, 375 to 2 heads, and 125 to 3 heads. The column of cumulative probabilities was added to facilitate the assignment of the random numbers. Observe that in each case the last random number is one less than the number formed by the three decimal digits of the corresponding cumulative probability.

EXAMPLE Use this scheme and the 16th, 17th, and 18th columns of the second page of Table XI, starting with the 6th row and going down the page, to simulate ten flips of three balanced coins.

Solution Reading the three-digit random numbers off that page, we get 974, 611, 345, 664, 041, 203, 531, 421, 031, and 925, and we interpret this as 3, 2, 1, 2, 0, 1, 2, 1, 0, and 3 heads.

Had we worked this example by using the computer software which yielded the printout of Figure 5.3, the command would have been BRANDOM 1Ø N=3 P=0.5.

The method we have illustrated here with reference to a game of chance can be used to simulate observations of any discrete random variable with a given probability distribution.

EXAMPLE   The probabilities that 0, 1, 2, 3, ..., or 13 cars will arrive at the tollbooth of a certain bridge in any one minute during the early afternoon are 0.008, 0.037, 0.089, 0.146, 0.179, 0.175, 0.143, 0.100, 0.062, 0.034, 0.016, 0.007, 0.003, and 0.001.

(a) Distribute the three-digit random numbers from 000 to 999 among the fourteen values of this random variable, so that they can be used to simulate the arrival of cars at the tollbooth.

(b) Use the 6th, 7th, and 8th columns of the first page of Table XI, starting with the 21st row and going down the page, to simulate the arrival of cars at the tollbooth during 20 one-minute intervals in the early afternoon.

Solution   (a) Calculating the cumulative probabilities and following the suggestion on page 207, we arrive at the following scheme:

Number of cars	Probability	Cumulative probability	Random numbers
0	0.008	0.008	000–007
1	0.037	0.045	008–044
2	0.089	0.134	045–133
3	0.146	0.280	134–279
4	0.179	0.459	280–458
5	0.175	0.634	459–633
6	0.143	0.777	634–776
7	0.100	0.877	777–876
8	0.062	0.939	877–938
9	0.034	0.973	939–972
10	0.016	0.989	973–988
11	0.007	0.996	989–995
12	0.003	0.999	996–998
13	0.001	1.000	999

(b) Reading off the three-digit random numbers, we get 836, 712, 524, 762, 325, 081, 960, 594, 473, 370, 305, 178, 523, 184, 368, 864, 676, 975, 553, and 618, and this means that 7, 6, 5, 6, 4, 2, 9, 5, 5, 4, 4, 3, 5, 3, 4, 7, 6, 10, 5, and 5 cars arrived at the tollbooth during 20 simulated one-minute intervals.

In all problems of this kind, the page in the table of random numbers, the columns, and the row with which we start, should all be selected at random; perhaps, the choice may be left to a table of random numbers.

```
MTB > PRANDOM 20 MU=4.9 C1
    20 POISSON OBS. WITH MEAN     4.9000
       3.       6.       4.       8.       5.       4.       3.       6.       5.       7.
       9.       3.       9.       3.      11.       7.       5.       9.       6.       2.
```

FIGURE 8.6 *Computer simulations of 20 values of a random variable having the Poisson distribution with* $\lambda = 4.9$.

The simulation of the preceding example was fairly easy and straightforward, but it might have been even easier if we had used a computer. Making use of the fact that the distribution we gave is that of a random variable having the Poisson distribution with $\lambda = 4.9$, we get the printout of Figure 8.6, which tells us that 3, 6, 4, 8, 5, 4, 3, 6, 5, 7, 9, 3, 9, 3, 11, 7, 5, 9, 6, and 2 cars arrived at the tollbooth during 20 simulated one-minute intervals.

It is important to remember that simulations are not necessarily exact re-placements for real experiments. Simulations are designed to exhibit the mathe-matical aspects of real situations, but they sometimes do a poor job of capturing behavioral peculiarities. The simulations of coin flips and traffic at a toll booth illustrated in this section are quite reliable. A simulation of an inventory manage-ment scheme, however, cannot deal with the possibility that persons managing an inventory scheme might invent new procedures as they go along or might fail to follow established rules.

EXERCISES

★ **8.77** Letting the digits 1, 2, 3, 4, 5, and 6 represent the corresponding faces of a die (and omitting 0, 7, 8, and 9), use Table XI to simulate 120 rolls of a balanced die. Also determine the accumulated proportion of 6's after each tenth roll and draw a diagram like that of Figure 5.4. Does it seem to support the Law of Large Numbers?

★ **8.78** As we have seen, the probabilities of getting 0, 1, or 2 heads in two flips of a balanced coin (or when flipping a pair of balanced coins) are 0.25, 0.50, and 0.25.
  (a) Distribute the two-digit random numbers from 00 to 99 among the three possibilities so that the corresponding random numbers can be used to simulate repeated flips of a pair of balanced coins.
  (b) Use Table XI to simulate 200 flips of a pair of balanced coins and determine the accumulated proportion of the time that both coins came up heads after each tenth pair of flips. Also draw

a diagram like that of Figure 5.4 and check whether the simulation supports the Law of Large Numbers.

★ **8.79** The probabilities that a real estate broker will sell 0, 1, 2, 3, 4, 5, or 6 houses in a week are 0.14, 0.27, 0.27, 0.18, 0.09, 0.04, and 0.01.
  (a) Distribute the two-digit random numbers from 00 through 99 among these seven possibilities so that the corresponding random numbers can be used to simulate the number of houses the real estate broker sells in a week.
  (b) Use the results of part (a) to simulate the real estate broker's weekly sales during 25 consecu-tive weeks.

★ **8.80** With reference to the preceding exercise, how big is the difference between the mean of the simulated numbers of sales and $\mu = 1.97$, the mean of the given probability distribution?

★ **8.81** Suppose that the probabilities are 0.41, 0.37, 0.16, 0.05, and 0.01 that there will be 0, 1, 2, 3, or 4 UFO sightings in a certain region on any one day.

  (a) Distribute the two-digit random numbers from 00 through 99 among the five values of this random variable, so that the corresponding random numbers can be used to simulate the sighting of UFO's in the given region.

  (b) Use the result of part (a) to simulate the sighting of UFO's in the given region on 30 days.

★ **8.82** Suppose the probabilities are 0.2466, 0.3452, 0.2417, 0.1128, 0.0395, 0.0111, 0.0026, and 0.0005 that there will be 0, 1, 2, 3, 4, 5, 6, or 7 polluting spills in the Great Lakes on any one day.

  (a) Distribute the four-digit random numbers from 0000 through 9999 among the eight values of this random variable, so that the corresponding random numbers can be used to simulate the occurrence of polluting spills in the Great Lakes.

  (b) Use the results of part (a) to simulate the number of polluting spills in the Great Lakes on 40 consecutive days.

★ **8.83** The owner of a bakery knows that the probabilities are 0.05, 0.15, 0.25, 0.25, 0.20, and 0.10 that the demand for a highly perishable cheese cake will be 0, 1, 2, 3, 4, or 5.

  (a) Distribute the two-digit random numbers from 00 through 99 among these six possibilities so that the corresponding random numbers can be used to simulate the daily demand.

  (b) Use the results of part (a) to simulate the demand for the cheese cake on 30 consecutive business days.

  (c) If the baker makes a profit of $2.00 on each cake that he sells, but loses $1.00 on each cake that goes to waste (namely, each cake that cannot be sold on the day it is baked), find the baker's profit or loss for each of the 30 days of part (b), assuming that each day he bakes 3 of these cakes. Also find the baker's average profit for the 30 days.

  (d) Repeat part (c), assuming that each day he bakes 4 rather than 3 of the cakes. Do not repeat part (b) in doing this. Which of the two appears to be more profitable?

★ **8.84** Depending on the availability of parts, a company can manufacture 3, 4, 5, or 6 units of a certain item per week with corresponding probabilities of 0.10, 0.40, 0.30, and 0.20. The probabilities that there will be a weekly demand for 0, 1, 2, 3,..., or 8 units are 0.05, 0.10, 0.30, 0.30, 0.10, 0.05, 0.05, 0.04, and 0.01. If a unit is sold during the week that it is made, it will yield a profit of $100; this profit is reduced by $20 for each week that a unit has to be stored. Use random numbers to simulate the operations of this company for 50 consecutive weeks and estimate its expected weekly profit.

8.10
CHECKLIST OF KEY TERMS
(with page references to their definitions)

8.11
REVIEW EXERCISES

8.85 If 12 percent of all medical students want to specialize in surgery, what are the probabilities that in a random sample of three medical students
  (a) none want to specialize in surgery;
  (b) only one wants to specialize in surgery?

8.86 A panel of 300 persons chosen for jury duty includes 30 under 25 years of age. Since the jury of 12 persons chosen from this panel to judge a narcotics violation does not include anyone under 25 years of age, the youthful defendant's attorney complains that this jury is not really representative. Indeed, he argues, if the selection were random, the probability of having one of the 12 jurors under 25 years of age should be *many times* the probability of having none of them under 25 years of age. Actually, what is the ratio of these two probabilities?

8.87 Referring to the computer printout of Figure 8.5 on page 198, find the probability that a random variable having the Poisson distribution with $\lambda = 0.6$ will take on a value greater than 2 by
  (a) adding the probabilities corresponding to 3, 4, ...;
  (b) working with the cumulative probabilities.

8.88 In each case determine whether the given values can be looked upon as the values of a probability distribution of a random variable which can take on the values 1, 2, 3, 4, and 5, and explain your answers:
  (a) $f(1) = 0.18$ (b) $f(1) = 0.10$
      $f(2) = 0.20$ $f(2) = 0.10$
      $f(3) = 0.22$ $f(3) = 0.80$
      $f(4) = 0.20$ $f(4) = 0.10$
      $f(5) = 0.18$ $f(5) = 0.10$
  (c) $f(1) = 0.15$
      $f(2) = 0.20$
      $f(3) = 0.25$
      $f(4) = 0.30$
      $f(5) = 0.35$

8.89 The probabilities are 0.22, 0.34, 0.25, 0.13, 0.05, and 0.01 that 0, 1, 2, 3, 4, or 5 of a doctor's patients will come down with the flu during the first week of January.
  (a) find the mean of this probability distribution;
  (b) use the computing formula to find the variance of this probability distribution.

★ 8.90 The probability that a burglar will get caught on any given job is 0.20. Use the formula for the geometric distribution to find the probability that he will get caught for the first time on his fifth job.

8.91 Among 25 workers on a picket line, 14 are men and 11 are women. If a television news reporter randomly picks four of them to be shown on camera, what are the probabilities that this will include
  (a) two men and two women;
  (b) one man and three women?

★ 8.92 Suppose that the probabilities are 0.46, 0.27, 0.15, 0.08, and 0.04 that it will take an accountant 1, 2, 3, 4, or 5 hours to complete a certain income tax form. Distribute the two-digit random numbers from 00 to 99 among the five values of this random variable, so that they can be used to simulate the amount of time it takes the accountant to complete such forms.

★ 8.93 A certain supermarket carries four grades of ground beef, and the probabilities that a shopper will buy the poorest, third best, second best, or best grade are, respectively, 0.10, 0.20, 0.40, and 0.30. Find the probability that among 12 (randomly chosen) shoppers buying ground beef at this supermarket, one will choose the poorest kind, two will choose the third best kind, six will buy the second best kind, and three will buy the best kind.

8.94 Check in each case whether the condition for the binomial approximation to the hypergeometric distribution is satisfied:
  (a) $a = 80$, $b = 120$, and $n = 8$;
  (b) $a = 75$, $b = 75$, and $n = 10$;
  (c) $a = 136$, $b = 164$, and $n = 24$.

8.95 Find the mean of the binomial distribution with $n = 9$ and $p = 0.50$, using
  (a) Table V and the formula which defines μ;
  (b) the special formula for the mean of a binomial distribution.

8.96 Find the standard deviation of the binomial distribution of the preceding exercise, using
  (a) Table V, the result of part (b) of the preceding exercise, and the computing formula for σ^2;
  (b) the special formula for the standard deviation of a binomial distribution.

8.97 Use the computer printout of Figure 8.7 to find the probabilities that a random variable having the binomial distribution with $n = 18$ and $p = 0.27$ will take on
(a) the value 4;
(b) a value less than 7;
(c) a value anywhere from 3 to 8.

```
MTB > BINOMIAL N=18 P=0.27

    BINOMIAL PROBABILITIES FOR N =  18  AND P =  .270000

      K            P( X = K)           P(X LESS OR = K)
      0              .0035               .0035
      1              .0231               .0265
      2              .0725               .0991
      3              .1431               .2422
      4              .1985               .4406
      5              .2055               .6462
      6              .1647               .8109
      7              .1044               .9153
      8              .0531               .9684
      9              .0218               .9903
     10              .0073               .9975
     11              .0020               .9995
     12              .0004               .9999
```

FIGURE 8.7 *Computer printout of the binomial distribution with* $n = 18$ *and* $p = 0.27$.

8.98 It is known that 4 percent of all rats carry a certain disease.
   (a) If we examine 200 rats, constituting a random sample, will we satisfy the condition in the text for using the Poisson approximation to the binomial distribution of the number carrying the disease?
   (b) Calculate the probability that exactly 6 of the 200 rats will carry the disease. Use both the exact binomial probability and the Poisson approximation, and decide whether the difference vindicates your solution to part (a).

8.99 During the month of August, the daily number of persons visiting a certain tourist attraction is a random variable with $\mu = 1,200$ and $\sigma = 80$.
   (a) What does Chebyshev's theorem with $k = 7$ tell us about the number of persons who will visit the tourist attraction on an August day?
   (b) According to Chebyshev's theorem, with what probability can we assert that between 1,000 and 1,400 persons will visit the tourist attraction on an August day?

★ **8.100** The size of an animal population is sometimes estimated by the **capture–recapture method**. In this method, n_1 of the animals are captured, marked, and released. Later, n_2 of the animals are captured, x of them are found to be marked, and all this information is used to estimate N, the size of the population. If $n_1 = 3$ rare owls are captured, marked, and released, and later $n_2 = 4$ such owls are captured and $x = 1$ of them is found to be marked, for what value of N is the probability of getting $x = 1$ a maximum? (*Hint*: Try $N = 9, 10, 11, 12, 13,$ and 14.)

8.101 A restaurant manager wants to review the calculations on the diner checks presented to his customers. During the previous week, 1,428 of these checks were written, and 35 of these have errors in prices or in addition. The manager selects 50 of these at random to review, and we would like to calculate the probability of finding exactly 2 with errors.
   (a) Write the expression which gives the exact probability, using the hypergeometric distribution. Do not complete the calculation.
   (b) Use the binomial distribution as an approximation to obtain the probability of finding exactly 2 errors.
   (c) Use the Poisson distribution as an approximation to obtain the probability of finding exactly 2 errors.

8.102 Among 600 plants exposed to excessive radiation, 90 show abnormal growth. If a scientist collects the seed of three of the plants chosen at random, find the probability that he will get the seed from one plant with abnormal growth and two plants with normal growth by using
(a) the hypergeometric distribution;
(b) the binomial distribution as an approximation.

8.103 The probabilities that 0, 1, 2, 3, 4, 5, or 6 fires will break out in a national forest on a July morning are 0.120, 0.207, 0.358, 0.162, 0.096, 0.043, and 0.014. Find the mean and the variance of this probability distribution.

⋆ **8.104** If 12 of 18 medieval gold coins are genuine, 4 are gold-plated counterfeits, and 2 are pure brass counterfeits, and if three of these coins are randomly selected to be sold at auction, what are the probabilities that
(a) all three coins will be genuine;
(b) two of the coins will be genuine;
(c) one coin will be genuine;
(d) one coin will be genuine, one will be a gold-plated counterfeit, and one will be a pure brass counterfeit?

⋆ **8.105** The probabilities are 0.70, 0.20, and 0.10 that an atuomobile using a certain road will have tires which can be classified as adequate, worn, and dangerous, respectively. Ten cars are stopped in a random safety check. Find the probabilities that
(a) at least 8 of the 10 will have adequate tires,
(b) at most 2 of the 10 will have dangerous tires;

(c) 6 will have adequate tires, 3 will have worn tires, and 1 will have dangerous tires.

8.106 Determine whether the following can be probability distributions (defined in each case only for the given values of x) and explain your answers:

(a) $f(x) = \dfrac{x}{10}$ for $x = 0, 1, 2, 3, 4$;

(b) $f(x) = \dfrac{x-1}{2}$ for $x = 0, 1, 2, 3$;

(c) $f(x) = \dfrac{1}{5}\dbinom{2}{x}$ for $x = 0, 1,$ and 2.

8.107 A study shows that 95 percent of the families in a certain suburban area have a color television set. Use Table V to find the probabilities that among 16 families randomly selected in this area
(a) 14 will have a color television set;
(b) at least 14 will have a color television set;
(c) at most 14 will have a color television set.

8.108 The number of blossoms on a rare plant is a random variable having the Poisson distribution with $\lambda = 2.4$. What are the probabilities that such a plant will have
(a) no blossoms;
(b) at most two blossoms?

8.109 Find the mean and the standard deviation of the binomial distribution with $n = 324$ and $p = 0.50$.

8.110 According to Chebyshev's theorem, with what probability can we assert that in 40,000 flips of a balanced coin the proportion of heads will be between 0.45 and 0.55?

8.12
REFERENCES

A great deal of information about various probability distributions may be found in

HASTINGS, N. A. J., and PEACOCK, J. B., *Statistical Distributions.* London: Butterworth & Company (Publishers) Ltd, 1975.

More detailed tables of binomial probabilities may be found in

ROMIG, H. G., *50–100 Binomial Tables.* New York: John Wiley & Sons, Inc., 1953.
Tables of the Binomial Probability Distribution, National Bureau of Standards Applied

Mathematics Series No. 6. Washington, D.C.: U.S. Government Printing Office, 1950.

and a detailed table of Poisson probabilities is given in

Molina, E. C., *Poisson's Exponential Binomial Limit*. Princeton, N.J.: D. Van Nostrand Company, Inc., 1947.

The wide availability of computer programs for binomial and Poisson probabilities makes it unlikely that the above-named tables will be extended or updated.

Among the many published tables of random numbers, one of the most widely used is

RAND Corporation, *A Million Random Digits with 100,000 Normal Deviates*. New York: Macmillan Publishing Co., Inc., third printing 1966.

There also exist calculators which are preprogrammed to generate random numbers and it is fairly easy to program a computer so that a person can generate his or her own random numbers. The following is one of many articles on this subject.

Kimberling, C., "Generate Your Own Random Numbers." *Mathematics Teacher*, February 1984.

A book discussing the use of random numbers in programming is

Hennefeld, J., *Using Microsoft and IBM BASIC*, Boston: PWS Publishers, 1987.

For more information about simulation and Monte Carlo methods, see

Rubinstein, R. Y., *Simulation and the Monte Carlo Method*. New York: John Wiley & Sons, Inc., 1981.

Travers, K. J., "Using Monte Carlo Methods to Teach Probability and Statistics." in *Teaching Statistics and Probability*, 1981 Yearbook of the National Council of Teachers of Mathematics. Reston, Va.: NCTM, 1981.

THE NORMAL DISTRIBUTION

Continuous sample spaces and **continuous random variables** arise when we deal with quantities that are measured on a continuous scale—for instance, when we measure the speed of a car, the amount of alcohol in a person's blood, the net weight of a package of frozen food, or the amount of tar in a cigarette. Although there exist continuums of possibilities in situations like these, we nearly always round measurements to the nearest whole unit or to a few decimals. In actual practice, we shall thus be interested in probabilities associated with intervals or regions of a sample space, and not in probabilities associated with individual points. For instance, we may want to know the probability that at a given time a car is moving between 50 and 55 miles per hour (not at exactly $16\pi = 50.26548246\ldots$ miles per hour), or that a package of frozen food weighs more than 5.95 ounces (not exactly $\sqrt{35.9} = 5.99166087\ldots$ ounces).

In this chapter we shall learn how to determine, and work with, probabilities relating to continuous random variables. The place of histograms will be taken by continuous curves, as in Figure 9.1; we can picture them mentally as being approached by histograms with narrower and narrower classes.

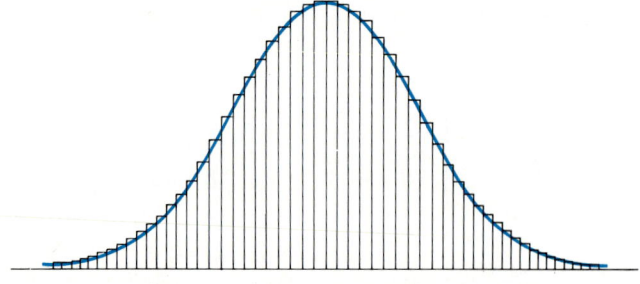

FIGURE 9.1 *Continuous distribution curve.*

After a general introduction to **continuous distributuons** in Section 9.1, we devote the remainder of the chapter to the **normal distribution**, which is basic to most of the "bread and butter" techniques of modern statistics. Various applications are discussed in Sections 9.4 and 9.5, and the optional material in Sections 9.3 and 9.6 cover a test to decide whether observed data follow the general pattern of the normal distribution and the simulation of normally distributed random variables.

9.1
CONTINUOUS DISTRIBUTIONS

In all the histograms we saw in previous chapters, the frequencies, percentages, proportions, or probabilities were represented by the heights of the rectangles, or by their areas. In the continuous case, we also represent probabilities by areas, but not by areas of rectangles—they are given by areas under continuous curves. This is illustrated by Figure 9.2, where the diagram on the left shows a histogram of the probability distribution of a discrete random variable which takes on only the

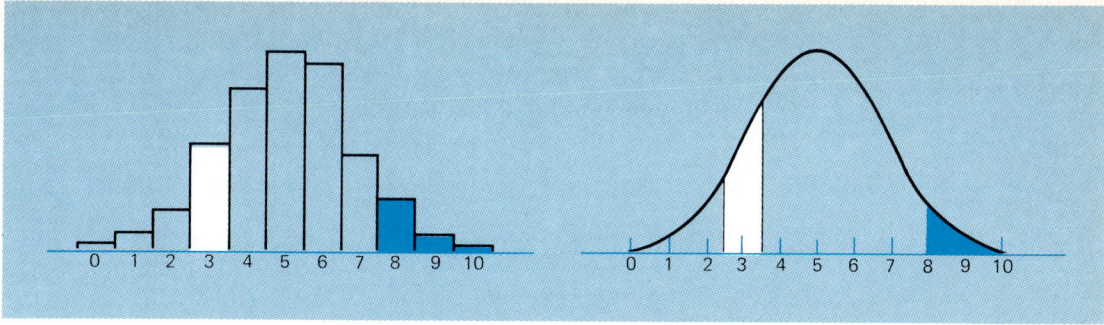

FIGURE 9.2 *Histogram of probability distribution and graph of continuous distribution.*

values 0, 1, 2, …, and 10. The probability that it will take on the value 3, for example, is given by the area of the white rectangle, and the probability that it will take on a value greater than or equal to 8 is given by the sum of the areas of the three tinted rectangles on the right-hand side. The other diagram of Figure 9.2 refers to a continuous random variable which can take on any value on the interval from 0 to 10. The probability that it will take on a value on the interval from 2.5 to 3.5, for example, is given by the area of the white region under the curve, and the probability that it will take on a value greater than or equal to 8 is given by the area of the tinted region under the curve on the right-hand side.

Continuous curves such as the one shown on the right in Figure 9.2 are the graphs of functions called **probability densities**, or informally, **continuous distributions**. The term "probability density" comes from physics, where the terms "weight" and "density" are used in just about the same way in which we use the terms "probability" and "probability density" in statistics. As is illustrated by Figure 9.3, probability densities are characterized by the fact that

The area under the curve between any two values *a* and *b* gives the probability that a random variable having this continuous distribution will take on a value on the interval from *a* to *b*.

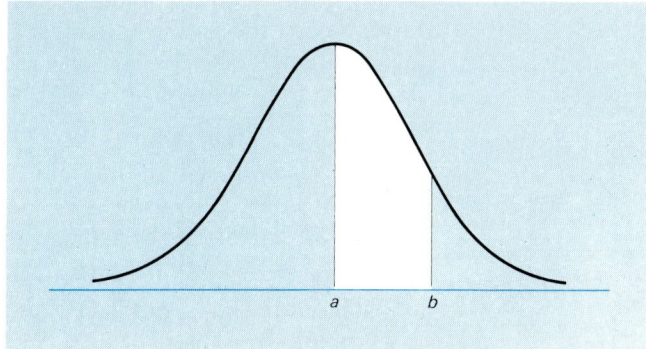

FIGURE 9.3 *Continuous distribution.*

It follows that the values of a continuous distribution must be non-negative, and that the total area under the curve, representing the certainty that a random variable must take on one of its values, is always equal to 1. This parallels the two rules about probability distributions which we gave on page 182.

EXAMPLE   Verify that $f(x) = \dfrac{x}{8}$ can serve as the probability density of a continuous random variable which can take on any value on the interval from 0 to 4.

Solution   The first condition is satisfied since $\dfrac{x}{8}$ is 0 for $x = 0$, and positive for any other value of x on the interval from 0 to 4. So far as the second condition is concerned, it can be seen from Figure 9.4 that the total area under the curve is that of a triangle, and that the base and the height of the triangle are, respectively, 4 and $\frac{4}{8} = \frac{1}{2}$. The usual formula for the area of a triangle yields $\frac{1}{2} \cdot 4 \cdot \frac{1}{2} = 1$.

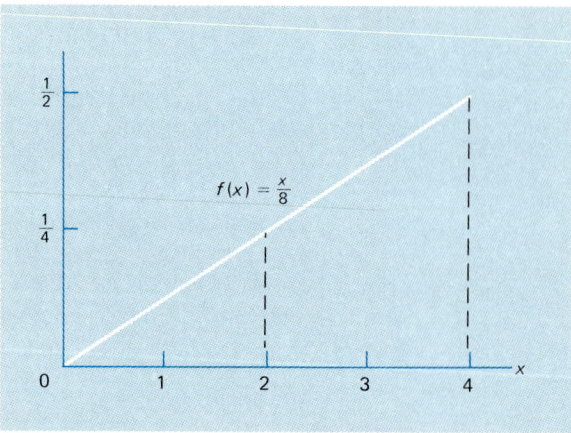

FIGURE 9.4 *A probability density.*

EXAMPLE   With reference to the preceding example, find the probabilities that a random variable having the given probability density will take on a value
   (a) less than 2;
   (b) less than or equal to 2.

Solution   **(a)** The probability is given by the area of the smaller triangle of Figure 9.4, which is bounded on the right by the dashed line at $x = 2$. Its base is 2, its height is $\frac{2}{8} = \frac{1}{4}$, so its area is $\frac{1}{2} \cdot 2 \cdot \frac{1}{4} = \frac{1}{4}$.
   (b) The probability is the same as that of part (a), namely, $\frac{1}{4}$.

This illustrates the important fact that in the continuous case the probability is zero that a random variable will take on any particular value. In our example, the probability is zero that the random variable will take on the value 2, and by 2 we mean *exactly* 2, which does not include nearby values such as 1.9999998 or 2.0000001.

It is a consequence of measuring (rather than counting) that we must assign probability zero to any particular outcome. We claim that the probability must be zero that an individual will have a weight of *exactly* 145.27 pounds or that a horse will run a race in *exactly* 58.442 seconds. Observe however, that even though every particular outcome has probability zero, the process will still produce a value (whether or not we can measure it with extra-fine precision); apparently then events of probability zero not only can occur but must occur when dealing with measured random quantities.

Statistical descriptions of continuous distributions are as important as descriptions of probability distributions or distributions of observed data, but most of them, including the mean and the standard deviation, cannot be defined without using calculus. Informally, though, we can always picture continuous distributions as being approximated by histograms of probability distributions (see Figure 9.1), whose mean and standard deviation we can calculate. Then, if we choose histograms with narrower and narrower classes, the means and the standard deviations of the corresponding probability distributions will approach the mean and the standard deviation of the continuous distribution. Actually, the mean and the standard deviation of a continuous distribution measure the same properties as the mean and the standard deviation of a probability distribution—the expected value of a random variable having the given distribution, and the expected value of its squared deviations from the mean. More intuitively, the mean μ of a continuous distribution is a measure of its "center" or "middle," and the standard deviation σ of a continuous distribution is a measure of its "dispersion" or "spread."

9.2
THE NORMAL DISTRIBUTION

Among the many different continuous distributions used in statistics, the most important one is the **normal distribution**. Its study dates back to eighteenth-century investigations into the nature of errors of measurement. It was observed that discrepancies among repeated measurements of the same physical quantity displayed a surprising degree of regularity; their pattern (distribution), it was found, could be closely approximated by a certain kind of continuous distribution curve, referred to as the "normal curve of errors" and attributed to the laws of chance. The mathematical equation of this kind of curve is

$$f(x) = \frac{1}{\sigma\sqrt{2\pi}} e^{-\frac{1}{2}\left(\frac{x-\mu}{\sigma}\right)^2}$$

for $-\infty < x < \infty$, where e is the irrational number $2.71828\ldots$ we met on page 195 in connection with the Poisson distribution. We have given this equation only to point out some of the key features of normal distributions; it will not be used in any of our calculations.

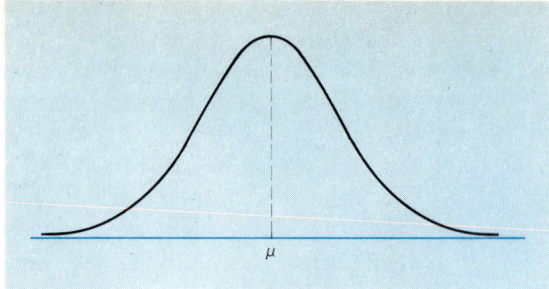

FIGURE 9.5 *Normal distribution curve.*

The graph of a normal distribution is a bell-shaped curve that extends indefinitely in both directions. Although this may not be apparent from a small drawing like that of Figure 9.5, the curve comes closer and closer to the horizontal axis without ever reaching it, no matter how far we go in either direction away from the mean. Fortunately, it is seldom necessary to extend the tails of a normal distribution very far because the area under the curve more than four or five standard deviations away from the mean is negligible for most practical purposes (see Exercise 9.12 on page 231).

An important feature of normal distributions, apparent from the equation above, is that they depend only on the two quantities μ and σ, which are, indeed, the mean and the standard deviation. In other words, there is one and only one normal distribution with a given mean μ and a given standard deviation σ. The fact that we will get different curves depending on the values of μ and σ is illustrated by Figure 9.6 on page 225. At the top there are two normal curves with unequal means but equal standard deviations; in the middle there are two normal curves with equal means but unequal standard deviations, and at the bottom there are two normal curves with unequal means and unequal standard deviations.

In all our work with normal distributions, we shall be concerned only with areas under their curves—so-called **normal-curve areas**—and such areas are found in practice from tables such as Table I at the end of the book. As it is physically impossible, but also unnecessary, to construct separate tables of normal-curve areas for all conceivable pairs of values of μ and σ, we tabulate these areas only for the normal distribution with $\mu = 0$ and $\sigma = 1$, called the **standard normal distribution**. Then, we obtain areas under any normal curve by performing the change of scale (see Figure 9.7 on page 220) which converts the units of measurement from the original scale, or x-scale, into **standard units, standard scores,** or **z-scores,** by means of the formula

Standard units

$$z = \frac{x - \mu}{\sigma}$$

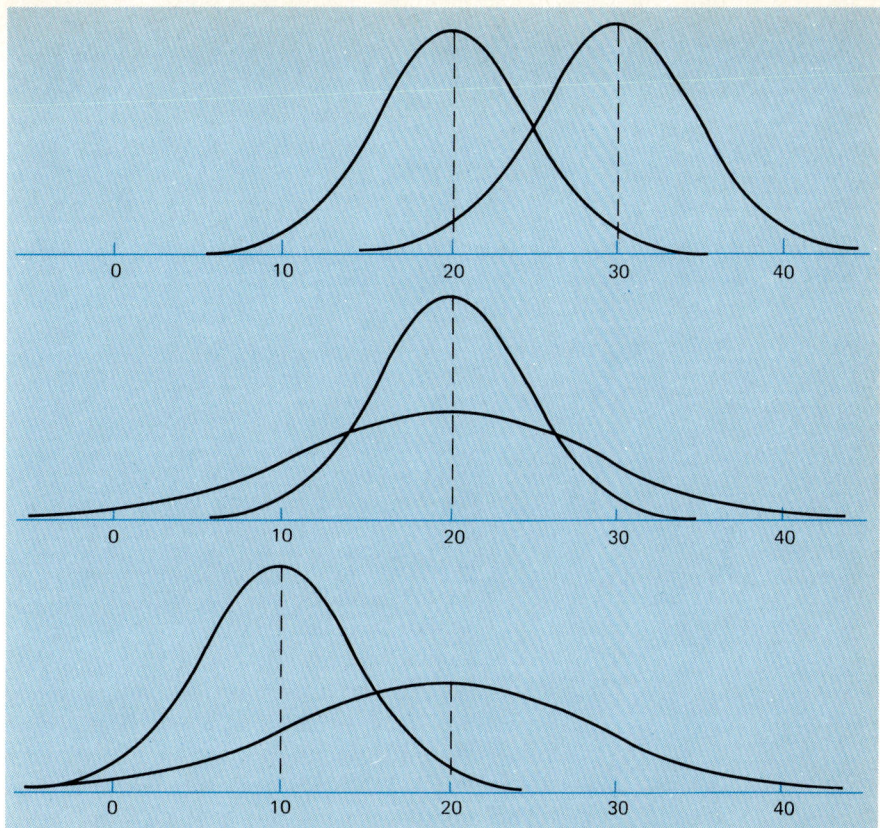

FIGURE 9.6 *Three pairs of normal distributions.*

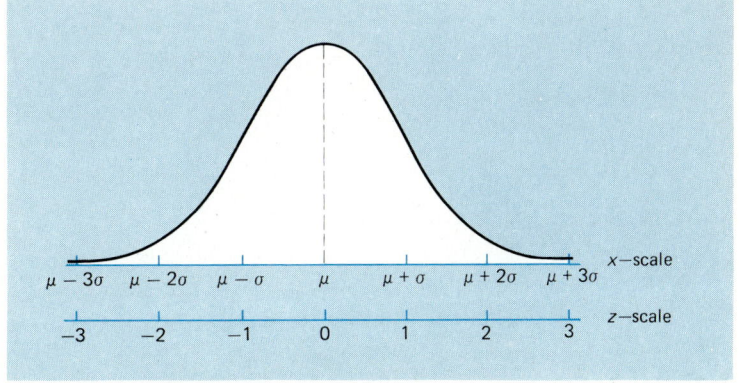

FIGURE 9.7 *Change of scale to standard units.*

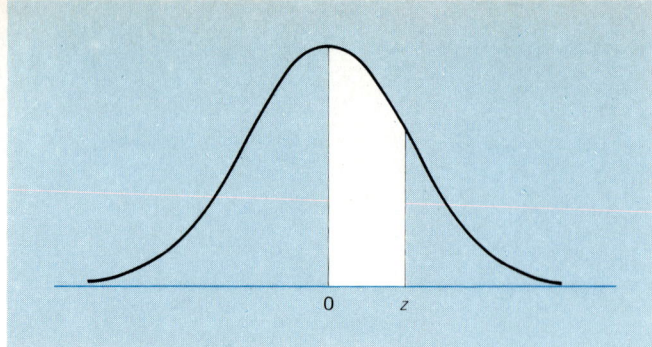

FIGURE 9.8 *Tabulated normal-curve areas.*

In this new scale, the z-scale, a value of z simply tells us how many standard deviations the corresponding value of x lies above or below the mean of its distribution.

The entries in Table I are the areas under the standard normal curve between the mean $z = 0$ and $z = 0.00, 0.01, 0.02, \ldots, 3.08$, and 3.09, and also $z = 4.00$, $z = 5.00$, and $z = 6.00$. In other words, the entries in Table I are normal-curve areas like that of the white region of Figure 9.8.

Table I has no entries corresponding to negative values of z, for these are not needed by virtue of the symmetry of any normal curve about its mean. [This follows from the equation on page 223, which remains unchanged when we substitute $-(x - \mu)$ for $x - \mu$. In other words, we get the same value for $f(x)$ when we go the distance $x - \mu$ to the left or to the right of μ.]

EXAMPLE   Find the standard-normal-curve area between $z = -1.20$ and $z = 0$.

Solution   As can be seen from Figure 9.9, the area under the curve between $z = -1.20$ and $z = 0$ equals the area under the curve between $z = 0$ and $z = 1.20$. So, we look up the entry corresponding to $z = 1.20$ in Table I and we get 0.3849.

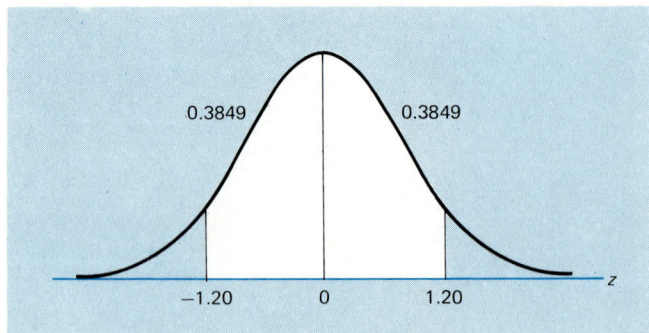

FIGURE 9.9 *Symmetry of normal distribution.*

Questions concerning areas under normal distributions arise in various ways, and the ability to find any desired area quickly can be a big help. Although the table gives only areas between $z = 0$ and selected positive values of z, we often have to find areas to the left or to the right of given positive or negative values of z, or areas between two given values of z. This is easy, provided we remember exactly what areas are represented by the entries in Table I, and also that the standard normal distribution is symmetrical about $z = 0$, so that the area under the curve to the left of $z = 0$ and that to the right of $z = 0$ are both equal to 0.5000.

EXAMPLE Find the standard-normal-curve area
  (a) to the left of $z = 0.94$;
  (b) to the right of $z = -0.65$;
  (c) to the right of $z = 1.76$;
  (d) to the left of $z = -0.85$;
  (e) between $z = 0.87$ and $z = 1.28$;
  (f) between $z = -0.34$ and $z = 0.62$.

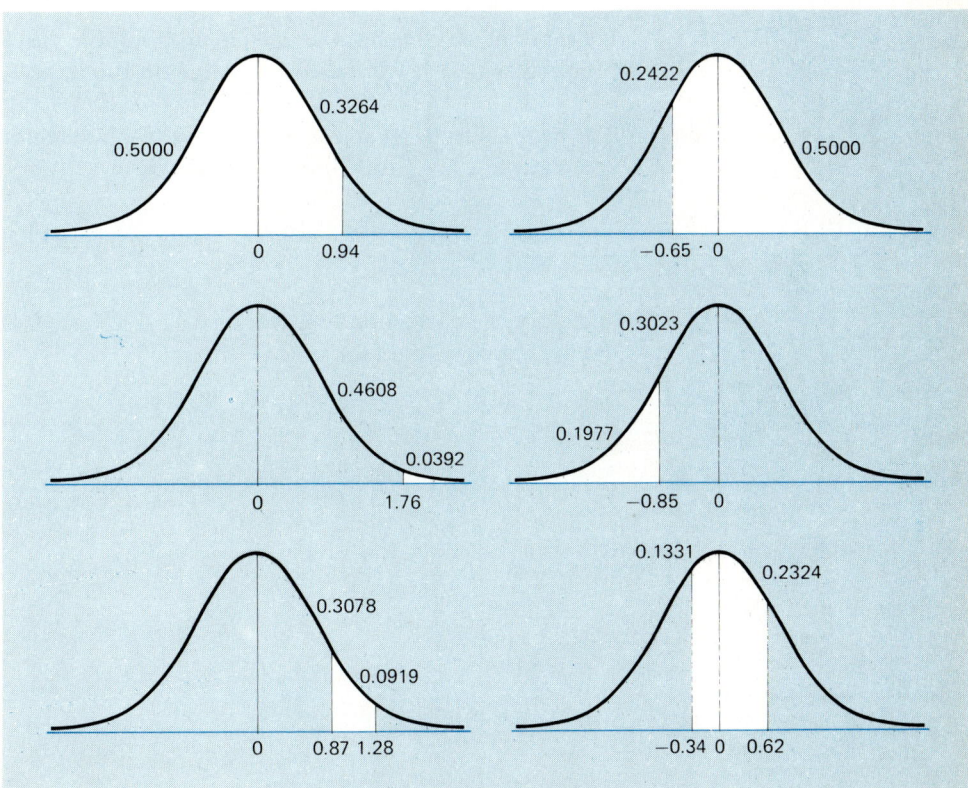

FIGURE 9.10 *Standard-normal-curve areas.*

Solution For each part refer to the corresponding diagram in Figure 9.10.

(a) The area to the left of $z = 0.94$ is 0.5000 plus the entry in Table I corresponding to $z = 0.94$, or $0.5000 + 0.3264 = 0.8264$.

(b) The area to the right of $z = -0.65$ is 0.5000 plus the entry in Table I corresponding to $z = 0.65$, or $0.5000 + 0.2422 = 0.7422$.

(c) The area to the right of $z = 1.76$ is 0.5000 minus the entry in Table I corresponding to $z = 1.76$, or $0.5000 - 0.4608 = 0.0392$.

(d) The area to the left of $z = -0.85$ is 0.5000 minus the entry in Table I corresponding to $z = 0.85$, or $0.5000 - 0.3023 = 0.1977$.

(e) The area between $z = 0.87$ and $z = 1.28$ is the difference between the entries in Table I corresponding to $z = 0.87$ and $z = 1.28$, or $0.3997 - 0.3078 = 0.0919$.

(f) The area between $z = -0.34$ and $z = 0.62$ is the sum of the entries in Table I corresponding to $z = 0.34$ and $z = 0.62$, or $0.1331 + 0.2324 = 0.3655$.

In both of the preceding examples we dealt directly with the standard normal distribution. Now let us consider an example where μ and σ are not 0 and 1, so that we must first convert to standard units.

EXAMPLE If a random variable has the normal distribution with $\mu = 10$ and $\sigma = 5$, what is the probability that it will take on a value on the interval from 12 to 15?

Solution The probability is given by the area of the white region of Figure 9.11. Converting $x = 12$ and $x = 15$ to standard units, we get

$$z = \frac{12 - 10}{5} = 0.40 \quad \text{and} \quad z = \frac{15 - 10}{5} = 1.00$$

and since the corresponding entries in Table I are 0.1554 and 0.3413, we find that the probability asked for in this example is $0.3413 - 0.1554 = 0.1859$.

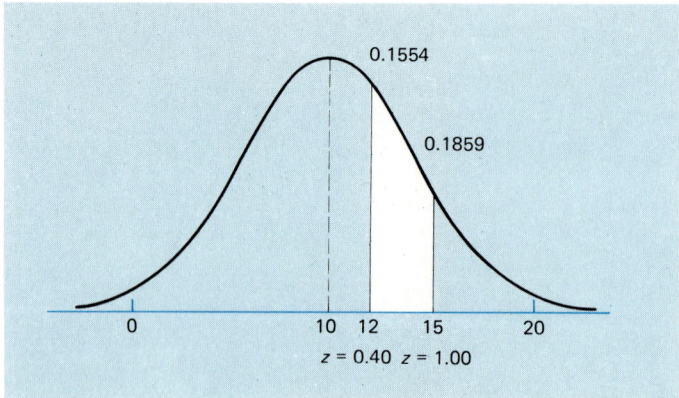

FIGURE 9.11 *Normal-curve area.*

There are also problems in which we are given areas under normal curves and asked to find the corresponding values of z. The results of the example which follows will be used extensively in subsequent chapters.

EXAMPLE If z_α denotes the value of z for which the standard-normal-curve area to its right is equal to α (Greek lowercase *alpha*), find

(a) $z_{0.01}$;

(b) $z_{0.05}$.

Solution For both parts refer to the diagram of Figure 9.12.

(a) Since $z_{0.01}$ corresponds to an entry of $0.5000 - 0.0100 = 0.4900$ in Table I, we look for the entry closest to 0.4900 and find 0.4901 corresponding to $z = 2.33$; thus, we let $z_{0.01} = 2.33$.

(b) Since $z_{0.05}$ corresponds to an entry of $0.5000 - 0.0500 = 0.4500$ in Table I, we look for the entry closest to 0.4500 and find 0.4495 and 0.4505 corresponding to $z = 1.64$ and $z = 1.65$. Since 0.4500 is precisely halfway between these values we let $z_{0.05} = 1.645$.

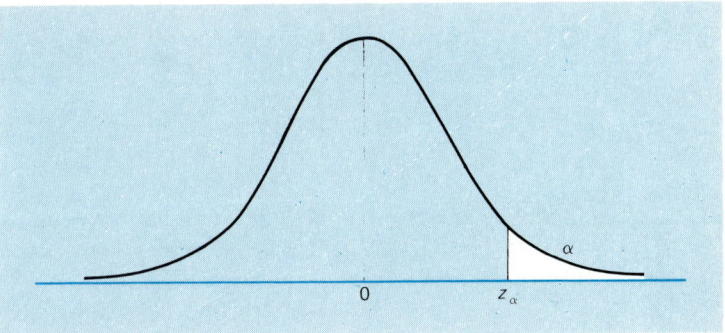

FIGURE 9.12 *Determination of z_α.*

Table I also enables us to verify the remark on page 79 that for frequency distributions having the general shape of the cross section of a bell, about 68 percent of the values will lie within one standard deviation of the mean, about 95 percent will lie within two standard deviations of the mean, and about 99.7 percent will lie within three standard deviations of the mean. These percentages apply to frequency distributions having the general shape of a normal distribution, and the reader will be asked to verify them in the first three parts of Exercise 9.12. The other two parts of that exercise show that, although the "tails" extend indefinitely in both directions, the standard-normal-curve area to the right of $z = 4$ or $z = 5$, or to the left of $z = -4$ or $z = -5$, is negligible.

Although this chapter is devoted to the normal distribution and its importance cannot be denied, it would be a serious mistake to think that the normal distribution is the only continuous distribution that matters in the study of statistics. In Chapter 11 and in subsequent chapters we shall meet other continuous distributions which play important roles in problems of statistical inference.

EXERCISES

9.1 Explain in each case why the given equation cannot serve as that of the probability density of a continuous random variable which takes on values on the interval from 1 to 4:

(a) $f(x) = \frac{1}{4}$ for $1 \leq x \leq 4$;

(b) $f(x) = \frac{2}{15}(4x - 7)$ for $1 \leq x \leq 4$.

9.2 Figure 9.13 shows the graph of the **uniform distribution** of a random variable which takes on values on the interval from 2 to 10. Find the probabilities that this random variable will take on a value

(a) less than 7;

(b) between 2.4 and 8.8.

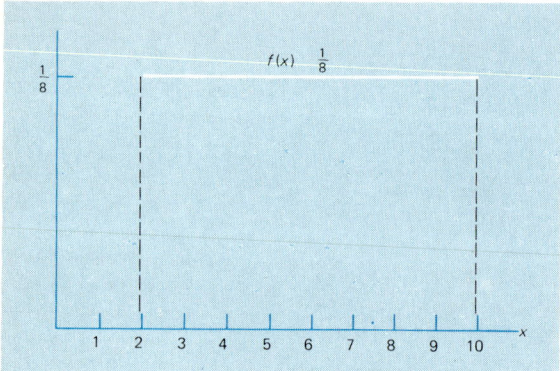

FIGURE 9.13 *Diagram for Exercise 9.2.*

9.3 With reference to the example on page 222 and Figure 9.4, what is the probability that the random variable will take on a value greater than 3?

9.4 Figure 9.14 shows the graph of the distribution of a continuous random variable which takes on values on

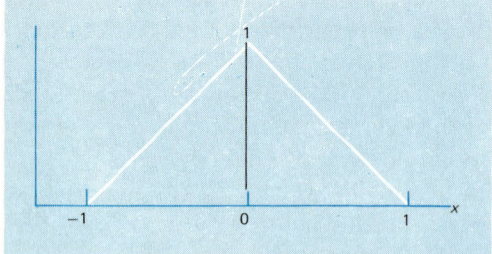

FIGURE 9.14 *Diagram for Exercise 9.4.*

the interval from -1 and 1. Find the probabilities that this random variable will take on a value

(a) between $\frac{1}{2}$ and 1;

(b) between $-\frac{1}{3}$ and $\frac{1}{3}$.

9.5 For each of the cases cited below, involving areas under the standard normal curve, decide whether the first area is bigger, the second area is bigger, or the two areas are equal:

(a) the area to the right of $z = 1.5$ or the area to the right of $z = 2$;

(b) the area to the left of $z = -1.5$ or the area to the left of $z = -2$;

(c) the area to the right of $z = 1$ or the area to the left of $z = -1.5$;

(d) the area to the right of $z = 2$ or the area to the left of $z = -2$;

(e) the area to the right of $z = -2.5$ or the area to the right of $z = -1.5$;

(f) the area to the left of $z = 0$ or the area to the right of $z = -0.1$;

(g) the area to the right of $z = 0$ or the area to the left of $z = 0$;

(h) the area to the right of $z = -1.4$ or the area to the left of $z = -1.4$.

9.6 For each of the cases cited below, involving areas under the standard normal curve, decide whether the first area is bigger, the second area is bigger, or the two areas are equal:

(a) the area between $z = 0$ and $z = 1.3$ or the area between $z = 0$ and $z = 1$;

(b) the area between $z = -0.2$ and $z = 0.2$ or the area between $z = -0.4$ and $z = 0.4$;

(c) the area between $z = -1$ and $z = 1$ or the area between $z = 0$ and $z = 2$;

(d) the area between $z = 0.1$ and $z = 0.2$ or the area between $z = 1.1$ and $z = 1.2$;

(e) the area between $z = -1$ and $z = -0.5$ or the area between $z = 0.5$ and $z = 1$;

(f) the area to the left of $z = -1.5$ or the area to the right of $z = -0.5$;

(g) the area between $z = -1$ and $z = 1.5$ or the area between $z = -1.5$ and $z = 1$;

(h) the area between $z = 1$ and $z = 2$ or the area between $z = 2$ and $z = 4$.

9.7 Find the standard-normal-curve area which lies
  (a) between $z = 0$ and $z = 0.87$;
  (b) between $z = -1.66$ and $z = 0$;
  (c) to the right of $z = 0.48$;
  (d) to the right of $z = -0.27$;
  (e) to the left of $z = 1.30$;
  (f) to the left of $z = -0.79$.

9.8 Find the standard-normal-curve area which lies
  (a) between $z = 0.55$ and $z = 1.12$;
  (b) between $z = -1.75$ and $z = -1.05$;
  (c) between $z = -1.95$ and $z = 0.44$.

9.9 Find the standard-normal-curve area which lies
  (a) between $z = -0.70$ and $z = 0.92$;
  (b) to the right of $z = 1.15$;
  (c) to the left of $z = 0.22$;
  (d) between $z = 0.24$ and $z = 1.82$;
  (e) to the left of $z = -1.14$;
  (f) to the right of $z = -0.76$;
  (g) between $z = -1.82$ and $z = -0.79$.

9.10 Find z if the standard-normal-curve area
  (a) is the area between 0 and z is 0.1915;
  (b) to the left of z is 0.8078;
  (c) to the left of z is 0.0132;
  (d) between $-z$ and z is 0.8502.

9.11 Find z if the standard-normal-curve area
  (a) between 0 and z is 0.4306;
  (b) to the right of z is 0.7704;
  (c) to the right of z is 0.1314;
  (d) between $-z$ and z is 0.9700.

9.12 Find the standard-normal-curve area between $-z$ and z if
  (a) $z = 1$;
  (b) $z = 2$;
  (c) $z = 3$;
  (d) $z = 4$;
  (e) $z = 5$.

9.13 Verify that
  (a) $z_{0.025} = 1.96$;
  (b) $z_{0.005} = 2.575$.

9.14 For each of the cases cited below, involving random variables with normal distributions, decide whether the first probability is bigger, the second probability is bigger, or the two probabilities are equal:
  (a) for a random variable with a normal distribution with $\mu = 100$ and $\sigma = 20$, the probability of a value greater than 140 or the probability of a value greater than 130;
  (b) for a random variable with a normal distribution with $\mu = 80$ and $\sigma = 20$, the probability of a value greater than 100 or the probability of a value less than 70;
  (c) for a random variable with a normal distribution with $\mu = 60$ and $\sigma = 12$, the probability of a value between 48 and 72 or the probability of a value between 60 and 84;
  (d) for a random variable with a normal distribution with $\mu = 200$ and $\sigma = 40$, the probability of a value greater than 250 or the probability of a value less than 140.

9.15 For each of the cases cited below, involving random variables with normal distributions, decide whether the first probability is bigger, the second probability is bigger, or the two probabilities are equal:
  (a) the probability that a random variable having the normal distribution with $\mu = 50$ and $\sigma = 10$ takes a value less than 60 or the probability that a random variable having the normal distribution with $\mu = 500$ and $\sigma = 100$ takes a value less than 600;
  (b) the probability that a random variable having the normal distribution with $\mu = 40$ and $\sigma = 5$ takes a value greater than 40 or the probability that a random variable having the normal distribution with $\mu = 50$ and $\sigma = 5$ takes a value greater than 40;
  (c) the probability that a random variable having the normal distribution with $\mu = 50$ and $\sigma = 10$ takes a value less than 60 or the probability that a random variable having the normal distribution with $\mu = 50$ and $\sigma = 20$ takes a value less than 60;
  (d) the probability that a random variable having the normal distribution with $\mu = 100$ and $\sigma = 5$ takes a value greater than 110 or the probability that a random variable having the normal distribution with $\mu = 108$ and $\sigma = 5$ takes a value greater than 110;
  (e) the probability that a random variable having the normal distribution with $\mu = 5,000$ and $\sigma = 1,000$ takes a value between 4,000 and 6,000 or the probability that a random variable having the normal distribution with $\mu = 7,000$ and $\sigma = 1,000$ takes a value between 4,000 and 6,000.

9.16 If a random variable has the normal distribution with $\mu = 80.0$ and $\sigma = 4.8$, find the probabilities that it will take on a value

    (a) less than 87.2;
    (b) greater than 76.4;
    (c) between 81.2 and 86.0;
    (d) between 71.6 and 88.4.

9.17 If a random variable has the normal distribution with $\mu = 62.5$ and $\sigma = 12.4$, find the probabilities that it will take on a value

    (a) less than 40.1;
    (b) greater than 69.3;
    (c) between 65.0 and 75.0;
    (d) between 57.4 and 67.6.

9.18 A normal distribution has the mean $\mu = 62.4$. Find its standard deviation if 20 percent of the area under the curve lies to the right of 79.2.

9.19 A random variable has a normal distribution with $\sigma = 10$. If the probability that the random variable will take on a value less than 82.5 is 0.8212, what is the probability that it will take on a value greater than 58.3?

★ **9.20** Another continuous distribution, called the **exponential distribution**, has many important applications. If a random variable has an exponential distribution with the mean μ, the probability that it will take on a value between 0 and any given nonnegative value of x is $1 - e^{-x/\mu}$; see Figure 9.15. Here e is the irrational number which appears also in the formula for the normal distribution. Many calculators have keys for computing expressions of the form $e^{-x/\mu}$, and selected values may be obtained from Table XIII. Find the probabilities that a random variable having the exponential distribution with the mean $\mu = 10$ will take on a value

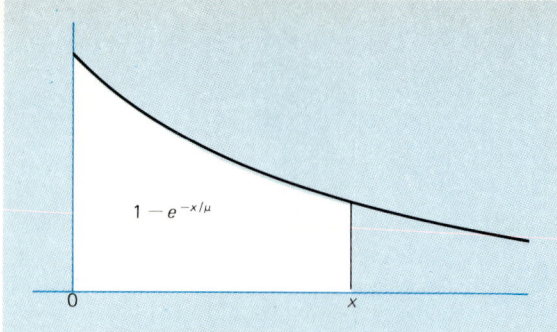

FIGURE 9.15 *Exponential distribution.*

    (a) less than 4;
    (b) between 5 and 9;
    (c) greater than 16.

★ **9.21** The lifetime of a certain electronic component is a random variable which has the exponential distribution with the mean $\mu = 2,000$ hours. Use the formula of Exercise 9.20 to find the probabilities that such a component will last

    (a) at most 2,400 hours;
    (b) at least 1,600 hours;
    (c) between 1,800 and 2,200 hours.

★ **9.22** According to medical research, the time between successive reports of a rare tropical disease is a random variable having the exponential distribution with the mean $\mu = 120$ days. Find the probabilities that the time between successive reports of the disease will

    (a) exceed 240 days;
    (b) exceed 360 days;
    (c) be less than 60 days.

9.3

A CHECK FOR "NORMALITY" ★

There are various ways in which we can test whether an observed distribution has roughly the shape of a normal distribution. The one we shall present here is crude and largely subjective, but it has the decided advantage that it is very easy to perform.

    To illustrate this technique, let us refer again to the sulfur oxides emission data which we used as an example in the early chapters of the book. First we convert the cumulative frequencies of the table on page 25 into cumulative percentages by dividing each one by 80, the total frequency, and then multiplying by 100. This yields

Tons of sulfur oxides	Cumulative percentage
Less than 4.95	0.00
Less than 8.95	3.75
Less than 12.95	16.25
Less than 16.95	33.75
Less than 20.95	65.00
Less than 24.95	86.25
Less than 28.95	97.50
Less than 32.95	100.00

where we show the class boundaries instead of the class limits, although this does not really matter unless we continue with the analysis as in Exercise 9.27.

Before we plot this cumulative percentage distribution on the special graph paper shown in Figure 9.16, let us briefly examine its scales. When such graph

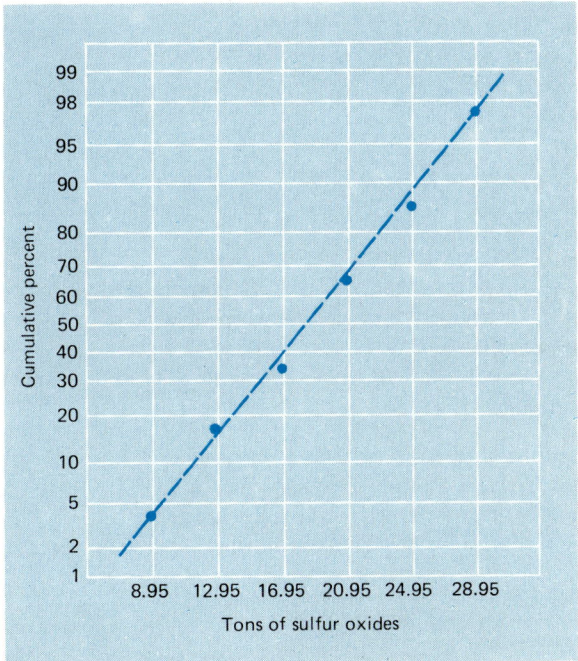

FIGURE 9.16 *Normal probability paper*

paper is purchased commercially, the cumulative percentage scale is already printed in the special way which makes it suitable for our purpose. The other scale consists of equal subdivisions. This kind of graph is called **normal probability paper**, or **arithmetic probability paper**, and it should be available at most college or university bookstores.

Once the cumulative "less than" percentages have been plotted as in Figure 9.16, we use the following criterion:

If the points follow pretty much the pattern of a straight line, we consider this as positive evidence that the distribution has roughly the shape of a normal distribution.

Of course, "pretty much" and "roughly" are not very precise terms, but we pointed out from the start that this is a crude and largely subjective, though easy-to-perform, technique. The most common pattern in which the distribution would be declared "not normal" is one where the rightmost points fall below the straight line determined by the balance of the points. A more rigorous way of checking the "normality" of a distribution of observed data is explained in Exercise 13.88 on page 386.

Returning to Figure 9.16, we find that the points are all close to the dashed line, and we conclude that the distribution of the sulfur oxides emission data has roughly the shape of a normal distribution. Note that in Figure 9.16 we did not plot the cumulative percentages corresponding to 4.95 and 32.95. As we pointed out on page 224, we never quite reach 0 or 100 percent of the area under a normal curve, no matter how far away from the mean we go in either direction.

EXERCISES

★ **9.23** Use normal probability paper to check whether the distribution of Exercise 2.25 on page 28, which deals with the numbers of customers a restaurant serves for lunch on 120 weekdays, has roughly the shape of a normal distribution.

★ **9.24** Use normal probability paper to check whether the distribution of Exercise 4.33 on page 88, which deals with the lengths of the downtimes of a certain machine, has roughly the shape of a normal distribution.

★ **9.25** The following is the distribution of the amounts of time that 200 persons required to complete a certain job application:

Time (minutes)	Number of persons
24 or less	15
25–29	50
30–34	75
35–39	40
40–44	15
45 or over	5

Use normal probability paper to check whether this distribution has roughly the shape of a normal distribution.

★ 9.26 The following is the distribution of the numbers of inquiries a realty firm received about 500 pieces of property:

Number of inquiries	Frequency
3– 6	55
7–10	227
11–14	170
15–18	42
19–22	6

Use normal probability paper to check whether this distribution has roughly the shape of a normal distribution.

★ 9.27 Normal probability paper can also be used to obtain crude estimates of the mean and the standard deviation of a distribution which has roughly the shape of a normal distribution. To estimate the mean, we observe that 50 percent of the area under any normal curve lies to the left of the mean. Hence, if we check the 50 percent mark on the cumulative percentage scale and go horizontally to the line we fit to the points (for instance, the dashed line of Figure 9.16), then the corresponding value on the horizontal scale provides an estimate of the mean of the distribution. To estimate the standard deviation, we observe that the areas under the curve to the left of $z = -1$ and $z = +1$ are roughly 0.16 and 0.84. Hence, if we check 16 percent and 84 percent on the vertical scale, we can judge by the straight line we have fitted to the points what values on the horizontal scale correspond to $z = -1$ and $z = +1$; their difference divided by 2 provides an estimate of the standard deviation of the distribution. Use this method to estimate the mean and the standard deviation of the sulfur oxides emission data from Figure 9.16. Compare with $\bar{x} = 18.85$ and $s = 5.55$ obtained in Chapters 3 and 4.

★ 9.28 Use the method of Exercise 9.27 to estimate the mean and the standard deviation of the distribution of Exercise 4.43 on page 88, and compare the results with the values of $\bar{x}$ and s obtained in that exercise.

★ 9.29 Use the method of Exercise 9.27 to estimate the mean and the standard deviation of the distribution of Exercise 9.25.

★ 9.30 Normal probability paper can also be used for ungrouped data. Begin by sorting the data values into increasing order. Then one point is plotted for each data value; the point corresponding to the ith largest of n values is given cumulative percentage $\dfrac{i}{n+1} \cdot 100$ on the vertical scale. For example, the 15th largest of 86 values is given a cumulative percentage of $\dfrac{15}{86+1} \cdot 100 = 17.24$.

If n is large, say 30 or more, then the points will be very close together, and thus not all the points need to be plotted. As a reasonable guideline, the points for the ten smallest and ten largest values should be plotted, along with every tenth value between these.

Examine the following set of $n = 39$ data values, which have been sorted, and use normal probability paper to assess whether these seem to have roughly the shape of a normal distribution.

51.5	57.9	63.4	63.6	63.9
64.9	66.6	69.8	69.8	70.1
70.2	72.9	73.1	74.0	74.4
75.4	75.6	76.5	77.6	77.9
79.3	79.7	79.8	80.0	82.0
82.0	82.5	83.0	83.9	84.4
85.0	86.4	89.3	91.4	93.6
94.0	94.0	95.3	102.6	

★ 9.31 The following $n = 49$ values, which have been sorted, were obtained as a random sample.

2.1	2.9	3.6	5.5	6.9
7.4	7.6	9.0	9.5	11.2
11.6	12.9	14.8	15.2	17.6
18.2	19.0	19.2	21.0	21.1
36.1	37.5	42.8	44.6	44.9
47.6	49.5	49.7	54.0	64.4
70.6	79.6	80.0	87.1	89.6
96.8	103.7	104.9	105.3	107.4
107.5	121.6	128.1	132.1	133.4
138.5	167.0	181.0	257.3	

Use normal probability paper to assess whether these data seem to have roughly the shape of a normal distribution. (See the previous exercise.)

9.4
APPLICATIONS OF THE NORMAL DISTRIBUTION

Let us now consider some applications, where it will be assumed in each case that the distribution of the data, or the distribution of the random variable under consideration, can be approximated closely with a normal distribution.

EXAMPLE   If the amount of cosmic radiation to which a person is exposed while flying by jet across the United States is a random variable having a normal distribution with $\mu = 4.35$ mrem and $\sigma = 0.59$ mrem, find the probabilities that a person on such a flight will be exposed to

    (a) more than 5.00 mrem of cosmic radiation;

    (b) anywhere from 3.00 to 4.00 mrem of cosmic radiation.

Solution   **(a)** This probability is given by the area of the white region of the upper diagram of Figure 9.17, namely, the area under the curve to the right of

$$z = \frac{5.00 - 4.35}{0.59} = 1.10$$

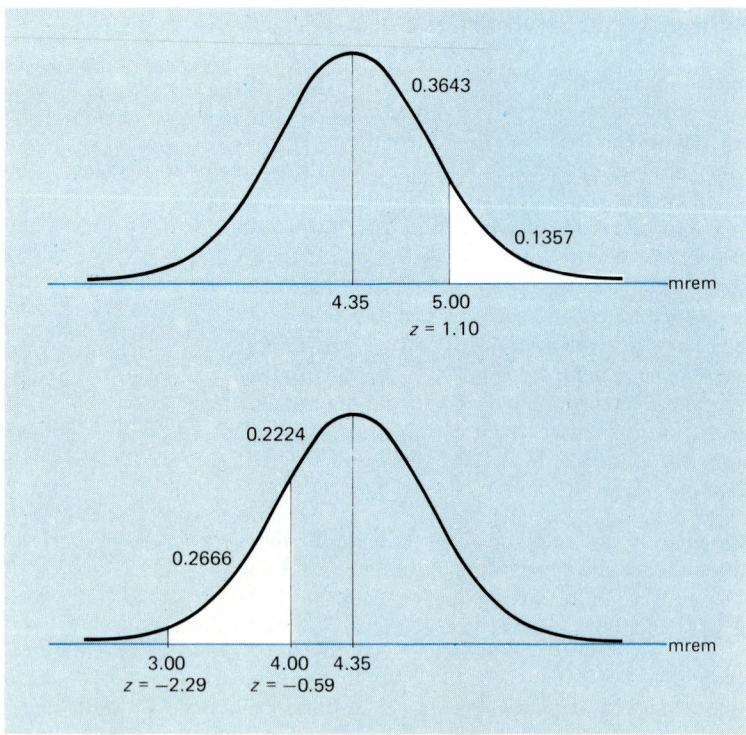

FIGURE 9.17 *Diagrams for cosmic radiation example.*

Since the entry in Table I corresponding to $z = 1.10$ is 0.3643, we find that the probability is $0.5000 - 0.3643 = 0.1357$ or approximately 0.14, that a person will be exposed to more than 5.00 mrem of cosmic radiation on such a flight.

(b) This probability is given by the area of the white region of the lower diagram of Figure 9.17, namely, the area under the curve between

$$z = \frac{3.00 - 4.35}{0.59} = -2.29 \quad \text{and} \quad z = \frac{4.00 - 4.35}{0.59} = -0.59$$

Since the entries in Table I corresponding to $z = 2.29$ and $z = 0.59$ are, respectively, 0.4890 and 0.2224, we find that the probability is $0.4890 - 0.2224 = 0.2666$, or approximately 0.27, that a person will be exposed to anywhere from 3.00 to 4.00 mrem of cosmic radiation on such a flight.

EXAMPLE   The actual amount of instant coffee which a filling machine puts into "6-ounce" jars varies from jar to jar, and it may be looked upon as a random variable having a normal distribution with a standard deviation of 0.04 ounce. If only 2 percent of the jars are to contain less than 6 ounces of coffee, what must be the mean fill of these jars?

Solution   Here we are given $\sigma = 0.04$, $x = 6.00$, a normal-curve area (that of the white region of Figure 9.18), and we are asked to find μ. Since the value of z for which the entry in Table I is closest to $0.5000 - 0.0200 = 0.4800$ is $z = 2.05$ corresponding to 0.4798, we have

$$-2.05 = \frac{6.00 - \mu}{0.04}$$

Then solving for μ, we get

$$6.00 - \mu = -2.05(0.04) = -0.082$$

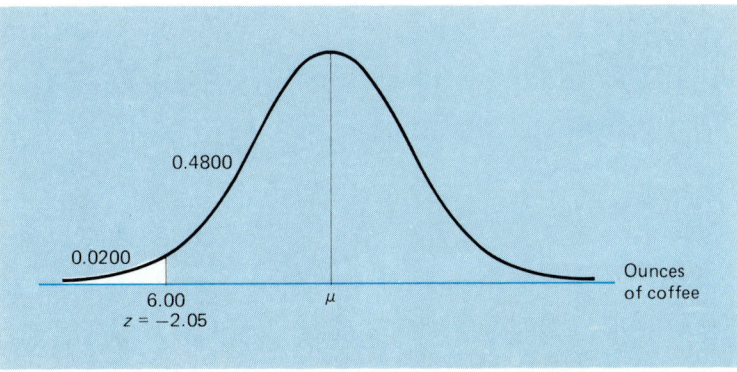

FIGURE 9.18   *Diagram for instant coffee filling example.*

and

$$\mu = 6.00 + 0.082 = 6.082 \text{ ounces}$$

or 6.08 ounces rounded to the nearest hundredth of an ounce.

Although the normal distribution is a continuous distribution which applies to continuous random variables, it is often used to approximate distributions of discrete random variables, which can take on only a finite number of values or as many values as there are positive integers. To do this, we must use the **continuity correction** illustrated in the following example:

EXAMPLE In a study of aggressive behavior, male white mice, returned to the group in which they live after four weeks of isolation, averaged 18.6 fights in the first five minutes with a standard deviation of 3.3 fights. If it can be assumed that the distribution of this random variable (the number of fights into which such a mouse gets under the stated conditions) can be approximated closely with a normal distribution, what is the probability that such a mouse will get into at least 15 fights in the first five minutes?

Solution The answer is given by the area of the white region of Figure 9.19 below; namely, by the area under the curve to the right of 14.5, and not 15. The reason for this is that the number of fights in which such a mouse gets involved is a whole number. Hence, if we want to approximate the distribution of this random variable with a normal distribution, we must "spread" its values over a continuous scale, and we do this by representing each whole number k by the interval from $k - \frac{1}{2}$ to $k + \frac{1}{2}$. For instance, 5 is represented by the interval from 4.5 to 5.5, 10 is represented by the interval from 9.5 to 10.5, 20 is represented by the interval from 19.5 to 20.5, and the probability of 15 or more is given by the area under the curve to the right of 14.5. Accordingly, we get

$$z = \frac{14.5 - 18.6}{3.3} = -1.24$$

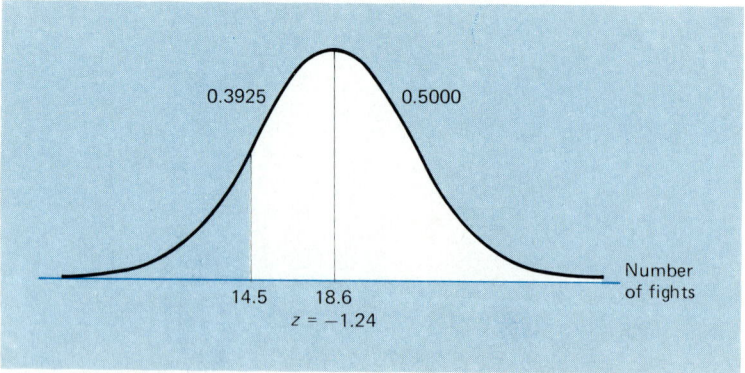

FIGURE 9.19 *Diagram for example dealing with aggresive behavior of mice.*

and it follows from Table I that the area of the white region of Figure 9.19—the probability that such a mouse will get into at least 15 fights in the first five minutes—is $0.5000 + 0.3925 = 0.8925$, or approximately 0.89.

All the examples of this section dealt with random variables having normal distributions, or distributions which can be approximated closely with normal curves. When we observe a value (or values) of a random variable having a normal distribution, we may say that we are sampling a **normal population**; this is consistent with the terminology introduced at the end of Section 8.3.

9.5
THE NORMAL APPROXIMATION TO THE BINOMIAL DISTRIBUTION

The normal distribution provides a close approximation to the binomial distribution when n, the number of trials, is large and p, the probability of a success on an individual trial, is close to $\frac{1}{2}$. Figure 9.20 shows the histograms of binomial distributions with $p = \frac{1}{2}$ and $n = 2, 5, 10,$ and 25, and it can be seen that with increasing n these distributions approach the symmetrical bell-shaped pattern of the normal distribution. In fact, normal distributions with the mean $\mu = np$ and the standard deviation $\sigma = \sqrt{np(1 - p)}$ can often be used to approximate binomial probabilities when n is not all that large and p differs quite a bit from $\frac{1}{2}$. Since "not

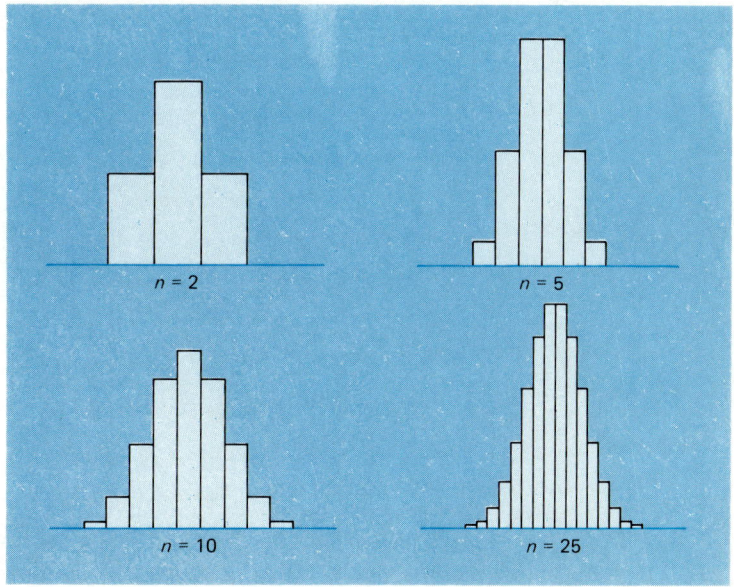

FIGURE 9.20 *Binomial distributions with $p = \frac{1}{2}$.*

all that large" and "differs quite a bit" are not very precise terms, let us state the following rule of thumb:

> **It is considered sound practice to use the normal approximation to the binomial distribution only when np and $n(1 − p)$ are both greater than 5; symbolically, when**
>
> $$np > 5 \quad \text{and} \quad n(1 − p) > 5$$

EXAMPLE Use the normal distribution to approximate the probability of getting 6 heads and 10 tails in 16 flips of a balanced coin, and compare the result with the corresponding value of Table V.

Solution It is reasonable to use the normal approximation since $np = 16 \cdot 0.5 = 8$ and $n(1 − p) = 16 \cdot (1 − 0.5) = 8$ are both larger than 5. To find the normal approximation to this probability, we use the continuity correction and represent 6 heads by the interval from 5.5 and 6.5. Hence, we must determine the area of the white region of Figure 9.21 below. Since $\mu = 16 \cdot \frac{1}{2} = 8$ and $\sigma = \sqrt{16 \cdot \frac{1}{2} \cdot \frac{1}{2}} = 2$, we get

$$z = \frac{5.5 − 8}{2} = −1.25 \quad \text{and} \quad z = \frac{6.5 − 8}{2} = −0.75$$

in standard units for $x = 5.5$ and $x = 6.5$. The entries corresponding to 1.25 and 0.75 in Table I are 0.3944 and 0.2734, and we get $0.3944 − 0.2734 = 0.1210$ for the normal approximation to the probability of getting 6 heads and 10 tails in 16 flips of a balanced coin. Since the corresponding entry in Table V is 0.122, we find that the error of the approximation is 0.001.

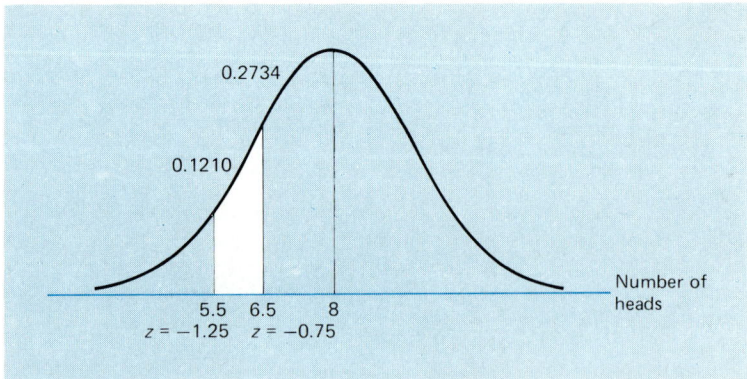

FIGURE 9.21 *Normal approximation to a binomial distribution.*

EXAMPLE A study shows that 40 percent of all self-employed writers in a certain income bracket have Keogh retirement accounts. Use the normal distribution to approximate the probability that among 20 such writers, randomly selected from IRS files, 10 will have Keogh retirement accounts. Also compare the result with the corresponding value in Table V.

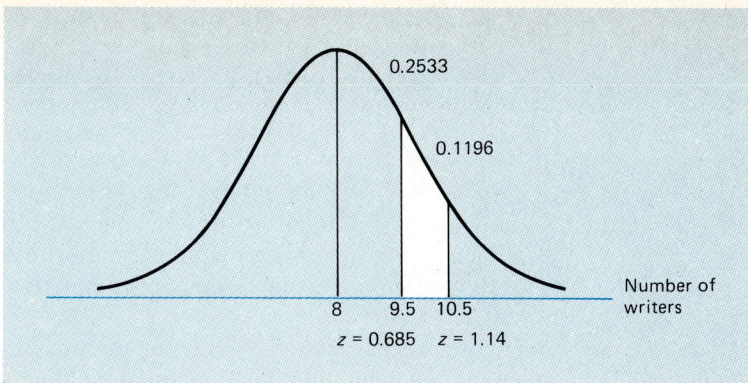

FIGURE 9.22 *Normal approximation to a binomial distribution.*

Solution It is reasonable to use the normal approximation since $np = 20 \cdot 0.4 = 8$ and $n(1 - p) = 20 \cdot (1 - 0.4) = 12$ are both larger than 5. Using the continuity correction, we must find the area of the white region of Figure 9.22, namely, that between 9.5 and 10.5. Since

$$\mu = 20(0.40) = 8 \quad \text{and} \quad \sigma = \sqrt{20(0.40)(0.60)} = 2.19$$

we get

$$z = \frac{9.5 - 8}{2.19} = 0.685 \quad \text{and} \quad z = \frac{10.5 - 8}{2.19} = 1.14$$

in standard units for $x = 9.5$ and $x = 10.5$. The entries corresponding to 0.68 and 0.69 in Table I are 0.2517 and 0.2549, so that by interpolation we find that the area corresponding to 0.685 is

$$\frac{0.2517 + 0.2549}{2} = 0.2533$$

Also the area corresponding to 1.14 is 0.3729. Thus, we get $0.3729 - 0.2533 = 0.1196$ for the normal approximation to the probability that 10 of the 20 self-employed writers will have Keogh retirement accounts. Since the corresponding entry in Table V is 0.117, we find that the error of the approximation is $0.1196 - 0.117$ or 0.003 rounded to three decimals.

To assess the size of the error of an approximation like this, it usually helps to determine the **percentage error**, namely, the error expressed as a percentage of the quantity we are trying to approximate. For the example above we get $\frac{0.003}{0.117} \cdot 100\% = 2.6\%$.

The normal approximation to the binomial distribution comes in very handy when we would otherwise have to use the formula for the binomial distribution repeatedly to calculate the values of many different terms.

EXAMPLE   Suppose that 5 percent of the adobe bricks shipped by a manufacturer have blemishes. Use the normal distribution to approximate the binomial probability that among 150 adobe bricks shipped by the manufacturer at least eight will have blemishes. Also use the computer printout of Figure 8.4 on page 197 to determine the percentage error of this approximation.

Solution   Using the continuity correction, we must find the area of the white region of Figure 9.23, namely, that to the right of 7.5. Since

$$\mu = 150(0.05) = 7.5 \quad \text{and} \quad \sigma = \sqrt{150(0.05)(0.95)} = 2.67$$

we get

$$z = \frac{7.5 - 7.5}{2.67} = 0$$

in standard units for $x = 7.5$. Thus, the area under the curve to the right of 7.5 is 0.5000, and this is the normal approximation to the binomial probability that at least eight of the 150 adobe bricks will have blemishes. Since the printout of Figure 8.4 shows that the corresponding binomial probability is $1 - 0.5228 = 0.4772$, we find that the error of the approximation is $0.5000 - 0.4772 = 0.0228$, and hence that the percentage error is $\frac{0.0228}{0.4772} \cdot 100\% = 4.8$ percent.

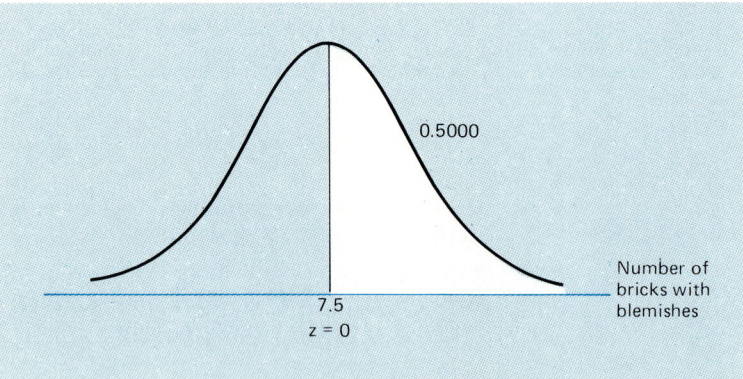

FIGURE 9.23 *Normal approximation to a binomial distribution.*

Whether the approximations of the preceding examples are good enough, depends on how they are to be used. Let us add, though, that besides the rule of thumb on page 240, use of the normal approximation to the binomial distribution requires a good deal of professional judgment. For instance, when p is quite small and we want to approximate the probability of a value that is not too close to the mean, the approximation may be poor even though the rule of thumb on page 240 is satisfied. As the reader will be asked to verify in Exercise 9.51 on page 244, if we had wanted to approximate the probability that there will be only one brick with blemishes among the 150 shipped by the manufacturer, the answer would have been more than twice the correct value. Since we do not expect the reader to become an instant expert at using the normal approximation to the binomial distribution, let us add that we have presented it here mainly because it will be needed in Chapter 13 for large-sample inferences concerning proportions.

EXERCISES

9.32 If the assembly time of an "easy to assemble" toy is a random variable having the normal distribution with $\mu = 15.4$ minutes and $\sigma = 4.4$ minutes, what are the probabilities that this kind of toy can be assembled in
    (a) less than 10.0 minutes;
    (b) anywhere from 12.0 minutes to 18.0 minutes;
    (c) anywhere from 11.0 minutes to 19.8 minutes?

9.33 With reference to the preceding exercise, for what value of x would it be fair to give 3 to 1 odds that it will take less than x minutes to assemble the given kind of toy?

9.34 The reduction of a person's oxygen consumption during periods of transcendental meditation may be looked upon as a random variable having the normal distribution with $\mu = 38.6$ cc per minute and $\sigma = 6.5$ cc per minute. Find the probabilities that during a period of transcendental meditation a person's oxygen consumption will be reduced by
    (a) at least 33.4 cc per minute;
    (b) at most 34.7 cc per minute.

9.35 The sardines processed by a cannery have a mean length of 4.54 inches with a standard deviation of 0.25 inch. If the distribution of the lengths of the sardines can be approximated closely with a normal distribution, what percentage of the sardines are
    (a) shorter than 4.00 inches;
    (b) between 4.40 and 4.60 inches long?

9.36 With reference to the preceding exercise, find x so that only 20 percent of the sardines processed by the cannery are longer than x.

9.37 In a certain photographic process, the developing time of prints may be looked upon as a random variable

having the normal distribution with $\mu = 16.20$ seconds and $\sigma = 0.52$ second. Find the probabilities that the time it takes to develop one of the prints will be
    (a) at least 17 seconds;
    (b) at most 16 seconds;
    (c) anywhere from 16 to 17 seconds.

9.38 With reference to the filling-machine example on page 237, show that 97.7 percent of the jars will contain at least 6 ounces of coffee if the machine is adjusted so that $\mu = 6.02$ ounces and $\sigma = 0.01$ ounce.

9.39 In a very large class in European history, the final examination grades have a mean of 71.6 and a standard deviation of 12.6. If it is reasonable to approximate the distribution of these grades with a normal distribution, what percentage of the grades should exceed 79?

9.40 The yearly number of major earthquakes, the world over, is a random variable having approximately the normal distribution with $\mu = 20.8$ and $\sigma = 4.5$. Find the probabilities that in any given year there will be
    (a) exactly 19 major earthquakes;
    (b) at most 19 major earthquakes;
    (c) at least 19 major earthquakes.

9.41 An airline knows that the number of suitcases it loses per week on a certain route is a random variable having approximately the normal distribution with $\mu = 26.2$ and $\sigma = 5.8$. Find the probabilities that in one week on the given route it will lose
    (a) exactly 22 suitcases;
    (b) at most 22 suitcases;
    (c) at least 22 suitcases.

9.42 With reference to the example on page 238, what is the probability that such a mouse will get into exactly 15 fights in the first five minutes?

9.43 Check in each case whether the conditions for the normal approximation to the binomial distribution are satisfied:

    (a) $n = 16$ and $p = \frac{1}{4}$;
    (b) $n = 65$ and $p = 0.10$;
    (c) $n = 120$ and $p = 0.98$.

9.44 Check in each case whether the conditions for the normal approximation to the binomial distribution are satisfied:

    (a) $n = 200$ and $p = 0.01$;
    (b) $n = 150$ and $p = 0.97$;
    (c) $n = 100$ and $p = \frac{1}{8}$.

9.45 Use the normal distribution to approximate the probability of getting 7 heads and 7 tails in 14 flips of a balanced coin. Also refer to Table V to find the error of this approximation.

9.46 Use the normal distribution to approximate the probability that at most 40 of 225 loan applications received by a bank will be refused, if the probability is 0.20 that any such loan application will be refused.

9.47 Use the normal distribution to approximate the probability that at most 20 of 40 clouds seeded with silver iodide will show spectacular growth, if the probability is 0.62 that any cloud seeded with silver iodide will show spectacular growth.

★ 9.48 Use a computer printout of the binomial distribution with $n = 40$ and $p = 0.62$ or the National Bureau of Standards Tables referred to on page 217, to find the error of the approximation of the preceding exercise.

9.49 Use the normal distribution to approximate the probability that at most 12 of 50 patients will get a headache from using a certain kind of medication, if the probability is 0.22 that any one patient will get a headache from using the medication.

9.50 Use a computer printout of the binomial distribution with $n = 50$ and $p = 0.22$, or the 50–100 Binomial Tables referred to on page 197, to find the percentage error of the approximation of the preceding exercise.

9.51 Use the printout of Figure 8.4 on page 197 to show that if we use the normal distribution to approximate the binomial probability with $x = 1$, $n = 150$, and $p = 0.05$, the percentage error is about 117 percent.

9.52 Use the normal distribution to approximate the probability that at least 55 of 90 persons flying across the Atlantic Ocean will feel the effect of the time difference for at least 24 hours, if the probability is 0.70 that any one person flying across the Atlantic Ocean will feel the effect of the time difference for at least 24 hours.

9.53 Use a computer printout of the binomial distribution with $n = 90$ and $p = 0.70$, or the 50–100 Binomial Tables referred to on page 217, to find the percentage error of the approximation of the preceding exercise.

9.54 Use the normal distribution to approximate the probability that at least 150 of 400 persons who have cable television will watch a certain movie, if the probability is 0.34 that any one person who has cable television will watch the movie.

★ 9.55 In some situations we can use either the normal approximation or the Poisson approximation to the binomial distribution, in some situations we can use one but not the other, and in some situations we can use neither. Check in each case whether we can use the normal approximation and/or the Poisson approximation to the binomial distribution:

    (a) $n = 50$ and $p = \frac{1}{12}$;
    (b) $n = 200$ and $p = 0.06$;
    (c) $n = 100$ and $p = 0.04$;
    (d) $n = 100$ and $p = 0.08$.

★ 9.56 For a certain brand of lawnmower, the probability is 0.07 that it will need in-store servicing before being placed on the sales floor. You wish to find the probability that exactly three of 120 of these lawnmowers will need in-store servicing. Either the normal approximation or the Poisson approximation may be used for this binomial distribution with $n = 120$ and $p = 0.07$.

    (a) Find the exact binomial probability that three of 120 of these lawnmowers will need in-store servicing.
    (b) Use the normal approximation to obtain a value for this probability.
    (c) Use the Poisson approximation to obtain a value for this probability.

★ 9.57 In Exercise 9.51 the reader was asked to verify that the normal distribution yields a very poor approximation to the binomial probability with $x = 1$, $n = 150$, and $p = 0.05$. Show that for the Poisson approximation the percentage error is only 15 percent (compared to 117 percent).

9.6

TECHNICAL NOTE (Simulation) ★

There are several ways in which one can simulate values of continuous random variables, including a graphical method which can be used for any continuous random variable. Limiting ourselves here to random variables having normal distributions, we shall demonstrate the use of published tables of **random normal numbers**, also called **random normal deviates**, and show some computer-generated results.

Tables of random normal numbers consist of many pages, on which numbers (rounded to three decimals in Table XII at the end of the book) are set down in a random fashion, much as they would appear if they were generated one at a time by a gambling device which somehow "produces" values of a random variable having the standard normal distribution. With suitable software, they can easily be produced as shown in Figure 9.24, where the appropriate command yielded twenty values of a random variable having the standard normal distribution.

```
MTB > NRANDOM 20 MU=0.0 SIGMA=1.0 C1
    20 NORMAL OBS. WITH MU =         .0000 AND SIGMA =       1.0000
    -.2820    -.5089      .7306      .2049      .1639      .6778
     .5455     .6587    -1.3667     -.0210    -1.9784     -.7907
    -.4938     .5346     -.3686     -.5387    -1.6011     -.6137
    -.2855    -.2206
```

FIGURE 9.24 *Computer simulation of 20 values of a random variable having the standard normal distribution.*

If we are using a computer, we can generate values of a random variable having a normal distribution other than the standard normal distribution by a simple change in command. To obtain values of such a random variable from a table like Table XII, we make use of the formula

Changing from standard units

$$x = \mu + \sigma z$$

which changes z's into x's, namely, values of a random variable having the standard normal distribution into values of a random variable having the normal distribution with given values of μ and σ. It follows by solving for x the formula $z = \dfrac{x - \mu}{\sigma}$ for converting to standard units. As in the use of "ordinary" random numbers tables, the choice of the page and the place from which to start should, in practice, be left to chance.

EXAMPLE In a certain city, the time it takes the police to respond to emergencies may be regarded as a random variable having the normal distribution with $\mu = 7.3$

minutes and $\sigma = 0.6$ minute. Use the eighth column of the second page of Table XII, starting with the sixth row and going down the page, to simulate the amounts of time it takes the police in this city to respond to five emergencies.

Solution The values we get from Table XII are $-0.745, 0.655, -1.115, 0.027,$ and -2.520, so that the simulated response times are

$$7.3 + 0.6(-0.745) = 6.9$$

$$7.3 + 0.6(0.655) = 7.7$$

$$7.3 + 0.6(-1.115) = 6.6$$

$$7.3 + 0.6(0.027) = 7.3$$

and

$$7.3 + 0.6(-2.520) = 5.8$$

minutes rounded to one decimal.

The exercises which follow may be done in this way, or, if suitable software is available, with the use of a computer. Using a computer to rework the preceding example, we obtained the printout shown in Figure 9.25.

```
MTB > NRANDOM 5 MU=7.3 SIGMA=0.6 C1
      5 NORMAL OBS. WITH MU =        7.3000 AND SIGMA =        .6000
    6.3195     6.9863     7.2877     7.3012     7.4496
```

FIGURE 9.25 *Computer simulation of 5 values of a random variable having the normal distribution with $\mu = 7.3$ and $\sigma = 0.6$.*

EXERCISES

★ **9.58** The time it takes a person to learn how to operate a certain piece of machinery is a random variable having the normal distribution with $\mu = 5.6$ hours and $\sigma = 1.2$ hours. Simulate the times it takes eight persons to learn how to operate the piece of machinery.

★ **9.59** The scores of college-bound high school seniors on a certain standardized test may be regarded as values of a random variable having the normal distribution with $\mu = 54.3$ and $\sigma = 6.2$. Rounding the results to the nearest whole numbers, simulate the scores of 12 college-bound high school seniors on this test.

★ **9.60** Suppose that the increase in the pulse rate of a person performing a certain strenuous task is a random variable having the normal distribution with $\mu = 28.40$ and $\sigma = 4.17$. Rounding the results to one decimal, simulate the increase in the pulse rate of 20 persons performing the given task.

★ **9.61** The weights of grapefruits shipped by a large orchard are values of a random variable having the normal distribution with $\mu = 19.6$ ounces and $\sigma = 2.2$ ounces. Rounding the results to one decimal, simulate the weight of 24 grapefruits shipped by the orchard.

9.7

CHECKLIST OF KEY TERMS
(with page references to their definitions)

★ Arithmetic probability paper, 234
Continuity correction, 238
Continuous distribution, 220
★ *Exponential distribution*, 232
Normal approximation to binomial distribution, 239
Normal distribution, 223
Normal population, 239
★ Normal probability paper, 234

Normal-curve areas, 224
Percentage error, 241
Probability density, 221
★ Random normal deviates, 245
★ Random normal numbers, 245
Standard normal distribution, 224
Standard units, 224
Uniform distribution, 230
z-score, 224

9.8

REVIEW EXERCISES

9.62 If the amount of time a tourist spends in a famous museum is a random variable having the normal distribution with $\mu = 43.4$ minutes and $\sigma = 6.8$ minutes, find the probabilities that a tourist will spend
  (a) at most 36.0 minutes in the museum;
  (b) anywhere from 40.0 to 50.0 minutes in the museum.

9.63 Use the normal distribution to approximate the probability of getting 12 heads in 20 flips of a balanced coin. Refer to Table V to determine the error of this approximation.

9.64 Find the standard-normal-curve area which lies
  (a) between $z = 0$ and $z = 1.83$;
  (b) to the left of $z = 2.50$;
  (c) to the right of $z = -0.64$;
  (d) to the right of $z = 1.24$;
  (e) to the left of $z = -0.71$.

9.65 If a random variable has the normal distribution with $\mu = 102.4$ and $\sigma = 3.6$, find the probabilities that it will take on a value
  (a) less than 107.8;
  (b) greater than 99.7;
  (c) between 106.9 and 110.5;
  (d) between 96.1 and 104.2.

9.66 Find the values of
  (a) $z_{0.02}$;
  (b) $z_{0.10}$.

★ **9.67** The following is the distribution of the miles per gallon obtained with 200 cars:

Miles per gallon	Frequency
18.0–19.9	2
20.0–21.9	10
22.0–23.9	19
24.0–25.9	39
26.0–27.9	62
28.0–29.9	37
30.0–31.9	20
32.0–33.9	8
34.0–35.9	3

Use normal probability paper to check whether this distribution has roughly the shape of a normal curve.

9.68 Use the normal distribution to approximate the probability that more than 90 of 100 scorpion stings will cause extensive discomfort, if the probability is 0.85 that any one of them will cause extensive discomfort.

★ **9.69** Suppose that the time it takes to complete a certain new tax form is a random variable having the normal distribution with $\mu = 32.6$ minutes and $\sigma = 4.5$ minutes. Simulate the time it takes 12 persons to complete the new tax form.

9.70 Figure 9.26 shows the probability density of a continuous random variable which takes on values on the interval from 0 to $1\frac{1}{2}$. Verify that the total area under the curve is equal to 1.

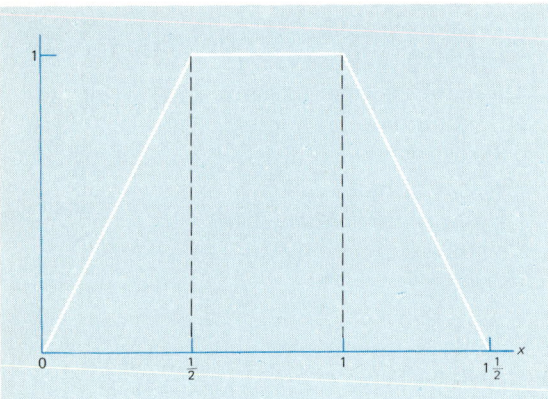

FIGURE 9.26 *Diagram for Exercise 9.70.*

9.71 With reference to the preceding exercise, find the probabilities that the random variable will take on a value

(a) less than $\frac{3}{4}$;

(b) greater than $\frac{1}{2}$;

(c) between $\frac{1}{4}$ and $1\frac{1}{4}$.

9.72 The number of complaints received by the complaint department of a department store per day is a random variable which has approximately the normal distribution with $\mu = 48.4$ and $\sigma = 7.5$. Approximately, what are the probabilities that on any one day they will receive

(a) at least 55 complaints;

(b) anywhere from 40 to 50 complaints?

9.73 A random variable has a normal distribution with $\sigma = 4.0$. If the probability is 0.9713 that this random variable will take on a value less than 87.6, what is the probability that it will take on a value in excess of 75.0?

9.74 The average time required to perform job A is 78.5 minutes with a standard deviation of 16.2 minutes, and the average time required to perform job B is 103.2 minutes with a standard deviation of 11.3 minutes. Assuming normal distributions, what proportion of the time will job A take longer than the average job B, and

what proportion of the time will job B take less time than the average job A?

9.75 Use the normal distribution to approximate the probabilities that a random variable having the binomial distribution with $n = 18$ and $p = 0.27$ will take on a value

(a) less than 7;

(b) anywhere from 3 to 8.

9.76 With reference to the preceding exercise, use the results of Exercise 8.97 on page 216 or the printout of Figure 8.7 to find the percentage errors of the two approximations.

9.77 Find the standard-normal-curve area which lies

(a) between $z = 0$ and $z = -1.11$;

(b) between $z = -0.63$ and $z = 0.63$;

(c) between $z = 0.40$ and $z = 0.55$;

(d) between $z = -1.18$ and $z = -0.68$;

(e) between $z = -1.22$ and $z = 1.82$.

9.78 Check in each case whether the conditions for the normal approximation to the binomial distribution are satisfied:

(a) $n = 40$ and $p = \frac{1}{5}$;

(b) $n = 135$ and $p = \frac{1}{45}$;

(c) $n = 150$ and $p = \frac{1}{25}$;

(d) $n = 50$ and $p = 0.95$.

9.79 If the weekly number of muggings reported in a certain precinct is a random variable having the normal distribution with $\mu = 25.3$ and $\sigma = 5.5$, find the probabilities that in any one week

(a) 27 muggings will be reported;

(b) at least 27 muggings will be reported.

9.80 Find z if the standard-normal curve area

(a) between 0 and z is 0.2019;

(b) to the right of z is 0.8810;

(c) to the right of z is 0.1788;

(d) between $-z$ and z is 0.4038;

(e) between $-z$ and $2z$ is 0.4530.

9.81 To illustrate the Law of Large Numbers which we mentioned in Sections 5.4 and 8.8, use the normal approximation to the binomial distribution to find the probabilities that the proportion of heads will be anywhere from 0.49 to 0.51 when a balanced coin is flipped

(a) 1,000 times;

(b) 10,000 times.

9.9
REFERENCES

More detailed tables of normal-curve areas, as well as tables for some other continuous distributions, may be found in

FISHER, R. A., and YATES, F., *Statistical Tables for Biological, Agricultural and Medical Research.* Cambridge: The University Press, 1954.

PEARSON, E. S., and HARTLEY, H. O., *Biometrika Tables for Statisticians*, *Vol. I.* New York: John Wiley & Sons, Inc., 1968.

and in other handbooks of statistical tables. Extensive tables of random normal numbers are given in the RAND Corporation tables listed on page 218. Detailed information about various continuous distributions may be found in

HASTINGS, N. A. J., and PEACOCK, J. B., *Statistical Distributions.* London, Butterworth & Company (Publishers) Ltd., 1975.

Further information about the normal approximation to the binomial distribution may be found in

GREEN, J., and ROUND-TURNER, J., "The Error in Approximating Cumulative Binomial and Poisson Probabilities," *Teaching Statistics*, May 1986.

10 SAMPLING AND SAMPLING DISTRIBUTIONS

The main objective of most statistical studies, analyses or investigations, is to make sound generalizations on the basis of samples about the populations from which the samples came. Note the word "sound," because the question of when and under what conditions samples permit such generalizations is not easily answered. For instance, if we want to estimate the average amount of money a person spends on a vacation, would we take as our sample the amounts spent by deluxe-class passengers on a four-week cruise; or would we attempt to estimate, or predict, the wholesale price of all farm products on the basis of the price of fresh asparagus alone? Obviously not, but just what vacationers and what farm products we should include in our samples, and how many of them, is neither intuitively clear nor self-evident.

In most of the methods we shall study in the remainder of this book, it will be assumed that we are dealing with so-called **random samples**. We pay this much attention to random samples, which are defined and discussed in Section 10.1, because they permit valid, or logical, generalizations. As we shall see, however, random sampling is not always feasible, or even desirable, and some alternative sampling procedures are mentioned in the optional Sections 10.2 through 10.5.

In Section 10.6 we introduce the related concept of a **sampling distribution**, which tells us how quantities determined from samples may vary from sample to sample. Then, in Sections 10.7 through 10.9 we learn how such chance variations can be measured, predicted, and perhaps even controlled.

10.1
RANDOM SAMPLING

In Section 3.1 we distinguished between populations and samples, stating that a population consists of all conceivably possible (or hypothetically possible) observations of a given phenomenon, while a sample is simply part of a population. In what follows, we shall distinguish further between two kinds of populations—**finite populations** and **infinite populations**.

A population is finite if it consists of a finite, or fixed, number of elements, measurements, or observations. Examples of finite populations are the net weights of the 3,000 cans of paint in a certain production lot, the SAT scores of all the freshmen admitted to a given college in the fall of 1991, and the daily high temperatures recorded at a weather station during the years 1987–1991.

In contrast to finite populations, a population is infinite if it contains, hypothetically at least, infinitely many elements. This is the case, for example, when we observe a value of a continuous random variable and there are infinitely many different outcomes. It is also the case when we observe the totals obtained in repeated rolls of a pair of dice, when we repeatedly measure the boiling point of a silicon compound, and when we sample with replacement from a finite population. There is no limit to the number of times that we can roll a pair of dice, no limit to the number of times that we can measure the boiling point of the silicon compound, and no limit to the number of times that we can draw a sample from a finite population and replace it before the next one is drawn.

To present the idea of **random sampling from a finite population**, let us first see how many different samples of size n can be drawn from a finite population of size N. Referring to the rule for the number of combinations of n objects taken r at a time on page 101, we find that, with a change of letters, the answer is $\binom{N}{n}$.

EXAMPLE How many different samples of size n can be drawn from a finite population of size N, when
(a) $n = 2$ and $N = 12$;
(b) $n = 3$ and $N = 100$?

Solution

(a) There are $\binom{12}{2} = \dfrac{12 \cdot 11}{2!} = 66$ different samples.

(b) There are $\binom{100}{3} = \dfrac{100 \cdot 99 \cdot 98}{3!} = 161{,}700$ different samples.

Based on the result that there are $\binom{N}{n}$ different samples of size n from a finite population of size N, let us now give the following definition of a **random sample** (sometimes referred to also as a **simple random sample**) from a finite population:

A sample of size n from a finite population of size N is random if it is chosen in such a way that each of the $\binom{N}{n}$ possible samples has the same probability, $\dfrac{1}{\binom{N}{n}}$, of being selected.

For instance, if a population consists of the $N = 5$ elements a, b, c, d, and e (which might be the annual incomes of five persons, the weights of five cows, or the prices of five commodities), there are $\binom{5}{3} = 10$ possible samples of size $n = 3$. They consist of the elements abc, abd, abe, acd, ace, ade, bcd, bce, bde, and cde. If we choose one of these samples in such a way that each sample has the probability $\frac{1}{10}$ of being selected, we call this sample a random sample.

Next comes the question of how random samples are drawn in actual practice. In a simple situation like the one described immediately above, we could write each of the ten samples on a slip of paper, put them in a hat, shuffle them thoroughly, and then draw one without looking. Obviously, though, this would be impractical in a more realistically complex situation where n and N, or only N, are large. For instance, for $n = 4$ and $N = 200$ we would have to label and draw one of $\binom{200}{4} = 64{,}684{,}950$ slips of paper.

Fortunately, we can draw a random sample from a finite population without listing all possible samples, which we mentioned here only to stress the point that the selection of a random sample must depend entirely on chance. Instead of listing all possible samples, we can write each of the N elements of the finite population on a slip of paper, and draw n of them one at a time without replacement, making sure in each of the successive drawings that all of the remaining elements of the population have the same chance of being selected. As the reader will be asked to verify in Exercise 10.14 on page 256, this procedure also leads to the same probability, $\dfrac{1}{\binom{N}{n}}$, for each possible sample.

This relatively easy procedure can be simplified further by choosing random numbers instead of drawing slips of paper, or by letting all the work be done by a computer. As we pointed out on page 210, published tables of random numbers (such as the one from which Table XI of this book is excerpted) consist of pages on which the digits $0, 1, 2, \ldots$, and 9 are set down in much the same fashion as they might appear if they had been generated by a chance or gambling device giving each digit the same probability, $\frac{1}{10}$, of appearing at any given place in the table.

EXAMPLE   Draw a random sample of size $n = 12$ from the population which consists of the amounts of sales tax collected by a city's 247 drugstores in December 1990, by numbering the stores $001, 002, 003, \ldots$, and 247 (say, in the order in which they are listed in the telephone directory), and reading three-digit random numbers off the second page of Table XI, using the 26th, 27th, and 28th columns, starting with the sixth row and going down the page.

Solution   Following these instructions, we get

$$046 \quad 230 \quad 079 \quad 022 \quad 119 \quad 150 \quad 056 \quad 064 \quad 193 \quad 232 \quad 040 \quad 146$$

where we ignored numbers greater than 247; had any numbers recurred, we would have ignored them too. The twelve numbers we got here are the numbers assigned to the drugstores—the corresponding sales tax figures constitute the desired random sample.

The procedure we used in this example was quite easy, but it would have been even easier if we had the software which leaves most of the work to a computer. For instance, the printout of Figure 10.1 shows a computer-generated random sample of size $n = 12$ from the finite population which consists of the numbers $1, 2, 3, \ldots, 246$, and 247. The sample values are 197, 147, 82, 171, 60, 39, 51, 129, 71, 45, 86, and 224.

```
MTB > GENE 247 C1
MTB > SAMPLE 12 C1 C2
    12 ROWS SELECTED OUT OF    247
    THE ROWS SELECTED FROM COLUMN    C1    CONTAIN
   197.0000   147.0000    82.0000   171.0000    60.0000    39.0000
    51.0000   129.0000    71.0000    45.0000    86.0000   224.0000
```

FIGURE 10.1 *Computer generated random sample.*

When lists are available so that items can readily be numbered, it is easy to draw random samples with the aid of random number tables or computers. Unfortunately, though, there are many situations where it is impossible to proceed in the way we have just described. For instance, if we want to use a sample to estimate the mean outside diameter of thousands of ball bearings packed in a large crate, or if we want to estimate the mean height of the trees in a forest, it would be impossible to number the ball bearings or the trees, choose random numbers, and then locate and measure the corresponding ball bearings or trees. In these and in many similar situations, all we can do is proceed according to the dictionary definition of the word "random," namely, "haphazard, without aim or purpose." That is, we must not select or reject any element of a population because of its seeming typicalness or lack of it, nor must we favor or ignore any part of a population because of its accessibility or lack of it, and so forth. With some reservations, such samples can often be treated as if they were, in fact, random samples.

So far we have discussed random sampling only in connection with finite populations. For infinite populations we say that

A sample of size *n* from an infinite population is random if it consists of values of independent random variables having the same distribution.

As we pointed out in connection with the binomial and normal distributions, it is this "same" distribution which we refer to as the population being sampled. Also, by "independent" we mean that the probabilities relating to any one of the random variables are the same regardless of what values may have been observed for the other random variables.

For instance, if we get 2, 5, 1, 3, 6, 4, 4, 5, 2, 4, 1, and 2 in twelve rolls of a die, these numbers constitute a random sample if they are values of independent random variables having the same probability distribution

$$f(x) = \tfrac{1}{6} \qquad \text{for } x = 1, 2, 3, 4, 5, \text{ or } 6$$

To give another example of a random sample from an infinite population, suppose that eight students obtained the following measurements of the boiling point of a silicon compound: 136, 153, 170, 148, 157, 152, 143, and 150 degrees Celsius. According to the definition, these values constitute a random sample if they are values of independent random variables having the same distribution, say, the normal distribution with $\mu = 152$ and $\sigma = 10$. To judge whether this is actually the case, we would have to ascertain, among other things, that the eight students' measuring techniques are equally precise (so that σ is the same for each of the random variables), that there was no collaboration (which might make the random variables dependent), and that there were no impurities in the raw material. In practice, it is not an easy task to judge whether a set of data may be looked upon as a random sample, and we shall go into this further in Section 17.9.

EXERCISES

10.1 How many different samples of size $n = 2$ can be selected from a finite population of size
    (a) $N = 6$;   (b) $N = 10$;   (c) $N = 25$?

10.2 How many different samples of size $n = 3$ can be selected from a finite population of size
    (a) $N = 15$;   (b) $N = 30$;   (c) $N = 60$?

10.3 What is the probability of each possible sample if
    (a) a random sample of size 4 is to be drawn from a finite population of size 12;
    (b) a random sample of size 5 is to be drawn from a finite population of size 22?

10.4 With reference to the example on page 252, where we listed all possible samples of size $n = 3$ from the finite population which consists of the elements $a, b, c, d,$ and e, what is the probability that a random sample of size $n = 3$ from this population will contain a specific element, say, the element c?

10.5 List the $\binom{8}{2} = 28$ possible samples of size $n = 2$ that can be drawn from a finite population whose elements are denoted by $a, b, c, d, e, f, g,$ and h.

10.6 With reference to the preceding exercise, what is the probability that a random sample of size $n = 2$ from the given population will include the element denoted by the letter d?

10.7 List all possible choices of two of the following six airlines: TWA, American, United, Eastern, Delta, and Western. If a person randomly selects two of these airlines to study their safety records, find the probability
    (a) of each possible sample;
    (b) that TWA will be included in the sample.

10.8 A bacteriologist wants to double-check a sample of $n = 10$ of the 812 blood specimens analyzed by a medical laboratory in a given month. If he numbers the specimens 001, 002, 003, . . . , 811, and 812, which ones will he select if he chooses them by using the eleventh, twelfth, and thirteenth columns of the second page of Table XI, starting with the eighth row and going down the page?

10.9 A county assessor wants to reassess a random sample of 15 of 8,019 one-family homes. If she numbers them 0001, 0002, . . . , 8,018, and 8,019, which ones (by number) will she select if she chooses them by using the 11th, 12th, 13th, and 14th columns of the first page of Table XI, starting with the fourth row and going down the page?

10.10 A sociologist wants to include ten of the 83 counties in Michigan in a survey. If he numbers these counties 01, 02, . . . , 82, and 83, which ones (by number) will he include in the survey if he selects them by using the 21st and 22nd columns of the second page of Table XI, starting with the tenth row and going down the page?

10.11 Suppose that you have a population of 40 items and you wish to take a sample of size 4. Suppose that you use a random number table to make the selection. What is the probability that the first four numbers you choose will involve no duplicates?

10.12 On page 253 we said that a random sample can be drawn from a finite population by choosing the elements to be included in the sample one at a time, making sure in each of the successive drawings that all of the remaining elements of the population have the same chance of being selected. Verify this for the example on page 252, which dealt with random samples of size $n = 3$ drawn from the finite population which consists of the elements a, b, c, d, and e, by showing that the probability of any particular sample drawn by this method (say, b, c, and e) is $\frac{1}{10}$. (*Hint*: Multiply the probabilities of getting one of the three letters on the first draw, one of the two remaining letters on the second draw, and the third letter on the third draw.)

10.13 Use the same kind of argument as in the preceding exercise to verify that

(a) each possible random sample of size $n = 3$, drawn one at a time from a finite population of size $N = 100$, has the probability

$$\frac{1}{\binom{100}{3}} = \frac{1}{161,700}$$

(b) each possible random sample of size n, drawn one at a time from a finite population of size $N = 200$, has the probability

$$\frac{1}{\binom{200}{3}} = \frac{1}{1,313,400}$$

10.14 Use the same kind of argument as in Exercise 10.12 to verify that each possible random sample of size n, drawn one at a time from a finite population of size N, has the probability $1 \Big/ \binom{N}{n}$.

10.15 Randy McGill and Susan Martin are both members of a population of 50 students. A researcher is going to select 5 students from this population.

(a) What is the probability that Randy will be in the sample?

(b) What is the probability that Susan will be in the sample?

(c) What is the probability that both Randy and Susan will be in the sample?

(d) Is your solution to (c) greater than or less than $\left(\frac{5}{50}\right)^2 = \frac{1}{100}$?

10.2

SAMPLE DESIGNS ★

The only kind of samples we have discussed so far are random samples, and we did not even consider the possibility that under certain conditions there may be samples which are better (say, easier to obtain, cheaper, or more informative) than random samples, and we did not go into any details about the question of what might be done when random sampling is impossible. Indeed, there are many other ways of selecting a sample from a population, and there is an extensive literature devoted to the subject of designing sampling procedures.

In statistics, a **sample design** is a definite plan, completely determined before any data are actually collected, for obtaining a sample from a given population. Thus, the plan to take a simple random sample of 12 of a city's 247 drugstores by using a table of random numbers in a prescribed way constitutes a sample design. In the next three sections we shall discuss briefly some of the most widely used kinds of sample designs.

10.3

SYSTEMATIC SAMPLING ★

In some instances, the most practical way of sampling is to select, say, every 20th name on a list, every 12th house on one side of a street, every 50th piece coming off an assembly line, and so on. This is called **systematic sampling**, and an element of randomness can be introduced into this kind of sampling by using random numbers to pick the unit with which to start. Although a systematic sample may not be a random sample in accordance with the definition, it is often reasonable to treat systematic samples as if they were random samples; indeed, in some instances, systematic samples actually provide an improvement over simple random samples inasmuch as the samples are spread more evenly over the entire populations.

If the members of the population appear sequentially in time, as with pieces coming off a production line or automobiles approaching a toll booth, systematic sampling will smoothly spread out the work of sampling over time. This desirable feature of systematic sampling helps reduce the number of clerical errors.

The real danger in systematic sampling lies in the possible presence of hidden periodicities. For instance, if we inspect every 40th piece made by a particular machine, the results would be very misleading if, because of a regularly recurring failure, every 10th piece produced by the machine has blemishes. Also, a systematic sample might yield biased results if we interview the residents of every 12th house along a certain route and it so happens that each 12th house along the route is a corner house on a double lot.

10.4

STRATIFIED SAMPLING ★

If we have information about the makeup of a population (that is, its composition) and this is of relevance to our investigation, we may be able to improve on random sampling by **stratification**. This is a procedure which consists of stratifying (or dividing) the population into a number of non-overlapping subpopulations, or **strata**, and then taking a sample from each stratum. If the items selected from each stratum constitute simple random samples, the entire procedure—first stratification and then random sampling—is called **stratified (simple) random sampling**.

Suppose, for instance, that we want to estimate the mean weight of four persons on the basis of a sample of size 2, and that the (unknown) weights of the four persons are 115, 135, 185, and 205 pounds. Thus, the mean weight we want to estimate is

$$\mu = \frac{115 + 135 + 185 + 205}{4} = 160 \text{ pounds}$$

If we take an ordinary random sample of size 2 from this population, the $\binom{4}{2} = 6$ possible samples are 115 and 135, 115 and 185, 115 and 205, 135 and 185, 135 and 205, and 185 and 205, and the corresponding means are 125, 150, 160, 160, 170, and 195. Observe that since each of these samples has the probability $\frac{1}{6}$, the probabilities are $\frac{1}{3}$, $\frac{1}{3}$, and $\frac{1}{3}$ that our error (the difference between the sample mean and $\mu = 160$) will be 0, 10, or 35. Now suppose that we know that two of these persons are men and two are women, and suppose that the (unknown) weights of the men are 185 and 205 pounds, while the (unknown) weights of the women are 115 and 135 pounds. Stratifying our sample (by sex) and randomly choosing one of the two men and one of the two women, we find that there are only the four stratified samples 115 and 185, 115 and 205, 135 and 185, and 135 and 205. The means of these samples are 150, 160, 160, and 170, and now the probabilities are $\frac{1}{2}$ and $\frac{1}{2}$ that our error will be 0 or 10. Clearly, stratification has greatly improved our chances of getting a good (close) estimate of the mean weight of the four persons. See, however, Exercise 10.19.

Essentially, the goal of stratification is to form strata in such a way that there is some relationship between being in a particular stratum and the answer sought in the statistical study, and that within the separate strata there is as much homogeneity (uniformity) as possible. In our example there is such a connection between sex and weight and there is much less variability in weight within each of the two groups than there is within the entire population.

In the example above, we used **proportional allocation**, which means that the sizes of the samples from the different strata are proportional to the sizes of the strata. In general, if we divide a population of size N into k strata of size N_1, $N_2, \ldots$, and N_k, and take a sample of size n_1 from the first stratum, a sample of size n_2 from the second stratum, $\ldots$, and a sample of size n_k from the kth stratum, we say that the allocation is proportional if

$$\frac{n_1}{N_1} = \frac{n_2}{N_2} = \cdots = \frac{n_k}{N_k}$$

or if these ratios are as nearly equal as possible. In the example dealing with the weights we had $N_1 = 2$, $N_2 = 2$, $n_1 = 1$, and $n_2 = 1$, so that

$$\frac{n_1}{N_1} = \frac{n_2}{N_2} = \frac{1}{2}$$

and the allocation was, indeed, proportional.

As the reader will be asked to verify in Exercise 10.22 on page 261, allocation is proportional if

Sample sizes for proportional allocation

$$n_i = \frac{N_i}{N} \cdot n \qquad for\ i = 1, 2, \ldots, and\ k$$

where $n = n_1 + n_2 + \cdots + n_k$ is the total size of the sample. When necessary, we use the integers closest to the values given by this formula. See Exercise 10.24 and the references at the end of the chapter.

EXAMPLE A stratified sample of size $n = 60$ is to be taken from a population of size $N = 4,000$, which consists of three strata of size $N_1 = 2,000$, $N_2 = 1,200$, and $N_3 = 800$. If the allocation is to be proportional, how large a sample must be taken from each stratum?

Solution Substituting into the formula, we get

$$n_1 = \frac{2,000}{4,000} \cdot 60 = 30 \qquad n_2 = \frac{1,200}{4,000} \cdot 60 = 18$$

and

$$n_3 = \frac{800}{4,000} \cdot 60 = 12$$

This illustrates proportional allocation, but we should add that there exist other ways of allocating portions of a sample to the different strata. One of these, called **optimum allocation**, is described in Exercise 10.26 on page 262. It accounts not only for the size of the strata, as in proportional allocation, but also for the variability (of whatever characteristic is of concern) within the strata.

Stratification is not restricted to a single variable of classification, or characteristic, and populations are often stratified according to several characteristics. For instance, in a system-wide survey designed to determine the attitude of its students, say, toward a new tuition plan, a state college system with 17 colleges might stratify its sample not only with respect to the colleges, but also with respect to students' class standing, sex, and major. So, part of the sample would be allocated to junior women in college A majoring in engineering, another part to sophomore men in college L majoring in English, and so on. Up to a point, stratification like this, called **cross stratification**, will increase the precision (reliability) of estimates and other generalizations, and it is widely used, particularly in opinion sampling and market research.

In stratified sampling, the cost of taking random samples from the individual strata is often so expensive that interviewers are simply given quotas to be filled from the different strata, with few (if any) restrictions on how they are to be filled. For instance, in determining voters' attitudes toward increased medical coverage for elderly persons, an interviewer working a certain area might be told to interview 6 male self-employed homeowners under 30 years of age, 10 female wage earners in the 45–60 age bracket who live in apartments, 3 retired males over 60 who live in trailers, and so on, with the actual selection of the individuals being left to the interviewer's discretion. This is called **quota sampling**, and it is a convenient, relatively inexpensive, and sometimes necessary procedure, but as it is often executed, the resulting samples do not have the essential features of random samples. In the absence of any controls on their choice, interviewers naturally tend to select individuals who are most readily available—persons who work in the same building, shop in the same store, or perhaps reside in the same general area. Quota samples are thus essentially **judgment samples**, and inferences based on such samples generally do not lend themselves to any sort of formal statistical evaluation.

10.5

CLUSTER SAMPLING ★

To illustrate another important kind of sampling, suppose that a large foundation wants to study the changing patterns of family expenditures in the San Diego area. In attempting to complete schedules for 1,200 families, the foundation finds that simple random sampling is practically impossible, since suitable lists are not available and the cost of contacting families scattered over a wide area (with possibly two or three callbacks for the not-at-homes) is very high. One way in which a sample can be taken in this situation is to divide the total area of interest into a number of smaller, non-overlapping areas, say, city blocks. A number of these blocks are then randomly selected, with the ultimate sample consisting of all (or samples of) the families residing in these blocks.

In this kind of sampling, called **cluster sampling**, the total population is divided into a number of relatively small subdivisions, and some of these subdivisions, or clusters, are randomly selected for inclusion in the overall sample. If the clusters are geographic subdivisions, as in the example above, this kind of sampling is also called **area sampling**. To give another example of cluster sampling, suppose that the Dean of Students of a university wants to know how fraternity men at the school feel about a certain new regulation. He can take a cluster sample by interviewing some or all of the members of several randomly selected fraternities.

Although estimates based on cluster samples are usually not as reliable as estimates based on simple random samples of the same size (see Exercise 10.25 on page 262), they are often more reliable per unit cost. Referring again to the survey of family expenditures in the San Diego area, it is easy to see that it may well be possible to take a cluster sample several times the size of a simple random sample for the same cost. It is much cheaper to visit and interview families living close together in clusters than families selected at random over a wide area.

In practice, several of the methods of sampling we have discussed may well be used in the same study. For instance, if government statisticians want to study the attitude of American elementary school teachers toward certain federal programs, they might first stratify the country by states or some other geographic subdivisions. To take a sample from each stratum, they might then use cluster sampling, subdividing each stratum into a number of smaller geographic subdivisions (say, school districts), and finally they might use simple random sampling or systematic sampling to select a sample of elementary school teachers within each cluster.

EXERCISES

★ **10.16** The following are average monthly AFDC (Aid to Families with Dependent Children) payments in a recent year by the fifty states, listed in alphabetical order.

```
111  539  216  128  462  283  413  237  193  177
406  257  290  213  325  306  184  168  310  266
279  393  450   92  241  302  319  193  281  313
295  402  183  310  257  257  302  315  353  128
244  116  127  348  418  232  400  166  451  315
335  707  266   91  703  380  618   79  588  199
```

List the five possible samples of size $n = 10$ that can be taken from this population by starting with one of the first five numbers and then taking each tenth number on the list.

★ **10.17** The following are monthly figures on the amount of mail (in millions of ton-miles) carried by domestic air operations in a recent year.

```
67  62  75  67  70   68
64  70  66  73  73   97
76  73  80  78  78   72
75  75  73  83  76  108
84  78  86  85  81   78
78  75  78  86  76  111
79  77  87  84  82   77
79  77  80  84  78  117
```

List the six possible systematic samples of size $n = 8$ that can be taken from this population by starting with one of the first six numbers and then taking each sixth number on the list.

★ **10.18** If one of the six samples of the preceding exercise is randomly chosen to estimate the average amount of mail (in millions of ton-miles) carried per month, explain why there is a serious risk of getting a very misleading result.

★ **10.19** To generalize the example on page 257, suppose that we want to estimate the mean weight of six persons, whose (unknown) weights are 115, 125, 135, 185, 195, and 205 pounds.

    (a) List all possible random samples of size 2 which can be taken from this population, calculate their means, and determine the probability that the mean of such a sample will differ by more

than 5 from 160, the actual mean weight of the six persons.

    (b) Suppose that the first three weights are those of women and the other three are those of men. List all possible stratified samples of size 2 which can be taken by randomly choosing one of the three women and one of the three men, calculate their means, and determine the probability that the mean of such a sample will differ by more than 5 from 160, the actual mean weight of the six persons.

    (c) Suppose that three of the persons, those with weights 125, 135, and 185 pounds, are under 25 years of age, while the remaining are over 25 years of age. List all possible stratified samples of size 2 which can be taken by randomly choosing one of the three younger persons and one of the three older persons, calculate their means, and determine the probability that the mean of such a sample will differ by more than 5 from 160, the actual mean weight of the six persons.

    (d) Compare the results of parts (a), (b), and (c).

★ **10.20** Based on their volumes of sales, 18 of the 24 clothing stores in a city are classified as being small, and the other six are classified as being large. How many different stratified samples of four of these clothing stores can we choose, if

    (a) half of the sample is to be allocated to each of the two strata;

    (b) the allocation is to be proportional?

★ **10.21** Among 240 persons empaneled for jury duty, 120 are white, 80 are black, and 40 are Hispanic. How many different stratified samples of six of the 240 persons can we choose, if

    (a) one third of the sample is to be allocated to each of the three strata;

    (b) the allocation is to be proportional?

★ **10.22** Verify that if the formula

$$n_i = \frac{N_i}{N} \cdot n \qquad \text{for } i = 1, 2, \ldots, \text{ and } k$$

is used to determine the sample sizes allocated to the k strata, then

    (a) the allocation is proportional, that is, the ratios n_i/N_i all equal the same constant;

    (b) the sum of the n_i is equal to n.

★ **10.23** A stratified sample of size $n = 80$ is to be taken from a population of size $N = 2,000$, which consists of four strata of size $N_1 = 500$, $N_2 = 1,200$, $N_3 = 200$, and $N_4 = 100$. If the allocation is to be proportional, how large a sample must be taken from each of the four strata?

★ **10.24** In general, the formula used for the sample sizes in proportional allocation, given on page 258, will not produce integers for the n_i values. Suppose that a population with $k = 3$ strata has sizes $N_1 = 62$, $N_2 = 20$, $N_3 = 18$, and that we desire to take a sample of size $n = 12$.
  (a) Use the formula for proportional allocation, rounding the n_i values to the nearest integer, and show that the resulting sample sizes sum to 11.
  (b) Revise the results of part (a) to make the total sample size equal to 12 by increasing that value of n_i for which the decimal part of $\dfrac{N_i}{N} \cdot n$ is largest. Then give the values of the ratios N_i/n_i.
  (c) Revise the results of part (a) to make the total sample size equal to 12 by increasing that value of n_i for which the ratio N_i/n_i is largest. Then give the values of the ratios N_i/n_i.

The rounding issues presented in this exercise are precisely those of computing the numbers of seats, given to each of the states, in the United States House of Representatives. The ratios N_i/n_i correspond to the number of people represented by one congressperson. See the list of references at the end of this chapter.

★ **10.25** With reference to Exercise 10.19, list all possible cluster samples of size 2 which can be taken by randomly choosing either two of the three women or two of the three men, calculate their means, and determine the probability that the mean of such a sample will differ by more than 5 from 160, the actual mean weight of the six persons. If we compare this probability with those obtained in parts (a) and (b) of Exercise 10.19, what does this tell us about the relative merits of simple random sampling, stratified sampling, and cluster sampling in the given situation?

★ **10.26** In stratified sampling with proportional allocation, the importance of differences in stratum size is accounted for by letting the larger strata contribute relatively more items to the sample. However, strata differ not only in size but also in variability, and it would seem reasonable to take larger samples from the more variable strata and smaller samples from the less variable strata. If we let $\sigma_1, \sigma_2, \ldots,$ and σ_k denote the standard deviations of the k strata, we can account for both differences in stratum size and differences in stratum variability, by requiring that

$$\frac{n_1}{N_1 \sigma_1} = \frac{n_2}{N_2 \sigma_2} = \cdots = \frac{n_k}{N_k \sigma_k}$$

The sample sizes for this kind of allocation, called **optimal allocation** or **Neyman allocation** (in honor of its inventor), are given by the formula

$$n_i = \frac{n \cdot N_i \sigma_i}{N_1 \sigma_1 + N_2 \sigma_2 + \cdots + N_k \sigma_k}$$

for $i = 1, 2, \ldots,$ *and* k, where, if necessary, we round to the nearest integer. Verify that
  (a) with this formula, the quantities $\dfrac{n_i}{N_i \sigma_i}$ all equal the same constant;
  (b) the sum of the n_i is equal to n.

★ **10.27** A sample of size $n = 100$ is to be taken from a population consisting of two strata for which $N_1 = 10,000$, $N_2 = 30,000$, $\sigma_1 = 45$, and $\sigma_2 = 60$. To attain optimum allocation, how large a sample must be taken from each of the two strata?

★ **10.28** A sample of size $n = 84$ is to be taken from a population consisting of three strata for which $N_1 = 5,000$, $N_2 = 2,000$, $N_3 = 3,000$, $\sigma_1 = 15$, $\sigma_2 = 18$, and $\sigma_3 = 5$. To attain optimum allocation, how large a sample must be taken from each of the three strata?

★ **10.29** To estimate the mean of a population on the basis of a stratified sample, we calculate the weighted mean of the means $\bar{x}_1, \bar{x}_2, \ldots,$ and $\bar{x}_k$ obtained for the k strata, using as weights the strata sizes N_i. Verify that for proportional allocation, the weighted mean thus obtained equals the mean of the values obtained for all the strata.

★ **10.30** The records of a casualty insurance company show that among 3,800 claims filed against the company over a period of time 2,600 were minor claims (under $200), while the other 1,200 were major claims ($200 or more). To estimate the average size of these claims, the company takes a 1 percent sample, proportionally allocated to the two strata, with the following results (rounded to the nearest dollar):

Minor claims:	42	115	63	78	45	148	195
	66	18	73	55	89	170	41
	92	103	22	138	49	62	88
	113	29	71	58	83		

| *Major claims*: | 246 | 355 | 872 | 649 | 253 | 338 |
| | 491 | 860 | 755 | 502 | 488 | 311 |

(a) Find the means of these two samples and then determine their weighted mean, using as weights the two strata sizes $N_1 = 2{,}600$ and $N_2 = 1{,}200$.

(b) Verify that the result of part (a) equals the mean of the combined sample data. This being the case, proportional allocation is said to be **self-weighting**.

10.6

SAMPLING DISTRIBUTIONS

Let us now introduce the concept of the **sampling distribution** of a statistic, probably the most basic concept of statistical inference. As we shall see, this concept is closely related to the idea of chance variation, or chance fluctuations, which we mentioned earlier to emphasize the need for measuring the variability of data. In this chapter we shall concentrate mainly on the sample mean and its sampling distribution, but in some of the exercises on page 277 and in later chapters we shall also consider the sampling distributions of other statistics.

To illustrate the concept of a sampling distribution, let us construct the one for the mean of a random sample of size $n = 2$ drawn without replacement from the finite population of size $N = 5$, whose elements are the numbers 3, 5, 7, 9, and 11. The mean of this population is

$$\mu = \frac{3 + 5 + 7 + 9 + 11}{5} = 7$$

and its standard deviation is

$$\sigma = \sqrt{\frac{(3 - 7)^2 + (5 - 7)^2 + (7 - 7)^2 + (9 - 7)^2 + (11 - 7)^2}{5}}$$

$$= \sqrt{8}$$

Now, if we take a random sample of size $n = 2$ from this population, there are the $\binom{5}{2} = 10$ possibilities

| 3 and 5 | 3 and 7 | 3 and 9 | 3 and 11 | 5 and 7 |
| 5 and 9 | 5 and 11 | 7 and 9 | 7 and 11 | 9 and 11 |

and their means are 4, 5, 6, 7, 6, 7, 8, 8, 9, and 10. Since each sample has the probability $\frac{1}{10}$, we thus get the following sampling distribution of the mean of a random sample of size $n = 2$ from the given population:

$\bar{x}$	Probability
4	$\frac{1}{10}$
5	$\frac{1}{10}$
6	$\frac{2}{10}$
7	$\frac{2}{10}$
8	$\frac{2}{10}$
9	$\frac{1}{10}$
10	$\frac{1}{10}$

A histogram of this probability distribution is shown in Figure 10.2.

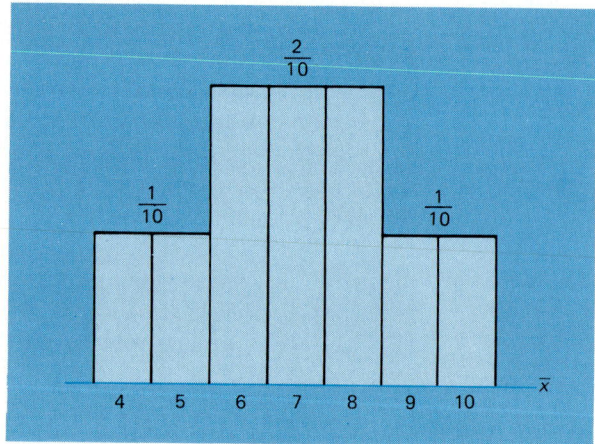

FIGURE 10.2 *Sampling distribution of the mean.*

An examination of this sampling distribution reveals some pertinent information relative to the problem of estimating the mean of the given population on the basis of a random sample of size 2. For instance, we see that corresponding to $\bar{x} = 6$, 7, or 8, the probability is $\frac{6}{10}$ that a sample mean will not differ from the population mean $\mu = 7$ by more than 1. Also, corresponding to $\bar{x} = 5$, 6, 7, 8, or 9, the probability is $\frac{8}{10}$ that a sample mean will not differ from the population mean $\mu = 7$ by more than 2. So, if we did not know the mean of the given population and wanted to estimate it with the mean of a random sample of size $n = 2$, the above would give us some idea about the possible size of our error.

Further useful information about this sampling distribution of the mean can be obtained by calculating its mean $\mu_{\bar{x}}$ and its standard deviation $\sigma_{\bar{x}}$, where the subscripts serve to distinguish between these parameters and those of the original population. Using the definitions of the mean and the variance of a probability distribution on pages 202 and 205, we get

$$\mu_{\bar{x}} = 4 \cdot \frac{1}{10} + 5 \cdot \frac{1}{10} + 6 \cdot \frac{2}{10} + 7 \cdot \frac{2}{10} + 8 \cdot \frac{2}{10} + 9 \cdot \frac{1}{10} + 10 \cdot \frac{1}{10}$$

$$= 7$$

and

$$\sigma_{\bar{x}}^2 = (4 - 7)^2 \cdot \frac{1}{10} + (5 - 7)^2 \cdot \frac{1}{10} + (6 - 7)^2 \cdot \frac{2}{10} + (7 - 7)^2 \cdot \frac{2}{10}$$

$$+ (8 - 7)^2 \cdot \frac{2}{10} + (9 - 7)^2 \cdot \frac{1}{10} + (10 - 7)^2 \cdot \frac{1}{10}$$

$$= 3$$

so that $\sigma_{\bar{x}} = \sqrt{3}$. Observe that, at least for this example, $\mu_{\bar{x}}$ equals μ and $\sigma_{\bar{x}}$ is smaller than σ. These relationships are of fundamental importance in statistics, and we shall return to them in Section 10.7.

To give an example of a sampling distribution, we took a very small sample of size $n = 2$ from a very small population of size $N = 5$, but it would be difficult to use the same method to construct the sampling distribution of the mean of a large sample from a large population. We would have to enumerate too many possibilities. For instance, for $n = 10$ and $N = 100$, we would have to list more than 17 trillion samples.

So, to get some idea about the sampling distribution of the mean of a somewhat larger sample from a large finite population, we shall use a **computer simulation**. In other words, we shall leave it to a computer to take repeated random samples from a given population, determine their means, and describe the distribution of these means in various ways. This will give us some idea about the overall shape and some of the key features of the real sampling distribution of the mean for random samples of the given size from the given population.

Figure 10.3 on page 266 shows the results of a computer simulation, in which one hundred random samples of size $n = 15$ are drawn without replacement from the finite population which consists of the integers from 1 to 1,000, $\bar{x}$ is calculated for each of the samples, and the distribution of the $\bar{x}$'s is suitably described. Note that although sampling is without replacement for each sample of size $n = 15$, the whole sample is, so to speak, replaced by the computer before the next sample is drawn.

Near the top of the printout we find that the mean and the standard deviation of the population are $\mu = 500.5$ and $\sigma = 289$ (see also Exercise 10.55 on page 276). The printout does not show the actual samples or their means, but further down it tells us that the mean of the hundred sample means is 506.3 and that their standard deviation is 63.9. Also, it shows the means grouped into a distribution with the class marks 25.5, 75.5, 125.5, ..., which were chosen so that the class boundaries are the impossible values 0.5, 50.5, 100.5, ...; these values cannot be taken on by a mean of 15 positive integers. Asked to describe this distribution, we

```
MTB > GENE 1 1ØØØ, C1
MTB > MEAN C1
    MEAN     =        5ØØ.5
MTB > STDEV C1
    STDEV    =         289.
MTB > SET C3
DATA> 4
DATA> END
MTB > STORE
STOR> SAMPLE 15 C1 C2
STOR> LET K1=AVER(C2)
STOR> JOIN K1 TO C3 PUT IN C3
STOR> ERASE C2
STOR> END
MTB > EXECUTE 1ØØ TIMES
MTB > OMIT ROW WITH 4 IN C3 AND PUT REST IN C4
MTB > MEAN C1
    MEAN     =        5Ø6.3
MTB > STDEV C1
    STDEV    =         63.9
MTB > HIST C4 25.5 5Ø

    C4

    MIDDLE OF     NUMBER OF
    INTERVAL      OBSERVATIONS
      25.5           Ø
      75.5           Ø
     125.5           Ø
     175.5           Ø
     225.5           Ø
     275.5           Ø
     325.5           1      *
     375.5           4      ****
     425.5          16      ****************
     475.5          23      ***********************
     525.5          33      *********************************
     575.5          16      ****************
     625.5           5      *****
     675.5           2      **
```

FIGURE 10.3 *Computer simulation of a sampling distribution of the mean.*

might say that it is fairly symmetrical and bell-shaped; in fact, the overall pattern is quite close to that of a normal curve. All this applies to the distribution which we obtained by means of the computer simulation involving only one hundred random samples of size $n = 15$ drawn without replacement from the finite population which consists of the integers from 1 to 1,000. It applies, we expect, also to the actual sampling distribution of the mean which pertains to all possible random samples of size $n = 15$ drawn without replacement from this population.

Note also that the results of the simulation support the relationships pointed out on page 265, where the mean and the standard deviation of a sampling distribution of the mean were, respectively, equal to and less than the corresponding parameters of the population. Although the mean of the hundred sample means is 506.3 and not exactly $\mu = 500.5$, this is very close; also, the standard deviation of the hundred sample means is 63.9, which is less than $\sigma = 289$.

10.7

THE STANDARD ERROR OF THE MEAN

In most practical situations we cannot proceed as in the two examples of the preceding section. That is, we cannot enumerate all possible samples or simulate a sampling distribution in order to judge how close a sample mean might be to the mean of the population from which the sample came. Fortunately, though, we can usually get the information we need from two theorems, which express essential facts about sampling distributions of the mean. One of these will be discussed below and the other in Section 10.8.

The first of these two theorems expresses formally what we discovered from both of the examples of the preceding section—the mean of the sampling distribution of $\bar{x}$ equals the mean of the population sampled, and the standard deviation of the sampling distribution of $\bar{x}$ is smaller than the standard deviation of the population sampled. It may be phrased as follows: For random samples of size n taken from a population with the mean μ and the standard deviation σ, the sampling distribution of $\bar{x}$ has the mean

Mean of sampling distribution of $\bar{x}$

$$\mu_{\bar{x}} = \mu$$

and the standard deviation

Standard error of the mean

$$\sigma_{\bar{x}} = \frac{\sigma}{\sqrt{n}} \quad or \quad \sigma_{\bar{x}} = \frac{\sigma}{\sqrt{n}} \cdot \sqrt{\frac{N-n}{N-1}}$$

depending on whether the population is infinite or finite of size N.

It is customary to refer to $\sigma_{\bar{x}}$ as the **standard error of the mean**, where "standard" is used in the sense of an average, as in "standard deviation." Its role in statistics is fundamental, as it measures the extent to which sample means can be expected to fluctuate, or vary, due to chance. If $\sigma_{\bar{x}}$ is small, the chances are good that the mean of a sample will be close to the mean of the population; if $\sigma_{\bar{x}}$ is large, we are more likely to get a sample mean which differs considerably from the mean of the population.

What determines the size of $\sigma_{\bar{x}}$ can be seen from the two formulas above. Both formulas (for infinite and finite populations) show that $\sigma_{\bar{x}}$ increases as the variability of the population increases, and that it decreases as the sample size increases. In fact, it is directly proportional to σ and inversely proportional to the square root of n. (For finite populations it decreases even faster due to the n appearing in $\sqrt{\dfrac{N-n}{N-1}}$.)

EXAMPLE When we take a sample from an infinite population, what happens to the standard error of the mean, and hence to the error we might expect when we use a sample mean to estimate the mean of a population, if the sample size is increased from 50 to 200?

Solution The ratio of the two standard errors is

$$\frac{\dfrac{\sigma}{\sqrt{200}}}{\dfrac{\sigma}{\sqrt{50}}} = \frac{\sqrt{50}}{\sqrt{200}} = \sqrt{\frac{50}{200}} = \sqrt{\frac{1}{4}} = \frac{1}{2}$$

and where n is quadrupled, the standard error of the mean is only divided by 2.

The factor $\sqrt{\dfrac{N-n}{N-1}}$ in the second formula for $\sigma_{\bar{x}}$ is called the **finite population correction factor**, for without it the two formulas for $\sigma_{\bar{x}}$ (for infinite and finite populations) are the same. In practice, it is omitted unless the sample constitutes at least 5 percent of the population, for otherwise it is so close to 1 that it has little effect on the value of $\sigma_{\bar{x}}$. This factor appears also in the standard deviation of the hypergeometric distribution, as in Exercise 8.69 on page 209.

EXAMPLE Find the value of the finite population correction factor for $n = 100$ and $N = 10,000$.

Solution Substituting $n = 100$ and $N = 10,000$, we get

$$\sqrt{\frac{N-n}{N-1}} = \sqrt{\frac{10,000 - 100}{10,000 - 1}} = 0.995$$

and this is so close to 1 that the correction factor can be omitted for all practical purposes.

Since we stated the two formulas for the standard error of the mean without proof, let us verify the one for finite populations by referring to the results of the two examples of Section 10.6.

EXAMPLE With reference to the example on page 263, where we had $n = 2$, $N = 5$, and $\sigma = \sqrt{8}$, verify that the second of the two formulas for $\sigma_{\bar{x}}$ will yield $\sqrt{3}$. This is the value we got on page 265.

Solution   Substituting $n = 2$, $N = 5$, and $\sigma = \sqrt{8}$ into the second of the two formulas for $\sigma_{\bar{x}}$, we get

$$\sigma_{\bar{x}} = \frac{\sqrt{8}}{\sqrt{2}} \cdot \sqrt{\frac{5-2}{5-1}} = \frac{\sqrt{8}}{\sqrt{2}} \cdot \sqrt{\frac{3}{4}} = \sqrt{\frac{8}{2} \cdot \frac{3}{4}} = \sqrt{3}$$

EXAMPLE   With reference to the computer simulation of Figure 10.3, where we had $n = 15$, $N = 1,000$, and $\sigma = 289$, what value might we have expected for the standard deviation of the hundred sample means?

Solution   Substituting $n = 15$, $N = 1,000$, and $\sigma = 289$ into the second of the two formulas for $\sigma_{\bar{x}}$, we get

$$\sigma_{\bar{x}} = \frac{289}{\sqrt{15}} \cdot \sqrt{\frac{1,000 - 15}{1,000 - 1}} = 74.1$$

This is not very close to 63.9, the value obtained in the computer simulation of Figure 10.3, but such things can happen due to chance; another simulation yielded 73.3, which is much closer to 74.1.

10.8
THE CENTRAL LIMIT THEOREM

When we use a sample mean to estimate the mean of a population, the uncertainties about our error can be expressed in various ways. When we know the exact sampling distribution of the mean, which we seldom do, we could proceed as in the example on page 264 and calculate the probabilities associated with errors of various size. Also, we always can, but in practice don't, use Chebyshev's theorem and assert with a probability of at least $1 - \frac{1}{k^2}$ that the mean of a random sample will differ from the mean of the population sampled by less than $k \cdot \sigma_{\bar{x}}$.

EXAMPLE   Based on Chebyshev's theorem with $k = 2$, what can we say about the size of our error, if we are going to use the mean of a random sample of size $n = 64$ to estimate the mean of an infinite population with $\sigma = 20$?

Solution   Substituting $n = 64$ and $\sigma = 20$ into the appropriate formula for the standard error of the mean, we get

$$\sigma_{\bar{x}} = \frac{20}{\sqrt{64}} = 2.5$$

and it follows that we can assert with a probability of at least $1 - \frac{1}{2^2} = 0.75$ that the error will be less than $k \cdot \sigma_{\bar{x}} = 2(2.5) = 5$.

Here, the trouble is that "at least 0.75" does not tell us enough when in reality the probability may be, say, 0.98 or even 0.999.

Chebyshev's theorem provides the logical connection between errors and the probabilities that they will be committed, but there exists another theorem which, in many instances, enables us to make much stronger probability statements about our potential errors. This theorem, which is the second of the two theorems referred to on page 267, is called the **central limit theorem** and it may be stated as follows:

Central limit theorem

> *For large samples, the sampling distribution of the mean can be approximated closely with a normal distribution.*

If we combine this theorem with the one of the preceding section, according to which $\mu_{\bar{x}} = \mu$ and $\sigma_{\bar{x}} = \dfrac{\sigma}{\sqrt{n}}$ for random samples from infinite populations, we find that if $\bar{x}$ is the mean of a random sample of size n from an infinite population with the mean μ and the standard deviation σ and n is large, then

$$z = \frac{\bar{x} - \mu}{\sigma/\sqrt{n}}$$

is a value of a random variable having approximately the standard normal distribution.

The central limit theorem is of fundamental importance in statistics, because it justifies the use of normal-curve methods in a wide range of problems; it applies to infinite populations, and also to finite populations when n, though large, constitutes but a small portion of the population. It is difficult to say precisely how large n must be so that the central limit theorem can be applied, but unless the population distribution has a very unusual shape, $n = 30$ is usually regarded as sufficiently large. Note that when we are actually sampling a normal population, the sampling distribution of the mean is a normal distribution regardless of the size of n.

Let us now see what probability will take the place of "at least 0.75," if we use the central limit theorem instead of Chebyshev's theorem in the example on page 269.

EXAMPLE Based on the central limit theorem, what is the probability that the error will be less than 5, when the mean of a random sample of size $n = 64$ is used to estimate the mean of an infinite population with $\sigma = 20$?

Solution The probability is given by the area of the white region under the curve in Figure 10.4, namely, by the standard-normal-curve area between

$$z = \frac{-5}{20/\sqrt{64}} = -2 \quad \text{and} \quad z = \frac{5}{20/\sqrt{64}} = 2$$

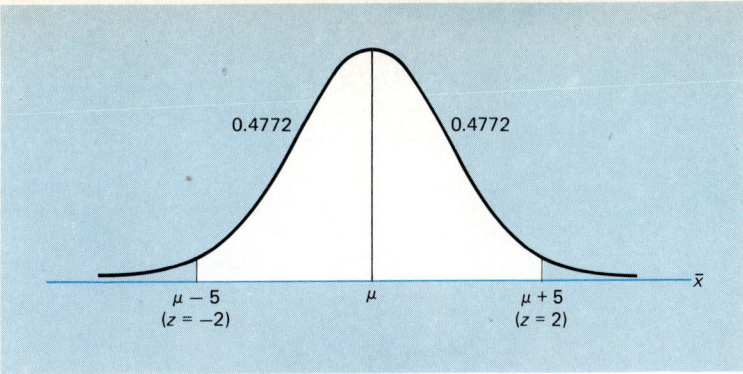

FIGURE 10.4 *Sampling distribution of the mean.*

Since the entry in Table I corresponding to $z = 2.00$ is 0.4772, the probability asked for is $0.4772 + 0.4772 = 0.9544$. Thus, the statement that the probability is "at least 0.75" is replaced by the much stronger statement that the probability is about 0.95 (that the mean of a random sample of size $n = 64$ from the given population will differ from the mean of the population by less than 5).

The central limit theorem can also be used for finite populations, but a precise description of the situations under which this can be done is rather complicated. The most common proper use is the case in which n is large while $\frac{n}{N}$ is small. This is the case in most political polls.

Since we presented the central limit theorem without proof, let us consider the following example:

EXAMPLE In the simulation of Figure 10.3, a hundred random samples of size $n = 15$ were drawn without replacement from the finite population which consists of the integers from 1 to 1,000, namely, from a finite population with the mean $\mu = 500.5$ and $\sigma = 289$. Based on the central limit theorem, what is the probability that the mean of such a sample will differ from $\mu = 500.5$ by 150 or less?

Solution The probability is given by the area of the white region under the curve in Figure 10.5 on page 272. Since

$$\sigma_{\bar{x}} = \frac{289}{\sqrt{15}} \cdot \sqrt{\frac{1,000 - 15}{1,000 - 1}} = 74.1$$

this area is the standard-normal-curve area between

$$z = \frac{-150}{74.1} = -2.02 \quad \text{and} \quad z = \frac{150}{74.1} = 2.02$$

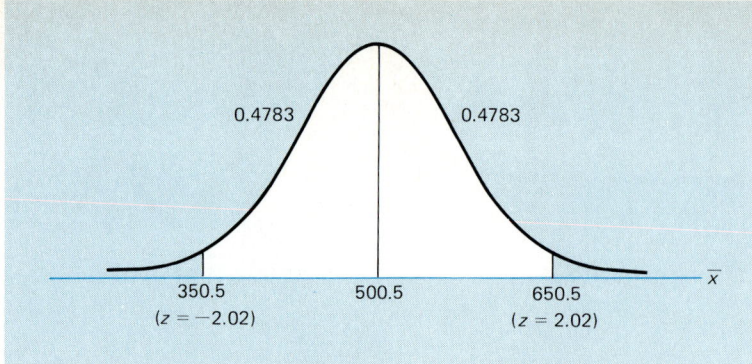

FIGURE 10.5 *Sampling distribution of the mean.*

Since the entry in Table I corresponding to $z = 2.02$ is 0.4783, the probability asked for is $0.4783 + 0.4783 = 0.9566$, or 0.96 rounded to two decimals. Now, from Figure 10.3 we find that 97 of the 100 sample means fall between $500.5 - 150 = 350.5$ and $500.5 + 150 = 650.5$, and this is very close, indeed, to the 96 out of 100 we might have expected on the basis of our calculations. Thus, the simulation provides definite support for the central limit theorem.

10.9

SOME FURTHER CONSIDERATIONS

In Sections 10.6 through 10.8, our main goal was to introduce the concept of a sampling distribution, and the one which we chose as an illustration was the sampling distribution of the mean. It should be clear, though, that instead of the mean we could have studied the median, the standard deviation, or some other statistic, and investigated its chance fluctuations. So far as the corresponding theory is concerned, this would have required a different formula for the standard error and theory analogous to, yet different from, the central limit theorem.

For instance, for large samples from continuous populations, the **standard error of the median** is approximately

$$\sigma_{\tilde{x}} = 1.25 \cdot \frac{\sigma}{\sqrt{n}}$$

where n is the size of the sample and σ is the population standard deviation. Note that comparison of the two formulas

$$\sigma_{\bar{x}} = \frac{\sigma}{\sqrt{n}} \quad \text{and} \quad \sigma_{\tilde{x}} = 1.25 \cdot \frac{\sigma}{\sqrt{n}}$$

reflects the fact that the mean is generally more reliable than the median (that is, it tends to expose us to smaller errors) when we estimate the mean of a symmetrical population. For symmetrical populations, the means of the sampling distributions of $\bar{x}$ and $\tilde{x}$ are both equal to the population mean μ.

EXAMPLE How large a random sample do we need so that its mean is as reliable an estimate of the mean of a symmetrical continuous population as the median of a random sample of size $n = 200$?

Solution Equating the two standard error formulas and substituting $n = 200$ into the one for the standard error of the median, we get

$$\frac{\sigma}{\sqrt{n}} = 1.25 \cdot \frac{\sigma}{\sqrt{200}}$$

which, solved for n, yields $n = 128$. Thus, for the stated purpose, the mean of a random sample of size $n = 128$ is as "good" as the median of a random sample of size $n = 200$.

Also, a point worth repeating is that the examples of Section 10.6 were used as teaching aids, designed to convey the idea of a sampling distribution, but they do not reflect what we do in actual practice. In practice, we can seldom list all possible samples, and ordinarily we base an inference on one sample and not on 100 samples. In Chapter 11 and in subsequent chapters we shall go further into the problem of translating theory about sampling distributions into methods of evaluating the merits and shortcomings of statistical procedures.

Another point worth repeating concerns the $\sqrt{n}$ which appears in the denominator of the formula for the standard error of the mean. It makes sense that when n becomes larger and larger, our generalizations will be subject to smaller errors, but the $\sqrt{n}$ in the formula for the standard error of the mean tells us that the gain in reliability is not proportional to the increase in the size of the sample. As we saw, quadrupling the size of the sample will only double the reliability of a sample mean as an estimate of the mean of a population. Indeed, to quadruple the reliability, we would have to multiply the sample size by 16. This relationship between reliability and sample size suggests that it seldom pays to take samples that are really large.

EXERCISES

10.31 Suppose that in the example on page 263, where random samples of size $n = 2$ were drawn without replacement from the finite population which consists of the numbers 3, 5, 7, 9, and 11, sampling had been with replacement.

(a) List the 25 ordered samples that can be drawn with replacement from the given population and calculate their means. (By "ordered" we mean that 3 and 7, for example, is a different sample than 7 and 3.)

(b) Assuming that sampling is random, namely, that each of the ordered samples of part (a) has the probability $\frac{1}{25}$, construct the sampling distribution of the mean for random samples of size $n = 2$ drawn with replacement from the given population.

10.32 With reference to the preceding exercise, find the probabilities that the mean of a random sample of size $n = 2$ drawn with replacement from the given population will differ from $\mu = 7$ by
(a) not more than 1;
(b) at most 2.

10.33 Calculate the standard deviation of the sampling distribution obtained in part (b) of Exercise 10.31, and verify the result by substituting $n = 2$ and $\sigma = \sqrt{8}$ into the first of the two standard error formulas on page 267.

10.34 A finite population consists of the $N = 6$ numbers 6, 9, 12, 15, 18, and 21, and it can be easily verified that its mean is $\mu = 13.5$ and its standard deviation is $\frac{1}{2}\sqrt{105}$, which is 5.12, rounded to two decimals.
(a) List the 20 possible samples of size $n = 3$ that can be drawn without replacement from this finite population.
(b) Calculate the means of the 20 samples obtained in (a).
(c) Assigning the probability $\frac{1}{20}$ to each of the samples obtained in part (a), construct the sampling distribution of the mean for random samples of size $n = 3$ drawn without replacement from the given finite population.
(d) What is the probability that the mean of a random sample of size $n = 3$ drawn without replacement from the given population will differ from $\mu = 13.5$ by less than 3?

10.35 With reference to the preceding exercise, calculate the mean and the standard deviation of the sampling distribution obtained in part (c) and verify the results using the theory of Section 10.7.

10.36 With reference to Exercise 10.34, determine the medians of the 20 samples obtained in part (a). Construct the sampling distribution of the median for random samples of size $n = 3$ drawn without replacement from the given population. Find the probability that the median of such a random sample will differ from $\mu = 13.5$ by less than 3. Compare this probability with the one obtained in part (d) of Exercise 10.34.

10.37 With reference to Exercise 10.34, determine the ranges of the 20 samples obtained in part (a) and construct the sampling distribution of the range for random samples of size $n = 3$ drawn without replacement from the given population.

10.38 When we sample an infinite population, what happens to the standard error of the mean if the sample size is
(a) increased from 60 to 240;
(b) increased from 200 to 450;
(c) increased from 25 to 225;
(d) decreased from 640 to 40?

10.39 What is the value of the finite population correction factor when
(a) $n = 10$ and $N = 200$;
(b) $n = 10$ and $N = 500$;
(c) $n = 10$ and $N = 2,000$;
(d) $n = 20$ and $N = 200$;
(e) $n = 40$ and $N = 400$;
(f) $n = 400$ and $N = 4,000$?

★ **10.40** Show that if the mean of a random sample of size n is used to estimate the mean of an infinite population with the standard deviation σ and n is large, there is a fifty-fifty chance that the magnitude of the error will be less than

$$0.6745 \cdot \frac{\sigma}{\sqrt{n}}$$

It has been the custom to refer to this quantity as the **probable error of the mean**; nowadays, it is used mainly in military applications.
(a) If a random sample of size $n = 64$ is drawn from an infinite population with $\sigma = 20.6$, what is the probable error of the mean?
(b) If a random sample of size $n = 100$ is drawn from a very large population (consisting of the fines paid for various traffic violations in a certain county in 1986) with $\sigma = \$18.26$, what is the probable error of the mean? Explain its significance.

★ **10.41** With reference to Exercise 10.16 on page 261, assign each of the systematic samples the probability $\frac{1}{10}$ and calculate the standard deviation of the corresponding sampling distribution of the mean. Compare the result with $\sigma_{\bar{x}} = 2.3$, the standard error of the mean for ordinary random samples of size $n = 5$ drawn from the given finite population.

★ **10.42** With reference to Exercise 10.17 on page 261, assign each of the systematic samples the probability $\frac{1}{6}$ and calculate the standard deviation of the corresponding sampling distribution of the mean. Compare the result with $\sigma_{\bar{x}} = 3.5$, the standard error of the mean for ordinary random samples of size $n = 8$ from the given finite population.

★ **10.43** In the example on page 258, we compared stratified samples from a population consisting of four weights with ordinary random samples of the same size.
  (a) Assigning each of the random samples on page 258 the probability $\frac{1}{6}$, show that the mean and the standard deviation of this sampling distribution of the mean are $\mu_x = 160$ and $\sigma_x = 21.0$.
  (b) Assigning each of the stratified samples on page 258 the probability $\frac{1}{4}$, show that the mean and the standard deviation of this sampling distribution of the mean are $\mu_{\bar{x}} = 160$ and $\sigma_{\bar{x}} = 7.1$.
  Compare the results of parts (a) and (b).

★ **10.44** If $\bar{x}$ is the mean of a stratified random sample of size n obtained by proportional allocation from a finite population of size N, which consists of k strata of size $N_1, N_2, \ldots,$ and N_k, then

$$\sigma_{\bar{x}}^2 = \sum_{i=1}^{k} \frac{(N-n)N_i^2}{nN^2(N_i-1)} \cdot \sigma_i^2$$

where $\sigma_1^2, \sigma_2^2, \ldots,$ and σ_k^2 are the corresponding variances for the individual strata. Use this formula to verify the result of part (b) of the preceding exercise.

★ **10.45** In the example on page 258, the two cluster samples, which consist of choosing and weighing either the two women or the two men, have the means 125 and 195. Assigning the probability $\frac{1}{2}$ to each of these means, calculate the standard deviation of this sampling distribution of the mean, and compare it with the corresponding values for ordinary random sampling and stratified sampling obtained in the two parts of Exercise 10.43.

10.46 The mean of a random sample of size $n = 36$ is used to estimate the mean of an infinite population having the standard deviation $\sigma = 9$. What can we assert about the probability that the error will be less than 4.5, if we use
  (a) Chebyshev's theorem;
  (b) the central limit theorem?

10.47 The mean of a random sample of size $n = 25$ is used to estimate the mean of a very large population (consisting of the attention spans of persons over 65), which

has a standard deviation of $\sigma = 2.4$ minutes. What can we assert about the probability that the error will be less than 1.2 minutes, if we use
  (a) Chebyshev's theorem;
  (b) the central limit theorem?

10.48 The mean of a random sample of size $n = 144$ is used to estimate the mean of a very large population (consisting of the weights of certain animals), which has the standard deviation $\sigma = 3.6$ ounces. Based on the central limit theorem, what can we say about the probability that the error will be
  (a) less than 0.42 ounce;
  (b) greater than 0.75 ounce?

10.49 If measurements of the specific gravity of a metal can be looked upon as a sample from a normal population having the standard deviation $\sigma = 0.025$, what is the probability that the mean of a random sample of size $n = 16$ will be "off" by at most 0.01?

10.50 If the distribution of the weights of all men traveling by air between Dallas and El Paso has a mean of 163 pounds and a standard deviation of 18 pounds, what is the probability that the combined weight of 36 men traveling on such a flight is more than 6,012 pounds?

10.51 The number of driving miles before a certain kind of tire begins to show wear is, on average, 16,800 miles with a standard deviation of 3,300 miles. A car rental agency buys 36 of these tires for replacement purposes and puts each one on a different car.
  (a) Find the probability that the 36 tires will average less than 16,000 miles until they begin to show wear.
  (b) Find the probability that the 36 tires will average more than 18,000 miles until they begin to show wear.
  (c) Explain why the probabilities in (a) and (b) could not be calculated from the given information if the car rental agency puts four of the 36 tires on each of nine cars.

10.52 Suppose that a population of jelly beans has a mean weight of $\mu = 4.5$ grams and a standard deviation of $\sigma = 0.4$ gram. What is the minimum number of jelly beans to select if you want your sample of jelly beans to have a total weight of at least 200 grams with a probability of 0.90 or more? (*Hint:* Suppose that you took a sample of $n = 60$. Getting a total weight in excess of 200 grams would require that the average weight in the sample exceeds $\frac{200}{60} = 3.33$ grams. The probability that the average of a sample of 60 will exceed 3.33 grams is larger than 0.90; therefore 60 jelly beans is too large a sample. Continue this logic to find the minimum sample size.)

10.53 Verify that the mean of a random sample of size $n = 256$ is as reliable an estimate of the mean of a symmetrical continuous population as the median of a random sample of size $n = 400$.

10.54 A gourmet chef would like to have 40 pounds of shad roe for a special banquet. Shad roe is sold in "sets" rather than by weight, but the chef knows that the sets come from a population that has the mean weight $\mu = 0.60$ pound and the standard deviation $\sigma = 0.20$ pound. How many sets should the chef order if he wants the probability of reaching the 40 pound objective to be at least 0.90?

★ **10.55** Using the result that the sum of the first n positive integers is $\dfrac{n(n + 1)}{2}$ and the sum of their squares is $\dfrac{n(n + 1)(2n + 1)}{6}$, it can be shown that the mean and the standard deviation of the finite population which consists of the first n positive integers are

$$\mu = \frac{n + 1}{2} \quad \text{and} \quad \sigma = \sqrt{\frac{n^2 - 1}{12}}$$

Verify that these formulas yield the population mean and standard deviation shown in the printout of Figure 10.3 on page 266.

10.10

TECHNICAL NOTE (Simulation)★

Simulation provides one of the most effective ways of illustrating, and thus teaching, some of the basic concepts of statistics. As we shall see in the chapters which follow, the evaluation and interpretation of statistical techniques will often require that we imagine what would happen if experiments are repeated over and over again. Since, most of the time, such repetitions are neither practical nor feasible, we can resort instead to simulations, preferably with the use of computers. Simulation also plays a role in the development of statistical theory, for there are situations where simulation is easier and cheaper than a detailed mathematical analysis.

In earlier chapters we used simulation to generate values of random variables having binomial, Poisson, and normal distributions, and in this chapter we used simulation to generate random samples from specified populations as well as a sampling distribution of the mean. It is hoped that the reader will have the opportunity to generate and study simulations of the sampling distribution of the median, the sampling distribution of s or s^2, or the sampling distributions of some other statistics.

Simulations of sampling distributions can also be performed without computers by referring to a table of random numbers, but to simplify matters for the reader we present here 40 simulated random samples, each consisting of $n = 5$ values of a random variable having the Poisson distribution with the mean $\lambda = 16$. (The reader may picture these figures as data on the number of emergency calls which an ambulance service receives in an afternoon, the number of cars that arrive at a tollbooth during a five-minute interval, the number of pieces of junk mail that a family receives in a week, and so forth.)

Sample						Sample					
1	22	13	15	18	17	21	19	10	15	16	18
2	20	10	15	15	19	22	16	18	26	14	20
3	18	12	13	19	9	23	15	14	21	17	11
4	12	13	13	14	12	24	18	8	21	14	15
5	20	17	16	18	19	25	19	17	16	13	15
6	10	11	9	11	11	26	12	12	20	11	15
7	16	18	14	9	11	27	17	9	18	16	9
8	11	17	19	20	17	28	10	18	19	13	20
9	14	19	16	13	21	29	18	14	23	23	14
10	17	16	11	17	11	30	16	13	10	14	20
11	12	21	10	15	16	31	15	11	21	8	17
12	15	26	19	20	15	32	12	11	19	17	16
13	13	15	15	13	18	33	15	16	17	17	16
14	15	16	16	15	17	34	20	17	11	19	15
15	22	15	13	19	11	35	15	11	14	14	18
16	18	13	15	11	12	36	17	22	21	16	20
17	11	23	12	20	14	37	22	21	16	15	13
18	20	8	17	16	13	38	16	17	20	17	7
19	16	10	9	19	15	39	21	18	17	26	25
20	14	19	16	18	20	40	12	18	15	14	16

In the exercises which follow, the reader will be asked to use these figures for the construction of various simulated sampling distributions.

EXERCISES

★ **10.56** Determine the means of the 40 samples given above. Give the mean and standard deviation of this list of 40 means, and compare the results with the corresponding values expected in accordance with the theory of Section 10.7.

★ **10.57** Determine the medians of the 40 samples given above, and give the standard deviation of this list of 40 means. Compare this to the standard deviation of the list of 40 means in the previous problem.

★ **10.58** Group the 40 sample means obtained in Exercise 10.56 and the 40 sample medians obtained in Exercise 10.57 into side-by-side histograms with the classes 9.5–10.5, 10.5–11.5, 11.5–12.5, …, and 20.5–21.5. Compare the two histograms.

★ **10.59** Determine the ranges of the 40 samples given above and construct a histogram showing how many times each value occurs. Based on these data, what would be reasonable odds that the range of such a sample will be greater than 10?

★ **10.60** For large samples, the formula $\dfrac{\sigma}{\sqrt{2n}}$ may be used to approximate the standard error of s, namely, the standard deviation of the sampling distribution of s. Calculate the standard deviation of 3.4, 4.0, 4.2, 0.8, 1.6, 0.9, 3.6, 3.5, 3.4, 3.1, 4.2, 4.5, 2.0, 0.8, 4.5, 2.8, 5.2, 4.5, 4.2, 2.4, 3.5, 4.6, 3.7, 4.9, 2.2, 3.7, 4.4, 4.3, 4.5, 3.7, 5.1, 3.4, 0.8, 3.6, 2.5, 2.6, 3.9, 4.9, 4.0, and 2.2, which are the standard deviations of the 40 samples given above. Compare the result with the corresponding value obtained by using the formula above with $n = 5$ and $\sigma = \sqrt{16} = 4$. Note that even though $n = 5$ is small, the approximation is very good.

★ **10.61** Use Table XII to simulate 50 random samples of size $n = 3$ from a standard normal population and calculate their means. Also calculate the mean and the standard deviation of the 50 means and compare them with the values we might expect in accordance with the theory on page 267.

10.11

CHECKLIST OF KEY TERMS
(with page references to their definitions)

★ Area sampling, 260
Central limit theorem, 270
★ Cluster sampling, 260
★ Computer simulation, 265
★ Cross stratification, 259
Finite population, 251
Finite population correction factor, 268
Infinite population, 251
★ Judgment sample, 260
★ *Neyman allocation*, 262
★ *Optimum allocation*, 262
★ *Probable error of the mean*, 274

★ Proportional allocation, 258
★ Quota sampling, 259
Random sample, 252
★ Sample design, 256
Sampling distribution, 263
★ *Self-weighting*, 263
Simple random sample, 252
Standard error of the mean, 267
Standard error of the median, 272
★ Stratified sampling, 257
★ Systematic sampling, 257

10.12

REVIEW EXERCISES

10.62 The mean of a random sample of size $n = 81$ is used to estimate the mean of a normal population with $\sigma = 18$. With what probability can we assert that the error of this estimate will be less than 9, if we use
   (a) Chebyshev's theorem;
   (b) the central limit theorem?

★ **10.63** Among 80 persons interviewed for certain jobs by a government agency, 40 are married, 20 are single, 10 are divorced, and 10 are widowed. In how many ways can a 10 percent stratified sample be chosen from among the persons interviewed, if
   (a) one-fourth of the sample is to be allocated to each group;
   (b) the allocation is proportional?

10.64 A quality control inspector wants to examine 15 of 625 computers which were received as part of a single shipment. Which ones (by number) will he examine, using the last three columns of the random numbers on the second page of Table XI, starting with the 6th row and going down the page, if the computers are numbered 001, 002, 003, . . . , and 625?

10.65 If a random sample of size $n = 3$ is to be chosen from a finite population of size $N = 80$, what is the probability of each possible sample?

10.66 Convert the 40 samples on page 277 into 20 samples of size $n = 10$ by combining samples 1 and 21, sam-

ples 2 and 22, . . . , and samples 20 and 40. Calculate the mean of each of these new samples and determine their mean and their standard deviation. Compare these values with those we might expect in accordance with the theorem on page 267.

10.67 Random samples of size $n = 2$ are drawn from the finite population which consists of the numbers 1, 3, 5, and 7. As can easily be verified, $\mu = 4$ for this population and $\sigma = \sqrt{5}$.
   (a) List the six possible samples of size $n = 2$ that can be drawn without replacement from the given population, and calculate their means.
   (b) Assigning each of the samples obtained in part (a) the probability $\frac{1}{6}$, construct the sampling distribution of the mean for random samples of size $n = 2$ drawn without replacement from the given population.
   (c) Calculate the mean and the standard deviation of the probability distribution obtained in part (b), and compare the results with the values we would expect in accordance with the formulas on page 267.

10.68 Rework the preceding exercise for sampling with replacement, so that there are 16 possible samples and each one has the probability $\frac{1}{16}$.

★ 10.69 A finite population consists of three strata for which $N_1 = 5{,}000$, $N_2 = 2{,}000$, $N_3 = 3{,}000$, $\sigma_1 = 15$, $\sigma_2 = 18$, and $\sigma_3 = 5$. If a stratified sample of size $n = 168$ is to be taken from this population using optimum allocation, what size samples must be taken from the three strata?

★ 10.70 In Exercise 10.19 on page 261, the reader was asked to compare stratified sampling from a population which consists of the weights of three women and the weights of three men with ordinary random sampling.
(a) Assigning each of the random samples the probability $\frac{1}{15}$, show that $\sigma_{\bar{x}} = 22.7$ for the means of random samples of size $n = 2$ from the given population.
(b) Assigning each of the stratified samples the probability $\frac{1}{9}$, show that $\sigma_{\bar{x}} = 5.8$ for the means of stratified samples of size $n = 2$ from the given population.
Compare the results of parts (a) and (b).

★ 10.71 Use the formula of Exercise 10.44 on page 275 to verify the result of part (b) of the preceding exercise.

★ 10.72 With reference to Exercise 10.19 on page 261, how many different cluster samples of size $n = 2$ can we choose from the given population, if we want to choose the weights of two of the women or the weights of two of the men?

10.73 What is the value of the finite population correction factor when
(a) $n = 30$ and $N = 120$;
(b) $n = 20$ and $N = 500$?

10.74 When we sample from an infinite population, what happens to the standard error of the mean if the sample size is
(a) increased from 20 to 500;
(b) decreased from 490 to 40?

10.75 The annual dollar value of homes sold by real estate salespersons in a certain community is 7.5 million dollars, on average, with a standard deviation of 4.4 million dollars. Find the probabilities that in a sample of 36 such salespersons the average annual dollar value of home sales will
(a) be less than 6.5 million dollars;
(b) exceed 9 million dollars.

10.76 The weights of persons who use a certain elevator are normally distributed, at least approximately, with mean $\mu = 130$ pounds and standard deviation $\sigma = 15$ pounds. The elevator contains a sign which says "capacity 3,000 pounds." What is the greatest number of persons that can ride in the elevator while keeping below 0.05 the probability that the capacity is violated?

10.77 How many different samples of size $n = 4$ can be selected from a finite population of size
(a) $N = 18$;
(b) $N = 30$;
(c) $N = 100$?

10.78 What is the probability of each possible sample if a random sample of size $n = 5$ is to be drawn from a finite population of size $N = 24$?

10.79 The mean of a random sample of size $n = 81$ is used to estimate the mean annual growth of certain plants. Assuming that $\sigma = 3.6$ mm for such data, use the central limit theorem to find the probabilities that this estimate will be off either way by
(a) less than 1.0 mm;
(b) less than 0.5 mm.

★ 10.80 With reference to Exercise 10.17 on page 261, list the eight possible systematic samples of size $n = 6$ that can be taken from the given population by starting with one of the first eight numbers and then taking each eighth number on the list.

10.13
REFERENCES

Derivations of the formulas for the standard error of the mean and more general formulations (and proofs) of the central limit theorem may be found in most textbooks on mathematical statistics; for instance, in

FREUND, J. E., and WALPOLE, R. E., *Mathematical Statistics*, *4th ed.* Englewood Cliffs, N.J.: Prentice-Hall, Inc., 1987.

All sorts of information about sampling may be found in

COCHRAN, W. G., *Sampling Techniques, 3rd ed.* New York: John Wiley & Sons, Inc., 1977.

SCHAFFER, R. L., MENDENHALL, W., and OTT, L., *Elementary Survey Sampling, 2nd ed.*, Boston: Duxbury Press, 1979.

SLONIN, M. J., *Sampling in a Nutshell.* New York: Simon and Schuster, 1973.

WILLIAMS, W. H., *A Sampler on Sampling.* New York: John Wiley & Sons, Inc., 1978.

Some books on simulation are referred to among the references at the end of Chapter 8, and the difficulties of drawing a random sample in practice are illustrated by "Was the 1970 Draft Lottery Fair?" in

HOLLANDER, M., and PROSCHAN, F., *The Statistical Exorcist: Dispelling Statistics Anxiety.* New York: Marcel Dekker, Inc., 1984.

The paradoxical problems of proportional allocation to the United States House of Representatives are discussed in

BALINSKI, M. L., and YOUNG, H. P., "The Apportionment of Representation," in *Proceedings of Symposia in Applied Mathematics*, American Mathematical Society, Providence, R.I.: 1985.

11 INFERENCES ABOUT MEANS

Traditionally, problems of statistical inference have been divided into problems of **estimation**, where we assign numerical values to population parameters, **tests of hypotheses**, where we accept or reject assertions about the parameters or the form of populations, and problems of **prediction**, where we forecast future values of random variables. In each case, the inferences are all based on sample data. In this chapter, and in those that immediately follow, we concentrate on problems of estimation and tests of hypotheses; methods of prediction will be taken up in Chapters 15 and 16.

Problems of estimation arise everywhere—in science, in business, and in everyday life. In science, a psychologist may want to determine the average time that it takes an adult person to react to a visual stimulus; in business, a union official may want to know how much variation there is in the time it takes members to get to work; and in everyday life we may want to find out what percentage of all one-car accidents are due to driver fatigue. These are all problems of estimation, but they would be tests of hypotheses if the psychologist wanted to decide whether the average time it takes an adult person to react to the stimulus is really 0.44 second, if the union official wanted to know whether the variation in the time it takes members to get to work exceeds company claims, and if we wanted to check whether it is true that 14.5 percent of all one-car accidents are due to driver fatigue.

Among the examples of the preceding paragraph, the psychologist's problem concerned a population mean, the union official's problem concerned a measure of variation, and our "everyday life" problem concerned a percentage. Conceptually, such problems are all treated in the same way, but there are differences in the methods which we employ.

In this chapter, Sections 11.1 through 11.3 are devoted to some general problems of estimation and, in particular, to the estimation of means. Then, after a general introduction to tests of hypotheses in Sections 11.4 and 11.5, Sections 11.6 through 11.10 are devoted to tests of hypotheses about means. Corresponding inferences about the variability of populations and percentages (or proportions) are taken up in Chapters 12 and 13.

11.1
THE ESTIMATION OF MEANS

To illustrate some of the problems we face when we estimate the mean of a population from sample data, let us refer to a study in which industrial designers wanted to determine the average (mean) time it takes an adult to assemble an "easy to assemble" toy. Using a random sample, they obtained the following data (in minutes) for 36 persons who assembled the toy:

17	13	18	19	17	21	29	22	16	28	21	15
26	23	24	20	8	17	17	21	32	18	25	22
16	10	20	22	19	14	30	22	12	24	28	11

The mean of this sample is $\bar{x} = 19.9$ minutes, and in the absence of any other information, this figure can be used as an estimate of μ, the "true" average time it takes an adult to assemble the given kind of toy.

An estimate like this is called a **point estimate**, since it consists of a single number, or a single point on the real number scale. Although this is the most common way in which estimates are expressed, it leaves room for many questions. For instance, it does not tell us on how much information the estimate is based, and it does not tell us anything about the possible size of the error. And of course, we must expect an error. This should be clear from our discussion of the sampling distribution of the mean in Chapter 10, where we saw that the chance fluctuations of the mean (and, hence, its reliability of an estimate of μ) depend on two things—the size of the sample and the size of the population standard deviation σ. Thus, we might supplement the estimate, $\bar{x} = 19.9$ minutes, with the information that it is the mean of a random sample of size $n = 36$, whose standard deviation is $s = 5.73$ minutes. Although this does not tell us the actual value of σ, the sample standard deviation can serve as an estimate of this quantity.

Scientific reports often present sample means in this way, together with the values of n and s, but this does not supply readers of the report with a coherent picture unless they have had some formal training in statistics. To take care of this, we refer to the theory of Sections 10.7 and 10.8, and the definition on page 229, according to which $z_{\alpha/2}$ is such that the area to its right under the standard normal curve is $\alpha/2$, and hence the area under the standard normal curve between $-z_{\alpha/2}$ and $z_{\alpha/2}$ is equal to $1 - \alpha$. Making use of the fact that, for large random samples from infinite populations, the sampling distribution of the mean is approximately a normal distribution with $\mu_{\bar{x}} = \mu$ and $\sigma_{\bar{x}} = \dfrac{\sigma}{\sqrt{n}}$, we find from Figure 11.1 that the probability is $1 - \alpha$ that the mean of a large random sample from an infinite population will differ from the mean of the population by at most $z_{\alpha/2} \cdot \dfrac{\sigma}{\sqrt{n}}$. In other words,

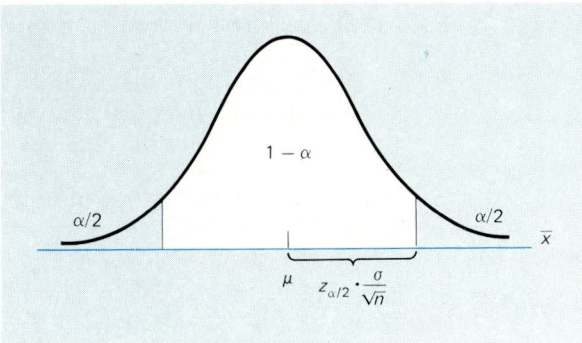

FIGURE 11.1 *Sampling distribution of the mean.*

If we are going to use $\bar{x}$ as an estimate of μ, the probability is $1 - \alpha$ that this estimate will be "off" either way by at most

Maximum error
of estimate

$$E = z_{\alpha/2} \cdot \frac{\sigma}{\sqrt{n}}$$

This result applies when n is large, $n \geq 30$, and the population is infinite (or large enough so that the finite population factor need not be used.) The two values which are most commonly, though not necessarily, used for $1 - \alpha$ are 0.95 and 0.99, and the corresponding values of $\alpha/2$ are 0.025 and 0.005. As the reader was asked to show in Exercise 9.13 on page 231, $z_{0.025} = 1.96$ and $z_{0.005} = 2.575$.

EXAMPLE A team of efficiency experts intends to use the mean of a random sample of size $n = 150$ to estimate the average mechanical aptitude of assembly-line workers in a large industry (as measured by a certain standardized test). If, based on experience, the efficiency experts can assume that $\sigma = 6.2$ for such data, what can they assert with probability 0.99 about the maximum error of their estimate?

Solution Substituting $n = 150$, $\sigma = 6.2$, and $z_{0.005} = 2.575$ into the formula for E, we get

$$E = 2.575 \cdot \frac{6.2}{\sqrt{150}} = 1.30$$

Thus, the efficiency experts can assert with probability 0.99 that their error will be at most 1.30.

Suppose now that these efficiency experts actually collect the necessary data and get $\bar{x} = 69.5$. Can they still assert with probability 0.99 that the error of their estimate, $\bar{x} = 69.5$, is at most 1.30? After all, $\bar{x} = 69.5$ differs from the true (population) mean by at most 1.30 or it does not, and they have no way of knowing whether it is one or the other. Actually, they can make this assertion, but it must be understood that the 0.99 probability applies to the *method* used (getting the sample data and using the formula for E) and not directly to the single problem at hand.

To make this distinction, it has become the custom to use the word "**confidence**" here instead of "probability."

In general, we make probability statements about future values of random variables (say, the potential error of an estimate) and confidence statements once the data have been obtained.

Accordingly, we would say in our example that the efficiency experts can be 99 percent confident that the error of their estimate, $\bar{x} = 69.5$, is at most 1.30.

Use of the formula for E involves one complication. To be able to judge the size of the error we might make when we use $\bar{x}$ as an estimate of μ, we must know the value of the population standard deviation σ. Since this is not the case in most practical situations, we have no choice but to replace σ with an estimate, usually the sample standard deviation s. In general, this is considered to be reasonable provided the sample is sufficiently large, and by sufficiently large we mean again $n \geq 30$.

EXAMPLE With reference to the example on page 282, what can we assert with 95 percent confidence about the maximum error, if we use $\bar{x} = 19.9$ minutes as an estimate of the average time it takes an adult to assemble the given kind of toy?

Solution Substituting $n = 36$, $s = 5.73$ for σ, and $z_{0.025} = 1.96$ into the formula for E, we find that we can assert with 95 percent confidence that the error is at most

$$E = 1.96 \cdot \frac{5.73}{\sqrt{36}} = 1.87 \text{ minutes}$$

Of course, the error is at most 1.87 minutes or it is not, and we do not know which is actually the case, but if we had to bet, 95 to 5 (or 19 to 1) would be fair odds that the error is at most 1.87 minutes.

The formula for E can also be used to determine the sample size that is needed to attain a desired degree of precision. Suppose that we want to use the mean of a large random sample to estimate the mean of a population, and we want to be able to assert with probability $1 - \alpha$ that the error of this estimate will be less than some prescribed quantity E. As before, we write $E = z_{\alpha/2} \cdot \dfrac{\sigma}{\sqrt{n}}$, and upon solving this equation for n we get

Sample size for estimating μ

$$n = \left[\frac{z_{\alpha/2} \cdot \sigma}{E} \right]^2$$

EXAMPLE The dean of a college wants to use the mean of a random sample to estimate the average amount of time students take to get from one class to the next, and she wants to be able to assert with probability 0.95 that her error will be at most 0.25 minute. If she knows from studies of a similar kind that it is reasonable to let $\sigma = 1.50$ minutes, how large a sample will she need?

Solution Substituting $z_{0.025} = 1.96$, $E = 0.25$, and $\sigma = 1.50$ into the formula for n, we get

$$n = \left[\frac{1.96 \cdot 1.50}{0.25} \right]^2 = 139$$

rounded up to the nearest integer. Thus, a random sample of size $n = 139$ is required for the estimate.

Note that the treatment would be the same if we replace "she wants to be able to assert with probability 0.95 that her error *will be* at most 0.25 minute" by "she wants to be able to assert with 95 percent confidence that her error *is* at most 0.25 minute." The difference is in when the assertion is to be made—before or after she collects the data.

As can be seen from the formula for n and also from the example, this method has the shortcoming that it cannot be used unless we know (at least approximately) the value of the population standard deviation. For this reason, we sometimes begin with a relatively small sample and then use the sample standard deviation as an estimate of σ to determine whether more data are required.

Let us now give a different way of presenting a sample mean together with an assessment of the error we might make if we use it to estimate the mean of the population from which the sample came. As on page 283, we shall make use of the fact that, for large random samples from infinite populations, the sampling distribution of the mean is approximately a normal distribution with the mean $\mu_{\bar{x}} = \mu$ and the standard deviation $\sigma_{\bar{x}} = \dfrac{\sigma}{\sqrt{n}}$, so that

$$z = \frac{\bar{x} - \mu}{\sigma/\sqrt{n}}$$

is a value of a random variable having approximately the standard normal distribution. Since the probability is $1 - \alpha$ that a random variable having this distribution will take on a value between $-z_{\alpha/2}$ and $z_{\alpha/2}$, namely, that

$$-z_{\alpha/2} < z < z_{\alpha/2}$$

we can substitute into this inequality the foregoing expression for z and get

$$-z_{\alpha/2} < \frac{\bar{x} - \mu}{\sigma/\sqrt{n}} < z_{\alpha/2}$$

If we now apply some relatively simple algebra, we can rewrite this inequality as

Large-sample confidence interval for μ

$$\bar{x} - z_{\alpha/2} \cdot \frac{\sigma}{\sqrt{n}} < \mu < \bar{x} + z_{\alpha/2} \cdot \frac{\sigma}{\sqrt{n}}$$

and we can assert with probability $1 - \alpha$ that it will be satisfied for any given sample.[†] In other words, we can assert with $(1 - \alpha)100$ percent confidence that the interval from $\bar{x} - z_{\alpha/2} \cdot \dfrac{\sigma}{\sqrt{n}}$ to $\bar{x} + z_{\alpha/2} \cdot \dfrac{\sigma}{\sqrt{n}}$, determined on the basis of a large random sample, contains the population mean we are trying to estimate. When σ is unknown and n is at least 30, we replace σ by the sample standard deviation s. If n is at least 30, then the use of s is recommended, even if the value of σ is claimed to be known.

An interval like this is called a **confidence interval**, its endpoints are called **confidence limits**, and $(1 - \alpha)100$ percent is called the **degree of confidence**. As before, the values most often used for the degree of confidence are 95 percent and 99 percent, and the corresponding values of $z_{\alpha/2}$ are 1.96 and 2.575. In contrast to

---

[†] Since the probability is also $1 - \alpha$ that $-z_{\alpha/2} \leq z \leq z_{\alpha/2}$, some authors substitute $\leq$ for $<$ in this confidence-interval formula. Practically speaking, this does not make any difference.

point estimates, estimates given in the form of a confidence interval are called **interval estimates**. They have the advantage that they require no further elaboration about their reliability—this is taken care of indirectly by their width and the degree of confidence.

EXAMPLE   With reference to the example on page 282, where we had $n = 36$, $\bar{x} = 19.9$, and $s = 5.73$, construct a 95 percent confidence interval for the average time it takes an adult to assemble the given kind of toy.

Solution   Substituting $n = 36$, $\bar{x} = 19.9$, $s = 5.73$ for σ, and $z_{0.025} = 1.96$ into the confidence interval formula, we get

$$19.9 - 1.96 \cdot \frac{5.73}{\sqrt{36}} < \mu < 19.9 + 1.96 \cdot \frac{5.73}{\sqrt{36}}$$

and

$$18.0 < \mu < 21.8$$

where the confidence limits are in minutes, rounded to one decimal. Of course the statement that the interval from 18.0 to 21.8 contains the true average time it takes an adult to assemble the given kind of toy is either true or false, and we do not know whether it is true or false, but we can be 95 percent confident that it is true. Why? Because the method we used works 95 percent of the time. To put it differently, the interval may contain μ or it may not, but if we had to bet, 95 to 5 (or 19 to 1) would be fair odds that it does.

Had we wanted to construct a 99 percent confidence interval in the preceding example, we would have substituted 2.575 instead of 1.96 for $z_{\alpha/2}$, and we would have obtained $17.4 < \mu < 22.4$. The 99 percent confidence interval is wider than the 95 percent confidence interval—it goes from 17.4 to 22.4 instead of 18.0 to 21.8—and this illustrates the important fact that **when we increase the degree of confidence, the confidence interval becomes wider and thus tells us less about the quantity we are trying to estimate**. Indeed, as is also illustrated by Exercise 11.8 on page 294, we might say that "the surer we want to be, the less we have to be sure of."

11.2

THE ESTIMATION OF MEANS (Small Samples)

In Section 11.1 we assumed that the samples were large enough, $n \geq 30$, to approximate the sampling distribution of the mean with a normal distribution and, where necessary, to replace σ with s. To develop corresponding methods which apply also to small samples, it will be necessary to assume that the populations we are sampling have roughly the shape of normal distributions. We can then base

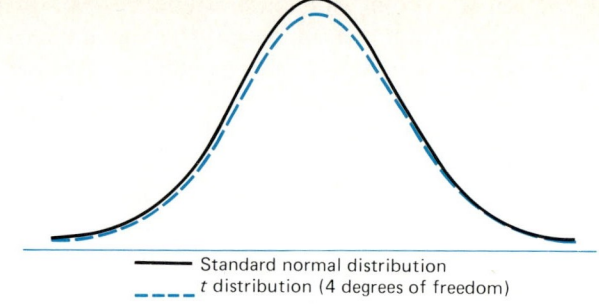

———— Standard normal distribution
– – – – t distribution (4 degrees of freedom)

FIGURE 11.2 *Standard normal distribution and t distribution.*

our methods on the statistic

$$t = \frac{\bar{x} - \mu}{s/\sqrt{n}}$$

which is a value of a random variable having the **t distribution**. More specifically, this distribution is called the **Student-t distribution** or **Student's t distribution**, as it was first developed by a statistician, W. S. Gosset, who published his work under the pen name "Student." As is shown in Figure 11.2, the shape of this continuous distribution is very similar to that of the standard normal distribution—like the standard normal distribution, it is bell-shaped and symmetrical with zero mean. The exact shape of the t distribution depends on a parameter called the **number of degrees of freedom**, or simply the **degrees of freedom**, which, for the methods of this section, equals $n - 1$, the sample size less one.

For the standard normal distribution, we defined $z_{\alpha/2}$ in such a way that the area under the curve to its right equals $\alpha/2$, and, hence, the area under the curve between $-z_{\alpha/2}$ and $z_{\alpha/2}$ equals $1 - \alpha$. As is shown in Figure 11.3, the corresponding values for the t distribution are $-t_{\alpha/2}$ and $t_{\alpha/2}$. Since these values depend on $n - 1$,

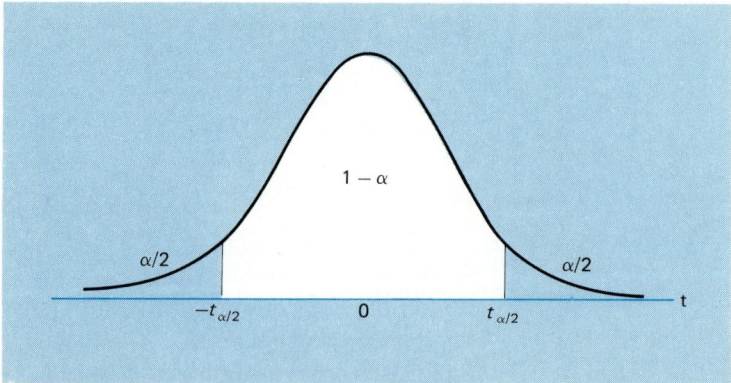

FIGURE 11.3 *t distribution.*

the number of degrees of freedom, they must be obtained from a special table, such as Table II at the end of this book, or perhaps a computer. Table II contains among others the values of $t_{0.025}$ and $t_{0.005}$ for 1 through 29 degrees of freedom, and it can be seen that $t_{0.025}$ and $t_{0.005}$ approach the corresponding values for the standard normal distribution as the number of degrees of freedom becomes large.

Now we can proceed as on page 286. Since the probability is $1 - \alpha$ that a random variable having the t distribution will take on a value between $-t_{\alpha/2}$ and $t_{\alpha/2}$, namely, that

$$-t_{\alpha/2} < t < t_{\alpha/2}$$

we can substitute into this inequality the expression for t on page 288 and get

$$-t_{\alpha/2} < \frac{\bar{x} - \mu}{s/\sqrt{n}} < t_{\alpha/2}$$

Applying some relatively simple algebra, we can rewrite this inequality as

Small-sample confidence interval for μ

$$\bar{x} - t_{\alpha/2} \cdot \frac{s}{\sqrt{n}} < \mu < \bar{x} + t_{\alpha/2} \cdot \frac{s}{\sqrt{n}}$$

and we can assert with probability $1 - \alpha$ that it will be satisfied for any given sample. The only difference between this confidence interval formula and the large-sample formula (with s substituted for σ) is that $t_{\alpha/2}$ takes the place of $z_{\alpha/2}$.

EXAMPLE   To test the durability of a new paint for white center lines, a highway department painted test strips across heavily traveled roads in eight different locations, and electronic counters showed that they deteriorated after having been crossed by (to the nearest hundred) 142,600, 167,800, 136,500, 108,300, 126,400, 133,700, 162,000, and 149,400 cars. Construct a 95 percent confidence interval for the average amount of traffic (car crossings) this paint can withstand before it deteriorates.

Solution   The mean and the standard deviation of these $n = 8$ values are $\bar{x} = 140,800$ and $s = 19,200$ to the nearest hundred, and since $t_{0.025}$ for $8 - 1 = 7$ degrees of freedom equals 2.365 (according to Table II), substitution into the formula yields

$$140,800 - 2.365 \cdot \frac{19,200}{\sqrt{8}} < \mu < 140,800 + 2.365 \cdot \frac{19,200}{\sqrt{8}}$$

$$124,700 < \mu < 156,900$$

This is the desired 95 percent confidence interval for the average amount of traffic (car crossings) the paint can withstand before it deteriorates. Again, we can't tell for sure whether the interval from 124,700 to 156,900 contains the true average number of car crossings the paint can withstand before it deteriorates, but 95 to 5 would be fair odds that it does. These odds are based on the fact that the method we used—taking a random sample from a normal population and using the formula given above—works 95 percent of the time

```
MTB > SET C1
DATA> 142600  167800  136500  108300  126400  133700  162000  149400
MTB > TINT 95 C1

                N      MEAN   STDEV  SE MEAN    95.0 PERCENT C.I.
C1              8    140837   19228     6798   ( 124758,  156917)
```

FIGURE 11.4 *Computer printout for small-sample confidence interval for μ.*

A computer printout showing the result we obtained in the preceding example is given in Figure 11.4. The differences between 124,700 and 124,758, and 156,900 and 156,917, are due to rounding. Indeed, since the original data were given to the nearest hundred, we rounded the mean, the standard deviation, and the confidence limits to the nearest hundred.

The method which we used earlier to determine the maximum error we risk with $(1 - \alpha)100$ percent confidence when we use a sample mean to estimate the mean of a population can easily be adapted to small samples, provided that the population we are sampling has roughly the shape of a normal distribution. All we have to do is substitute s for σ and $t_{\alpha/2}$ for $z_{\alpha/2}$ in the formula for the maximum error E on page 284.

EXAMPLE   While performing a certain task under simulated weightlessness, the pulse rate of twelve astronauts increased on the average by 27.33 beats per minute with a standard deviation of 4.28 beats per minute. If we use $\bar{x} = 27.33$ as an estimate of the true average increase of the pulse rate of astronauts performing the given task, what can we assert with 99 percent confidence about the maximum error?

Solution   Substituting $s = 4.28$, $n = 12$, and $t_{0.005} = 3.106$ (the entry in Table II for $12 - 1 = 11$ degrees of freedom) into the new formula for E, we get

$$E = t_{\alpha/2} \cdot \frac{s}{\sqrt{n}} = 3.106 \cdot \frac{4.28}{\sqrt{12}} = 3.84$$

Thus, if we use $\bar{x} = 27.33$ beats per minute as an estimate of the true average increase in the pulse rate of astronauts performing the given task (under the stated conditions), we can assert with 99 percent confidence that our error is at most 3.84 beats per minute.

11.3

THE ESTIMATION OF MEANS (a Bayesian Method) ★

In recent years there has been mounting interest in methods of inference in which parameters (for example, the population mean μ or the population standard deviation σ) are looked upon as random variables having **prior distributions** which reflect how a person feels about the different values that a parameter can take on. Such prior considerations are then combined with direct sample evidence to obtain **posterior distributions** of the parameters, on which subsequent inferences are based. Since the method used to combine the prior considerations with the direct sample evidence is based on a generalization of Bayes' theorem of Section 6.7, we refer to such inferences as **Bayesian**.

In this section we shall present a Bayesian method of estimating the mean of a population. As we said, our prior feelings about the possible values of μ are expressed in the form of a prior distribution, and like any distribution, this kind of distribution has a mean and a standard deviation. We shall denote these values by μ_0 and σ_0 and call them the **prior mean** and the **prior standard deviation**.

The method we shall describe here involves quite a few assumptions, but without them it is nearly impossible to get results that are in an easy-to-use form. So, if we are sampling a population with the mean μ (which we want to estimate) and the standard deviation σ, if the sample is large enough to apply the central limit theorem, and if the prior distribution of μ has roughly the shape of a normal distribution, it can be shown that the posterior distribution of μ is also a normal distribution with the mean

Posterior mean

$$\mu_1 = \frac{\dfrac{n}{\sigma^2} \cdot \bar{x} + \dfrac{1}{\sigma_0^2} \cdot \mu_0}{\dfrac{n}{\sigma^2} + \dfrac{1}{\sigma_0^2}}$$

and the standard deviation σ_1 given by the formula

Posterior standard deviation

$$\frac{1}{\sigma_1^2} = \frac{n}{\sigma^2} + \frac{1}{\sigma_0^2}$$

in which we can solve for σ_1.

Since the posterior mean, μ_1, can be used as an estimate of the population mean μ, let us examine some of its most important features. First we note that μ_1 is a weighted mean of $\bar{x}$ and μ_0, and that the weights are $\dfrac{n}{\sigma^2}$ and $\dfrac{1}{\sigma_0^2}$, namely, the reciprocals of the variances of the distribution of $\bar{x}$ and the prior distribution of μ. We see also that when no direct information is available and $n = 0$, the weight assigned $\bar{x}$ is 0, the formula reduces to $\mu_1 = \mu_0$, and the estimate is based entirely on the prior distribution. Without sample data, our estimate will correspond precisely to the prior distribution. When $\sigma_0^2 = \dfrac{\sigma^2}{n}$, our estimate μ_1 is a 50–50 average of μ_0 and $\bar{x}$. As more and more direct evidence becomes available (that is, as n becomes larger and larger), the weight shifts more and more toward the direct sample evidence, the sample mean $\bar{x}$. Finally, we see that when the prior feelings about the possible values of μ are vague (that is, when σ_0 is relatively large), the estimate will be based to a greater extent on $\bar{x}$. However, when there is a great deal of variability in the population which yields the direct sample evidence (that is, when σ is relatively large), the estimate will be based to a greater extent on μ_0.

Once we have determined μ_1 and σ_1, we can actually make probability statements about the population mean μ, which we are trying to estimate. However, with subjective prior feelings and objective data, such probabilities reflect a combination of the corresponding probability concepts, and must be interpreted with care.

This procedure requires the specification of μ_0 and σ_0, and it also requires that we know the population standard deviation σ or be willing to estimate it with s. In the case where σ is not known, which is usually the case, some Bayesians will formulate a prior distribution for σ; the resulting estimate for μ is rather complex and has not found wide acceptance.

EXAMPLE An investor who is planning to open a new travel agency feels most strongly that he should net on the average $\mu_0 = \$3,600$ per month; also, the subjective prior distribution which he attaches to the various possible values of μ has roughly the shape of a normal distribution with the standard deviation $\sigma_0 = \$130$. If during nine months the operation of the travel agency nets $\$3,810$, $\$3,690$, $\$3,350$, $\$3,400$, $\$3,320$, $\$3,250$, $\$3,430$, $\$3,600$, and $\$3,670$, what is the posterior probability that the travel agency will net on the average between $\$3,500$ and $\$3,600$ per month?

Solution The mean and the standard deviation of the sample data are $\bar{x} = 3,502$ and $s = 195$. Using this sample standard deviation to estimate σ, and substituting $n = 9$, $\bar{x} = 3,502$, $s = 195$ for σ, $\mu_0 = 3,600$, and $\sigma_0 = 130$ into the formulas for the posterior mean and the posterior standard deviation, we get

$$\mu_1 = \frac{\dfrac{9}{195^2} \cdot 3,502 + \dfrac{1}{130^2} \cdot 3,600}{\dfrac{9}{195^2} + \dfrac{1}{130^2}} = \$3,522$$

and

$$\frac{1}{\sigma_1^2} = \frac{9}{195^2} + \frac{1}{130^2} \quad \text{and} \quad \sigma_1 = 58.1$$

Having found the mean and the standard deviation of the posterior distribution of μ pictured in Figure 11.5, we must now determine the area of the white region under the curve, namely, that between

$$z = \frac{3,500 - 3,522}{58.1} = -0.38 \quad \text{and} \quad z = \frac{3,600 - 3,522}{58.1} = 1.34$$

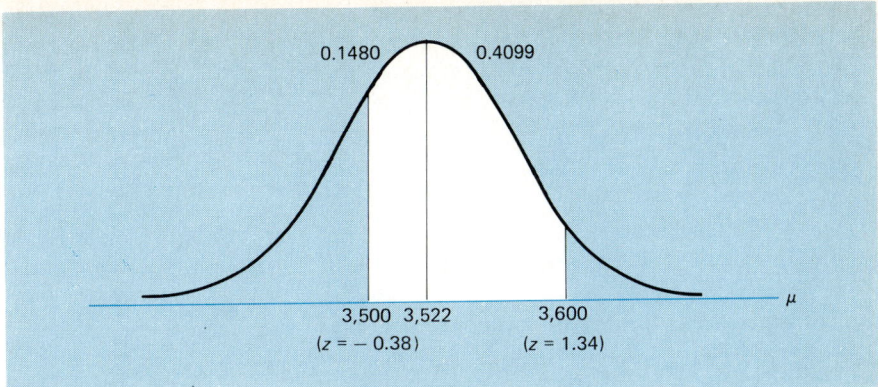

FIGURE 11.5 *Posterior distribution of μ.*

Since the entries corresponding to $z = 0.38$ and $z = 1.34$ in Table I are 0.1480 and 0.4099, we get

$$0.1480 + 0.4099 = 0.5579 \quad \text{or approximately } 0.56$$

for the (mostly subjective but partly objective) posterior probability that the travel agency will net on the average between $3,500 and $3,600 per month.

This brief introduction to **Bayesian inference** serves to bring out the following points: (1) In Bayesian statistics the parameter about which an inference is made is looked upon as a random variable having a distribution of its own, and (2) this kind of inference permits the use of direct as well as collateral information. When we referred to the result of our example as being "mostly subjective but partly objective," we were trying to express the fact that the prior distribution of the investor was probably based on a subjective evaluation of various factors (business conditions in general, for instance, and indirect information about other travel agencies), and that this evaluation was combined with the figures which were actually observed for the nine months. It must also be noted that two persons performing this analysis will likely have different prior distributions and therefore produce different results.

EXERCISES

11.1 To estimate the average service time at a hamburger fast-food restaurant, a management consultant noted the times that it took for 35 counter persons, a random sample, to complete a standard order (consisting of two hamburgers, two packages of French-fried potatoes, and two drinks). It took these persons, on average, 72.2 seconds with a standard deviation of 12.8 seconds to complete the orders. What can the consultant assert with 95 percent confidence about the maximum error, if he uses $\bar{x} = 72.2$ seconds as an estimate of the true average time required to complete this standard order?

11.2 With reference to the preceding exercise, construct a 95 percent confidence interval for the true average time that it takes a counter person to complete the standard order.

11.3 A study of the annual growth of certain cacti showed that 64 of them, selected at random in a desert region, grew on the average 52.80 mm with a standard deviation of 4.5 mm. Construct a 99 percent confidence interval for the true average annual growth of the given kind of cactus.

11.4 With reference to the preceding exercise, what can we assert with 99 percent confidence about the maximum error, if we use $\bar{x} = 52.8$ mm as an estimate of the true average annual growth of the given kind of cactus?

11.5 A study conducted by an airline showed that a random sample of 120 of its passengers disembarking at Kennedy airport on flights from Europe took on the average 24.15 minutes with a standard deviation of 3.29 minutes to claim their luggage and get through customs. What can they assert with 95 percent confidence about the maximum error, if they use $\bar{x} = 24.15$ minutes as an estimate of the true average time it takes one of their passengers disembarking at Kennedy airport on a flight from Europe to claim his or her luggage and get through customs?

11.6 With reference to the preceding exercise, by how much is the maximum error increased if they want to be 99 percent instead of 95 percent confident?

11.7 With reference to Exercise 11.5, construct a confidence interval for the true average time it takes on of the airline's passengers disembarking at Kennedy airport on a flight from Europe to claim his or her luggage and get through customs, using
    (a) 95 percent confidence;
    (b) 98 percent confidence;
    (c) 99 percent confidence;
    (d) 99.9 percent confidence.

11.8 In the example on page 287, the 95 percent confidence interval had a width of $21.8 - 18.0 = 3.8$ minutes and the 99 percent confidence interval had a width of $22.4 - 17.4 = 5.0$ minutes. What would be the width of a corresponding 99.99994 percent confidence interval? Note how greatly the width has increased from the 99 percent confidence interval to the 99.99994 percent confidence interval even though the confidence is increased by less than 1 percent.

11.9 A district official intends to use the mean of a random sample of 150 sixth graders from a very large school district to estimate the mean score which all the sixth graders in the district would get if they took a certain arithmetic achievement test. If, based on experience, the official knows that $\sigma = 9.4$ for such data, what can she assert with probability 0.95 about the maximum error? Why did we use the word "probability" here instead of the word "confidence?"

11.10 A sample survey conducted in a large city in 1986 showed that 200 families spent on the average $218.67 per week on food with a standard deviation of $14.93. As it is desired to pin the true average weekly food expenditure of families in that city to a fairly narrow interval, construct a 90 percent confidence interval. Explain a possible disadvantage of this procedure.

11.11 A distributor of soft-drink vending machines plans to use the mean number of drinks dispensed during one week by 60 of her machines to estimate the average number dispensed by any one of her machines during one week. If the 60 randomly selected machines had a mean of 255.3 drinks, with a standard deviation of 48.2 drinks, give a 95 percent confidence interval for the true average number dispensed by any one of her machines during one week.

11.12 A random sample of 60 cans of pear halves has a mean weight of 16.1 ounces and a standard deviation of 0.3 ounce. If $\bar{x} = 16.1$ ounces is used as an estimate of the mean weight of all the cans of pear halves in the large lot from which the sample came, with what confidence can we assert that the error of this estimate is at most 0.1 ounce?

⋆ **11.13** If a sample constitutes an appreciable portion of a finite population (say, 5 percent or more), the various formulas of Section 11.1 must be modified by appending the finite population correction factor to the standard error of the mean. For instance, the formula for the maximum error on page 284 becomes

$$E = z_{\alpha/2} \cdot \frac{\sigma}{\sqrt{n}} \cdot \sqrt{\frac{N - n}{N - 1}}$$

A random sample of 50 scores on a law school's admission test is drawn from the scores of the 420 persons who applied to the law school in 1985. If the sample mean and standard deviation are $\bar{x} = 541$ and $s = 87$, what can we assert with 95 percent confidence about the maximum error, when $\bar{x} = 541$ is used as an estimate of the average score of all 420 applicants?

★ 11.14 Use the method of the preceding exercise on the data of Exercise 11.9, given that there are 1,200 sixth graders in the district.

★ 11.15 A random sample of 40 drums of a chemical, drawn at random from among 200 such drums whose weights can be assumed to have the standard deviation $\sigma = 10.2$ pounds, has a mean weight of 240.80 pounds. Using the same modification as in Exercise 11.13, construct a 95 percent confidence interval for the mean weight of the 200 drums.

★ 11.16 Suppose that there are only 3,200 families in the city referred to in Exercise 11.10. Using the same modification as in Exercise 11.13, construct a 90 percent confidence interval for the true average weekly food expenditures of families in that city. Compare the result with that of Exercise 11.10.

★ 11.17 A stratified random sample of size $n = 336$ is allocated proportionally to the four strata of a finite population for which $N_1 = 4,000$, $N_2 = 6,000$, $N_3 = 2,000$, $N_4 = 12,000$, $\sigma_1 = 20$, $\sigma_2 = 20$, $\sigma_3 = 16$, and $\sigma_4 = 30$ cm. If the mean of this stratified sample is used to estimate the mean of the population,
   (a) what can we assert with 95 percent confidence about the maximum error;
   (b) what can we assert with 99 percent confidence about the maximum error?
(*Hint:* Use the standard error formula of Exercise 10.44 on page 275.)

11.18 The manager of a restaurant plans to introduce child-size portions for a number of menu items. As an experiment, she would like to give hamburger patties to a number of six-year-olds in an "all you can eat" situation in order to estimate the average preferred quantity. She believes that the population of children would choose, on the average, somewhere between 2 ounces and 6 ounces, but she has no idea about the standard deviation. She needs to know what to say with 95 percent confidence about the maximum error in using the sample mean to estimate the population mean, if $n = 40$ children are used in the experiment.
   (a) Suggest a plausible value for the standard deviation σ and then provide a value for the maximum error, using 95 percent confidence.
   (b) Suggest a pessimistic value for the standard deviation σ and then provide a value for the maximum error, using 95 percent confidence.

11.19 A candy distributor wishes to determine the average water content of bottles of maple syrup from a partic-

ular New Hampshire producer. The bottles contain 12 ounces of liquid, and she decides to determine the water content of 10 of these bottles, using the sample mean as an estimate of the true population average. What can she say, with probability 0.95, about the maximum error, if the largest possible standard deviation that she is willing to believe is 2.0 ounces?

11.20 It is desired to estimate the mean number of hours of continuous use until a certain kind of computer will first require repairs. If it can be assumed that $\sigma = 48$ hours, how large a sample is needed so that it can be asserted with probability 0.99 that the sample mean will be off by at most 10 hours? Why did we use the word "probability" here, whereas in Exercise 11.19 we used the word "confidence"?

11.21 Suppose that we want to estimate the mean score of senior citizens on a current events test, and that we want to be able to assert with probability 0.98 that the error of our estimate, the mean of a random sample, will be at most 2.5. How large a sample will we need, if it can be assumed that $\sigma = 7.3$?

11.22 Before bidding on a contract, a contractor wants to be 95 percent confident that he is in error by at most 5 minutes in using the mean of a random sample to estimate the average time it takes a certain kind of adobe brick to harden.
   (a) How large a sample will he need, if he can assume that $\sigma = 24$ minutes for such bricks?
   (b) Why did we use the word "confident" here, whereas in Exercise 11.21 we used the word "probability"?
   (c) Would the solution to (a) be different if we substitute "less than 5 minutes" for "at most 5 minutes"?

11.23 Before purchasing a large shipment of ground meat, a sausage manufacturer wants to be 95 percent confident that he is in error by no more than 2.5 grams in estimating the fat content (per 100 grams of meat). If the standard deviation of the fat content (per 100 grams of meat) is assumed to be 8 grams, on how large a sample should he base his estimate?

11.24 The ratio $E^* = E/\sigma$ is the error in standard deviation units. In some situations, lack of knowledge about σ makes it difficult to establish realistic values for E. In such cases, specifying E^* is a useful alternative. For example, requiring that E^* be 0.25 or less asks for an error of 0.25σ or less, whatever the value of σ. If a school administrator wants to estimate the average reading achievement score for his whole school by using a sample of children, how large a sample should he take if he wishes to assert, with 95 percent confidence, that E^* is at most 0.25?

11.25 A computer is programmed to yield values of a random variable having a normal distribution, whose mean and standard deviation are known only to the programmer. Each of 30 students is asked to use the computer to simulate a random sample of size $n = 5$ and then construct a small-sample 90 percent confidence interval for μ. The following are their results:

$6.30 < \mu < 8.26,$ $6.50 < \mu < 7.72,$ $6.93 < \mu < 8.01,$
$6.60 < \mu < 8.00,$ $6.51 < \mu < 7.51,$ $6.82 < \mu < 8.66,$
$7.02 < \mu < 8.11,$ $6.94 < \mu < 7.64,$ $6.24 < \mu < 7.26,$
$6.87 < \mu < 8.17,$ $6.77 < \mu < 8.13,$ $6.14 < \mu < 6.82,$
$6.83 < \mu < 7.93,$ $6.66 < \mu < 8.10,$ $6.73 < \mu < 7.49,$
$6.41 < \mu < 7.67,$ $6.76 < \mu < 7.57,$ $6.97 < \mu < 7.47,$
$6.01 < \mu < 7.43,$ $7.15 < \mu < 7.89,$ $6.87 < \mu < 7.81,$
$7.35 < \mu < 7.99,$ $6.60 < \mu < 8.16,$ $6.47 < \mu < 7.81,$
$7.01 < \mu < 8.33,$ $6.97 < \mu < 7.55,$ $6.56 < \mu < 7.48,$
$7.13 < \mu < 8.03,$ $7.39 < \mu < 8.01,$ $5.98 < \mu < 7.68.$

(a) How many of these 30 confidence intervals would we expect to contain the actual mean of the population sampled?

(b) Given that the computer was programmed so that $\mu = 7.30$, how many of the 30 confidence intervals actually contain the mean of the population sampled? Discuss this result.

11.26 A major truck-stop owner has kept extensive records on various transactions with his customers. If a random sample of 18 of these records show average sales of 58.22 gallons of diesel fuel with a standard deviation of 4.80 gallons,

(a) construct a 95 percent confidence interval for the mean of the population sampled;

(b) construct a 99 percent confidence interval for the mean of the population sampled;

(c) indicate the maximum error that we can assert with 95 percent confidence if we use $\bar{x} = 58.22$ as an estimate of the truck stop's average sales of diesel fuel;

(d) indicate the maximum error that we can assert with 99 percent confidence if we use $\bar{x} = 58.22$ as an estimate of the truck stop's average sales of diesel fuel.

11.27 There is often some uncertainty about the level of confidence that should be used in constructing intervals. It is claimed that, in general, 99 percent con-

fidence intervals are about 30% longer than the corresponding 95 percent interval. Investigate this claim with the data given in Exercise 11.26, where $n = 18$.

11.28 The excess length of the 99 percent confidence over the corresponding 95 percent interval, expressed as a percentage of the length of the 95 percent invertal (see Exercise 11.27), is, in fact, determined only by the sample size n.

(a) What is the excess length when $n = 10$?

(b) What is the excess length when $n = 20$?

(c) What is the excess length when $n = 30$?

11.29 Nine bearings manufactured by a certain process have a mean diameter of 0.404 cm and a standard deviation of 0.003 cm. What can be said about the maximum error if $\bar{x} = 0.404$ cm is used as an estimate of the mean diameter of bearings made by the process:

(a) with 95 percent confidence;

(b) with 99 percent confidence?

11.30 In an air pollution study, a random sample of $n = 8$ specimens collected within a mile downwind from a certain factory contained on the average 2.26 micrograms of suspended benzene-soluble organic matter per cubic meter with a standard deviation of 0.56. If $\bar{x} = 2.26$ is used as an estimate of the mean of the population sampled, what can be said with 99 percent confidence about the maximum error?

11.31 With reference to the preceding exercise, construct a 95 percent confidence interval for the mean of the population sampled.

11.32 With reference to the example on page 289, suppose that $\bar{x} = 140,800$ is used as a point estimate of the average amount of traffic (car crossings) the paint can withstand before it deteriorates. What can we assert with 90 percent confidence about the maximum error?

11.33 With reference to the example on page 290, construct confidence intervals for the true average increase in the pulse rate of astronauts performing the given task under simulated weightlessness, using the

(a) 95 percent degree of confidence;

(b) 99 percent degree of confidence.

11.34 On six occasions, presumably a random sample, it took 21, 26, 24, 22, 23, and 22 minutes to clean up a school cafeteria. Construct a 95 percent confidence interval for the average time it takes to clean up this cafeteria.

11.35 To establish the authenticity of an ancient coin, its weight is often of critical importance. If four experts independently weighed a Phoenician tetradrachm and got 14.28, 14.34, 14.26, and 14.32 grams, what can we assert with 99 percent confidence about the maximum error, when the mean of these four values is used as an estimate of the actual weight of the coin?

11.36 A dentist finds in a routine check that six prison inmates, a random sample, require 2, 3, 6, 0, 4, and 3 fillings. Construct a 90 percent confidence interval for the average number of fillings that are required by the inmates of this (very large) prison.

★ **11.37** With reference to the example on page 292, where an investor is planning to open a new travel agency, what prior probability does the investor assign to the agency's netting between $3,500 and $3,600 per month? Compare this prior probability with the corresponding posterior probability obtained on page 293.

★ **11.38** An actuary feels that the prior distribution of the average annual claims for a certain kind of health insurance coverage is a normal distribution with the mean $\mu_0 = \$352$ and the standard deviation $\sigma_0 = \$12$. She also knows that the claims for this policy vary from one person to another with the standard deviation $\sigma = \$120$. If the claims for ten policies like this

average $440.22, find the posterior mean as a Bayesian estimate of its true average annual claim amount.

★ **11.39** With reference to the preceding exercise, what is the posterior probability that the policy's true average annual claim amounts are between $360 and $400?

★ **11.40** A college professor is making up a final examination in history which is to be given to a large group of students. His feelings about the average grade they should get is expressed subjectively by a normal distribution which has the mean $\mu_0 = 65.2$ and the standard deviation $\sigma_0 = 1.5$. What prior probability does the professor assign to the event that the average grade of all the students will lie on the interval from 63.0 to 68.0?

★ **11.41** With reference to the preceding exercise, suppose that the examination is given to a random sample of 40 of the students and that their grades have the mean $\bar{x} = 72.9$ and the standard deviation $s = 7.4$.
  (a) Calculate the posterior mean as a point estimate of the mean grade which all the students will get on the test.
  (b) Find the posterior probability which the professor assigns to the event that the average grade of all the students will lie on the interval from 63.0 and 68.0.

11.4
TESTS OF HYPOTHESES

In the introduction to this chapter we gave several examples of tests of hypotheses. We mentioned a psychologist who wanted to know whether an adult person's mean reaction time to a visual stimulus is really 0.44 second, a union official who wondered whether the variation in the time it takes members to get to work might exceed company claims, and we asked whether it is really true that 14.5 percent of all one-car accidents are due to driver fatigue. Since the psychologist was concerned with a value of μ, the union official wondered about variability, say, a value of σ, and we questioned a value of p, the parameter of a binomial population, these are all tests of **statistical hypotheses**. In general,

A statistical hypothesis is an assertion or conjecture about a parameter, or parameters, of a population; it may also concern the type, or nature, of the population.

With regard to the second part of this definition, we shall see in Section 13.6 how we can test whether it is reasonable to treat a random variable as having a binomial distribution, or perhaps a Poisson distribution, and whether it is reasonable to treat a set of data as coming from a normal population. In this chapter we shall be concerned only with hypotheses about means—the mean of one population or the means of two populations.

To develop procedures for testing statistical hypotheses, we must always know exactly what to expect when a hypothesis is true, and it is for this reason that we often hypothesize the opposite of what we hope to prove. Suppose, for instance, that we suspect that a dice game is not honest. If we formulate the hypothesis that the dice are crooked, everything would depend on how crooked they are, but if we assume that they are perfectly balanced, we could calculate all the necessary probabilities and take it from there. Also, if we want to show that one method of teaching computer programming is more effective than another, we would hypothesize that the two methods are equally effective; if we want to show that one diet is healthier than another, we hypothesize that they are equally healthy; and if we want to show that a new copper-bearing steel has a higher yield strength than ordinary steel, we hypothesize that the two yield strengths are the same. Since we hypothesize that there is no difference in the effectiveness of the two teaching methods, medically no difference between the two diets, and no difference in the yield strength of the two kinds of steel, we call hypotheses like these **null hypotheses** and denote them by H_0. Nowadays, the term "null hypothesis" is used for any hypothesis set up primarily to see whether it can be rejected.

The idea of setting up a null hypothesis is common even in nonstatistical thinking. It is precisely what we do in criminal proceedings, where an accused is presumed to be innocent until his guilt has been established beyond a reasonable doubt. The presumption of innocence is a null hypothesis.

The hypothesis which we use as an alternative to the null hypothesis, namely, the hypothesis which we accept when the null hypothesis is rejected, is appropriately called the **alternative hypothesis**, and it is denoted by H_A. It must always be formulated together with the null hypothesis, for otherwise we would not know when to reject H_0. For instance, if the psychologist of the example on page 282 tests the null hypothesis $\mu = 0.44$ second against the alternative hypothesis $\mu > 0.44$ second, he would reject the null hypothesis only if he gets a sample mean which is much greater than 0.44 second. On the other hand, if he uses the alternative hypothesis $\mu \neq 0.44$ second, he would reject the null hypothesis if he gets a sample mean which is much greater than, or much less than, 0.44 second.

As in the preceding illustration, alternative hypotheses usually specify that the population mean (or whatever other parameter may be of concern) is less than, greater than, or not equal to the value assumed under the null hypothesis. For any given problem, the choice of one of these alternatives depends on what we hope to be able to show, or perhaps on where we want to put the burden of proof.

EXAMPLE An appliance manufacturer is considering the purchase of a new machine for stamping out sheet metal parts. If μ_0 is the average number of good parts stamped out per hour by his old machine and μ is the corresponding average for the new machine, the manufacturer wants to test the null hypothesis $\mu = \mu_0$ against a suitable alternative. What should the alternative be if

(a) he does not want to buy the new machine unless it is more productive than the old one;

(b) he wants to buy the new machine (which has some other nice features) unless it is less productive than the old one.

Solution
(a) The manufacturer should use the alternative hypothesis $\mu > \mu_0$ and purchase the new machine only if the null hypothesis can be rejected.

(b) The manufacturer should use the alternative hypothesis $\mu < \mu_0$ and purchase the new machine unless the null hypothesis is rejected.

In the example above, we treat μ_0 as a known number. We wish to make an assertion about μ, the parameter representing the unknown mean for the new machine.

To illustrate in detail the problems we face when testing a statistical hypothesis, suppose that the members of an airport's planning commission are considering the possibility of redesigning the parking facilities. In connection with this they first want to check the claim that on the average cars remain in the short-term parking area for 42.5 minutes, or whether this figure is too high or too low. Thus, they shall want to test the null hypothesis

$$H_0: \quad \mu = 42.5 \text{ minutes}$$

against the alternative hypothesis

$$H_A: \quad \mu \neq 42.5 \text{ minutes}$$

To perform this test, they instruct a member of their staff to take a random sample of 50 ticket stubs showing time of arrival and time of departure, with the intention of accepting the claim if the mean of the sample falls anywhere from 40.5 to 44.5 minutes. Otherwise, they will reject the claim, and in either case they will take whatever action is thus called for in their plans.

This provides a clear-cut criterion for accepting or rejecting the claim, but unfortunately it is not infallible. Since the decision is based on a sample, there is the possibility that the sample mean may be less than or equal to 40.5 minutes or greater than or equal to 44.5 minutes even though the true mean is $\mu = 42.5$ minutes, and there is also the possibility that the sample mean may fall on the interval from 40.5 minutes to 44.5 minutes even though the true mean is, say, $\mu = 45.5$ minutes. Thus, before adopting the criterion, it would seem wise to investigate the chances that it may lead to a wrong decision.

Assuming that it is known from similar studies that $\sigma = 7.6$ minutes for this kind of data, let us first investigate the possibility that the sample mean may be less than or equal to 40.5 minutes or greater than or equal to 44.5 minutes even though the true mean is $\mu = 42.5$ minutes. The probability that this will happen purely due to chance is given by the sum of the areas of the white regions of Figure 11.6, and it can easily be determined by approximating the sampling distribution of the mean with a normal distribution. Assuming that the population sampled is large enough to be treated as infinite, we have

$$\sigma_{\bar{x}} = \frac{\sigma}{\sqrt{n}} = \frac{7.6}{\sqrt{50}} = 1.07$$

and the dividing lines of the criterion, in standard units, are

$$z = \frac{40.5 - 42.5}{1.07} = -1.87 \quad \text{and} \quad z = \frac{44.5 - 42.5}{1.07} = 1.87$$

It follows from Table I that the area in each "tail" of the sampling distribution of Figure 11.6 is $0.5000 - 0.4693 = 0.0307$, and hence that the probability of getting a value in either tail of the sampling distribution (and erroneously rejecting the hypothesis $\mu = 42.5$ minutes) is $0.0307 + 0.0307 = 0.0614$, or approximately 0.06. Whether this is an acceptable risk is for the members of the planning commission to decide; it would have to depend on the consequences of their making such an error.

Let us now consider the other possibility, where the test fails to detect that μ is not equal 42.5 minutes. Suppose again, for the sake of argument, that the true mean is $\mu = 45.5$ minutes, so that the probability of getting a sample mean on the interval from 40.5 minutes to 44.5 minutes (and, hence, erroneously accepting the claim that $\mu = 42.5$ minutes) is given by the area of the white region of Figure 11.7. The mean of the sampling distribution is now 45.5, its standard deviation is as before

$$\sigma_{\bar{x}} = \frac{7.6}{\sqrt{50}} = 1.07$$

and the dividing lines of the criterion, in standard units, are

$$z = \frac{40.5 - 45.5}{1.07} = -4.67 \quad \text{and} \quad z = \frac{44.5 - 45.5}{1.07} = -0.93$$

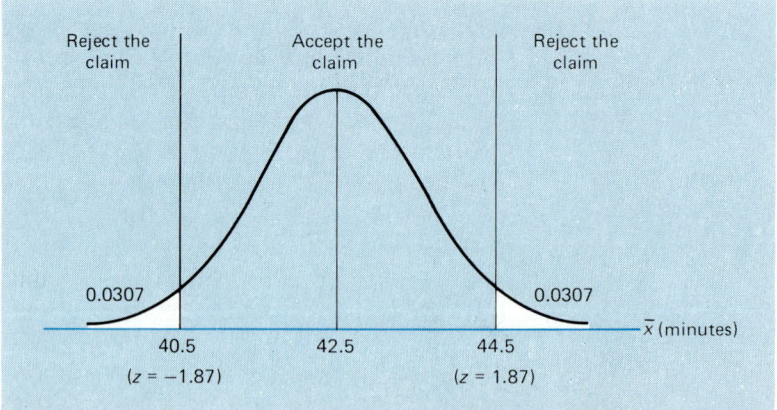

FIGURE 11.6 *Test criterion.*

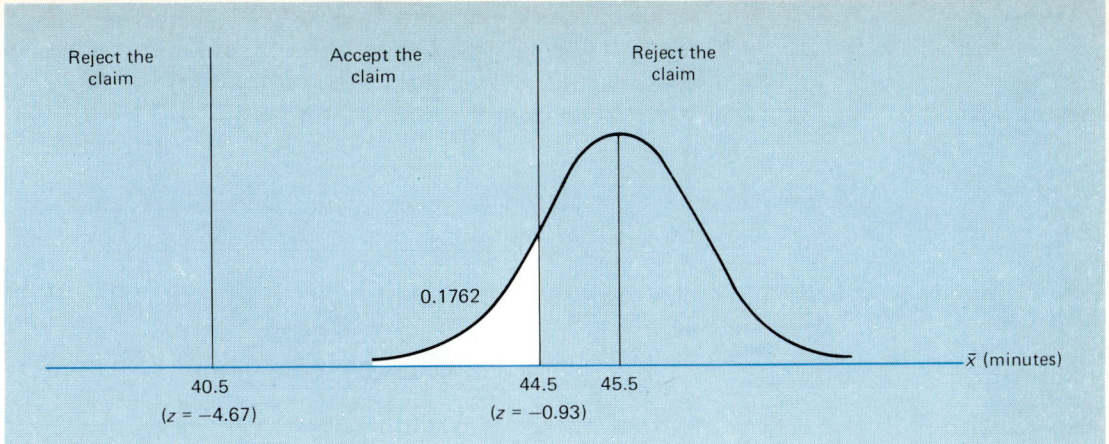

FIGURE 11.7 *Test criterion.*

Since the area under the curve to the left of $z = -4.67$ is negligible, it follows from Table I that the white region area of Figure 11.7 is $0.5000 - 0.3238 = 0.1762$, or approximately 0.18. This is the probability of erroneously accepting the null hypothesis when actually $\mu = 45.5$ minutes. Again, it is up to the members of the planning commission to decide whether this represents an acceptable risk.

The situation described here is typical of tests of hypotheses, and it may be summarized as in the following table:

	Accept H_0	*Reject* H_0
H_0 is true	Correct decision	Type I error
H_0 is false	Type II error	Correct decision

If the null hypothesis H_0 is true and accepted or false and rejected, the decision is in either case correct; if it is true and rejected or false and accepted, the decision is in either case in error. The first of these errors is called a **Type I error** and the probability of committing it is designated by the Greek letter α (*alpha*); the second is called a **Type II error** and the probability of committing it is designated by the Greek letter β (*beta*). Thus, in our example we showed that for the given test criterion $\alpha = 0.06$, and $\beta = 0.18$ when $\mu = 45.5$ minutes.

The scheme outlined above is reminiscent of what we did in Section 7.2. Analogous to the decision which the furniture manufacturer had to make in the example on page 168, we now have to decide whether to accept or reject H_0. It is difficult to carry this analogy much further, though, because in actual practice we can seldom associate cash payoffs with the various possibilities, as we did in that example.

EXAMPLE Suppose that the planning commission's staff member actually takes the sample and gets $\bar{x} = 41.8$ minutes. What decision will the members of the planning commission make and will it be in error if in reality

(a) $\mu = 42.5$ minutes;

(b) $\mu = 41.2$ minutes?

Solution Since $\bar{x} = 41.8$ falls in the interval from 40.5 to 44.4, they will accept the null hypothesis that on the average cars remain in the short-term parking area for 42.5 minutes.

(a) Since the null hypothesis is true and accepted, they will not be making an error.

(b) Since the null hypothesis is false but accepted, they will be making a Type II error.

In calculating the probability of a Type II error in our example, we arbitrarily chose the alternative value $\mu = 45.5$ minutes. However, in this problem, as in most others, there are infinitely many other alternatives, and for each one of them there is a positive probability β of erroneously accepting H_0. So, in practice we choose some key alternative values and calculate the corresponding probabilities β of committing a Type II error, or we sidestep the issue by proceeding in a way which will be explained in Section 11.5.

If we do calculate β for various alternative values of μ and plot these probabilities as in Figure 11.8, we obtain a curve which is called the **operating characteristic curve**, or simply the **OC-curve**, of the test criterion. Since the probability of a Type II error is the probability of accepting H_0 when it is false, we "completed the picture" in Figure 11.8 by labeling the vertical scale "Probability of

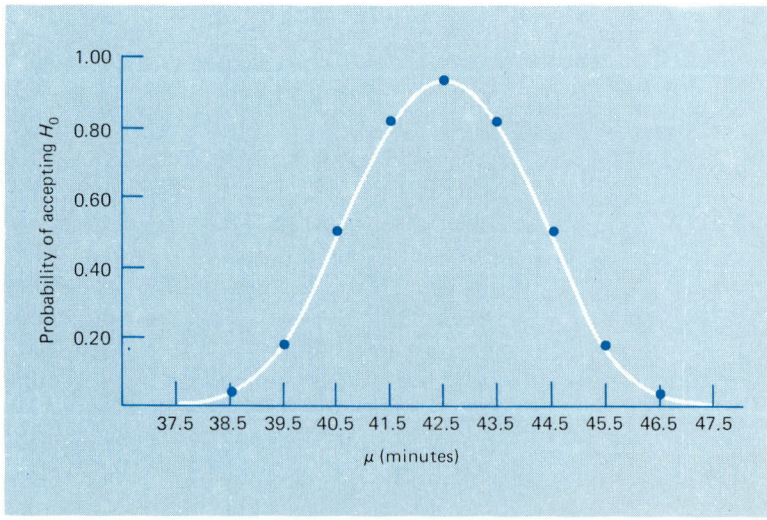

FIGURE 11.8 *Operating characteristic curve.*

accepting H_0" and plotting at $\mu = 42.5$ minutes the probability of accepting H_0 when it is true, namely, $1 - \alpha = 1 - 0.06 = 0.94$.

Examination of the curve of Figure 11.8 shows that the probability of accepting H_0 is greatest when it is true, and that it is still high for small departures from $\mu = 42.5$ minutes. However, for larger and larger departures from $\mu = 42.5$ minutes in either direction, the probabilities of failing to detect them and accepting H_0 become smaller and smaller. In Exercise 11.54 the reader will be asked to verify some of the probabilities plotted in Figure 11.8.

If we had plotted the probabilities of rejecting H_0 instead of those of accepting H_0, we would have obtained the graph of the **power function** of the test criterion instead of its operating characteristic curve. The concept of an OC-curve is used more widely in applications, especially in industrial applications, while the concept of a power function is used more widely in matters that are of theoretical interest. A detailed study of operating characteristic curves and power functions would go considerably beyond the scope of this text, and the purpose of our example was mainly to show how statistical methods can be used to measure and control the risks to which one is exposed when testing hypotheses. Of course, the methods we have discussed here are not limited to the particular problem concerning the average length of time that cars are parked in the given area—H_0 could be the hypothesis that the average age of divorced women at the time of their divorce is 35.6 years, the hypothesis that an antibiotic is 83 percent effective, the hypothesis that a computer-assisted method of instruction will on the average raise students' scores on a standard achievement test by 8.4 points, and so forth.

EXERCISES

11.42 The average drying time of a manufacturer's paint is 20 minutes. Investigating the effectiveness of a modification in the chemical composition of his paint, the manufacturer wants to test the null hypothesis $\mu = 20$ minutes against a suitable alternative, where μ is the average drying time of the modified paint.
  (a) What alternative hypothesis should the manufacturer use if he does not want to make the modification in the chemical composition of the paint unless it decreases the drying time?
  (b) What alternative hypothesis should the manufacturer use if the new process is actually cheaper and he wants to make the modification unless it increases the drying time of the paint?

11.43 A city police department is considering replacing the tires on its cars with radial tires. If μ_1 is the average number of miles the old tires last and μ_2 is the average number of miles the new tires will last, the null hypothesis to be tested is $\mu_1 = \mu_2$.

  (a) What alternative hypothesis should the department use if it does not want to buy the radial tires unless they are definitely proved to give better mileage? In other words, the burden of proof is put on the radial tires and the old tires are to be kept unless the null hypothesis can be rejected.
  (b) What alternative hypothesis should the department use if it is anxious to get the new tires (which have some other good features) unless they actually give poorer mileage than the old tires? Note that now the burden of proof is on the old tires, which will be kept only if the null hypothesis can be rejected.
  (c) What alternative hypothesis should the department use so that the rejection of the null hypothesis can lead either to keeping the old tires or to buying the new ones?

11.44 Rework the example on page 302, supposing that $\bar{x} = 45.2$ minutes instead of $\bar{x} = 41.8$ minutes.

11.45 A botanist wishes to test the null hypothesis that the average diameter of the flowers of a particular plant is 9.6 cm. He decides to take a random sample of size $n = 80$ and accept the null hypothesis if the mean of the sample falls between 9.3 cm and 9.9 cm; if the mean of the sample falls outside this interval he will reject the null hypothesis. What decision will he make and will it be in error if
    (a) he gets $\bar{x} = 10.2$ cm and $\mu = 9.6$ cm;
    (b) he gets $\bar{x} = 10.2$ cm and $\mu = 9.8$ cm;
    (c) he gets $\bar{x} = 9.2$ cm and $\mu = 9.6$ cm;
    (d) he gets $\bar{x} = 9.2$ cm and $\mu = 9.8$ cm?

11.46 An education specialist is considering the use instructional material on audiocasettes for a special class of third-grade students with reading difficulties. Students in this class are given a standardized reading test in May of the school year, and μ_1 is the average score obtained after many years of experience. Let μ_2 be the average score for students using the audiocasettes, and assume that high scores are desirable.
    (a) What null hypothesis should the education specialist use?
    (b) What alternative hypothesis should be used if the specialist does not want to adopt the new cassettes unless they improve the standardized test scores?
    (c) What alternative hypothesis should be used if the specialist wants to adopt the new casettes unless they worsen the standardized test scores?

11.47 Suppose that a psychological testing service is asked to check whether an executive is emotionally fit to assume the presidency of a large corporation. What type of error would it commit if it erroneously rejects the null hypothesis that the executive is fit for the job? What type of error would it commit if it erroneously accepts the null hypothesis that the executive is fit for the job?

11.48 Suppose we want to test the null hypothesis that an antipollution device for cars is effective. Explain under what conditions we would Commit a Type I error and under what conditions we would commit a Type II error.

11.49 Whether an error is a Type I error or a Type II error depends on how we formulate the null hypothesis. To illustrate this, rephrase the null hypothesis of the preceding exercise so that the Type I error becomes a Type II error, and vice versa.

11.50 For a given population with $\sigma = 8.4$ inches we want to test the null hypothesis $\mu = 80.0$ inches against the alternative hypothesis $\mu < 80.0$ inches on the basis of

a random sample of size $n = 100$. If the null hypothesis is rejected when $\bar{x} < 78.0$ inches and otherwise it is accepted, what is the probability of a Type I error?

11.51 With reference to the preceding exercise, what is the probability of a Type II error when in reality $\mu = 77.5$ inches?

11.52 For a given population with $\sigma = 21.0$ cm we want to test the null hypothesis $\mu = 255$ cm against the alternative hypothesis $\mu \neq 255$ cm on the basis of a sample of size $n = 36$. If the null hypothesis is rejected when $\bar{x} < 248$ cm or $\bar{x} > 262$ cm,
    (a) find the probability of a Type I error;
    (b) find the probability of a Type II error when in when in reality $\mu = 258.5$ cm.

11.53 For a given population with $\sigma = \$8$ we want to test the null hypothesis $\mu = \$65$ against the alternative hypothesis $\mu > \$65$ on the basis of a sample of size $n = 100$. If the null hypothesis is rejected when $\bar{x} > \$66$,
    (a) find the probability of a Type I error;
    (b) find the probability of a Type II error when in reality $\mu = \$65.50$;
    (c) find the probability of a Type II error when in reality $\mu = \$66.50$.

11.54 With reference to the operating characteristic curve of Figure 11.8, verify that the probabilities of Type II errors are
    (a) 0.82 when $\mu = 41.5$ minutes or $\mu = 43.5$ minutes;
    (b) 0.50 when $\mu = 40.5$ minutes or $\mu = 44.5$ minutes;
    (c) 0.18 when $\mu = 39.5$ minutes or $\mu = 45.5$ minutes;
    (d) 0.03 when $\mu = 38.5$ minutes or $\mu = 46.5$ minutes.

11.55 Suppose that in the airport parking example on page 299 the criterion is changed so that the null hypothesis $\mu = 42.5$ minutes is rejected if the sample mean is less than 41.0 minutes or greater than 44.0 minutes; otherwise, the null hypothesis is accepted.
    (a) What is the probability of a Type I error?
    (b) Find the probabilities of Type II errors when $\mu = 38.5, 39.5, 40.5, 41.5, 43.5, 44.5, 45.5,$ or 46.5 minutes, and plot the operating characteristic curve.

11.56 Suppose that in the airport parking example on page 299 the sample size is increased from 50 to 60, while the criterion remains as stated on page 299.
    (a) What is the probability of a Type I error?

(b) Find the probabilities of Type II errors when $\mu = 38.5, 39.5, 40.5, 41.5, 43.5, 44.5, 45.5,$ or 46.5 minutes, and plot the operating characteristic curve.

11.57 To reduce the probability of a Type I error in the airport parking example, the criterion is modified so that the null hypothesis $\mu = 42.5$ minutes is rejected if the sample mean is less than 40.0 minutes or greater than 45.0 minutes; otherwise, the null hypothesis is accepted.

(a) Show that this reduces the probability of a Type I error from 0.06 to 0.02.

(b) Show that for $\mu = 45.5$ the probability of a Type II error is increased from 0.18 to 0.32.

(c) Show that for $\mu = 46.5$ the probability of a Type II error is increased from 0.03 to 0.08.

11.5

SIGNIFICANCE TESTS

In the airport parking example of the preceding section, we had less trouble with Type I errors than with Type II errors, because we formulated the null hypothesis as a **simple hypothesis** about the parameter μ; that is, we formulated it so that μ took on a single value and the corresponding probability of a Type I error could be calculated.[†] Had we formulated instead a **composite hypothesis** about the parameter μ, say, $\mu \neq 42.5$ minutes, $\mu < 42.5$ minutes, or $\mu > 42.5$ minutes, where in each case μ can take on more than one possible value, we could not have calculated the probability of a Type I error without specifying how much μ differs from, is less than, or is greater than 42.5 minutes.

In the same example, the alternative hypothesis was the composite hypothesis $\mu \neq 42.5$ minutes, and it took quite some work to calculate the probabilities of Type II errors (for various alternative values of μ) shown in the OC-curve of Figure 11.8. Since this is typical of most practical situations (that is, alternative hypotheses are usually composite), let us illustrate how Type II errors can often be sidestepped altogether.

Experience shows that in a given city licensed drivers average 1.4 traffic tickets per year with a standard deviation of 0.4. To confirm her suspicion that licensed drivers over 65 average more than 1.4 traffic tickets per year, a social scientist checks the 1987 records of 40 randomly selected licensed drivers over 65 in the given city, and bases her decision on the following criterion:

Reject the null hypothesis $\mu = 1.4$ (and accept the alternative hypothesis $\mu > 1.4$) if the 40 licensed drivers over 65 average 1.5 or more traffic tickets per year; otherwise, reserve judgment (perhaps, pending further checks.)

_____

[†] Note that we are applying the term "simple hypothesis" to hypotheses about one specific parameter. Some statisticians use the term "simple hypothesis" only when there are no other unknown parameters and the hypothesis completely specifies the population. The hypothesis $H_0: \mu = 72$ regarding a normally distributed population with unknown standard deviation will be described here as a "simple hypothesis" even though some statisticians would not use this description.

The probability of committing a Type I error with this criterion is given by the area of the white region of Figure 11.9, and it equals 0.0571.

If we reserve judgment as in this criterion, there is no possibility of committing a Type II error—no matter what happens, the null hypothesis is never accepted. This would seem to be all right in our example, where the social scientist wants to see primarily whether her suspicion is justified, namely, whether the null hypothesis can be rejected. If it cannot be rejected, this does not mean that she must necessarily accept it; her suspicion may not be completely resolved.

The procedure we have outlined here is called a **significance test**, or a **test of significance**. If the difference between what we expect under the null hypothesis and what we observe in a sample is too large to be reasonably attributed to chance, we reject the null hypothesis. If the difference between what we expect and what we observe is so small that it may well be attributed to chance, we say that the result is **not statistically significant**, or simply that it is **not significant**. We then accept the null hypothesis or reserve judgment, depending on whether a definite decision one way or the other must be reached.

Since "significant" is often used interchangeably with "meaningful" or "important" in everyday language, it must be understood that we are using it here as a technical term. Specifically, the word "significant" is used in situations in which a null hypothesis is rejected. If a result is statistically significant, this does not mean that it is necessarily of any great importance, or that it is of any practical value. Suppose, for instance, that the social scientist of our example takes her sample and gets $\bar{x} = 1.525$. According to the criterion on page 305 this result is statistically significant—the difference between $\bar{x} = 1.525$ and $\mu = 1.4$ is too big to be attributed to chance—but then nobody may care. Even an insurance company which ought to be interested in such a result may well feel that it is not worth bothering about.

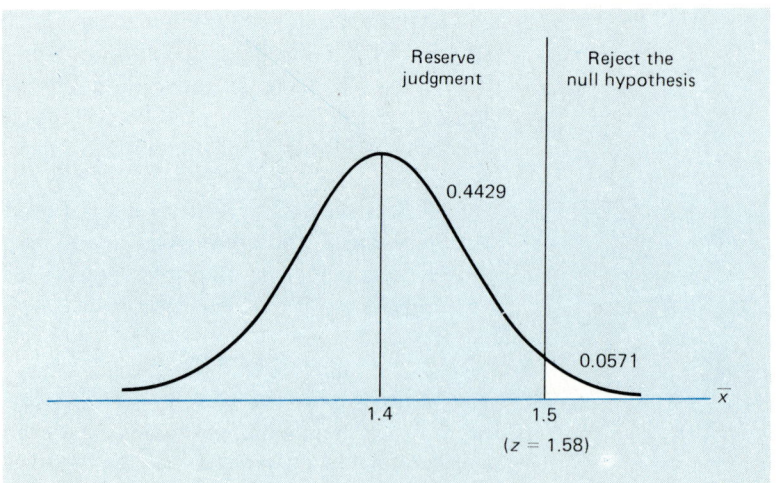

FIGURE 11.9 *Test criterion.*

Returning to the airport parking example, we could convert the criterion on page 299 into that of a significance test by writing

> **Reject the hypothesis $\mu = 42.5$ minutes (and accept the alternative $\mu \neq 42.5$ minutes) if the mean of the 50 sample values is less than or equal to 40.5 minutes or greater than or equal to 44.5 minutes; reserve judgment if the mean falls anywhere from 40.5 to 44.5 minutes.**

So far as the rejection of the null hypothesis is concerned, the criterion has remained unchanged and the probability of a Type I error is still 0.06. However, so far as its acceptance is concerned, the members of the planning commission are now playing it safe by reserving judgment.

Reserving judgment in a significance test is similar to what happens in court proceedings where the prosecution does not have sufficient evidence to get a conviction, but where it would be going too far to say that the defendent definitely did not commit the crime. In general, whether one can afford the luxury of reserving judgment in any given situation depends entirely on the nature of the situation. If a decision must be reached one way or the other, there is no way of avoiding the risk of committing a Type II error.

Since most of the remainder of this book will be devoted to significance tests—indeed, most statistical problems which are not problems of estimation or prediction deal with tests of this kind—it will help to perform such tests by proceeding systematically as outlined in the following five steps. The first of these may look simple and straightforward, yet it presents the greatest difficulties to most beginners.

1. We formulate the null hypothesis and an appropriate alternative.

In the airport parking example the null hypothesis was $\mu = 42.5$ minutes, and the alternative hypothesis was $\mu \neq 42.5$ minutes (presumably because the planning commission wanted to protect itself against the possibility that $\mu = 42.5$ minutes may be too high or too low). We refer to this kind of alternative as a **two-sided alternative**. In the traffic ticket example the null hypothesis was $\mu = 1.4$, and the alternative hypothesis was $\mu > 1.4$ (because the social scientist suspected that licensed drivers over 65 average more than 1.4 tickets per year). This is called a **one-sided alternative**. We can also write a one-sided alternative with the inequality going the other way. For instance, if we hope to be able to show that the average time required to do a certain job is less than 15 minutes, we would test the null hypothesis $\mu = 15$ minutes against the alternative hypothesis $\mu < 15$ minutes.

This is not the first time that we concerned ourselves with the formulation of hypotheses. In Section 11.4 we mentioned some of the things we must take into account when choosing H_A, but in the example on page 299 as well as in Exercises 11.42 and 11.43 on page 303 the null hypothesis was always specified. So, let us see how we go about choosing H_0.

Basically, there are two things we must watch. First, whenever possible we formulate the null hypothesis as a simple hypothesis about the parameter with which we are concerned; second, we formulate the null hypothesis in such a way that its rejection proves the point we hope to make. The reason for choosing H_0 as a simple hypothesis is that it enables us to calculate, or specify, the probability of a Type I error, and we saw how this works in the airport parking example. The reason for choosing the null hypothesis so that its rejection proves the point we hope to make is that in general it is much easier to prove that something is false than to prove that it is true.

Suppose, for instance, that somebody claims that all 6,000 male students attending a certain college weigh at least 145 pounds. To show that this is true, we literally have to weigh each of the 6,000 students; however, to show that it is false, we have only to find one student who weighs less than 145 pounds, and that should not be too hard.

EXAMPLE   A bakery uses a machine to fill boxes with crackers, averaging 454 grams (roughly one pound) of crackers per box.

   (a) If the management of the bakery is concerned about the possibility that the actual average is different from 454 grams, what null hypothesis and what alternative hypothesis should they use to put this to a test?

   (b) If the management of the bakery is concerned about the possibility that the actual average is less than 454 grams, what null hypothesis and what alternative hypothesis should it use to put this to a test?

Solution   **(a)** The words "different from" suggest that the hypothesis $\mu \neq 454$ grams is needed together with the only other possibility, namely, the hypothesis $\mu = 454$ grams. Since the second of these hypotheses is a simple hypothesis, and its rejection (and the acceptance of the other hypothesis) confirms the management's concern, we follow the two rules above by writing

$$H_0: \quad \mu = 454 \text{ grams}$$

$$H_A: \quad \mu \neq 454 \text{ grams}$$

   (b) The words "less than" suggest that we need the hypothesis $\mu < 454$ grams, but for the other hypothesis there are many possibilities, including $\mu = 454$ grams, $\mu \geq 454$ grams, and, say, $\mu = 456$ grams. Two of these (and many others) are simple hypotheses, but since it would be to the bakery's disadvantage to put too many crackers into the boxes, a sensible choice would be

$$H_0: \quad \mu = 454 \text{ grams}$$

$$H_A: \quad \mu < 454 \text{ grams}$$

Note that the null hypothesis is a simple hypothesis and that its rejection (and the acceptance of the alternative) confirms the management's suspicion.

Like the first step, the second step looks simple and straightforward, yet it is not without complications.

2. We specify the probability of a Type I error.

When H_0 is a simple hypothesis this can always be done, and we usually set the probability of a Type I error, also called the **level of significance**, at $\alpha = 0.05$ or $\alpha = 0.01$. Testing a simple hypothesis at the 0.05 (or 0.01) level of significance simply means that we are fixing the probability of rejecting H_0 when it is true at 0.05 (or 0.01).

The decision to use 0.05, 0.01, or some other value, depends mostly on the consequences of committing a Type I error. Although it may seem desirable to make the probability of a Type I error small, we cannot make it too small, since this would tend to make the probabilities of serious Type II errors too large, and make it difficult, perhaps too difficult, to get significant results. To some extent, the choice of 0.05 or 0.01, and not, say, 0.08 and 0.03, is dictated by the availability of statistical tables. For this reason, the choice of 0.05 or 0.01 has been criticized by some statisticians in recent years (see references on pages 332 and 333). Anyhow, in most of the exercises in this text, the level of significance will be specified.

Let us now see what can be done when the null hypothesis is not a simple hypothesis. Suppose, for instance, that the social scientist of the example on page 305 wants to allow for the possibility that licensed drivers over 65 in the given city might average fewer than 1.4 tickets per year, and hence test the null hypothesis $\mu \leq 1.4$ against the alternative hypothesis $\mu > 1.4$. This means that there is no unique value for the probability of a Type I error, but observe that if μ is less than 1.4, the normal curve of Figure 11.9 is shifted to the left, and the area under the curve to the right of 1.5 becomes less than 0.0571. Thus, if the null hypothesis is $\mu \leq 1.4$, we say that the probability of a Type I error is at most 0.0571, and we write $\alpha \leq 0.0571$. To allow for this possibility, we rewrite the second step as follows:

2. **We specify the probability of a Type I error, or the maximum probability of a Type I error.**

(This still does not take care of the case where the social scientist wants to test the null hypothesis $\mu < 1.4$ against the alternative hypothesis $\mu \geq 1.4$, but we rarely use null hypotheses given by *strict* inequalities.)

Another difficulty arises in connection with the second step when we cannot, or do not want to, specify the probability of a Type I error, or the maximum probability of a Type I error. This could happen when we do not have enough information about the consequences of Type I errors, or when one person processes the data while another person makes the decisions. What can be done in that case, is discussed on page 314.

After the null hypothesis, the alternative hypothesis, and the probability (or maximum probability) of a Type I error have been specified, the remaining steps are

3. **Based on the sampling distribution of an appropriate statistic, we construct a criterion for testing the null hypothesis against the given alternative hypothesis at the chosen level of significance.**
4. **We calculate from the data the value of the statistic on which the decision is to be based.**
5. **We decide whether to reject the null hypothesis, whether to accept it, or whether to reserve judgment.**

In the airport parking example we rejected the null hypothesis $\mu = 42.5$ minutes for values of $\bar{x}$ less than 40.5 and also for values of $\bar{x}$ greater than 44.5, and we refer to a criterion like this as a **two-sided test** or as a **two-tailed test**. As these names imply, we reject the null hypothesis for values of $\bar{x}$ falling into either tail of its sampling distribution. In the traffic ticket example we rejected the null hypothesis $\mu = 1.4$ for values of $\bar{x}$ greater than 1.5, and we refer to a criterion like this as a **one-sided test** or as a **one-tailed test**. As these names imply, we reject the null hypothesis only for values of $\bar{x}$ falling into one of the two tails of its sampling distribution. In general, a test is said to be two-sided if the null hypothesis is rejected for values of the **test statistic** falling into either tail of its sampling distribution, and it is said to be one-sided if the null hypothesis is rejected only for values of the test statistic falling into one specified tail of its sampling distribution. By "test statistic" we mean the statistic (for instance, the sample mean) on which the test is based. Although there are exceptions, two-sided test criteria are usually used in conjunction with two-sided alternative hypotheses, and one-sided test criteria are usually used in conjunction with one-sided alternative hypotheses.

As part of the third step we must also specifiy whether the alternative to rejecting the null hypothesis is to accept it or to reserve judgment. This, as we have said, depends on whether we must make a decision one way or the other, or whether the circumstances permit that we delay a decision pending further study. In exercises and examples, the phrase "whether or not" will sometimes be used to indicate that a decision must be reached one way or the other.

In connection with the fifth step, let us point out that we often accept null hypotheses with the tacit hope that we are not exposed to overly high risks of committing serious Type II errors. Of course, if it is necessary we can calculate enough probabilities of Type II errors to get an overall picture from the operating characteristic curve of the test criterion.

Before we discuss various special tests about means in the remainder of this chapter, let us point out that the concepts we have introduced here are not limited to tests concerning population means; they apply equally to tests concerning other parameters, or tests concerning the nature, or form, of populations.

EXERCISES

11.58 In a study designed to compare the IQ of persons of two different ethnic backgrounds, an educator gets a large difference between the respective sample means, and he concludes that the difference between the corresponding population means is statistically significant. Comment.

11.59 In a public opinion poll, 58 percent of the persons interviewed favored a certain candidate for the U.S. Senate, and this enabled the pollster to reject the null hypothesis that the candidate will get at least 60 percent of the vote at the 0.05 level of significance. Since the result is statistically significant, should the candidate be advised to change her campaign strategy?

11.60 The mean age of Mr. and Mrs. Green's three children is 16.3 years while the mean age of Mr. and Mrs. Brown's four children is 12.8 years. Does it make any sense to ask whether the difference between these two means is significant?

11.61 In a study of extrasensory perception, 250 persons were asked to predict patterns on cards drawn at random from a deck, and two of them scored better than could be expected at the 0.01 level of significance. Comment on the conclusion that these two persons must have extraordinary powers.

11.62 In the production of certain springs, samples are taken at regular intervals of time to check at the 0.05 level of significance whether the process is in control, namely, whether the mean length of the springs meets specifications. Is there reason for concern if, in forty such samples, the null hypothesis that the process is in control had to be rejected six times?

11.63 In a certain experiment, a null hypothesis is rejected at the 0.001 level of significance. Does this mean that the probability is at most 0.001 that the null hypothesis is true?

11.64 It has been claimed that, on the average, the students attending a certain university miss 3.4 lectures per semester per course. One of the deans, concerned that this figure might not be right, plans to put this claim to a test.
   (a) What null hypothesis and what alternative should he use?
   (b) If the dean plans to base the decision on the mean of a random sample, should he use a one-tailed test or a two-tailed test?

11.65 The manufacturer of a blood-pressure medication claims that, on the average, the medication will lower patients' blood pressure by at least 20 mm. A medical research worker suspects this claim and wishes to put it to a test.
   (a) What null hypothesis and what alternative should she use?
   (b) If she bases her decision on the mean of a random sample, should she use a one-tailed test or a two-tailed test?

11.66 A fire department claims that it takes on the average 6.2 minutes from the moment it receives a call until it reaches the scene of the fire. If it is suspected that this figure may be too low, what null hypothesis and what alternative hypothesis should be used to put this to a test?

11.67 A researcher develops a new topical treatment for acne and wants to show that its performance is not significantly different from that of a popular product currently on the market. Accordingly, he tests the null hypothesis that the population averages for the two products are the same against the two-sided alternative that they are different. He designs an experiment involving six subjects (three using each product) and standard measures of acne severity. The result of the experiment is that the products are not statistically significantly different. Comment.

11.68 A product line inspector is concerned that the average "fill" going into boxes of cornflakes might not be the desired 24 ounces. Letting μ be the relevant population average, she tests the null hypothesis that $\mu = 24$ ounces against the alternative that $\mu \neq 24$ ounces. Based on a random sample of 280 boxes with a sample mean of $\bar{x} = 24.08$ ounces she concludes that the null hypothesis must be rejected and that μ is significantly different from 24 ounces. Comment.

11.69 A lake is repeatedly monitored for concentration levels of a particular pollutant, for which the acceptable level is 0.0082 mg/liter. Letting μ be the average pollution level, a researcher tests the null hypothesis $\mu = 0.0082$ against the alternative $\mu > 0.0082$. Based on a sample of 120 readings with a sample mean of $\bar{x} = 0.0264$ and a sample standard deviation of $s = 0.0022$, he decides that the null hypothesis must be rejected. Comment.

11.6
TESTS CONCERNING MEANS

Having used tests concerning means to illustrate the basic principles of hypothesis testing, let us now demonstrate how we proceed in actual practice. In what follows, we shall depart somewhat from the procedure used in the preceding sections. In the airport parking example as well as in the traffic ticket example we stated the test criterion in terms of $\bar{x}$—in the first case we rejected the null hypothesis for $\bar{x} < 40.5$ or $\bar{x} > 44.5$, and in the second case we rejected it for $\bar{x} > 1.5$. Now we shall base it on the statistic

Statistic for test concerning mean

$$z = \frac{\bar{x} - \mu_0}{\sigma/\sqrt{n}}$$

where μ_0 is the value of the mean assumed under the null hypothesis. The reason for working with standard units, or z-values, is that it enables us to formulate criteria which are applicable to a great variety of problems, not just one.

The test of this section is essentially a **large-sample test**; that is, we require that the samples are large enough, $n \geq 30$, so that the sampling distribution of the mean can be approximated closely with a normal distribution and z is a value of a random variable having approximately the standard normal distribution. (In the special case where we sample a normal population, z is a value of a random variable having the standard normal distribution regardless of the size of n.)

Thus, we can use the test criteria shown in Figure 11.10 on page 313; depending on the alternative hypothesis, the dividing lines, or **critical values**, of the criteria are $-z_\alpha$ or z_α for the one-sided alternatives, and they are $-z_{\alpha/2}$ and $z_{\alpha/2}$ for the two-sided alternative. As before, z_α and $z_{\alpha/2}$ are such that the area to their right under the standard normal curve is α and $\alpha/2$. Symbolically, we can formulate these criteria as follows:

Alternative hypothesis	*Reject the null hypothesis if*	*Accept the null hypothesis or reserve judgment if*
$\mu < \mu_0$	$z \leq -z_\alpha$	$z > -z_\alpha$
$\mu > \mu_0$	$z \geq z_\alpha$	$z < z_\alpha$
$\mu \neq \mu_0$	*or* $\begin{array}{l} z \leq -z_{\alpha/2} \\ z \geq z_{\alpha/2} \end{array}$	$-z_{\alpha/2} < z < z_{\alpha/2}$

If the level of significance is 0.05, the dividing lines are -1.645 or 1.645 for the one-sided alternatives, and -1.96 and 1.96 for the two-sided alternative; if the level of significance is 0.01, the dividing lines are -2.33 or 2.33 for the one-sided alternatives, and -2.575 and 2.575 for the two-sided alternative. All these values come directly from Table I.[†]

[†] Some authors exclude the dividing lines from the values for which the null hypothesis is rejected, but this is of no practical consequence since the standard normal distribution is a continuous distribution.

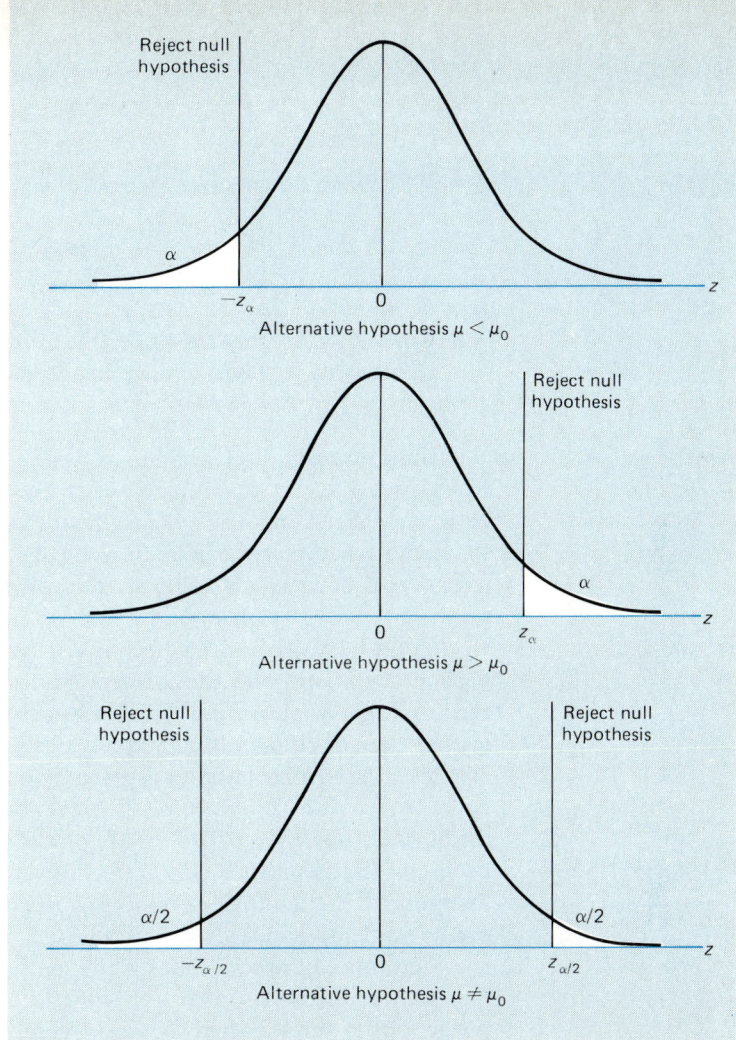

FIGURE 11.10 *Test criteria for large-sample tests concerning mean.*

EXAMPLE An oceanographer wants to test, on the basis of the mean of a random sample of size $n = 35$ and at the 0.05 level of significance, whether the average depth of the ocean in a certain area is 72.4 fathoms, as has been recorded. What will she decide if she gets $\bar{x} = 73.2$ fathoms, and she can assume from information gathered in similar studies that $\sigma = 2.1$ fathoms?

Solution 1. H_0: $\mu = 72.4$ fathoms
   H_A: $\mu \neq 72.4$ fathoms
2. $\alpha = 0.05$

3. Reject the null hypothesis if $z \leq -1.96$ or $z \geq 1.96$, where

$$z = \frac{\bar{x} - \mu_0}{\sigma/\sqrt{n}}$$

and otherwise accept it (or reserve judgment).

4. Substituting $\mu_0 = 72.4$, $\sigma = 2.1$, $n = 35$, and $\bar{x} = 73.2$ into the formula for z, we get

$$z = \frac{73.2 - 72.4}{2.1/\sqrt{35}} = 2.25$$

5. Since $z = 2.25$ exceeds 1.96, the null hypothesis must be rejected; to put it another way, the difference between $\bar{x} = 73.2$ and $\mu_0 = 72.4$ is significant.

If the oceanographer had used the 0.01 level of significance in this example, she would not have been able to reject the null hypothesis because $z = 2.25$ falls between -2.575 and 2.575. This illustrates the importance of specifying the level of significance before any calculations have actually been made. This will spare us the temptation of later choosing a level of significance which happens to suit our purpose.

In problems like this, some research workers accompany the calculated value of the test statistic with the corresponding **tail probability**, or ***p*-value**, namely, with the probability of getting a difference between $\bar{x}$ and μ_0 which is numerically greater than or equal to that actually observed. For instance, in the oceanographer example, the tail probability is given by the total area under the standard normal curve to the left of $z = -2.25$ and to the right of $z = 2.25$, and it equals $2(0.5000 - 0.4878) = 0.0244$. This practice is not new by any means, but it has been advocated more widely in recent years because of the general availability of computers. For some distributions, computers can provide tail probabilities which are not directly available from tables.

Quoting tail probabilities is the method referred to on page 309 for problems in which we cannot, or do not want to, specify the level of significance. This is proper, for example, when someone merely studies a set of data without having to reach a decision, or when someone merely processes a set of data for somebody else to make the decision; also, tail probabilities are given in many statistical programs, so that the same program can be used regardless of the level of significance and there is no need to compare results with tabular values. Be that as it may, if a decision must be reached, quoting tail probabilities does not relieve us of the responsibility of specifying the level of significance before the test is actually performed.

In general, *p*-values may be defined as follows: Corresponding to an observed value of a test statistic, the *p*-value is the lowest level of significance at which the null hypothesis could have been rejected. In our example, the *p*-value was 0.0244, and we could have rejected the null hypothesis at the 0.0244 level of significance. We certainly would have rejected the null hypothesis at the actual specified level of significance $\alpha = 0.05$.

If we wish to base tests of significance on p-values, steps 1 and 2 remain the same, but steps 3, 4, and 5 must be modified as follows:

3′. We specify the test statistic.
4′. We calculate the value of the specified test statistic and the corresponding p-value from the sample data.
5′. We compare the p-value obtained in step 4′ with the level of significance α specified in step 2. If the p-value is *less than or equal to* α, the null hypothesis must be rejected; otherwise we accept the null hypothesis or reserve judgment.[†]

The test we have described in this section requires that σ is known. When σ is not known, we substitute for σ the sample standard deviation s. If n is large, say $n \geq 30$, we recommend the use of s even if σ is claimed known.

EXAMPLE   A trucking firm suspects the claim that the average lifetime of certain tires is at least 28,000 miles. To check the claim, the firm puts 40 of these tires on its trucks and gets a mean lifetime of 27,563 miles with a standard deviation of 1,348 miles. What can it conclude if the probability of a Type I error is to be at most 0.01?

Solution
1. H_0: $\mu \geq 28{,}000$ miles
   H_A: $\mu < 28{,}000$ miles
2. $\alpha \leq 0.01$
3. Since the probability of a Type I error is greatest when $\mu = 28{,}000$ miles, we proceed as if we were testing the null hypothesis $\mu = 28{,}000$ miles against the alternative hypothesis. $\mu < 28{,}000$ miles at the 0.01 level of significance. Thus, the null hypothesis must be rejected if $z \leq -2.33$, where

$$z = \frac{\bar{x} - \mu_0}{\sigma/\sqrt{n}}$$

with σ replaced by s; otherwise, accept the null hypothesis or reserve judgment.
4. Substituting $\mu_0 = 28{,}000$, $\bar{x} = 27{,}563$, $n = 40$, and $s = 1{,}348$ for σ into the formula for z, we get

$$z = \frac{27{,}563 - 28{,}000}{1{,}348/\sqrt{40}} = -2.05$$

5. Since $z = -2.05$ is not less than -2.33, the null hypothesis cannot be rejected; in other words, the trucking firm's suspicion that $\mu < 28{,}000$ miles is not confirmed by its data. Whether it goes any further and accepts the null hypothesis or reserves judgment, is not a matter of statistics but an executive decision.

_____

[†]If test criteria are modified as in the footnote on page 312, p-values must be defined in a different way.

It would be acceptable to use $H_0: \mu = 28,000$ miles in step 1 and $\alpha = 0.01$ in step 2. Steps 3, 4, and 5 would not be affected by this change.

Let us repeat this procedure using the p-value. In the solution, steps 1 and 2 remain the same, but steps 3, 4, and 5 are replaced by

3'. The test statistic is

$$z = \frac{\bar{x} - \mu_0}{\sigma/\sqrt{n}}$$

with σ replaced by s.

4'. Substituting $n = 40$, $\bar{x} = 27,563$, $\mu_0 = 28,000$, and $s = 1,348$ for σ into the formula for z, we get

$$z = \frac{27,563 - 28,000}{1,348/\sqrt{40}} = -2.05$$

and from Table II we find that the p-value, the area under the curve to the left of $z = -2.05$, is 0.0202.

5'. Since 0.0202 is greater than $\alpha = 0.01$, the null hypothesis cannot be rejected. As before, the trucking firm's suspicion is not confirmed.

The p-value calculation is sometimes used to avoid the formalism of hypothesis testing, especially in situations where there is no immediate consequence to the accept-or-reject decision. In such cases the p-value may be used as a measure of how interesting the data are.

In the truck-tire example above, an actual decision is required: the firm must decide what brand of tires to use on its trucks. The end result must be a decision about adopting the specified brand of tires.

Consider next the plight of a social scientist exploring the relationships between family economics and school performance. He could be testing hundreds of hypotheses involving dozens of variables. The work is very complicated, and there are no immediate policy consequences. In this situation, the social scientist can tabulate the hypothesis tests according to their p-values. Those tests leading to the lowest p-values are the most provocative, and they will certainly be the subject of future discussion. The social scientist need not actually accept or reject the hypotheses, and the use of the p-value furnishes a convenient summary.

11.7

TESTS CONCERNING MEANS (Small Samples)

When we do not know the value of the population standard deviation and the sample is small, $n < 30$, we must assume as on page 287 that the population we are sampling has roughly the shape of a normal distribution. We can then base our decision on the statistic

*Statistic for
small-sample test
concerning mean*

$$t = \frac{\bar{x} - \mu_0}{s/\sqrt{n}}$$

which is a value of a random variable having the t distribution (see page 280) with $n - 1$ degrees of freedom. (If the assumption about the population cannot be met, we may have to use one of the alternative tests described in Chapter 17.)

The criteria for the **one-sample t test**, as this small-sample test is called, are similar to those shown in Figure 11.10 and in the table on page 312. Now the curves represent t distributions instead of normal distributions, and z, z_α, and $z_{\alpha/2}$ are replaced by t, t_α, and $t_{\alpha/2}$. As defined on page 288, t_α and $t_{\alpha/2}$ are values for which the area to their right under the t distribution are α and $\alpha/2$. All the dividing lines of the criteria may be read from Table II, with the number of degrees of freedom equal to $n - 1$.

EXAMPLE The yield of alfalfa from six test plots is 1.4, 1.8, 1.1, 1.9, 2.2, and 1.2 tons per acre. Test at the 0.05 level of significance whether this supports the contention that the average yield for this kind of alfalfa is 1.5 tons per acre.

Solution
1. H_0: $\mu = 1.5$
   H_A: $\mu \neq 1.5$
2. $\alpha = 0.05$
3. Reject the null hypothesis if $t \leq -2.571$ or $t \geq 2.571$, where

$$t = \frac{\bar{x} - \mu_0}{s/\sqrt{n}}$$

and 2.571 is the value of $t_{0.025}$ for $6 - 1 = 5$ degrees of freedom; otherwise, state that the data support the contention.
4. First calculating the mean and the standard deviation of the given data, we get $\bar{x} = 1.6$ and $s = 0.434$. Then, substituting these values together with $n = 6$ and $\mu_0 = 1.5$ into the formula for t, we get

$$t = \frac{1.6 - 1.5}{0.434/\sqrt{6}} = 0.56$$

5. Since $t = 0.56$ falls on the interval from -2.571 to 2.571, the null hypothesis cannot be rejected; in other words, the data support the contention that the average yield of the given kind of alfalfa is 1.5 tons per acre.

A computer prinout of the solution of the preceding example is shown in Figure 11.11. It confirms the values we got for $\bar{x}$, s, and t, and it shows that the tail probability is 0.60. Since 0.60 exceeds, 0.05, we conclude, as before, that the null hypothesis cannot be rejected.

```
MTB > SET C1
DATA> 1.4  1.8  1.1  1.9  2.2  1.2
MTB > TTEST MU=1.5 C1

TEST OF MU = 1.50 VS MU N.E. 1.50

              N      MEAN     STDEV    SE MEAN       T     P VALUE
C1            6     1.600     0.434      0.18     0.56        0.60
```

FIGURE 11.11 *Computer printout for one-sample t test.*

It is still possible, using Table II for the t distribution, to make an interval statement about the p-value. Here we note that $p > 2(0.10) = 0.20$. This is consistent with the 0.60 value given by the computer program—and with the conclusion that H_0 cannot be rejected.

EXERCISES

11.70 A law student, who wants to check a professor's claim that convicted embezzlers spend on the average 12.8 months in jail, takes a random sample of 60 such cases from court files. Using his results, namely $\bar{x} = 11.2$ months and $s = 3.5$ months, test the null hypothesis $\mu = 12.8$ months against the alternative hypothesis $\mu \neq 12.8$ months at the 0.01 level of significance.

11.71 Find the p-value in the previous problem. If you base the hypothesis test on the p-value, do you reach the same conclusion with regard to significance at the 0.01 level?

11.72 According to the norms established for a reading comprehension test, eighth graders should average 73.2 with a standard deviation of 8.6. If 45 randomly selected eighth graders from a certain school district average 76.7,
   (a) test the null hypothesis $\mu = 73.2$ against the alternative hypothesis $\mu > 73.2$ at the 0.01 level of significance;
   (b) comment on the district superintendent's claim that her eighth graders are above average;
   (c) find the p-value corresponding to the value of z obtained in the test in (a);
   (d) use the p-value to perform the hypothesis test.

11.73 The security department of a factory wants to know whether or not the true average time required by the night watchman to walk his round is 25 minutes. If, in a random sample of 32 rounds, the night watchman averaged 25.8 minutes with a standard deviation of 1.5 minutes, determine at the 0.01 level of significance whether this is sufficient evidence to reject the null hypothesis $\mu = 25$ minutes.

11.74 With reference to the preceding exercise, find the tail probability (meaning p-value) corresponding to the value obtained for z, and compare it with the specified level of significance.

11.75 A horticulturist wishes to determine whether the honeybees visiting a peach orchard this year differ from the typical weight of 0.85 gram. From previous work in weighing this species of honeybee, the standard deviation of weights is believed to be $\sigma = 0.18$ gram. A sample of 100 honeybees has a mean weight of 0.92 gram with a sample standard deviation of 0.24 gram. At the level of significance 0.05, can we conclude that this year's bees differ from the typical weight?

11.76 A large-sample test concerning a mean yielded a tail probability of 0.005. Can the null hypothesis be rejected at the 0.01 level of significance?

11.77 Tests performed with a random sample of 40 diesel engines produced by a large manufacturer showed that they have a mean thermal efficiency of 31.8 percent with a standard deviation of 2.2 percent. Based on this information and with a probability of a Type I error no greater than 0.05, should the person performing the tests accept the null hypothesis $\mu \geq 32.3$ percent or the alternative hypothesis $\mu < 32.3$ percent?

11.78 With reference to the preceding exercise, find the tail probability (meaning p-value) corresponding to the value obtained for z, and compare it with the specified maximum probability of committing a Type I error.

11.79 In a study of new sources of food, it is reported that a pound of a certain kind of fish yields on the average 2.41 ounces of FPC (fish-protein concentrate), which is used to enrich various food products (including flour). Yes or no, is this figure supported by a study in which 30 samples of this kind of fish yielded on the average 2.44 ounces of FPC (per pound of fish) with a standard deviation of 0.07 ounce, if we use the
 (a) 0.05 level of significance;
 (b) 0.01 level of significance?

11.80 With reference to the preceding exercise, find the tail probability corresponding to the value obtained for z, and use it to decide whether the null hypothesis could have been rejected at the 0.02 level of significance.

★ **11.81** If we wish to test the null hypothesis $\mu = \mu_0$ in such a way that the probability of a Type I error is α, and the probability of a Type II error is β for the specified alternative value $\mu = \mu_A$, we must take a random sample of size n, where

$$n = \frac{\sigma^2(z_\alpha + z_\beta)^2}{(\mu_A - \mu_0)^2}$$

if the alternative hypothesis is one-sided, and

$$n = \frac{\sigma^2(z_{\alpha/2} + z_\beta)^2}{(\mu_A - \mu_0)^2}$$

if the alternative hypothesis is two-sided.
 Suppose, for instance, that for a population with $\sigma = 6$ we want to test the null hypothesis $\mu = 200$ pounds against the alternative hypothesis $\mu < 200$ pounds. How large a sample will we need, if the probability of a Type I error is to be 0.05, and the probability of a Type II error is to be 0.05 when $\mu = 198$ pounds? Substituting into the first of the two

formulas for n above, we get

$$n = \frac{6^2(1.645 + 1.645)^2}{(198 - 200)^2} = 97.4$$

or 98 rounded up to the nearest integer.
 (a) Suppose that we want to test the null hypothesis $\mu = 540$ mm against the alternative hypothesis $\mu > 540$ mm for a population whose standard deviation is $\sigma = 88$ mm. How large a sample will we need if the probability of a Type I error is to be 0.05 and the probability of a Type II error is to be 0.10 when $\mu = 560$ mm? Determine also for what values of $\bar{x}$ the null hypothesis will be rejected
 (b) Suppose that we want to test the null hypotheses $\mu = \$650$ against the alternative hypotheses $\mu \neq \$650$ for a population whose standard deviation is $\sigma = \$26$. How large a sample will we need if the probability of a Type I error is to be 0.05 and the probability of a Type II error is to be 0.20 for $\mu = \$670$? Determine also for what values of $\bar{x}$ the null hypothesis will be rejected.

11.82 A random sample of 12 graduates of a secretarial school averaged 73.2 words per minute with a standard deviation of 7.9 words per minute on a typing test. Use the 0.05 level of significance to test the null hypothesis that graduates of this secretarial school average 75.0 words per minute on the given test against the alternative that they average less.

11.83 A soft-drink machine is set to dispense 6.0 ounces per cup. If the machine is tested nine times, yielding a mean cup fill of 6.2 ounces with a standard deviation of 0.15 ounce, can we reject the null hypothesis $\mu = 6.0$ ounces against the alternative hypothesis $\mu > 6.0$ ounces at the 0.01 level of significance?

11.84 The manufacturer of a new tranquilizer claims that it will reduce a person's pulse rate by at least 1.5 beats per minute. If the probability of a Type I error is to be at most 0.10, can we reject the manufacturer's claim on the basis of a random sample of 16 persons whose pulse rate was reduced on the average by 1.36 beats per minute with a standard deviation of 0.38 beats per minute?

11.85 A random sample of five-quart cartons is taken from a large production lot of ice cream. If their mean fat content is 12.6 percent with a standard deviation of 0.51 percent, can we reject the null hypothesis

$\mu = 12.0$ percent against the alternative hypothesis $\mu > 12.0$ percent at the 0.01 level of significance?

11.86 A random sample from a company's very extensive files shows that orders for a certain piece of machinery were filled in 12, 10, 17, 14, 13, 18, 11, and 9 days. At the 0.05 level of significance, can we reject the claim that on the average such orders are filled in 10.0 days?

11.87 An English teacher wants to determine whether the mean reading speed of a certain student is at least 600 words per minute. What can he conclude, if in six 1-minute intervals the student reads 606, 622, 617, 572, 570, and 605 words, and the probability of a Type I error is to be at most 0.05?

11.88 Five measurements of the tar content of a certain kind of cigarette yielded 14.5, 14.2, 14.4, 14.8, and 14.1 mg/cig (milligrams per cigarette). Is the difference between the mean of this sample and $\mu = 14.1$ (as claimed on the packages) significant at $\alpha = 0.05$, if the null hypothesis $\mu = 14.1$ mg/cig is tested against the alternative hypothesis
  (a) $\mu \neq 14.1$ mg/cig;
  (b) $\mu > 14.1$ mg/cig?

11.89 Five mature dogs of a certain breed weigh 66, 63, 64, 62, and 65 ounces. A kennel club claims that the average for this breed is 60 ounces.
  (a) Show that the difference between the mean of this sample and the claim $\mu = 60$ ounces is significant at the 0.05 level of significance.
  (b) Suppose that the fifth measurement were recorded as 80 instead of 65. Show that the difference between the mean of this sample and the claimed $\mu = 60$ is no longer significant.
  (c) With regard to (b), explain the apparent paradox that while the difference between $\bar{x}$ and 60 has increased, this difference is no longer significant.

11.90 Suppose that an unscrupulous manufacturer wants "scientific proof" that a totally useless chemical additive will improve the mileage yield of gasoline.
  (a) If a research group investigates this additive with one experiment, what is the probability that they will come up with a "significant result" using $\alpha = 0.05$ (to promote the additive with "scientific claims") even though the additive is totally ineffective?
  (b) If two independent research groups investigate the additive, what is the probability that at least one of them will come up with a "significant

result" even though the additive is totally ineffective?
  (c) If 32 independent research groups investigate the additive, what is the probability that at least one of them will come up with a "significant result" even though the additive is totally ineffective?

11.91 Suppose that a manufacturer of pharmaceuticals would like to find a new ointment to reduce swellings. It tries 20 different medications and tests each one for whether it reduces swellings at the 0.10 level of significance.
  (a) What is the probability that at least one of them will "prove" effective even though all of them are totally useless?
  (b) What is the probability that more than one will "prove" effective even though all of them are totally useless?

11.92 With reference to Exercise 11.86, use a suitable computer program to determine the tail probability corresponding to the observed value of t. Could we have rejected the claim at the 0.015 level of significance?

11.93 With reference to Exercise 11.87, use a suitable computer program to determine the tail probability corresponding to the observed value of t. What can the teacher conclude, if the probability of a Type I error is to be at most 0.03?

11.94 The standard formulation used by a certain company for making wall paint will enable a 1 gallon can to cover 250 square feet. The company's research laboratory has produced a new formulation, and the management is anxious to show that it does not cover significantly less area. Accordingly, they test the hypothesis H_0: $\mu = 250$ square feet against the alternative H_A: $\mu < 250$ square feet. The experiment involves 20 cans of the new formulation, and this sample gets mean coverage 237 square feet and a standard deviation of 25 square feet.
  (a) Give the value of the test statistic.
  (b) What is the p-value associated with this statistic?
  (c) After examining the p-value, the management decides to use 0.01 as the level of significance. Do the data support the claim, at level of significance 0.01, that the new formulation is not significantly different from the old?
  (d) Comment on the appropriateness of selecting the level of significance after observing the p-value.

11.8

DIFFERENCES BETWEEN MEANS

There are many problems in which we must decide whether an observed difference between two sample means can be attributed to chance. For instance, we may want to know whether there is really a difference in the average gasoline consumption of two kinds of cars, if sample data show that one kind averaged 24.6 miles per gallon while, under the same conditions, the other kind averaged 25.7 miles per gallon. Similarly, we may want to decide on the basis of samples whether men can perform a certain task faster than women, whether one kind of ceramic insulator is more brittle than another, whether the average diet in one country is more nutritious than that in another country, and so on.

The method we shall use to test whether an observed difference between two sample means can be attributed to chance, or whether it is statistically significant, is based on the following theory: If $\bar{x}_1$ and $\bar{x}_2$ are the means of two independent random samples, then the mean and the standard deviation of the sampling distribution of the statistic $\bar{x}_1 - \bar{x}_2$ are

$$\mu_{\bar{x}_1 - \bar{x}_2} = \mu_1 - \mu_2 \quad \text{and} \quad \sigma_{\bar{x}_1 - \bar{x}_2} = \sqrt{\frac{\sigma_1^2}{n_1} + \frac{\sigma_2^2}{n_2}}$$

where μ_1, μ_2, σ_1, and σ_2 are the means and the standard deviations of the two populations sampled. It is customary to refer to the standard deviation of this sampling distribution as the **standard error of the difference between two means**.

By "independent" samples we mean that the selection of one sample is in no way affected by the selection of the other. Thus, the theory does not apply to "before and after" kinds of comparisons, nor does it apply, say, if we want to compare the IQ's of husbands and wives. A special method for comparing the means of dependent samples is explained in Section 11.10.

Then, if we limit ourselves to large samples, $n_1 \geq 30$ and $n_2 \geq 30$, we can base tests of the null hypothesis $\mu_1 - \mu_2 = \delta$ (*delta*, the Greek letter for lowercase *d*) on the statistic

Statistic for test concerning difference between two means

$$z = \frac{(\bar{x}_1 - \bar{x}_2) - \delta}{\sqrt{\frac{\sigma_1^2}{n_1} + \frac{\sigma_2^2}{n_2}}}$$

which is a value of a random variable having approximately the standard normal distribution. Note that we obtained this formula for z by converting to standard units, namely, by subtracting from $\bar{x}_1 - \bar{x}_2$ the mean of its sampling distribution,

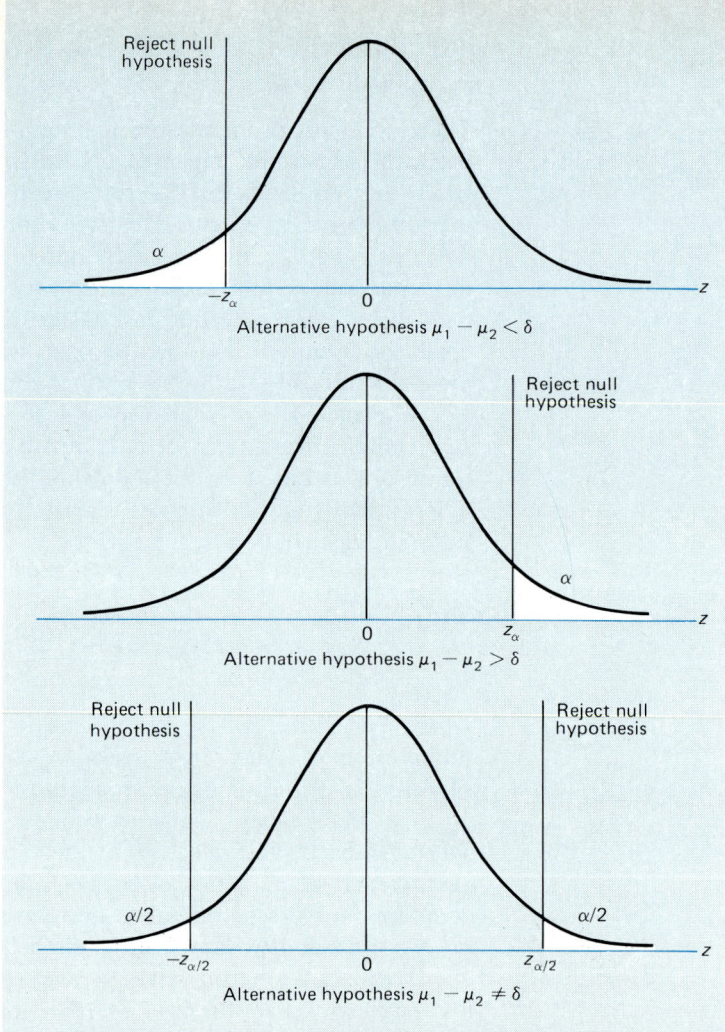

FIGURE 11.12 *Test criteria for large-sample tests concerning difference between means.*

which under the null hypothesis is $\mu_1 - \mu_2 = \delta$, and then dividing by the standard deviation of its sampling distribution.

Depending on whether the alternative hypothesis is $\mu_1 - \mu_2 < \delta$, $\mu_1 - \mu_2 > \delta$, or $\mu_1 - \mu_2 \neq \delta$, the criteria we use for the corresponding tests are shown in Figure 11.12 above. Note that they are like those shown in Figure 11.10 with $\mu_1 - \mu_2$ substituted for μ and δ substituted for μ_0. Analogous to the table on page 312, the criteria for tests of the null hypothesis $\mu_1 - \mu_2 = \delta$ are as follows:

Alternative hypothesis	Reject the null hypothesis if	Accept the null hypothesis or reserve judgment if
$\mu_1 - \mu_2 < \delta$	$z \leq -z_\alpha$	$z > -z_\alpha$
$\mu_1 - \mu_2 > \delta$	$z \geq z_\alpha$	$z < z_\alpha$
$\mu_1 - \mu_2 \neq \delta$	or $\begin{array}{c} z \leq -z_{\alpha/2} \\ z \geq z_{\alpha/2} \end{array}$	$-z_{\alpha/2} < z < z_{\alpha/2}$

Although δ can be any constant, it is worth noting that in the great majority of problems its value is zero and we test the null hypothesis of "no difference," namely, the null hypothesis $\mu_1 - \mu_2 = 0$ (or simply $\mu_1 = \mu_2$).

The test we have described here is essentially an approximate large-sample test; it is exact only when both of the populations we are sampling are normal. In most practical situations where σ_1 and σ_2 are unknown, we must make the further approximation of substituting for them the sample standard deviations s_1 and s_2.

EXAMPLE   In a study to test whether or not there is a difference between the average heights of adult females born in two different countries, random samples yielded the following results:

$$n_1 = 120 \qquad \bar{x}_1 = 62.7 \qquad s_1 = 2.50$$
$$n_2 = 150 \qquad \bar{x}_2 = 61.8 \qquad s_2 = 2.62$$

where the measurements are in inches. Use the 0.05 level of significance to test the null hypothesis that the corresponding population means are equal against the alternative hypothesis that they are not equal.

Solution   1. H_0: $\delta = 0$
   H_A: $\delta \neq 0$
2. $\alpha = 0.05$
3. Reject the null hypothesis if $z \leq -1.96$ or $z \geq 1.96$, where

$$z = \frac{\bar{x}_1 - \bar{x}_2 - \delta}{\sqrt{\dfrac{\sigma_1^2}{n_1} + \dfrac{\sigma_2^2}{n_2}}}$$

with s_1 and s_2 substituted for σ_1 and σ_2 and using $\delta = 0$; otherwise, except the null hypothesis or reserve judgment.
4. Substituting $n_1 = 120$, $n_2 = 150$, $\bar{x}_1 = 62.7$, $\bar{x}_2 = 61.8$, $s_1 = 2.50$ for σ_1, $s_2 = 2.62$ for σ_2, and $\delta = 0$ into the formula for z, we get

$$z = \frac{62.7 - 61.8}{\sqrt{\dfrac{(2.50)^2}{120} + \dfrac{(2.62)^2}{150}}} = 2.88$$

5. Since $z = 2.88$ exceeds 1.96, the null hypothesis must be rejected; in other words, the difference between $\bar{x}_1 = 62.7$ and $\bar{x}_2 = 61.8$ is significant.

Let us now rework this example using the p-value, which requires that we modify steps 3, 4, and 5.

3′.
$$z = \frac{\bar{x}_1 - \bar{x}_2}{\sqrt{\dfrac{\sigma_1^2}{n_1} + \dfrac{\sigma_2^2}{n_2}}}$$

4′.
$$z = \frac{62.7 - 61.8}{\sqrt{\dfrac{(2.50)^2}{120} + \dfrac{(2.62)^2}{150}}} = 2.88$$

The area under the normal curve to the right of $z = 2.88$ is 0.0020. This is also the area under the curve to the left of $z = -2.88$, so that the p-value is $0.0020 + 0.0020 = 0.0040$.

5′. The p-value is smaller than $\alpha = 0.05$, so that the null hypothesis must be rejected.

It must be noted that there is a certain awkwardness about comparing means when σ_1 and σ_2 have different values (or when the sample values s_1 and s_2 are not close). Consider, for example, two normal populations with means $\mu_1 = 50$ and $\mu_2 = 52$ and standard deviations $\sigma_1 = 5$ and $\sigma_2 = 15$. Though the second population has a larger mean, it is much more likely to produce a value below 40. An investigator faced with situations like this must decide whether the comparison of μ_1 and μ_2 really addresses the relevant problem.

11.9

DIFFERENCES BETWEEN MEANS (Small Samples)

As in Section 11.7, a small-sample test of the significance of the difference between two means may be based on an appropriate t statistic. For this test, which is used when $n_1 < 30$ or $n_2 < 30$, we must assume that our two samples are independent random samples from populations having roughly the shape of normal distributions with the same standard deviation. Then, we can base tests of the null hypothesis $\mu_1 - \mu_2 = \delta$ on the statistic

Statistic for small-sample test concerning difference between two means

$$t = \frac{\bar{x}_1 - \bar{x}_2 - \delta}{\sqrt{\dfrac{(n_1 - 1)s_1^2 + (n_2 - 1)s_2^2}{n_1 + n_2 - 2} \cdot \left(\dfrac{1}{n_1} + \dfrac{1}{n_2}\right)}}$$

which is a value of a random variable having the t distribution with $n_1 + n_2 - 2$ degrees of freedom.

The criteria for the **two-sample t test**, as this small-sample test is called, are similar to those shown in Figure 11.12 and in the table on page 323. Now the curves represent t distributions instead of normal distributions, and z, z_α, and $z_{\alpha/2}$ are replaced by t, t_α, and $t_{\alpha/2}$. Like the test of Section 11.8, the two-sample t test is used mainly to test the null hypothesis of "no difference," namely, the null hypothesis $\mu_1 - \mu_2 = 0$ (or simply $\mu_1 = \mu_2$).

EXAMPLE   The following random samples are measurements of the heat-producing capacity (in millions of calories per ton) of specimens of coal from two mines:

| Mine 1: | 8,400 | 8,230 | 8,380 | 7,860 | 7,930 |
| Mine 2: | 7,510 | 7,690 | 7,720 | 8,070 | 7,660 |

Use the 0.05 level of significance to test whether the difference between the means of these two samples is significant.

Solution   1. H_0: $\delta = 0$
      H_A: $\delta \neq 0$
2. $\alpha = 0.05$
3. Reject the null hypothesis if $t \leq -2.306$ or $t \geq 2.306$, where t is given by the formula on page 324 with $\delta = 0$, and 2.306 is the value of $t_{0.025}$ for $5 + 5 - 2 = 8$ degrees of freedom; otherwise, state that the difference between the means of the two samples is not significant.
4. The means and the variances of the two samples are $\bar{x}_1 = 8,160$, $\bar{x}_2 = 7,730$, $s_1^2 = 63,450$, and $s_2^2 = 42,650$. Substituting these values together with $n_1 = 5$, $n_2 = 5$, and $\delta = 0$ into the formula for t, we get

$$t = \frac{8,160 - 7,730}{\sqrt{\dfrac{4(63,450) + 4(42,650)}{5 + 5 - 2} \cdot \left(\dfrac{1}{5} + \dfrac{1}{5}\right)}}$$

$$= 2.95$$

5. Since $t = 2.95$ exceeds 2.306, the null hypothesis must be rejected; in other words, we conclude that the average heat-producing capacity of coal from the two mines is not the same.

A computer printout of the solution of the preceding example is shown in Figure 11.13. It confirms the value we got for t, and it shows that the tail probability is 0.018. Since 0.018 is less than 0.05, we conclude, as before, that the null hypothesis must be rejected.

The procedure of this section is used much more frequently than the procedure of Section 11.8. As note in the last paragraph of that section, statisticians try to avoid a direct comparison of means when the standard deviations are not equal.

```
MTB > SET C1
DATA> 8400    8230    8380    7860    7930
MTB > SET C2
DATA> 7510    7690    7720    8070    7660
MTB > POOL C1 .C2

TWOSAMPLE T FOR C1 VS C2
        N       MEAN       STDEV      SE MEAN
C1      5       8160        252         113
C2      5       7730        207          92

95 PCT CI FOR MU C1 - MU C2: (94, 766)
TTEST MU C1 = MU C2 (VS NE): T=2.95 P=0.018 DF=8.0
```

FIGURE 11.13 *Computer printout for two-sample t test.*

11.10
DIFFERENCES BETWEEN MEANS (Paired Data)

The methods of Sections 11.8 and 11.9 do not apply when the samples are not independent. For instance, they cannot be used when we deal with "before and after" kinds of comparisons, the weights of husbands and wives, and numerous other situations where the data are naturally paired. To handle data of this kind, we work with the (signed) differences of the paired data and test whether these differences may be looked upon as a random sample from a population which has the mean $\mu = \delta$. In most situations we are asking whether $\mu = 0$. If the sample is small, we use the one-sample t test of Section 11.7; otherwise, we use the corresponding large-sample test described in Section 11.6.

EXAMPLE The following are the average weekly losses of man-hours due to accidents in 10 industrial plants before and after a certain safety program was put into operation:

45 and 36   73 and 60   46 and 44   124 and 119   33 and 35,
57 and 51   83 and 77   34 and 29   26 and 24   17 and 11

Use the 0.05 level of significance to test whether the safety program is effective.

Solution The differences are 9, 13, 2, 5, -2, 6, 6, 5, 2, and 6, and for these data we perform the following test:

  1. H_0: $\mu = 0$
     H_A: $\mu > 0$ (The hypothesis that on the average there are more accidents "before" than "after.")
  2. $\alpha = 0.05$

3. Reject the null hypothesis if $t \geq 1.833$, where

$$t = \frac{\bar{x} - \mu_0}{s/\sqrt{n}}$$

and 1.833 is the value of $t_{0.05}$ for $10 - 1 = 9$ degrees of freedom; otherwise, accept the null hypothesis or reserve judgment (as the situation may demand).

4. First calculating the mean and the standard deviation of the ten differences, we get $\bar{x} = 5.2$ and $s = 4.08$. Then, substituting these values together with $n = 10$ and $\mu_0 = 0$ into the formula for t, we get

$$t = \frac{5.2 - 0}{4.08/\sqrt{10}} = 4.03$$

5. Since $t = 4.03$ exceeds 1.833, the null hypothesis must be rejected; in other words, we have shown that the industrial safety program is effective.

In connection with problems like this, the one-sample t test is usually referred to as the **paired-sample t test**.

EXERCISES

11.95 A sample study was made of the number of business lunches that executives claim as deductible expenses per month. If 40 executives in the insurance industry averaged 9.4 such deductions with a standard deviation of 3.3 in a given month, while 50 bank executives averaged 7.9 with a standard deviation of 2.9, test at the 0.05 level of significance whether the difference between these two sample means is significant.

11.96 An investigation of repair times for two kinds of photocopying equipment obtained these data values:

Equipment type	Number of repair jobs	Repair times Mean	Standard deviation
1	60	84.2	19.4
2	60	91.6	18.8

All times are in minutes.

(a) Test at the 0.01 level of significance whether the difference between these two sample means is significant.

(b) Find the p-value (meaning the tail probability) corresponding to the observed value of z in part (a).

(c) Use the p-value found in part (b) to decide whether the null hypothesis could also have been rejected at the 0.02 level of significance.

(d) Use the p-value found in part (b) to decide whether the null hypothesis could also have been rejected at the 0.03 level of significance.

(e) Use the p-value found in part (b) to decide whether the null hypothesis could also have been rejected at the 0.04 level of significance.

11.97 Suppose that we want to check whether it is true that on the average men earn in excess of \$30 more per week than women in a certain job classification. Sample data indicates the following:

Sex	Number of workers	Weekly salary Mean	Standard deviation
Male	75	\$422.18	\$35.20
Female	60	\$381.66	\$32.65

Test the null hypothesis $\mu_1 - \mu_2 = 30$ against a suitable alternative at the 0.01 level of significance.

11.98 The following are scores on the state social-studies exam for independent random samples of teenagers from two high schools.

School A: 78 84 81 78 76 83 79
          75 85 81
School B: 85 75 83 87 80 79 88
          94 87 82

Use the level of significance $\alpha = 0.05$ to test the claim that teenagers from school A have a lower average than teenagers from school B. Obtain the p-value as well.

11.99 To test the claim that the resistance of electric wire can be reduced by at most 0.050 ohm by alloying, 35 values obtained for standard wire yielded $\bar{x}_1 = 0.135$ ohm and $s_1 = 0.004$ ohm, and 35 values obtained for alloyed wire yielded $\bar{x}_2 = 0.082$ ohm and $s_2 = 0.005$ ohm. What can we conclude about the claim, if the probability of a Type I error is to be at most 0.01?

11.100 Sample surveys conducted in a large county in 1960 and again in 1990 showed that in 1960 the average height of 400 men in the age group from 18 to 24 was 68.4 inches with a standard deviation of 2.6 inches, while in 1990 the average height of 400 men in the same age group was 69.8 inches with a standard deviation of 2.5 inches. Use the 0.05 level of significance to test whether the true average increase in height was 0.6 inch or whether it was greater than that.

★ 11.101 If we substitute the formula for z on page 321 into $-z_{\alpha/2} < z < z_{\alpha/2}$, and manipulate the inequality algebraically so that the middle term is δ, we obtain a $(1 - \alpha)100$ percent confidence interval for $\delta = \mu_1 - \mu_2$. Use this method and the data of Exercise 11.97 to construct a 99 percent confidence interval for the difference between the true average weekly earnings of men and women in the given job classification.

★ 11.102 Use the method of the preceding exercise and the data of Exercise 11.99 to construct a 95 percent confidence interval for the true average amount by which the resistance of electric wire can be reduced by alloying.

11.103 Twelve measurements each of the hydrogen content (in percent number of atoms) of gases collected from the eruption of two volcanos yielded $\bar{x}_1 = 41.2$, $\bar{x}_2 = 45.8$, $s_1 = 5.2$, and $s_2 = 6.7$. Decide, at the 0.05 level of significance, whether to accept or reject the null hypothesis that there is no difference (with regard to hydrogen content) in the composition of the gases in the two eruptions.

11.104 Six guinea pigs injected with 0.5 mg of a medication took on the average 15.4 seconds to fall asleep with a standard deviation of 2.2 seconds, while six other guinea pigs injected with 1.5 mg of the medication took on the average 10.6 seconds to fall asleep with a standard deviation of 2.6 seconds. Use the 0.05 level of significance to test whether or not the increase in dosage from 0.5 mg to 1.5 mg reduces the average amount of time it takes a guinea pig to fall asleep by 2.0 seconds.

11.105 In the comparison of two kinds of paint, a consumer-testing service found that four one-gallon cans of one brand covered on the average 514 square feet with a standard deviation of 32 square feet, while four one-gallon cans of another brand covered on the average 487 square feet with a standard deviation of 27 square feet. Use the 0.05 level of significance to test whether the difference between the two sample means is significant.

11.106 The following are the numbers of sales which a random sample of nine salesmen of industrial chemicals in California and a random sample of six salesmen of industrial chemicals in Oregon made over a fixed period of time:

California: 41 47 62 39 56 64 37 61 52
Oregon: 34 63 45 55 24 43

Use the 0.01 level of significance to test whether the difference between the means of these two samples is significant.

11.107 As part of an industrial training program, some trainees are instructed by method 1, which is by teaching machine as well as some personal attention by an instructor, and some are instructed by method 2, which is straight teaching-machine instruction. Random samples of size ten are taken from large groups of trainees instructed by the two methods, and the following are the scores which they obtained on an appropriate achievement test:

Method 1:	81	71	79	83	76	75	84	90		
	83	78								
Method 2:	69	75	72	69	67	74	70	66		
	76	72								

What can we conclude about the claim that $\delta \leq 5$, where δ is the average amount by which the personal attention of an instructor will improve a trainee's score, if the probability of a Type I error is to be at most 0.05?

11.108 To compare two kinds of bumper guards, six of each kind were mounted on a certain make compact car. Then each car was run into a concrete wall at 5 miles per hour, and the following are the costs of the repairs (in dollars):

Bumper guard 1:	127	168	143	165	122
	139				
Bumper guard 2:	154	135	132	171	153
	149				

Test at the 0.01 level of significance whether the difference between the means of these two samples is significant.

★ **11.109** If we substitute the formula for t on page 324 into $-t_{\alpha/2} < t < t_{\alpha/2}$, and manipulate the inequality algebraically so that the middle term is δ, we obtain a $(1 - \alpha)100$ percent confidence interval for $\delta = \mu_1 - \mu_2$. Use this method and the data of Exercise 11.105 to construct a 95 percent confidence interval for the difference between the true average areas covered with a gallon of the two kinds of paint.

11.110 The following are the weights of 32 persons when they began a weight-reducing diet and two weeks later: 212 and 195, 193 and 185, 241 and 225, 218 and 199, 205 and 194, 216 and 193, 215 and 205, 198 and 176, 200 and 188, 233 and 224, 258 and 240, 186 and 174, 289 and 263, 250 and 238, 225 and 213, 244 and 241, 260 and 249, 209 and 201, 198 and 195, 211 and 196, 220 and 203, 245 and 236, 185 and 169, 206 and 195, 189 and 185, 202 and 195, 219 and 214, 263 and 255, 241 and 228, 235 and 229, 200 and 188, and 207 and 193. Use the 0.05 level of significance to test the null hypothesis that the mean weight loss of persons two weeks on this diet is 10.0 pounds against the alternative hypothesis that the weight loss is greater than that.

11.111 The following data were obtained in an experiment designed to check whether there is a systematic difference in the weights (in grams) obtained with two different scales:

Rock specimen	Scale I	Scale II
1	12.13	12.17
2	17.56	17.61
3	9.33	9.35
4	11.40	11.42
5	28.62	28.61
6	10.25	10.27
7	23.37	23.42
8	16.27	16.26
9	12.40	12.45
10	24.78	24.75

Use the 0.01 level of significance to test whether the difference between the means of the weights obtained with the two scales is significant.

11.11
CHECKLIST OF KEY TERMS
(With page references to their definitions)

Alternative hypothesis, 298
★ Bayesian inference, 293
Composite hypothesis, 305
Confidence, 284
Confidence interval, 286
Confidence limits, 286

Critical values, 312
Degree of confidence, 286
Degrees of freedom, 288
Estimation, 282
Interval estimate, 287
Level of significance, 309

11.12

REVIEW EXERCISES

11.112 A random sample of 40 cans of pineapple slices has a mean weight of 15.85 ounces and a standard deviation of 0.23 ounce. If this mean is used as an estimate of the mean weight of all the cans of pineapple slices in the large production lot from which the 40 cans came, with what confidence can we assert that this estimate is "off" by at most 0.06 ounce?

11.113 A random sample of 54 shirts worn by soldiers in a tropical climate has an average useful life of 63.9 washings with a standard deviation of 4.5. Under moderate weather conditions, such shirts are known to have an average useful life of 81.6 washings. At the 0.01 level of significance, can we conclude that their use in a tropical climate reduces the average useful life of such shirts?

11.114 Suppose that we want to perform a significance test regarding μ_A and μ_B, which are the average efficiencies of two brands of solar heaters, A and B. Indicate the null and alternative hypotheses which you would use for each of the following situations:

    (a) You would like to show that brand A is superior to brand B.

    (b) You would like to show that the two brands are not equally efficient.

    (c) You would like to show that brand A is not significantly worse than brand B.

    (d) You would like to show that brand A is at least as efficient as brand B.

    (e) You would like to show that the two brands are equally efficient.

11.115 In a French restaurant, the chef received 26, 21, 14, 22, 18, and 20 orders for coq au vin on six different nights. Construct a 95 percent confidence interval for the number of orders for coq au vin the chef can expect per night.

11.116 A general achievement test is standardized so that eighth graders should average 79.4 with a standard deviation of 4.8. If the superintendent of a large school district hopes to show that eighth graders in her district average better than that, and she has the test given to 32 eighth graders in her district (presumably a random sample), by how much must their average exceed 79.4 to make the difference significant at the 0.01 level of significance?

11.117 It is desired to test the null hypothesis $\mu = 0$ against the alternative hypothesis $\mu > 0$ on the basis of the mean of a random sample of size $n = 9$ from a normal population with $\sigma = 5$. If the probability of a Type I error is to be $\alpha = 0.05$,

   (a) verify that the null hypothesis must be rejected when $\bar{x} > 2.74$;

   (b) calculate β for $\mu = 2.50$, 5.00, and 7.50, and draw a rough sketch of the operating characteristic curve.

★ 11.118 A sales manager's feelings about the average monthly demand for a certain product may be described by means of a normal distribution with $\mu_0 = 4,800$ units and $\sigma_0 = 260$ units.

   (a) What probability does she, thus, assign to the true average monthly demand being somewhere on the interval from 4,500 to 5,000 units?

   (b) If data for ten months show an average demand of 4,702 units with a standard deviation of 380 units, find the mean and the standard deviation of the posterior distribution.

   (c) How would the probability of part (a) be modified in the light of the information given in part (b)?

★ 11.119 The assistant to the sales manager in the previous exercise feels that the average monthly demand may be described by means of a normal distribution with $\mu_0 = 4,500$ units and $\sigma_0 = 280$ units.

   (a) What probability does he assign to the true average monthly demand being somewhere on the interval from 4,500 to 5,000 units?

   (b) If data for ten months show an average demand of 4,702 units with a standard deviation of 380 units, find the mean and the standard deviation of the posterior distribution.

   (c) How would the probability of part (a) be modified with the information given in part (b)?

   (d) Why, even though based on the same ten months of data, are the solutions to part (c) of this exercise and part (c) of the previous exercise different?

11.120 An efficiency expert wants to determine the average time it takes a person to buy a week's groceries at a supermarket. If preliminary studies show that it is reasonable to let $\sigma = 4.2$ minutes, how large a sample will be needed to be able to assert with probability 0.95 that the mean of the sample will be "off" by at most 0.4 minute?

11.121 In fifteen rounds on various golf courses, two professional golfers had these scores:

Golfer 1	Golfer 2	Golfer 1	Golfer 2
71	72	70	73
69	71	71	72
70	70	69	73
72	75	70	74
69	70	73	72
71	70	68	74
74	72	70	78
68	71		

If the probability of a Type I error is to be at most 0.05, can we reject the null hypothesis that on the average the first golfer is at most 1.5 strokes better than the second?

11.122 In a study of the relationship between family size and intelligence, 40 "only children" had an average IQ of 101.5 with standard deviation of 6.7 and 50 "first-borns" in two-child families had an average IQ of 105.9 with a standard deviation of 5.8. Use the 0.05 level of significance to test whether the difference between these two means is significant.

11.123 Five burlap sacks of potatoes randomly selected from a large lot prepared by a particular distributor in Maine weigh 99.7, 99.5, 100.2, 99.2, and 99.9 pounds. What can we assert with 95 percent confidence about the maximum error, if we use the mean of this sample, $\bar{x} = 99.7$ pounds, to estimate the mean of the population sampled?

11.124 Consider the confidence interval formula on page 286 and the test described on page 312. Suppose that a random sample of size n is taken from a normal population with the mean μ and the standard deviation σ.

   (a) Find the set of values covered by the $(1 - \alpha) \cdot 100$ percent confidence interval for μ.

   (b) Consider the test at significance level α of the null hypothesis $\mu = \mu_0$ against the alternative $\mu \neq \mu_0$. Find the set of values for μ_0 for which the null hypothesis cannot be rejected.

   (c) Compare the sets obtained in (a) and (b).

11.125 In a national election, one candidate got 46 percent of the vote and the other candidate got 54 percent of the vote. Is it reasonable to ask whether the difference between these two percentages is statistically significant?

11.126 To check whether the true average wing span of a certain kind of insect is anywhere from 18.2 mm to 18.6 mm, a biologist decides to take a random sample of size $n = 60$ with the intention of rejecting the null hypothesis $18.2 \le \mu \le 18.6$ if the sample mean is less than 18.0 mm or greater than 18.8 mm; otherwise, he will accept it. If it is known from similar studies that $\sigma = 0.8$ mm for this kind of data, what is the maximum probability of a Type I error?

11.127 During the investigation of an alleged unfair trade practice, the Federal Trade Commission takes a random sample of 49 "3-ounce" candy bars from a large shipment. If the mean and the standard deviation of the sample are 2.96 ounces and 0.11 ounce, show that, at the 0.01 level of significance, the commission has grounds upon which to proceed against the manufacturer on the unfair practice of shortweight selling.

11.128 With reference to the preceding exercise, find the tail probability corresponding to the observed value of z, and use it to determine whether the null hypothesis could have been rejected at the 0.002 level of significance.

11.129 To compare freshmen's knowledge of history, samples of 50 freshmen from each of two universities were given a special test. If those from the first university have an average score of 67.4 with a standard deviation of 5.0, while those from the second university have an average score of 62.8 with a standard deviation of 4.6, test at the 0.05 level of significance whether the difference between the two sample means is significant.

11.130 Suppose that in the preceding exercise we had wanted to test whether freshmen from the first university score on the average 3.0 points higher on the test than freshman from the second university. Use the data of that exercise to test this conjecture at the 0.05 level of significance.

11.131 The operating rule for a quality control inspection is that "average weights in a sample must be not significantly worse than 90 percent of the standard of 50 pounds." A sample of 180 units shows a mean of 44.6 pounds and a standard deviation of 2.2 pounds. State the null and alternative hypotheses and then perform the test at the 5 percent level of significance.

11.132 In ten trials, a car ran for 28, 27, 21, 26, 29, 26, 29, 28, 29, and 27 miles with a gallon of a certain kind of gasoline. Does this support or refute the claim that with this kind of gasoline the car averages at least 28 miles per gallon, if the probability of a Type I error is to be at most 0.05?

11.133 Suppose that a law firm has one secretary whom it suspects of making more mistakes than the average of all its secretaries.
(a) If the law firm decides that it will let the secretary go, provided this suspicion is confirmed on the basis of observations made on the secretary's performance, what hypothesis and alternative should the law firm set up?
(b) If the law firm decides to let the secretary go unless he (she) can prove himself (herself) better than the average of all its secretaries, what hypothesis and alternative should the law firm set up?

11.134 A social scientist is studying the ages of executives in top management positions. What null hypothesis and what alternative hypothesis should she use, and under what conditions (accept or reject the null hypothesis) would she prove her point, if she hopes to be able to show that their mean age is
(a) greater than 48;
(b) at least 48?

11.13

REFERENCES

An informal introduction to interval estimation is given under the heading of "How to be precise though vague." in

MORONEY, M. J., *Facts from Figures.* London: Penguin Books, Ltd., 1956.

and some easy reading on estimation and tests of hypotheses may be found in Chapters 12 through 17 of

BROOK, R. J., ARNOLD, G. C., HASSARD, T. H., and PRINGLE, R. M., eds., *The Fascination of Statistics.* New York: Marcel Dekker, Inc., 1986.

A discussion of the theoretical foundation of the t distribution as well as other mathematical details omitted in this book may be found in most textbooks on mathematical statistics. Some of the theory of Bayesian estimation pertaining to Section 11.3 is discussed in Chapter 10 of the book by Freund and Walpole listed on page 279.

A detailed treatment of tests of significance, the choice of the level of significance, tail probabilities, and so forth, may be found in Chapters 26 and 29 of

FREEDMAN, D., PISANI, R., and PURVES, R., *Statistics.* New York: W. W. Norton & Company, Inc., 1978.

12

INFERENCES ABOUT STANDARD DEVIATIONS

In Chapter 11 we learned how to judge the size of the error when estimating the mean of a population, how to construct confidence intervals for population means, and how to perform tests of hypotheses about the mean of one population or the means of two populations. All these are very useful statistical techniques, but even more important are the concepts on which they are based—interval estimation, degree of confidence, the Bayesian approach, null and alternative hypotheses, Type I and Type II errors, test of significance, level of significance and tail probability, and above all the concept of statistical significance. Also, we learned about small-sample and large-sample techniques.

As we shall see in this chapter and in subsequent chapters, all these ideas carry over to inferences about population parameters other than the mean, to inferences about the nature of populations, and even to inferences about the randomness of samples. In this chapter we shall concentrate on population standard deviations, or population variances, which are not only important in their own right, but which must sometimes be estimated or compared before inferences about other parameters can be made. This was the case, for example, in Sections 11.1 and 11.6, where we needed an estimate of the population standard deviation to make inferences about the population mean, and in Section 11.9, where the two-sample t test required that the two populations sampled have equal standard deviations.

In this chapter, Section 12.1 is devoted to the estimation of σ and σ^2, and Sections 12.2 and 12.3 deal with tests of hypotheses about these population parameters.

12.1

THE ESTIMATION OF σ

There are various ways in which we can estimate the standard deviation of a population. We can base such an estimate on special graph paper as in Exercise 9.27 on page 235, or on the sample range as in Exercises 4.16 and 4.17 on pages 82 and 83 and in Exercise 12.15, but usually we base it on the sample standard deviation s, introduced in Chapter 4. To illustrate, let us begin with confidence intervals for σ based on s, which require that the population we are sampling has roughly the shape of a normal distribution. In that case, the statistic

Chi-square statistic

$$\chi^2 = \frac{(n-1)s^2}{\sigma^2}$$

called **chi square** (χ is the lowercase Greek letter *chi*), is a value of a random variable having approximately the **chi-square distribution**. The parameter of this important continuous distribution is called the **number of degrees of freedom**, just like the parameter of the t distribution, and as the chi-square distribution is used here, the number of degrees of freedom is $n-1$. An example of a chi-square distribution is

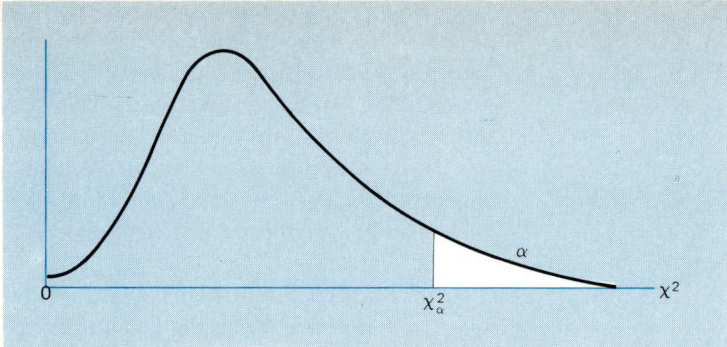

FIGURE 12.1 *Chi-square distribution.*

shown in Figure 12.1; unlike the normal and t distributions, its domain consists of the non-negative real numbers.

Analogous to z_α and t_α, we now define χ^2_α as the value for which the area under the curve to its right (see Figure 12.1) is equal to α; like t_α, this value depends on the number of degrees of freedom and must be obtained from a table, or perhaps a computer. Thus, $\chi^2_{\alpha/2}$ is such that the area under the curve to its right is $\alpha/2$, while $\chi^2_{1-\alpha/2}$ is such that the area under the curve to its left is $\alpha/2$ (see Figure 12.2). For instance, $\chi^2_{0.975}$ is the value for which the area under the curve to its left is 0.025. We made this distinction because the chi-square distribution, unlike the normal and t distributions, is not symmetrical. Values of $\chi^2_{0.995}$, $\chi^2_{0.975}$, $\chi^2_{0.025}$, and $\chi^2_{0.005}$, among others, are given in Table III at the end of the book for $1, 2, 3, \ldots,$ and 30 degrees of freedom.

We can now proceed as on pages 286 and 289. Since the probability is $1 - \alpha$ that a random variable having a chi-square distribution will take on a value between $\chi^2_{1-\alpha/2}$ and $\chi^2_{\alpha/2}$, namely, that

$$\chi^2_{1-\alpha/2} < \chi^2 < \chi^2_{\alpha/2}$$

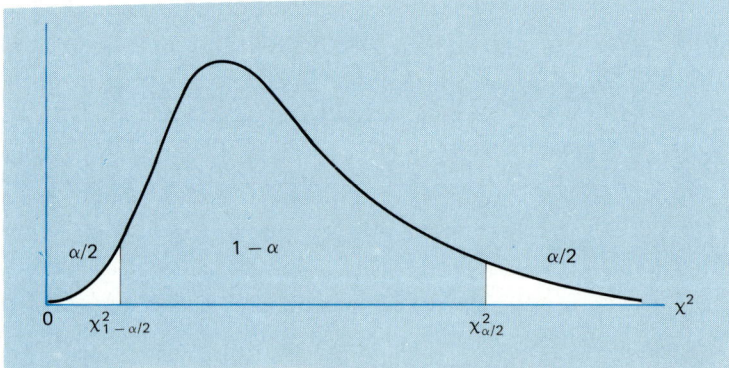

FIGURE 12.2 *Chi-square distribution.*

we can substitute into this inequality the expression for χ^2 on page 335 and get

$$\chi^2_{1-\alpha/2} < \frac{(n-1)s^2}{\sigma^2} < \chi^2_{\alpha/2}$$

Then applying some relatively simple algebra, we can rewrite this inequality as

Confidence interval for σ^2

$$\frac{(n-1)s^2}{\chi^2_{\alpha/2}} < \sigma^2 < \frac{(n-1)s^2}{\chi^2_{1-\alpha/2}}$$

This is a $(1-\alpha)100$ percent confidence interval formula for σ^2, the population variance, and if we take square roots, we get a corresponding $(1-\alpha)100$ percent confidence interval formula for σ, the population standard deviation. It should be emphasized that the use of this interval requires the assumption that the population we are sampling has roughly the shape of a normal distribution.

EXAMPLE A random sample of five hot dogs taken from a large production lot yielded the following data on their fat content: 32.4, 30.9, 33.2, 31.5, and 34.3 percent. Construct a 95 percent confidence interval for σ, which measures the variability of the percentage of fat in the given kind of hot dog.

Solution Since the sample standard deviation is $s = 1.35$ and since the values of $\chi^2_{0.975}$ and $\chi^2_{0.025}$ for $5 - 1 = 4$ degrees of freedom are 0.484 and 11.143 according to Table III, substitution into the confidence interval formula for σ^2 yields

$$\frac{4(1.35)^2}{11.143} < \sigma^2 < \frac{4(1.35)^2}{0.484}$$

$$0.65 < \sigma^2 < 15.06$$

Finally, taking square roots, we get

$$0.81 < \sigma < 3.88$$

The kind of confidence interval we have calculated here is often referred to as a **small-sample confidence interval**, because it is used mainly when n is small, $n < 30$, and, of course, only when it can be assumed that the population sampled has roughly the shape of a normal distribution. For large samples, $n \geq 30$, we make use of the theory that the sampling distribution of s can be approximated with a normal distribution having the mean σ and the standard deviation $\frac{\sigma}{\sqrt{2n}}$ (see Exercise 10.60 on page 277). Converting to standard units, we can thus assert with probability $1 - \alpha$ that

$$-z_{\alpha/2} < \frac{s - \sigma}{\dfrac{\sigma}{\sqrt{2n}}} < z_{\alpha/2}$$

and simple algebra leads to the following large-sample confidence interval formula for σ:

Large-sample confidence interval for σ

$$\frac{s}{1 + \dfrac{z_{\alpha/2}}{\sqrt{2n}}} < \sigma < \frac{s}{1 - \dfrac{z_{\alpha/2}}{\sqrt{2n}}}$$

EXAMPLE With reference to the example on page 23, in which $s = 5.55$ tons for a large industrial plant's emission of sulfur oxides on $n = 80$ days, construct a 95 percent confidence interval for the standard deviation of the population sampled.

Solution Substituting $n = 80$, $s = 5.55$, and $z_{\alpha/2} = 1.96$ into the confidence interval formula above, we get

$$\frac{5.55}{1 + \dfrac{1.96}{\sqrt{160}}} < \sigma < \frac{5.55}{1 - \dfrac{1.96}{\sqrt{160}}}$$

and

$$4.80 < \sigma < 6.57$$

This means that we are 95 percent confident that the interval from 4.80 tons to 6.57 tons contains σ, the true standard deviation of the plant's daily emission of sulfur oxides.

EXERCISES

12.1 The refractive indices of 18 pieces of glass, randomly selected from a large lot purchased by an optical firm, have a standard deviation of 0.014. Construct a 95 percent confidence interval for σ, the standard deviation of the population sampled.

12.2 A professional bowler bowled 15 games in a tournament and averaged 238 with a standard deviation of 14.8. Construct a 95 percent confidence interval for σ, which is a measure of this professional's consistency.

12.3 In the example on page 289 we referred to an experiment with a new paint for white center lines, and in the solution we obtained $s = 19,200$ car crossings for the eight locations. Construct a 95 percent confidence interval for σ, which here might measure the consistency of the paint, the consistency with which it is applied, or perhaps the uniformity (say, percentage of heavy trucks) of the traffic.

12.4 With reference to the exercises noted below, construct 95 percent confidence intervals for σ, the standard deviation of the population sampled.
    (a) Exercise 11.26, on page 296, in which $n = 18$ and $s = 4.80$ gallons;
    (b) Exercise 11.29, on page 296, in which $n = 9$ and $s = 0.003$ cm.

12.5 With reference to the exercises noted below, construct 99 percent confidence intervals for σ, the standard deviation of the population sampled.
   (a) Exercise 11.30, on page 296, in which $n = 8$ and $s = 0.56$ microgram;
   (b) Exercise 11.83, on page 319, in which $n = 9$ and $s = 0.15$ ounce.

12.6 With reference to the exercises noted below, construct 95 percent confidence intervals for σ^2, the variance of the population sampled.
   (a) Exercise 11.86, on page 320, in which $n = 8$ and $s = 3.21$ days;
   (b) Exercise 11.87, on page 320, in which $n = 6$ and $s = 22.39$ words.

12.7 With reference to the exercises noted below, construct 95 percent confidence intervals for σ, the standard deviation of the population sampled.
   (a) Exercise 11.1, on page 293, in which $n = 35$ and $s = 12.8$ seconds;
   (b) Exercise 11.3, on page 294, in which $n = 64$ and $s = 4.5$ mm.

12.8 With reference to the exercises noted below, construct 99 percent confidence intervals for σ, the standard deviation of the population sampled.
   (a) Exercise 11.5, on page 294, in which $n = 120$ and $s = 3.29$ minutes;
   (b) Exercise 11.10, on page 294, in which $n = 200$ and $s = \$14.93$.

12.9 With reference to Exercise 11.12 on page 294, where we had $n = 60$ and $s = 0.3$ ounce, construct a 95 percent confidence interval for σ^2, the variance of the weight of all the cans of pear halves in the given lot.

12.10 Reasonable estimates of the population standard deviation can often be obtained as $\dfrac{\text{IQR}}{1.35}$, where IQR is the interquartile range $Q_3 - Q_1$. In practice, one can use the hinges in place of the quartiles. Use this method on the power failure data of Exercise 3.27, and compare the value of $\dfrac{\text{IQR}}{1.35}$ with the actual value of the standard deviation. (The quartiles for the power failure data were obtained in Exercise 3.42 on page 59.)

12.11 When we deal with very small samples, good estimates of the population standard deviation can often be obtained on the basis of the sample range (the largest sample value minus the smallest). Such quick estimates of σ are given by the sample range divided by the divisor d, which depends on the size of the sample; for samples from populations having roughly the shape of a normal distribution, its values are shown in the following table for $n = 2, 3, \ldots,$ and 12:

n	2	3	4	5	6	7	8	9	10	11	12
d	1.13	1.69	2.06	2.33	2.53	2.70	2.85	2.97	3.08	3.17	3.26

For instance, if a psychiatrist saw 11, 9, 14, 8, 7, 11, 15, and 10 patients on eight days, we find that $s = 2.77$, $r = 15 - 7 = 8$, and $d = 2.85$ for $n = 8$. Thus, $\dfrac{8}{2.85} = 2.81$ and 2.77 are both estimates of the true standard deviation of the number of patients the psychiatrist sees per day. As can be seen, the difference between the two estimates is quite small. With reference to Exercise 11.34 on page 296, use this method to estimate the true standard deviation of the time it takes to clean up the cafeteria, and compare the result with the sample standard deviation s.

12.12 With reference to Exercise 11.35 on page 297, use the method of the preceding exercise to estimate σ, which here is an indication of the precision of such measurements, and compare the result with the sample standard deviation s.

12.13 With reference to Exercise 11.89 on page 320, use the method of Exercise 12.11 to estimate σ, which here measures the variation in the weight of such dogs, and compare the result with the sample standard deviation s.

12.2

TESTS CONCERNING STANDARD DEVIATIONS

In this section we shall consider the problem of testing whether a population standard deviation equals a specified constant σ_0, or whether a population variance equals σ_0^2. This kind of test may be required whenever we study the uniformity of a product, process, or operation. For instance, if we want to test whether a certain kind of glass is sufficiently homogeneous for making delicate optical equipment, whether the intelligence of a group of students is sufficiently uniform so that they can be taught in one class, whether a lack of uniformity in certain workers' performance may call for stricter supervision, and so forth.

The test of the null hypothesis $\sigma = \sigma_0$, that a population standard deviation equals a specified constant, is based on the same assumptions, the same statistic, and the same sampling theory as the small-sample confidence interval for σ^2 on page 337. Again assuming that our sample is random and comes from a population having roughly the shape of a normal distribution, we use the chi-square statistic

Statistic for test concerning standard deviation

$$\chi^2 = \frac{(n-1)s^2}{\sigma_0^2}$$

which is like the one on page 335 with σ replaced by σ_0. As before, the sampling distribution of the statistic is the chi-square distribution with $n-1$ degrees of freedom.

The test criteria are shown in Figure 12.3; depending on the alternative hypothesis, the dividing lines, or critical values, are $\chi_{1-\alpha}^2$ and χ_α^2 for the one-sided alternatives, and they are $\chi_{1-\alpha/2}^2$ and $\chi_{\alpha/2}^2$ for the two-sided alternative. Symbolically, we can formulate these criteria for testing the null hypothesis $\sigma = \sigma_0$ as follows:

Alternative hypothesis	Reject the null hypothesis if	Accept the null hypothesis or reserve judgment if
$\sigma < \sigma_0$	$\chi^2 \leq \chi_{1-\alpha}^2$	$\chi^2 > \chi_{1-\alpha}^2$
$\sigma > \sigma_0$	$\chi^2 \geq \chi_\alpha^2$	$\chi^2 < \chi_\alpha^2$
$\sigma \neq \sigma_0$	or $\begin{array}{c} \chi^2 \leq \chi_{1-\alpha/2}^2 \\ \chi^2 \geq \chi_{\alpha/2}^2 \end{array}$	$\chi_{1-\alpha/2}^2 < \chi^2 < \chi_{\alpha/2}^2$

The values of $\chi_{0.995}^2$, $\chi_{0.99}^2$, $\chi_{0.975}^2$, $\chi_{0.95}^2$, $\chi_{0.05}^2$, $\chi_{0.025}^2$, $\chi_{0.01}^2$, and $\chi_{0.005}^2$ are given in Table III at the end of the book for 1, 2, 3, ..., and 30 degrees of freedom.

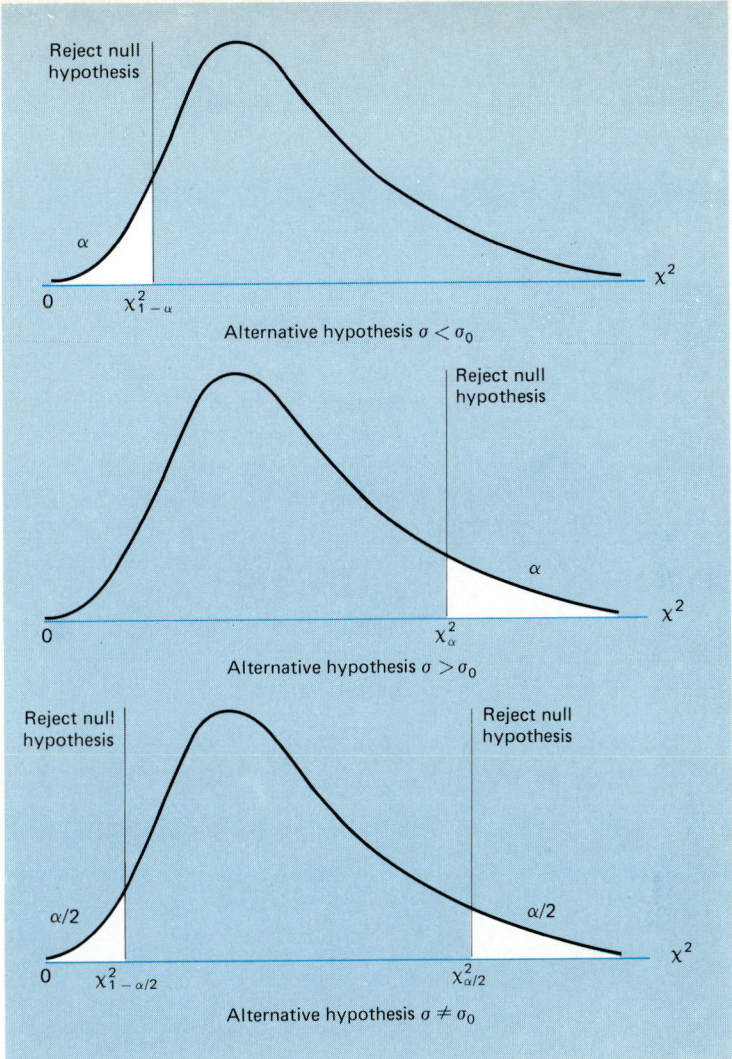

FIGURE 12.3 *Test criteria for tests concerning standard deviations.*

EXAMPLE   To judge certain safety features of a car, an engineer must know whether the reaction time of drivers to a given emergency situation has a standard deviation of 0.010 second, or whether it is greater than 0.010 second. What can she conclude at the 0.05 level of significance, if she gets $s = 0.014$ for a sample of size $n = 15$?

Solution   **1.** H_0: $\sigma = 0.010$
          H_A: $\sigma > 0.010$
      2. $\alpha = 0.05$

3. Reject the null hypothesis if $\chi^2 > 23.685$, where

$$\chi^2 = \frac{(n-1)s^2}{\sigma_0^2}$$

and 23.685 is the value of $\chi_{0.05}^2$ for $15 - 1 = 14$ degrees of freedom; otherwise, accept it.

4. Substituting $n = 15$, $s = 0.014$ and $\sigma_0 = 0.010$ into the formula for χ^2, we get

$$\chi^2 = \frac{14(0.014)^2}{(0.010)^2} = 27.44$$

5. Since $\chi^2 = 27.44$ exceeds 23.685, the null hypothesis must be rejected; in other words, the engineer can conclude that the standard deviation of the reaction time of drivers to the given emergency situation is greater than 0.010 second.

The test can also be done through the use of tail probabilities, meaning p-values. Table III gives only selected values for the chi-square distribution, so that we must make an interval statement for the p-value. The value 27.44 falls between $\chi_{0.025}^2 = 26.119$ and $\chi_{0.01}^2 = 29.141$, and we would state that $0.01 < p < 0.025$. The test using p-values would replace steps 3, 4, and 5 by these:

3′. The test statistic is

$$\chi^2 = \frac{(n-1)s^2}{\sigma_0^2}$$

4′. Substituting $n = 15$, $s = 0.014$, and $\sigma_0 = 0.010$ into the formula for χ^2, we get

$$\chi^2 = \frac{14(0.014)^2}{(0.010)^2} = 27.44$$

and from Table III we determine that $0.01 < p < 0.025$.

5′. Since $p < \alpha = 0.05$, the null hypothesis must be rejected.

When n is large, $n \geq 30$, tests of the null hypothesis $\sigma = \sigma_0$ can be based on the same theory as the large-sample confidence intervals of the preceding section. That is, we use the statistic

Statistic for large-sample test concerning standard deviation

$$z = \frac{s - \sigma_0}{\sigma_0/\sqrt{2n}}$$

which is a value of a random variable having the standard normal distribution.

Thus, the criteria for tests of the null hypothesis $\sigma = \sigma_0$ are like those shown in Figure 11.10 and in the table on page 312, the only difference is that μ and μ_0 are replaced by σ and σ_0.

EXAMPLE The specifications for the mass production of certain springs require, among other things, that the standard deviation of their compressed lengths should not exceed 0.040 cm. If a random sample of size $n = 35$ from a certain production lot yields $s = 0.053$ and the probability of a Type I error is not to exceed 0.01, does this constitute evidence for the null hypothesis $\sigma \le 0.040$ or for the alternative hypothesis $\sigma > 0.040$?

Solution
1. H_0: $\sigma \le 0.040$
   H_A: $\sigma > 0.040$
2. $\alpha \le 0.01$
3. The null hypothesis must be rejected if $z \ge 2.33$, where

$$z = \frac{s - \sigma_0}{\sigma_0/\sqrt{2n}}$$

and otherwise it must be accepted.
4. Substituting $n = 35$, $s = 0.053$, and $\sigma_0 = 0.040$ into the formula for z, we get

$$z = \frac{0.053 - 0.040}{0.040/\sqrt{70}} = 2.72$$

5. Since $z = 2.72$ exceeds 2.33, the null hypothesis must be rejected; in other words, the data show that the production lot does not meet specifications.

The tail probability for this example is 0.0033; since this probability is below 0.01, we would certainly reject the null hypothesis.

12.3

TESTS CONCERNING TWO STANDARD DEVIATIONS

In this section we shall discuss tests concerning the equality of the standard deviations of two populations. Among other applications, it is sometimes used in connection with the two-sample t test of Section 11.9, where it had to be assumed that the two populations sampled have the same standard deviation. For instance, in the example on page 325 dealing with the heat-producing capacity of coal from two mines, we had $s_1^2 = 63,450$ and $s_2^2 = 42,650$, and despite what may seem to be a large difference, we assumed that the corresponding population standard deviations were equal. Now we shall put this to a rigorous test.

Given independent random samples of size n_1 and n_2 from two populations with the standard deviations σ_1 and σ_2, we usually base tests of the null hypothesis $\sigma_1 = \sigma_2$ on the **F statistic**

Statistics for test concerning the equality of two standard deviations

$$F = \frac{s_1^2}{s_2^2} \quad \text{or} \quad F = \frac{s_2^2}{s_1^2}, \quad \text{whichever is large}$$

where s_1 and s_2 are the corresponding sample standard deviations. If the two populations sampled have roughly the shape of normal distributions, it can be shown that such ratios, appropriately called **variance ratios**, are values of random variables having the **F distribution**. This important continuous distribution depends on two parameters called the **numerator** and **denominator degrees of freedom**, which are $n_1 - 1$ and $n_2 - 1$ or $n_2 - 1$ and $n_1 - 1$, depending on which of the two sample variances go into the numerator and the denominator of the F statistic.

If we based all the tests on the statistic

$$F = \frac{s_1^2}{s_2^2}$$

we could reject the null hypothesis $\sigma_1 = \sigma_2$ for $F \leq F_{1-z}$ when the alternative hypothesis is $\sigma_1 < \sigma_2$, and for $F \geq F_\alpha$ when the alternative hypothesis is $\sigma_1 > \sigma_2$. In this notation, $F_{1-\alpha}$ and F_α are defined in the same way in which we defined the critical values $\chi_{1-\alpha}^2$ and χ_α^2 for the chi-square distribution. Unfortunately, things are not as simple as that. Since there exists a fairly straightforward mathematical relationship between $F_{1-\alpha}$ and F_α (see Exercise 12.28 on page 348), most F tables give only values corresponding to right-hand tails with α less than 0.50; for instance, Table IV at the end of the book contains only values of $F_{0.05}$ and $F_{0.01}$.

For this reason, we use

$$F = \frac{s_2^2}{s_1^2} \quad \text{or} \quad F = \frac{s_1^2}{s_2^2}$$

depending on whether the alternative hypothesis is $\sigma_1 < \sigma_2$ or $\sigma_1 > \sigma_2$, and in either case we reject the null hypothesis for $F > F_\alpha$ (see Figure 12.4). When the alternative hypothesis is $\sigma_1 \neq \sigma_2$, we use the greater of the two variance ratios,

$$F = \frac{s_1^2}{s_2^2} \quad \text{or} \quad F = \frac{s_2^2}{s_1^2}$$

and reject the null hypothesis for $F > F_{\alpha/2}$ (see Figure 12.4). In all these tests, the degrees of freedom are $n_1 - 1$ and $n_2 - 1$ or $n_2 - 1$ and $n_1 - 1$, depending on which sample variance goes into the numerator and which one goes into the denominator. Symbolically, these criteria for testing the null hypothesis $\sigma_1 = \sigma_2$

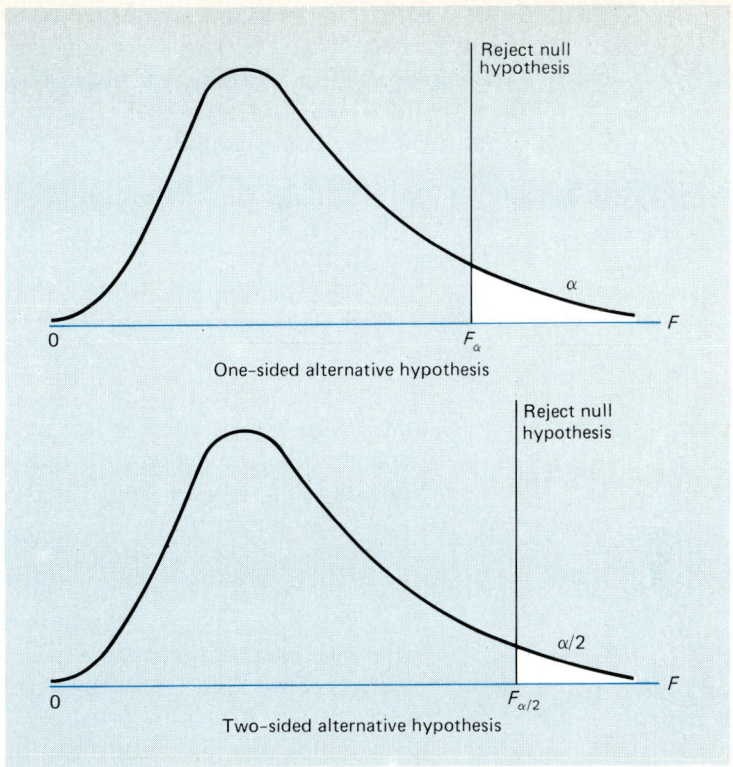

FIGURE 12.4 *Test criteria for tests concerning the equality of two standard deviations.*

are summarized in the following table:

Alternative hypothesis	Test statistic	Reject the null hypothesis if	Accept the null hypothesis or reserve judgment if
$\sigma_1 < \sigma_2$	$F = \dfrac{s_2^2}{s_1^2}$	$F \geq F_\alpha$	$F < F_\alpha$
$\sigma_1 > \sigma_2$	$F = \dfrac{s_1^2}{s_2^2}$	$F \geq F_\alpha$	$F < F_\alpha$
$\sigma_1 \neq \sigma_2$	*The larger of the two ratios*	$F \geq F_{\alpha/2}$	$F < F_{\alpha/2}$

The degrees of freedom are as indicated above.

EXAMPLE It is desired to determine whether there is less variability in the gold plating done by company 1 than in the gold plating done by company 2. If independent random samples yielded $s_1 = 0.033$ mil (based on $n_1 = 12$) and $s_2 = 0.061$ mil (based on $n_2 = 10$), test the null hypothesis $\sigma_1 = \sigma_2$ against the alternative hypothesis $\sigma_1 < \sigma_2$ at the 0.05 level of significance.

Solution

1. H_0: $\sigma_1 = \sigma_2$
   H_A: $\sigma_1 < \sigma_2$
2. $\alpha = 0.05$
3. Reject the null hypothesis if $F \geq 2.90$, where

$$F = \frac{s_2^2}{s_1^2}$$

and 2.90 is the value of $F_{0.05}$ for $10 - 1 = 9$ and $12 - 1 = 11$ degrees of freedom; otherwise, accept the null hypothesis or reserve judgment.
4. Substituting $s_1 = 0.033$ and $s_2 = 0.061$ into the formula for F, we get

$$F = \frac{(0.061)^2}{(0.033)^2} = 3.42$$

5. Since $F = 3.42$ exceeds 2.90, the null hypothesis must be rejected; in other words, we conclude that there is less variability in the gold plating done by company 1.

We can also provide the *p*-value for this example. Since $F_{0.01} = 4.63$ for 9 and 11 degrees of freedom, we note that $0.01 < p < 0.05$. The use of the *p*-value still allows us to reject the null hypothesis at the 5 percent level of significance.

EXAMPLE In the example dealing with the coal from two mines on page 325, we had $s_1^2 = 63,450$ and $s_2^2 = 42,650$ for two independent random samples of size $n = 5$. Use the 0.02 level of significance to test whether there is any evidence that the standard deviations of the two populations sampled are not equal.

Solution

1. H_0: $\sigma_1 = \sigma_2$
   H_A: $\sigma_1 \neq \sigma_2$
2. $\alpha = 0.02$
3. Reject the null hypothesis if $F \geq 16.0$, where

$$F = \frac{s_1^2}{s_2^2} \quad \text{or} \quad \frac{s_2^2}{s_1^2}$$

whichever is larger, and 16.0 is the value of $F_{0.01}$ for $5 - 1 = 4$ and $5 - 1 = 4$ degrees of freedom; otherwise, accept the null hypothesis.
4. Since s_1^2 is greater than s_2^2, we substitute $s_1^2 = 63,450$ and $s_2^2 = 42,650$ into the first of the two variance ratios and we get

$$F = \frac{63,450}{42,650} = 1.49$$

5. Since $F = 1.49$ does not exceed 16.0, the null hypothesis cannot be rejected; there is no reason why we cannot go ahead with the two-sample t test which requires that $\sigma_1 = \sigma_2$.

In a problem like this where the alternative hypothesis is $\sigma_1 \neq \sigma_2$, Table IV limits us to the 0.02 and 0.10 levels of significance. Had we wanted to use the 0.05 level of significance in the example immediately above, the table listed on page 349 would have yielded $F_{0.025} = 9.60$ and the conclusion would have been the same.

Since the test described here is very sensitive to departures from the underlying assumptions, it must be used with considerable caution. We say that the test is not **robust** and, if the circumstances permit, replace it with another procedure which, hopefully, is more robust.

EXERCISES

12.14 In a random sample the times that 18 women took to complete the written test for their driver's licenses had standard deviation $s = 2.4$ minutes. Use the 0.05 level of significance to test the null hypothesis $\sigma = 2.8$ minutes against the alternative hypothesis $\sigma \neq 2.8$ minutes.

12.15 In a laboratory experiment, $s = 0.0086$ for ten determinations of the specific heat of iron. Use the 0.05 level of significance to test the null hypothesis $\sigma = 0.0100$ for such determinations against the alternative hypothesis $\sigma < 0.0100$.

12.16 With reference to Exercise 11.82 on page 319, where we had $n = 12$ and $s = 7.9$ words per minute, test the null hypothesis $\sigma = 10.0$ words per minute against the alternative $\sigma < 10.0$ words per minute at the 0.01 level of significance.

12.17 With reference to Exercise 11.88 on page 320, what can we conclude about the null hypothesis that σ is at least 0.40 mg/cig for such determinations of the tar content of cigarettes, if the probability of a Type I error is to be at most 0.05?

12.18 Past data indicate that the standard deviation of measurements made on sheet metal stampings by experienced inspectors is 0.41 square inch. If a new inspector measures 60 stampings with a standard deviation of 0.48 square inch, test at the 0.05 level of significance whether he is making satisfactory measurements or whether the variability of his measurements is excessive.

12.19 With reference to the preceding exercise, find the tail probability corresponding to the value obtained for z.

12.20 With reference to Exercise 11.70 on page 318, where we had $n = 60$ and $s = 3.5$ months, test at the 0.01 level of significance whether $\sigma = 4.5$ months for the amount of time that convicted embezzlers spend in jail.

12.21 With reference to Exercise 11.73 on page 318, where we had $n = 32$ and $s = 1.5$ minutes, test at the 0.05 level of significance whether or not $\sigma = 1.9$ minutes for the amount of time it takes the night watchman to walk his round.

12.22 With reference to Exercise 11.77 on page 319, where we had $n = 40$ and $s = 2.2$ percent, test at the 0.05 level of significance whether $\sigma = 1.8$ percent for the thermal efficiency of the given kind of engine, or whether it is greater than that.

12.23 Two different lighting techniques are compared by measuring the intensity of light at selected locations in areas lighted by the two methods. If 12 measurements of the first technique have a standard deviation of 2.6 foot-candles and 16 measurements of the second technique have a standard deviation of 4.4 foot-candles, test at the 0.05 level of significance whether the two lighting techniques are equally variable or whether the first technique is less variable than the second.

12.24 The amounts of time required by Dr. L to do routine insurance checkups on 25 patients have a standard deviation of 4.2 minutes, while the amounts of time required by Dr. M to do the same procedure on 21 patients have a standard deviation of 3.0 minutes. Assuming that these data constitute independent random samples, test at the 0.05 level of significance whether the amounts of time for this procedure by these two doctors are equally variable or whether they are more variable for Dr. L.

12.25 With reference to Exercise 11.106 on page 328, test at the 0.02 level of significance whether there is any reason to doubt the assumption that $\sigma_1 = \sigma_2$, as required for the two-sample t test.

12.26 With reference to Exercise 11.107 on page 328, test at the 0.10 level of significance whether there is any reason to doubt the assumption that $\sigma_1 = \sigma_2$, as required for the two-sample t test.

12.27 With reference to Exercise 11.108 on page 328, test at the 0.02 level of significance whether there is any reason to doubt the assumption that $\sigma_1 = \sigma_2$, as required for the two-sample t test.

★ **12.28** If $F_{1-\alpha}$ is a critical value of F for m and n degrees of freedom and F_α is a critical value of F for n and m

degrees of freedom, then

$$F_{1-\alpha} = \frac{1}{F_\alpha}$$

Note that the degrees of freedom are interchanged. If, in the first example on page 346, we had wanted to reject the null hypothesis for

$$F = \frac{s_1^2}{s_2^2} < F_{0.95}$$

what would have been the critical value of $F_{0.95}$?

★ **12.29** To compare the variability of the lifetimes of two kinds of batteries, we want to test the null hypothesis $\sigma_1 = \sigma_2$ against the alternative hypothesis $\sigma_1 < \sigma_2$. If the sample sizes are $n_1 = 10$ and $n_2 = 8$ and we want to reject the null hypothesis when

$$F = \frac{s_1^2}{s_2^2} < F_{1-\alpha}$$

use the method of the preceding exercise to find the value of $F_{1-\alpha}$ for
(a) $\alpha = 0.05$;
(b) $\alpha = 0.01$.

12.4

CHECKLIST OF KEY TERMS
(with page references to their definitions)

Chi-square distribution, 335
Chi-square statistic, 335
Degrees of freedom, 335
Denominator degrees of freedom, 344
F distribution, 344

F statistic, 344
Number of degrees of freedom, 335
Numerator degrees of freedom, 344
Robust, 347
Variance ratio, 344

12.5

REVIEW EXERCISES

12.30 In a random sample of $n = 12$ rounds of golf played on her home course, a golf professional averaged 71.6 with a standard deviation of 2.48. Use the 0.01 level of significance to test the null hypothesis $\sigma = 2.00$ against the alternative hypothesis that her game is actually less consistent.

12.31 The following are the pull strengths (in pounds) required to break the bond of two kinds of glue:

Glue 1: 25.3 20.2 21.1 27.0 16.9 30.1
           17.8 22.9 27.2 20.0

Glue 2:   24.9   22.5   21.8   23.6   19.8   21.6
           20.4   22.1

As a preliminary to the two-sample t test, use the 0.02 level of significance to test whether it is reasonable to assume that the corresponding population standard deviations are equal.

12.32 In a random sample of 25 servings of a breakfast cereal, the sugar content averaged 10.40 grams with a standard deviation of 1.76 grams. Construct a 95 percent confidence interval for the standard deviation of the population sampled; that is, for the standard deviation of the sugar content in a serving of the given kind of breakfast cereal.

12.33 Test runs with 12 models of an experimental engine showed that they operated for 25, 31, 19, 31, 26, 22, 28, 25, 20, 21, 27, and 30 minutes with a gallon of fuel. Estimate the standard deviation of the population sampled using
    (a) the sample standard deviation;
    (b) the sample quartiles and the method of Exercise 12.10 on page 339;
    (c) the sample range and the method of Exercise 12.11 on page 339.

12.34 In a random sample of 50 income tax returns, the amount due after an audit averaged $880.22 with a standard deviation of $210.71. Construct a 99 percent confidence interval for the true standard deviation of the amount due on an income tax return after such an audit.

12.35 With reference to Exercise 11.103 on page 328, where we had $n_1 = 12$, $n_2 = 12$, $s_1 = 5.2$, and $s_2 = 6.7$, test at

the 0.05 level of significance whether there is more variation in the hydrogen content of gases from the second volcano than there is in the hydrogen content of gases from the first.

12.36 In a random sample, 40 lawyers had an average annual income of $93,520 with a standard deviation of $8,370. If the probability of a Type I error is to be at most 0.05, what can we conclude about the null hypothesis that the standard deviation of the population sampled is at least $10,000?

12.37 With reference to Exercise 11.84 on page 319, where we had $n = 16$ and $s = 0.38$, test the null hypothesis $\sigma = 0.3$ beats against the alternative $\sigma > 0.3$ beats at the 0.05 level of significance?

12.38 In a study of two topical treatments for sunburn, subjects were scored according to the fraction of rash area that was relieved. The relevant data values are these:

Treatment	Number of subjects	Fraction of area relieved by treatment	
		Mean	Standard deviation
G	12	0.68	0.21
R	11	0.62	0.09

Using the 10 percent level of significance, test the null hypothesis that the standard deviations of the populations sampled are equal.

12.6

REFERENCES

Theoretical discussions of the chi-square and F distributions may be found in most textbooks on mathematical statistics; for instance, in

FREUND, J. E., and WALPOLE, R. E., *Mathematical Statistics*, 4th ed. Englewood Cliffs, N.J.: Prentice-Hall, Inc., 1987.

For more detailed tables of the chi-square and F distributions, see for example

PEARSON, E. S., and HARTLEY, H. O., *Biometrika Tables for Statisticians, Vol. I.* New York: John Wiley & Sons, Inc., 1968.

INFERENCES
ABOUT
PROPORTIONS

In this chapter we shall deal with **count data**, namely, with data obtained by counting rather than measuring. For instance, we shall concern ourselves with the number of persons who experience some discomfort after they are given a flu vaccine, the number of defectives in a shipment of a manufactured product, the number of television viewers who enjoy a certain situation comedy, the number of tires which last more than 30,000 miles, the number of persons who like their hamburgers with onions, and so forth.

For the most part, we shall deal with values of random variables having binomial distributions, so we can use what we learned about this distribution in Chapter 8 and about its approximation by the normal distribution in Section 9.5. In particular, we shall concern ourselves with the binomial parameter p, which is the probability of a success on an individual trial. However, when it comes to applications other than games of chance, we usually interpret p as the proportion of the time a given kind of event will occur, or can be expected to occur. Correspondingly, we interpret $100p$ as the percentage of the time that such an event will occur.

Later in this chapter we shall study methods which apply when each trial has more than two possible outcomes. This is the case, for example, when we count how many persons favor a piece of legislation, are against it, or do not care, or when we count the number of times that a pair of dice comes up 2, 3, 4, ... , 11, or 12. In situations like this we shall assume that we are dealing with multinomial random variables, but we will use approximations rather than the formula of Section 8.6.

In this chapter, Sections 13.1 through 13.4 parallel the work of the two preceding chapters, but they deal with inferences about p instead of σ or μ. First we concern ourselves with the estimation of p (that is, the estimation of proportions, percentages, or probabilities), then we perform tests about the parameter p of a binomial population, and finally we perform tests about the corresponding parameters of two binomial populations. Sections 13.5 and 13.6 deal with tests pertaining to the multinomial case, where each trial permits more than two possible outcomes.

13.1
THE ESTIMATION OF PROPORTIONS

The information that is usually available for the estimation of a population proportion (percentage, or probability) is a **sample proportion**, $\frac{x}{n}$, where x is the number of times that an event has occurred in n trials. For instance, if a study shows that 54 of 120 cheerleaders, presumably a random sample, suffered what auditory experts called "moderate to severe damage" to their voices, then $\frac{x}{n} = \frac{54}{120} = 0.45$, and we can use this figure as a point estimate of the true proportion of cheerleaders who are afflicted in this way. Similarly, a supermarket chain might estimate the proportion of its shoppers who regularly use cents-off coupons as 0.68, if a random sample of 300 shoppers included 204 who regularly use cents-off coupons.

Throughout this section it will be assumed that the situations satisfy (at least approximately) the conditions underlying the binomial distribution; that is, our information will consist of the number of successes observed in a given number of independent trials, and it will be assumed that for each trial the probability of a success—the parameter we want to estimate—has the constant value p. Thus, the sampling distribution of the counts on which our methods will be based is the binomial distribution with the mean $\mu = np$ and the standard deviation $\sigma = \sqrt{np(1-p)}$, and we know from Section 9.5 that this distribution can be approximated with a normal distribution when np and $n(1-p)$ are both greater than 5. Usually, this requires that n must be large. For $n = 50$, for example, the normal-curve methods discussed here and later in this chapter may be used so long as $50p > 5$ and $50(1-p) > 5$; namely, so long as p lies between 0.10 and 0.90. Similarly, for $n = 100$ they may be used so long as p lies between 0.05 and 0.95, and for $n = 200$ they may be used so long as p lies between 0.025 and 0.975. This should give the reader some idea of what we mean here by "n being large."

Now, if we convert to standard units, it follows that, for large values of n, the statistic

$$z = \frac{x - np}{\sqrt{np(1-p)}}$$

is a value of a random variable having approximately the standard normal distribution. If we substitute this expression for z into the inequality

$$-z_{\alpha/2} < z < z_{\alpha/2}$$

(as on page 286) and use some relatively simple algebra, we arrive at the inequality

$$\frac{x}{n} - z_{\alpha/2}\sqrt{\frac{p(1-p)}{n}} < p < \frac{x}{n} + z_{\alpha/2}\sqrt{\frac{p(1-p)}{n}}$$

which looks like a confidence-interval formula for p. Indeed, if we use it repeatedly, the inequality should be satisfied $(1 - \alpha)100$ percent of the time, but observe that the unknown parameter p appears not only in the middle, but also in $\sqrt{\frac{p(1-p)}{n}}$ to the left of the first inequality sign and to the right of the other. The quantity $\sqrt{\frac{p(1-p)}{n}}$ is called the **standard error of a proportion**, as it is, in fact, the standard deviation of the sampling distribution of a sample proportion (see Exercise 13.27 on page 360).

To get around this difficulty and, at the same time, simplify the resulting formula, we substitute $\hat{p} = \frac{x}{n}$ for p in $\sqrt{\frac{p(1-p)}{n}}$, where $\hat{p}$ reads "p-hat." (This kind of notation is widely used in statistics. For instance, when we use the mean of a sample to estimate the mean of a population, we might denote it by $\hat{\mu}$, and when we use the standard deviation of a sample to estimate the standard deviation of a population, we might denote it by $\hat{\sigma}$.) Thus, we get the following $(1 - \alpha)100$ **percent large-sample confidence interval for p**:

Large-sample confidence interval for p

$$\hat{p} - z_{\alpha/2} \cdot \sqrt{\frac{\hat{p}(1 - \hat{p})}{n}} < p < \hat{p} + z_{\alpha/2} \cdot \sqrt{\frac{\hat{p}(1 - \hat{p})}{n}}$$

EXAMPLE   In a random sample, 136 of 400 persons given a flu vaccine experienced some discomfort. Construct a 95 percent confidence interval for the true proportion of persons who will experience some discomfort from the vaccine.

Solution   Substituting $n = 400$, $\hat{p} = \frac{136}{400} = 0.34$, and $z_{0.025} = 1.96$ into the confidence-interval formula, we get

$$0.34 - 1.96\sqrt{\frac{(0.34)(0.66)}{400}} < p < 0.34 + 1.96\sqrt{\frac{(0.34)(0.66)}{400}}$$

$$0.294 < p < 0.386$$

or, rounding to two decimals, $0.29 < p < 0.39$. To say again what we have said before more than once, an interval like this contains the parameter it is intended to estimate or it does not, and we really do not know which is actually the case, but the 95 percent confidence implies that the interval was obtained by a method which works 95 percent of the time. Note also that for $n = 400$ and p on the interval from 0.29 to 0.39, np and $n(1 - p)$ are both much greater than 5, so that there can be no question about n being large enough to use the normal approximation to the binomial distribution.

When it comes to small samples, we can construct confidence intervals for p by using a special table, but the resulting intervals are usually so wide that they are not of much value. For example, for $x = 4$ and $n = 10$, the 95 percent confidence interval is $0.12 < p < 0.75$. Clearly, this interval is so wide that it does not tell us very much about the actual value of p.

The large-sample theory presented here can also be used to assess the error we may be making when we use a sample proportion to estimate a population proportion, namely, the binomial parameter p. Proceeding as on pages 283 and 284, we can assert with probability $1 - \alpha$ that the difference between a sample proportion and p will be at most

$$E = z_{\alpha/2}\sqrt{\frac{p(1 - p)}{n}}$$

However, since p is unknown, we substitute for it again the sample proportion $\hat{p}$, and we arrive at the result that

If $\hat{p}$ is used as an estimate of p, we can assert with $(1 - \alpha)100$ percent confidence that the error is at most

Maximum error of estimate

$$E = z_{\alpha/2} \cdot \sqrt{\frac{\hat{p}(1 - \hat{p})}{n}}$$

Like the confidence-interval formula on page 353, this formula requires that n must be large.

EXAMPLE In a random sample of 250 persons interviewed while exiting from polling places all over a state, 145 said that they voted for the reelection of the incumbent governor. With 99 percent confidence, what can we say about the maximum error, if we use $\hat{p} = \frac{145}{250} = 0.58$ as an estimate of the actual proportion of the vote which the incumbent governor will get?

Solution Substituting $n = 250$, $\hat{p} = 0.58$, and $z_{0.005} = 2.575$ into the formula for E, we get

$$E = 2.575 \cdot \sqrt{\frac{(0.58)(0.42)}{250}} = 0.08$$

rounded to two decimals. Thus, if we use $\hat{p} = 0.58$ as an estimate of the actual proportion of the vote the incumbent governor will get, we can assert with 99 percent confidence that our error is at most 0.08.

As in Section 11.1, we can use the formula for the maximum error to determine how large a sample is needed to attain a desired degree of precision. If we want to use a sample proportion to estimate a population proportion p, and we want to be able to assert with probability $1 - \alpha$ that our error will not exceed some prescribed quantity E, we write as before

$$E = z_{\alpha/2} \cdot \sqrt{\frac{p(1 - p)}{n}}$$

Upon solving this equation for n, we get

Sample size for estimating p (with some information about p)

$$n = p(1 - p)\left[\frac{z_{\alpha/2}}{E}\right]^2$$

Since this formula involves p, it cannot be used unless we have some information about the possible values that p might assume. In that case, we substitute for p whichever of its values is closest to $\frac{1}{2}$. Without such information, we make use of the fact that $p(1-p)$ cannot exceed $\frac{1}{4}$ (which it equals for $p = \frac{1}{2}$) and use the formula

Sample size for estimating p (without information about p)

$$n = \frac{1}{4}\left[\frac{z_{\alpha/2}}{E}\right]^2$$

In either case, since the value we obtain for n may well be larger than necessary, we can say that the probability is *at least* $1 - \alpha$ that our error will not exceed E.

EXAMPLE   Suppose that a state highway department wants to estimate what proportion of all trucks operating between two cities carry too heavy a load, and it wants to be able to assert with a probability of at least 0.95 that its error will not exceed 0.04. How large a sample will it need if

(a) it knows that the true proportion lies somewhere on the interval from 0.10 to 0.25;

(b) it has no idea what the true value might be?

Solution   (a) Substituting $z_{0.025} = 1.96$, $E = 0.04$, and $p = 0.25$ into the formula for n, we get

$$n = (0.25)(0.75)\left[\frac{1.96}{0.04}\right]^2 = 451$$

rounded up to the nearest integer.

(b) Substituting $z_{0.025} = 1.96$, $E = 0.04$, and $p(1-p) = \frac{1}{4}$ into the formula for n, we get

$$n = \frac{1}{4}\left[\frac{1.96}{0.04}\right]^2 = 601$$

rounded up to the nearest integer.

The preceding example illustrates how some knowledge about the values p might assume can substantially reduce the sample size needed to attain a desired degree of precision. Note also that in a problem like this we round up, if necessary, to the nearest integer.

13.2

THE ESTIMATION OF PROPORTIONS
(a Bayesian Method)★

In Section 13.1 we looked upon the proportion p (which we tried to estimate) as an unknown constant; in Bayesian estimation this parameter is looked upon as a random variable having a prior distribution which reflects one's belief about the values it can assume. As in the Bayesian estimation of means, we are thus faced with the problem of combining prior information with direct sample evidence.

To give an example, consider a large company which routinely pays thousands of invoices submitted by its suppliers. Of course, it is of interest to know what proportion of these invoices might contain errors, and interviews of three of the company's executives reveal the following information: Mr. Martin feels that only 0.005 (or a half of 1 percent) of the invoices contain errors; Mr. Green, who is generally regarded to be as reliable in his estimates as Mr. Martin, feels that 0.01 (or 1 percent) of the invoices contain errors; and Mr. Jones, who is generally regarded to be twice as reliable in his estimates as either Mr. Martin or Mr. Green, feels that 0.02 (or 2 percent) of the invoices contain errors. Assuming for the sake of simplicity that $p = 0.005$, $p = 0.01$, and $p = 0.02$ are the only possibilities, we thus have the following **prior probabilities** for the proportion of invoices containing errors:

p	Prior probability
0.005	0.25
0.010	0.25
0.020	0.50

Here the prior probabilities assigned to Mr. Martin's and Mr. Green's estimates are the same, and that assigned to Mr. Jones's estimate is twice as large. The prior probabilities must sum to 1, of course.

Now suppose that a random sample of 200 invoices is carefully checked, and that only one of them contains an error. The probabilities of this happening when $p = 0.005$, $p = 0.010$, or $p = 0.020$ are

$$\binom{200}{1}(0.005)^1(0.995)^{199} = 0.37$$

$$\binom{200}{1}(0.010)^1(0.990)^{199} = 0.27$$

and

$$\binom{200}{1}(0.020)^1(0.980)^{199} = 0.07$$

where we used the formula for the binomial distribution. Combining these probabilities with the prior probabilities by using the formula for Bayes' theorem (see Section 6.7), we find that the **posterior probability** of $p = 0.005$ is

$$\frac{(0.25)(0.37)}{(0.25)(0.37) + (0.25)(0.27) + (0.50)(0.07)} = 0.47$$

while the corresponding posterior probabilities of $p = 0.010$ and $p = 0.020$ are

$$\frac{(0.25)(0.27)}{(0.25)(0.37) + (0.25)(0.27) + (0.50)(0.07)} = 0.35$$

and

$$\frac{(0.50)(0.07)}{(0.25)(0.37) + (0.25)(0.27) + (0.50)(0.07)} = 0.18$$

We have thus arrived at the following posterior distribution for the proportion of invoices containing errors:

p	Posterior probability
0.005	0.47
0.010	0.35
0.020	0.18

This distribution reflects the prior judgments as well as the direct sample evidence, and it should not come as a surprise that the highest posterior probability goes to $p = 0.005$ — after all, the sample proportion actually equaled $\dfrac{x}{n} = \dfrac{1}{200} = 0.005$.

To continue, we could use the mean of the posterior distribution, namely, $(0.005)(0.47) + (0.010)(0.35) + (0.020)(0.18) = 0.009$ as a point estimate of the true proportion of invoices containing errors. In contrast to the mean of the prior distribution, which was $(0.005)(0.25) + (0.010)(0.25) + (0.020)(0.50) = 0.014$, the mean of the posterior distribution reflects the prior judgments as well as the direct sample evidence.

In the preceding example it was assumed that p had to be 0.005, 0.010, or 0.020 to simplify the calculations. The method of analysis would have been the same, however, if we had considered ten different values of p or even a hundred. In fact, there exist Bayesian techniques, similar to that of Section 11.3, in which p can take on any value on the continuous interval from 0 to 1, and its prior distribution is a continuous curve. The corresponding theory goes considerably beyond the scope of this text, but there is a reference to it on page 391.

EXERCISES

13.1 In using the rule that both np and $n(1 - p)$ must be greater than 5 to use the normal approximation to the binomial distribution,
(a) what values of p are permitted if $n = 250$;
(b) what values of p are permitted if $n = 800$;
(c) what values of n are permitted if $p = 0.04$;
(d) what values of n are permitted if $p = 0.98$?

13.2 In a random sample of 400 persons interviewed in a large city, 288 said that they oppose the construction of any more freeways.

(a) Construct a 95 percent confidence interval for the corresponding proportion of the population sampled.

(b) What can we say with 95 percent confidence about the maximum error if we use the sample proportion as an estimate of the population proportion?

13.3 In a random sample of 200 students at a large university, 144 opposed an increase in the humanities requirement and 56 favored an increase in the humanities requirement.

(a) Give a 95 percent confidence interval for the population proportion opposed to an increase in the humanities requirement.

(b) Give a 95 percent confidence interval for the population proportion in favor of an increase in the humanities requirement.

(c) Indicate whether the endpoints of the interval in (b) can be found from the endpoints of the interval in (a).

13.4 In a random sample of 500 registered voters in a large city, 214 were registered as Democrats, 208 were registered as Republicans, and 78 were registered Independent.

(a) Give a 95 percent confidence interval for the proportion of registered voters in the city who are registered as Republicans.

(b) Give a 95 percent confidence interval for the proportion of registered voters in the city who are registered as Democrats.

(c) Give a 95 percent confidence interval for the proportion of registered voters in the city who are registered as Independent.

(d) Indicate whether there are any simple relationships among the endpoints of the intervals found in (a), (b), and (c).

13.5 Among 180 fish caught in a certain large lake, 24 were inedible as a result of the chemical pollution of the environment.

(a) Construct a 99 percent confidence interval for the corresponding true proportion.

(b) What can we say with 99 percent confidence about the maximum error if we use the sample proportion as an estimate of the population proportion?

13.6 A random sample of 300 shoppers at a supermarket includes 204 who regularly use cents-off coupons.

Construct a 98 percent confidence interval for the true proportion of shoppers at the supermarket who use cents-off coupons.

13.7 In a random sample of 120 cheerleaders, 54 had suffered moderate to severe damage to their voices. With 90 percent confidence, what can we say about the maximum error, if we use the sample proportion, $\frac{54}{120} = 0.45$, as an estimate of the true proportion of cheerleaders who are afflicted in this way?

13.8 In a random sample of 1,200 voters interviewed nationwide, only 324 felt that salaries of certain government officials should be raised. Construct a 95 percent confidence interval of the corresponding true proportion.

13.9 In a random sample of 400 television viewers interviewed in a certain area, 152 had seen a certain controversial program. With 99 percent confidence, what can we say about the maximum error, if we use the sample proportion $\frac{152}{400} = 0.38$, as an estimate of the corresponding true proportion?

13.10 In a random sample of 360 persons eating lunch at a department store cafeteria, only 126 had dessert. If we use $\frac{126}{360} = 0.35$ as an estimate of the corresponding true proportion, with what confidence can we assert that our error is at most 0.10?

13.11 In a random sample of 240 high school seniors in a Western state, 168 said that they expect to continue their education at an in-state college or university. Construct a 95 percent confidence interval for the corresponding true percentage.

13.12 With reference to the preceding exercise, what can we say with 98 percent confidence about the maximum error, if we use $\frac{168}{240} \cdot 100\% = 70\%$ as an estimate of the actual percentage of high school seniors in the state who expect to continue their education at an in-state college or university?

13.13 In a random sample of 140 supposed UFO sightings, 119 could easily be explained in terms of natural phenomena. Construct a 99 percent confidence interval for the probability that a supposed UFO sighting will easily be explained in terms of natural phenomena.

13.14 In a random sample of 80 persons convicted in U.S. District Courts on narcotics charges, 36 received probation. With 95 percent confidence, what can we say about the maximum error, if we use $\frac{36}{80} = 0.45$ as an estimate of the probability that a person convicted in a U.S. District Court on narcotics charges will receive probation?

13.15 In a random sample of 240 undergraduate biology students, 84 said that they intend to go on to graduate school. If we use $\frac{84}{240} \cdot 100\% = 35\%$ as an estimate of the true percentage of undergraduate biology students who intend to go on to graduate school, with what confidence can we assert that our error does not exceed 6 percentage points?

★ **13.16** When a sample constitutes more than 5 percent of a finite population, and the sample itself is large, we use the same finite population correction factor as in Section 10.7, and hence the following confidence limits for p:

Large-sample confidence limits for p (finite population)

$$\hat{p} \pm z_{\alpha/2} \cdot \sqrt{\frac{\hat{p}(1 - \hat{p})(N - n)}{n(N - 1)}}$$

where N is, as before, the size of the population sampled.

Among the 360 families in an apartment complex a random sample of 100 is interviewed, and it is found that 34 of the families have children of college age. Use the formula above to construct a 95 percent confidence interval for the proportion of all the families living in the apartment complex who have children of college age.

★ **13.17** A show manufacturer feels that unless its employees agree to a 10 percent wage reduction it cannot stay in business. If in a random sample of 80 of the manufacturer's 400 employees only 24 felt "kindly disposed" to the reduction, use the formula of the preceding exercise to construct a 99 percent confidence interval for the corresponding proportion for all the employees.

13.18 A private opinion pollster is engaged by a politician to estimate what proportion of her constituents favor the decriminalization of certain narcotics violations.
   (a) How large a sample will the pollster have to take to be at least 95 percent sure that the sample proportion is off by at most 0.02?
   (b) How large a sample will the pollster have to take if it is believed that the true proportion in favor of this decriminalization is approximately 0.35?

13.19 Many problems involve the construction of 95 percent confidence intervals, and the value $z_{0.025} = 1.96$ is used in the calculations. Some persons make an approximation by replacing 1.96 by 2. Show that the formula for sample size for estimating p (without information about p) on page 355 for 95 percent confidence intervals, replacing 1.96 by 2, reduces to

$$n = \frac{1}{E^2}.$$

   (a) Find the sample size needed to have 95 percent confidence that the error is at most 5 percentage points.
   (b) Find the sample size needed to have 95 percent confidence that the error is at most 4 percentage points.
   (c) Find the sample size needed to have 95 percent confidence that the error is at most 2 percentage points.

13.20 A political pollster is engaged by a politician to estimate the proportion of registered voters in his district who plan to vote for him in the next election. Find the sample size needed if he wants the poll to be accurate to within
   (a) 8 percentage points;
   (b) 4 percentage points;
   (c) 2 percentage points;
   (d) 1 percentage point.

13.21 Suppose that we want to estimate what proportion of all drivers exceed the maximum speed limit on a stretch of road between Los Angeles and Bakersfield. How large a sample will we need so that the error of our estimate is at most 0.04 if
   (a) the desired level of confidence is 90 percent;
   (b) the desired level of confidence is 95 percent;
   (c) the desired level of confidence is 99 percent?

13.22 A national manufacturer wants to determine what percentage of purchases of razor blades for use by men is actually made by women. How large a sample of men will the manufacturer need to be at least 98 percent confident that the sample percentage is not off by more than 2.5 percentage points if
   (a) nothing is known about the true proportion;
   (b) there is good reason to believe that the true proportion is at most 0.30?

13.23 A large finance company wants to estimate from a sample the probability that any one of its many customers will make a major appliance purchase on credit during the coming year. How large a sample will it need to be at least 95 percent confident that the difference between the sample proportion and actual probability does not exceed 0.03 if

  (a) nothing is known about the true probability;

  (b) there is good reason to believe that the true probability is between 0.15 and 0.40?

★ **13.24** In planning the operation of a new school, one school board member claims that 4 out of 5 newly hired teachers will stay with the school for more than a year, while another school board member claims that it would be correct to say 7 out of 10. In the past, the two board members have been about equally reliable in their predictions, so that in the absence of direct information we would assign their judgments equal weight. What posterior probabilities would we assign to their claims if it is found that 11 of 12 newly hired teachers stayed with the school for more than a year?

★ **13.25** The landscaping plans for a new hotel call for a row of palm trees along the driveway. The landscape designer tells the owner that if he plants *Washingtonia filifera*, 20 percent of the trees will fail to survive the first heavy frost, the manager of the nursery which supplies the trees tells the owner that 10 percent of the trees will fail to survive the first heavy frost, and the owner's wife tells him that 30 percent of the trees will fail to survive the first heavy frost.

  (a) If the owner feels that in this matter the landscape designer is 10 times as reliable as his wife and the manager of the nursery is 9 times as reliable as his wife, what prior probabilities should he assign to these percentages?

  (b) If 13 of these palm trees are planted and 2 fail to survive the first heavy frost, what posterior probabilities should the manager assign to the three percentages?

★ **13.26** The method of Section 13.2 can also be used to find the posterior distribution of the number of "successes" in a finite population. Suppose, for instance, that a coin dealer receives a shipment of five ancient coins from abroad, and that, on the basis of past experience, he feels that the probabilities that 0, 1, 2, 3, 4, or all 5 of them are counterfeits are 0.74, 0.11, 0.02, 0.01, 0.02, and 0.10. Since modern methods of coun-

terfeiting have become greatly refined, the cost of authentication has risen sharply and the dealer decides to select one of the coins at random and send it away for authentication. If it turns out that the coin is a forgery, find

  (a) the posterior probabilities that 0, 1, 2, 3, or all 4 of the remaining coins are counterfeits;

  (b) the mean of the posterior distribution obtained in part (a) and use it to estimate what proportion of the remaining coins are counterfeits.

13.27 Since the proportion of successes is simply the number of successes divided by n, the mean and the standard deviation of the sampling distribution of the proportion of successes may be obtained by dividing by n the mean and the standard deviation of the sampling distribution of the number of successes. Use this argument to verify the standard error formula given on page 352.

13.28 Can it ever happen that the lower end of the large-sample confidence interval for p is less than zero? *Hint*: Consider what happens if $x = 0$ or $x = 1$ or $x = 2$.

★ **13.29** The inequality $-z_{\alpha/2} < z < z_{\alpha/2}$ on page 352 is certainly equivalent to the inequality $z^2 < (z_{\alpha/2})^2$.

  (a) Show that the substitution of

$$z = \frac{x - np}{\sqrt{np(1 - p)}} \qquad \text{into} \qquad z^2 < (z_{\alpha/2})^2$$

  results in an inequality which can be written as

$$p^2[n^2 + n(z_{\alpha/2})^2] - p[2nx + n(z_{\alpha/2})^2] + x^2 < 0$$

  (b) Express the inequality obtained in (a) using numbers given in Exercise 13.3, namely $n = 200$, $x = 144$, and $z_{\alpha/2} = 1.96$.

  (c) Regard the left side of the inequality obtained in (b) as a quadratic expression in p. Find the values of p for which this quadratic is equal to zero; a 95% confidence interval for p is the interval between these values.

  (d) Compare the confidence interval found in (c) with the interval given in Exercise 13.3, part (a). *Note*: The confidence intervals should be very similar because the sample size n is large. With smaller values of n the two methods will differ noticeably.

13.3

TESTS CONCERNING PROPORTIONS

Now we shall concern ourselves with tests of hypotheses which enable us to decide, on the basis of sample data, whether the true value of a proportion (percentage, or probability) equals, is greater than, or is less than a given constant. These tests make it possible, for example, to determine whether the actual proportion of fifth graders who can name the governor of their state is 0.35, whether it is true that 10 percent of the answers which the IRS gives to taxpayers' telephone inquiries are in error, or whether the probability is really 0.25 that the downtime of a new computer will exceed two hours in any given week.

Questions of this kind are usually decided on the basis of the observed number of successes in n trials, or the observed proportion of successes, and it will be assumed throughout this section that these trials are independent and that the probability of a success is the same for each trial. In other words, we shall assume that we can use the binomial distribution and that we are, in fact, testing hypotheses about the parameter p of binomial populations.

When n is small, tests concerning proportions can be based directly on tables of binomial probability, as is illustrated by the following example:

EXAMPLE It has been claimed that more than 70 percent of the students attending a large university are opposed to a plan to increase student fees in order to build new parking facilities. If 15 of 18 students selected at random at that university are opposed to the plan, test the claim at the 0.05 level of significance.

Solution The structure of binomial probabilities in Table V makes it easiest to organize this test through the use of p-values, as described on pages 314–315.

1. H_0: $p = 0.70$
   H_A: $p > 0.70$
2. $\alpha = 0.05$
3'. The test statistic is the number of students in the sample opposed to the plan, namely $x = 15$.
4'. Table V shows that for $n = 18$ and $p = 0.70$ the probability of 15 or more successes is

$$0.105 + 0.046 + 0.013 + 0.002 = 0.166$$

Thus, $p = 0.166$.

5'. Since 0.166 is greater than 0.05, the null hypothesis cannot be rejected; in other words, the data do not support the claim that more than 70 percent of the students at the given university are opposed to the plan. We accept the null hypothesis or defer judgment.

We used the procedure based on p-values because, for tests based on a binomial table, this alternative procedure can save a good deal of work. As the reader will be asked to verify in Exercise 13.37, this is true, particularly, when a test is two-tailed, as in the example which follows.

EXAMPLE It has been claimed that 40 percent of all shoppers can identify a highly advertised trade mark. If, in a random sample, 13 of 20 shoppers were able to identify the trade mark, test at the 0.05 level of significance whether to accept the null hypothesis $p = 0.40$ or the alternative hypothesis $p \neq 0.40$.

Solution
1. H_0: $p = 0.40$
   H_A: $p \neq 0.40$
2. $\alpha = 0.05$
3'. The test statistic is x, the number of shoppers who can identify the trade mark; here $x = 13$.
4'. Table V shows that for $n = 20$ and $p = 0.40$ the probability of 13 or fewer successes is

$$1 - (0.005 + 0.001) = 0.994$$

and that the probability of 13 or more successes is

$$0.015 + 0.005 + 0.001 = 0.021$$

In a two-tailed test, the p-value is twice the smaller of these two numbers. Here, $p = 2(0.021) = 0.042$.
5'. Since 0.042 is less than 0.05, the null hypothesis must be rejected; in other words, the data show that the true percentage of shoppers who can identify the trade mark is not 40 percent.

When n is large, tests concerning proportions (percentages, or probabilities) may be based on the normal-curve approximation to the binomial distribution. Using the same z-statistic which led to the large-sample confidence interval formula for p on page 353, we base tests of the null hypothesis $p = p_0$ on the statistic[†]

Statistic for large-sample test concerning proportion

$$z = \frac{x - np_0}{\sqrt{np_0(1 - p_0)}}$$

which is a value of a random variable having approximately the standard normal distribution. The test criteria are like those shown in Figure 11.10 with p and p_0

[†] Some statisticians make a continuity correction here by substituting $x - \frac{1}{2}$ or $x + \frac{1}{2}$ for x in the formula for z, whichever makes z closer to zero. However, when n is large the effect of this correction is usually negligible.

substituted for μ and μ_0; analogous to the table on page 312, the criteria for tests of the null hypothesis $p = p_0$ are as follows:

Alternative hypothesis	Reject the null hypothesis if	Accept the null hypothesis or reserve judgment if
$p < p_0$	$z \leq -z_\alpha$	$z > -z_\alpha$
$p > p_0$	$z \geq z_\alpha$	$z < z_\alpha$
$p \neq p_0$	$or \begin{array}{l} z \leq -z_{\alpha/2} \\ z \geq z_{\alpha/2} \end{array}$	$-z_{\alpha/2} < z < z_{\alpha/2}$

EXAMPLE   A television critic claims that 80 percent of all viewers find the noise level of a certain commercial objectionable. If a random sample of 320 television viewers includes 245 who find the noise level of the commercial objectionable, test at the 0.05 level of significance whether the difference between the sample proportion, $\frac{245}{320} = 0.766$, and $p_0 = 0.80$ is significant.

Solution   1. H_0: $p = 0.80$
H_A: $p \neq 0.80$
2. $\alpha = 0.05$
3. Reject the null hypothesis if $z \leq -1.96$ or $z \geq 1.96$, where

$$z = \frac{x - np_0}{\sqrt{np_0(1 - p_0)}}$$

Otherwise, state that the difference between the sample proportion and $p = 0.80$ is not significant.
4. Substituting $n = 320$, $x = 245$, and $p_0 = 0.80$ into the formula for z, we get

$$z = \frac{245 - 320(0.80)}{\sqrt{320(0.80)(0.20)}} = -1.54$$

5. Since $z = -1.54$ falls on the interval from -1.96 to 1.96, the null hypothesis cannot be rejected; in other words, we conclude that the difference between $\frac{245}{320} = 0.766$ and $p = 0.80$ is not significant.

EXERCISES

13.30 A doctor claims that only 10 percent of all persons exposed to a certain amount of radiation will feel any ill effects. If, in a random sample, 5 of 18 persons exposed to such radiation feel some ill effects, test the null hypothesis $p = 0.10$ against the alternative hypothesis $p > 0.10$ at the 0.05 level of significance.

13.31 The manufacturer of a spot remover claims that his product removes 90 percent of all spots. If, in a random sample, the spot remover removes 11 of 16 spots, test the null hypothesis $p = 0.90$ against the alternative hypothesis $p < 0.90$ at the 0.05 level of significance.

13.32 It has been claimed that 20 percent of all undergraduate business students will take advanced work in accounting. In a random sample of 12 undergraduate business students, 5 say that they will take advanced work in accounting. Test the claimed value of 20 percent, using the alternative hypothesis $p > 0.20$ and the 0.05 level of significance.

13.33 In a random sample, 12 of 14 industrial accidents were due to unsafe working conditions. Use the 0.05 level of significance to test the hypothesis that 50 percent of all industrial accidents are due to unsafe working conditions.

13.34 A food processor wants to know whether the probability is really 0.60 that a customer will prefer a new kind of packaging to the old kind. If, in a random sample, 7 of 18 customers prefer the new kind of packaging to the old kind, test the null hypothesis $p = 0.60$ against the alternative hypothesis $p \neq 0.60$ at the 0.05 level of significance.

13.35 In a random sample, 5 of 12 medical students said that they will go into private practice soon after they finish their internships. If the probability of a Type I error is to be at most 0.05, does this support or refute the claim that at least 70 percent of all medical students plan to go into private practice soon after they finish their internships?

13.36 A random sample of size $n = 15$ is to be used to test the null hypothesis $p = 0.70$ against the alternative $p < 0.70$. For what values of x, the observed number of successes, will the null hypothesis be rejected if
(a) the level of significance is 0.10;
(b) the level of significance is 0.05;
(c) the level of significance is 0.01?

13.37 In the example on page 362, we tested the null hypothesis $p = 0.40$ against the alternative hypothesis $p \neq 0.40$. For what values of x, the number of shoppers in the sample of size $n = 20$ who can identify the

trademark, would the null hypothesis be rejected if
(a) the level of significance is 0.10;
(b) the level of significance is 0.05;
(c) the level of significance is 0.01?

13.38 In a random sample of 200 automobile accidents, it was found that 64 were due at least in part to driver fatigue. Use the 0.05 level of significance to test whether or not this supports the claim that 35 percent of all automobile accidents are due at least in part to driver fatigue.

13.39 In a study of aviophobia, a psychologist claims that 30 percent of all women are afraid of flying. If, in a random sample, 41 of 150 women are afraid of flying, test the null hypothesis $p = 0.30$ against the alternative hypothesis $p \neq 0.30$ at the 0.05 level of significance.

13.40 In a random sample of 600 cars making right turns at a certain intersection, 157 pulled into the wrong lane. Test the claim that 30 percent of all drivers make this mistake at the given intersection, using the level of significance
(a) $\alpha = 0.05$;
(b) $\alpha = 0.01$.

13.41 An airline claims that at most 6 percent of all of its lost luggage is never found. Test this claim, if, in a random sample, 17 of 200 pieces of the airline's lost luggage are not found and the probability of a Type I error is not to exceed 0.01.

13.42 To check an ambulance service's claim that at least 40 percent of its calls are life-threatening emergencies, a random sample was taken from its files, and it was found that 49 of 150 calls were life-threatening emergencies. If the probability of a Type I error is not to exceed 0.05, what can we conclude about the ambulance service's claim?

13.43 A food chemist notes that 90 percent of the bricks of a certain cheddar cheese will remain mold-free for three months under standard refrigeration. A variation is made in the salt content of this cheese, and a sample of 15 bricks of the modified cheese are kept under standard refrigeration for three months.
(a) Suppose that only 11 of the bricks remain mold-free. At the 5 percent level of significance,

test the null hypothesis $p = 0.90$ against the alternative hypothesis $p \neq 0.90$.

(b) Suppose that all 15 of the bricks remain mold-free. At the 5 percent level of significance, test the null hypothesis $p = 0.90$ against the alternative hypothesis $p \neq 0.90$. What can you say about this experiment's ability to detect p larger than 0.90?

13.44 For each of 400 simulated random samples, a statistics class determined a 90 percent confidence interval for μ, and it found that 368 of them contained the mean of the population sampled, the other 32 did not. At the 0.05 level of significance, is there evidence to doubt whether the method employed actually yields 90 percent confidence intervals?

13.4

DIFFERENCES BETWEEN PROPORTIONS

There are many problems in which we must decide whether an observed difference between two sample proportions can be attributed to chance, or whether it is indicative of the fact that the corresponding population proportions are not equal. For instance, we may want to decide on the basis of sample data whether there is a difference between the actual proportions of persons with and without flu shots who catch the disease, or we may want to test on the basis of samples whether two manufacturers of electronic equipment ship equal proportions of defectives.

The method we shall use here to test whether an observed difference between two sample proportions is significant, or whether it can be attributed to chance, is based on the following theory: If x_1 and x_2 are the numbers of successes obtained in n_1 trials of one kind and n_2 of another, the trials are all independent, and the corresponding probabilities of a success are, respectively, p_1 and p_2, then the sampling distribution of $\dfrac{x_1}{n_1} - \dfrac{x_2}{n_2}$ has the mean $p_1 - p_2$ and the standard deviation

$$\sqrt{\frac{p_1(1 - p_1)}{n_1} + \frac{p_2(1 - p_2)}{n_2}}$$

It is customary to refer to this standard deviation as the **standard error of the difference between two proportions**.

When we test the null hypothesis $p_1 = p_2 (= p)$ against an appropriate alternative hypothesis, the mean of the sampling distribution of the difference between two sample proportions is $p_1 - p_2 = 0$, and the standard error formula can be written

$$\sqrt{p(1 - p)\left(\frac{1}{n_1} + \frac{1}{n_2}\right)}$$

where p is usually estimated by **pooling** the data and substituting for p the combined sample proportion $\hat{p} = \dfrac{x_1 + x_2}{n_1 + n_2}$, which, as before, reads "p-hat." Then, converting to standard units, we obtain the statistic

Statistic for test concerning difference between two proportions

$$z = \frac{\dfrac{x_1}{n_1} - \dfrac{x_2}{n_2}}{\sqrt{\hat{p}(1 - \hat{p})\left(\dfrac{1}{n_1} + \dfrac{1}{n_2}\right)}} \quad \text{with} \quad \hat{p} = \frac{x_1 + x_2}{n_1 + n_2}$$

which, for large samples, is a value of a random variable having approximately the standard normal distribution. To make this formula appear more compact, we can substitute in the numerator $\hat{p}_1$ for $\dfrac{x_1}{n_1}$ and $\hat{p}_2$ for $\dfrac{x_2}{n_2}$.

The test criteria are again like those shown in Figure 11.10 with p_1 and p_2 substituted for μ and μ_0; analogous to the table on page 312, the criteria for tests of the null hypothesis $p_1 = p_2$ are as follows:

Alternative hypothesis	Reject the null hypothesis if	Accept the null hypothesis or reserve judgment if
$p_1 < p_2$	$z \leq -z_\alpha$	$z > -z_\alpha$
$p_1 > p_2$	$z \geq z_\alpha$	$z < z_\alpha$
$p_1 \neq p_2$	$z \leq -z_{\alpha/2}$ or $z \geq z_{\alpha/2}$	$-z_{\alpha/2} < z < z_{\alpha/2}$

EXAMPLE   To test the effectiveness of a new pain-relieving drug, 80 patients at a clinic were given a pill containing the drug and 80 others were given a placebo containing only powdered sugar. If 56 of the patients in the first group and 38 of those in the second group felt a beneficial effect, what can we conclude at the 0.01 level of significance about the effectiveness of the new drug?

Solution   1. H_0: $p_1 = p_2$
  H_A: $p_1 > p_2$
2. $\alpha = 0.01$
3. Reject the null hypothesis if $z \geq 2.33$, where z is given by the formula above; otherwise, accept it or reserve judgment.
4. Substituting

$$x_1 = 56, \quad x_2 = 38, \quad n_1 = 80, \quad n_2 = 80, \quad \text{and} \quad \hat{p} = \frac{56 + 38}{80 + 80} = 0.5875$$

into the formula for z, we get

$$z = \frac{\dfrac{56}{80} - \dfrac{38}{80}}{\sqrt{(0.5875)(0.4125)\left(\dfrac{1}{80} + \dfrac{1}{80}\right)}} = 2.89$$

5. Since $z = 2.89$ exceeds 2.33, the null hypothesis must be rejected; in other words, we conclude that the new drug is effective.

The method we have described here applies to tests of the null hypothesis $p_1 = p_2$, but it can easily be modified (see Exercise 13.52 on page 368) so that it applies also to tests of the null hypothesis $p_1 - p_2 = \delta$ (*delta*, the Greek letter for lowercase *d*). Also, when the alternative hypothesis is $p_1 \neq p_2$, but not when it is $p_1 < p_2$ or $p_1 > p_2$, we can base the test on a different statistic, which is somewhat easier to calculate. This method is described on page 376.

There are also many problems in which we must decide whether differences among more than two sample proportions are significant, or whether they can be attributed to chance. For instance, if 24 of 200 brand *A* tires, 21 of 200 brand *B* tires, 18 of 200 brand *C* tires, and 33 of 200 brand *D* tires failed to last 20,000 miles, we may want to decide whether the differences among $\frac{24}{200} = 0.120$, $\frac{21}{200} = 0.105$, $\frac{18}{200} = 0.090$, and $\frac{33}{200} = 0.165$ are significant, or whether they may be due to chance. Problems of this type can be treated separately, but we will consider them as a special case of the problem treated in Section 13.5.

EXERCISES

13.45 One method of seeding clouds was successful in 57 of 150 attempts, while another method was successful in 33 of 100 attempts. At the 0.05 level of significance, can we conclude that the first method is better than the second?

13.46 One mail solicitation for a charity brought 412 responses to 5,000 letters and another, more expensive, mail solicitation brought 311 responses to 3,000 letters. Use the 0.01 level of significance to test the null hypothesis that the two solicitations are equally effective against the alternative that the more expensive one is more effective.

13.47 In a random sample of 250 persons who skipped breakfast, 102 reported that they experienced midmorning fatigue, and in a random sample of 250 persons who ate breakfast, 73 reported that they experienced midmorning fatigue. Use the 0.01 level of significance to test the null hypothesis that there is no difference between the corresponding population proportions against the alternative hypothesis that midmorning fatigue is more prevalent among persons who skip breakfast.

13.48 A study showed that 74 of 200 persons who saw a deodorant advertised during the telecast of a baseball game and 86 of 200 other persons who saw it advertised on a variety show remembered two hours later the name of the deodorant. Use the 0.05 level of significance to test whether or not the difference between the two sample proportions, $\frac{74}{200} = 0.37$ and $\frac{86}{200} = 0.43$, is significant.

13.49 A random sample of 100 high school students were asked whether they would turn to their parents for help with a homework assignment in mathematics, and another random sample of 100 high school students were asked the same question with regard to a homework assignment in English. If 62 students in the first sample and 44 students in the second sample would turn to their parents for help, test at the 0.05 level of significance whether the difference between the two sample proportions, $\frac{62}{100}$ and $\frac{44}{100}$, may be attributed to chance.

13.50 The owner of an automobile repair shop must decide which of two soft-drink vending machines to install in his customer waiting room. Each machine was tested 200 times, and the first machine failed to work (neither delivered the drink nor returned the money) 11 times, while the second machine failed to work 6 times. Test at the 0.05 level of significance whether the difference between the corresponding sample proportions is significant.

13.51 Among 500 marriage license applications, chosen at random twelve years ago, 48 of the women were at least one year older than the men, and among 500 marriage license applications, chosen at random eight years later, 85 of the women were at least one year older than the men. Use the 0.05 level of significance to test whether or not there was an actual increase in the proportion of women on marriage license applications who were at least one year older than the men.

13.52 To test the null hypothesis that the difference between two population proportions equals some constant δ (*delta*, the Greek letter for lowercase *d*), we can use the statistic

Statistic for test concerning difference between two proportions
$$z = \frac{\hat{p}_1 - \hat{p}_2 - \delta}{\sqrt{\dfrac{\hat{p}_1(1 - \hat{p}_1)}{n_1} + \dfrac{\hat{p}_2(1 - \hat{p}_2)}{n_2}}}$$

which, for large samples, is a value of a random variable having approximately the standard normal distribution. In this formula, $\hat{p}_1$ and $\hat{p}_2$ denote the two sample proportions, $\dfrac{x_1}{n_1}$ and $\dfrac{x_2}{n_2}$. The test criteria are like those of Figure 11.12 on page 322 and the table on page 323 with p_1 and p_2 substituted for μ_1 and μ_2.

(a) In a true–false test, a test item is considered to be good if it discriminates between well-prepared and poorly prepared students. If 205 of 250 well-prepared students and 137 of 250 poorly prepared students answer a certain test item correctly, test at the 0.05 level of significance whether to accept the null hypothesis $p_1 - p_2 = 0.20$ or the alternative hypothesis $p_1 - p_2 > 0.20$.

(b) To check on regional preferences, a chain of donut houses interviews random samples of shoppers in North Carolina and in Vermont. Asked whether they prefer the chain's donuts to those of all its competitors, 177 of 500 persons interviewed in North Carolina and 110 of 400 persons interviewed in Vermont answered "Yes." Test at the 0.05 level of significance whether the corresponding true proportion for North Carolina exceeds that for Vermont by more than 0.06.

13.53 A medical research worker wants to study how male and female rats react to the injection of a certain toxic substance. If 72 of 200 male rats and 49 of 200 female rats reacted strongly to the injection, use the method of Exercise 13.52 to test at the 0.01 level of significance whether the corresponding true percentage for male rats exceeds that for female rats by 10 percent.

★ **13.54** If we substitute the formula for z of Exercise 13.52 into $-z_{\alpha/2} < z < z_{\alpha/2}$, and manipulate the inequality algebraically so that the middle term is δ, we obtain a $(1 - \alpha)100$ percent confidence interval for

$$\delta = p_1 - p_2$$

Use this method and the data of part (b) of Exercise 13.52 to construct a 99 percent confidence interval for the difference between the true proportions of shoppers in North Carolina and in Vermont who prefer the donut chain's donuts to those of all its competititors.

★ **13.55** Use the method of the preceding exercise and the data of Exercise 13.53 to construct a 98 percent confidence interval for the difference between the true percentages of male and female rats which react strongly to such an injection.

13.5

THE ANALYSIS OF AN $r \times c$ TABLE

The method we shall describe in this section applies to two kinds of problems, which differ conceptually but are analyzed in the same way. In the first kind of problem we deal with trials permitting two or more possible outcomes. This includes the binomial case, where there are always two possible outcomes, but it also includes situations where the outcomes may be that the weather has improved, remained the same, or got worse, that an undergraduate is a freshman, a sophomore, a junior, or a senior, or, say, that a person resides in Ohio, Indiana, Illinois, Michigan, or Wisconsin.

To give an example, let us consider a survey in which independent random samples of 80 single persons, 120 married persons, and 100 persons who are widowed or divorced were asked whether "friends and social life," "job or primary activity," or "health and physical condition" contribute most to their general happiness. The results are shown in the following table, called a 3×3 table (where 3×3 reads "3 by 3"), as it contains three rows and three columns:

	Single	Married	Widowed or divorced
Friends and social life	41	49	42
Job or primary activity	27	50	33
Health and physical condition	12	21	25
	80	120	100

There are three separate samples, the column totals are the fixed sample sizes, and each trial (each person interviewed) permits three different outcomes. Note that the row totals, $41 + 49 + 42 = 132$, $27 + 50 + 33 = 110$, and $12 + 21 + 25 = 58$, depend on the responses of the persons interviewed, and hence on chance. In general, a table like this, with r rows and c columns, is called an $r \times c$ ("r by c") **table**.

In the second kind of problem where the method of this section applies, the column totals as well as the row totals are left to chance. Suppose, for instance, that we want to investigate whether there is a relationship between the qualification test scores of persons who have gone through a certain job-training program and their subsequent performance on the job. Suppose, furthermore, that a random sample of 400 cases taken from very extensive files yielded the following results:

| | Performance | | | |
	Poor	Fair	Good	
Below average	67	64	25	156
Average	42	76	56	174
Above average	10	23	37	70
	119	163	118	400

Qualification test Scores labels the rows (Below average, Average, Above average).

Here there is only one sample, the **grand total** of 400 is its fixed size, and each trial (each case chosen from the files) permits nine different outcomes. It is mainly in connection with problems like this that $r \times c$ tables are referred to as **contingency tables**.

Before we demonstrate how $r \times c$ tables are analyzed, let us examine what null hypotheses we shall want to test. In the problem dealing with the different factors which contribute most to one's happiness, we want to test the null hypothesis that the probabilities of the three choices ("friends and social life," "job or primary activity," or "health and physical condition") are the same for each group. Symbolically, if p_{ij} is the probability of obtaining a response belonging to the ith row and the jth column, the null hypothesis is

$$H_0: \quad p_{11} = p_{12} = p_{13}$$
$$p_{21} = p_{22} = p_{23}$$
$$p_{31} = p_{32} = p_{33}$$

where the p's must add up to 1 for each column.[†] The alternative hypothesis is that the p's are not all equal for at least one row, namely,

$$H_A: \quad p_{i1}, p_{i2}, \text{ and } p_{i3} \text{ are not all equal for at least one value of } i.$$

In the other kind of problem, the one immediately above, we are concerned with the null hypothesis that on the job performance of persons who have gone through the training program is independent of their qualification test scores. In general, the null and alternative hypotheses are

$$H_0: \quad \text{The two variables under consideration are independent.}$$

$$H_A: \quad \text{The two variables are not independent.}$$

[†] We could also say that we are testing the null hypothesis that we are sampling three **multinomial populations** (see Section 8.6) with identical probabilities for the three alternatives.

In spite of the differences we have described, the analysis of an $r \times c$ table is the same for both kinds of situations, and we shall illustrate it here by analyzing the second example. If the null hypothesis of independence is true, the probability of randomly choosing a person whose qualification test score is below average and whose on the job performance is poor is given by the product of the probability of choosing a person whose qualification test score is below average and the probability of choosing a person whose on the job performance is poor. Using the total of the first row, the total of the first column, and the grand total for the entire table to estimate these probabilities, we get

$$\frac{67 + 64 + 25}{400} = \frac{156}{400}$$

for the probability of choosing a person whose qualification test score is below average, and

$$\frac{67 + 42 + 10}{400} = \frac{119}{400}$$

for the probability of choosing a person whose on the job performance is poor. Hence, we estimate the probability of choosing a person whose qualification test score is below average and whose on the job performance is poor as $\frac{156}{400} \cdot \frac{119}{400}$, and in a sample of size 400 we would expect to find

$$400 \cdot \frac{156}{400} \cdot \frac{119}{400} = \frac{156 \cdot 119}{400} = 46.4$$

persons who fit this description.

In the final step of the preceding paragraph, $\frac{156 \cdot 119}{400}$ is just the product of the total of the first row and the total of the first column divided by the grand total for the entire table. Indeed, the argument which led to this result can be used to show that in general

The expected frequency for any cell of an $r \times c$ table may be obtained by multiplying the total of the row to which it belongs by the total of the column to which it belongs and then dividing by the grand total for the entire table.

With this rule we get an expected frequency of

$$\frac{156 \cdot 163}{400} = 63.6$$

for the second cell of the first row, and

$$\frac{174 \cdot 119}{400} = 51.8 \quad \text{and} \quad \frac{174 \cdot 163}{400} = 70.9$$

for the first two cells of the second row.

It is not necessary to calculate all the expected frequencies in this way, as it can be shown that the sum of the expected frequencies for any row or column must equal the sum of the corresponding observed frequencies (see Exercises 13.79 and 13.80). Therefore, we can get some of the expected frequencies by subtraction from row or column totals. For instance, for our example we get

$$156 - 46.4 - 63.6 = 46.0$$

for the third cell of the first row,

$$174 - 51.8 - 70.9 = 51.3$$

for the third cell of the second row, and

$$119 - 46.4 - 51.8 = 20.8$$
$$163 - 63.6 - 70.9 = 28.5$$

and

$$118 - 46.0 - 51.3 = 20.7$$

for the three cells of the third row. These results are summarized in the following table, where the expected frequencies are shown in parentheses below the corresponding observed frequencies:

		Performance		
		Poor	Fair	Good
Qualification test score	Below average	67 (46.4)	64 (63.6)	25 (46.0)
	Average	42 (51.8)	76 (70.9)	56 (51.3)
	Above average	10 (20.8)	23 (28.5)	37 (20.7)

To test the null hypothesis under which the **expected cell frequencies** were calculated, we compare them with the **observed cell frequencies**. It stands to reason that the null hypothesis should be rejected if the discrepancies between the observed and expected frequencies are large, and that it should be accepted (or at least that we reserve judgment) if the discrepancies between the observed and expected frequencies are small.

Denoting the observed frequencies by the letter o and the expected frequencies by the letter e, we base this comparison on the following chi-square statistic:

Statistic for analysis of $r \times c$ table

$$\chi^2 = \sum \frac{(o - e)^2}{e}$$

which, if the null hypothesis is true, is a value of a random variable having approximately the chi-square distribution (see page 335) with $(r-1)(c-1)$ degree of freedom. When $r=3$ and $c=3$ as in our example, the number of degrees of freedom is $(3-1)(3-1)=4$, and it should be observed that after we had calculated four of the expected frequencies with the rule on page 371 all the others were automatically determined; that is, they were obtained by subtraction from row or column totals. Some statisticians prefer the alternative formula

Alternative formula for chi-square statistic

$$\chi^2 = \sum \frac{o^2}{e} - n$$

where n is the grand total of the frequencies for the entire table. This alternative formula does simplify the calculations, but the original formula shows more clearly how χ^2 is actually affected by the discrepancies between the o's and the e's.

Since we shall want to reject the null hypothesis when the discrepancies between the o's and e's are large, we use the test criterion shown in Figure 13.1; symbolically, we reject the null hypothesis at the α level of significance if $\chi^2 \geq \chi^2_\alpha$. Remember, though, that this test is only an approximate large-sample test, and it is recommended that it not be used when one (or more) of the expected frequencies is less than 5. The values of χ^2_α may be read from Table III at the end of the book, where, for the most part, they are given to three decimals. In practice, there is seldom any need to carry more than two decimals in calculating the value of the χ^2 statistic.

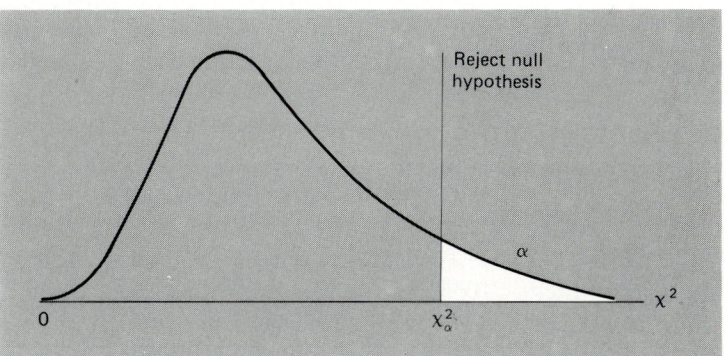

FIGURE 13.1 *Test criterion.*

EXAMPLE   With reference to the illustration on page 370, test at the 0.01 level of significance whether the on the job performance of persons who have gone through the training program is independent of their qualification test score.

Solution   **1.** H_0:   On the job performance and qualification test score are independent.
   H_A: On the job performance and qualification test score are not independent.

2. $\alpha = 0.01$

3. Reject the null hypothesis if $\chi^2 \geq 13.277$, where

$$\chi^2 = \sum \frac{(o - e)^2}{e}$$

and 13.277 is the value of $\chi^2_{0.01}$ for $(3 - 1)(3 - 1) = 4$ degrees of freedom; otherwise, accept the null hypothesis or reserve judgment.

4. Substituting the observed and expected frequencies from the table on page 372 into the formula for χ^2, we get

$$\chi^2 = \frac{(67 - 46.4)^2}{46.4} + \frac{(64 - 63.6)^2}{63.6} + \frac{25 - 46.0)^2}{46.0}$$

$$+ \frac{(42 - 51.8)^2}{51.8} + \frac{(76 - 70.9)^2}{70.9} + \frac{(56 - 51.3)^2}{51.3}$$

$$+ \frac{(10 - 20.8)^2}{20.8} + \frac{(23 - 28.5)^2}{28.5} + \frac{(37 - 20.7)^2}{20.7}$$

$$= 40.89$$

5. Since $\chi^2 = 40.89$ exceeds 13.277, the null hypothesis must be rejected; that is, we conclude that there is a relationship between qualification test score and on the job performance.

A computer printout of the preceding chi-square analysis is shown in Figure 13.2. The difference between the values of χ^2 obtained above and in Figure 13.2 is due to rounding. Some computer programs also give the tail probability corresponding to the observed value of χ^2 when the null hypothesis is true; for our example it is about 0.00000003.

The method which we used here to analyze the $r \times c$ table applies also when the column totals are fixed sample sizes, as in the "greatest contribution to happiness" example on page 369. In that case, the rule by which we multiply the row total by the column total and then divide by the grand total has to be justified in a different way (see Exercise 13.78 on page 380), but this is of no consequence—the expected frequencies are determined in exactly the same way. Actually, there exists an alternative way of analyzing $r \times c$ tables, which is preferable when we are dealing with ordinal data. Clearly, the three performance ratings in our example (poor, fair, and good) were ordered, and so were the three levels of test score (below average, average, and above average). The alternate method is discussed in Section 16.4, beginning on page 481. Other relevant references are given on page 391.

In the analysis of $r \times c$ tables, the special case where $r = 2$ and the column totals are fixed sample sizes has many important applications. Here we are testing, in fact, for significant differences among c sample proportions, and we can simplify our notation by letting $p_1, p_2, \ldots$, and p_c denote the corresponding true proportions.

```
MTB > READ C1 C2 C3
DATA> 67  64   25
DATA> 42  76   56
DATA> 10  23   37
MTB > CHIS C1 C2 C3

EXPECTED FREQUENCIES ARE PRINTED BELOW OBSERVED FREQUENCIES
          I   C1    I   C2    I   C3    ITOTALS
-------I-------I-------I-------I-------
     1 I    67  I    64  I    25  I     156
       I  46.4I  63.6I  46.0I
-------I-------I-------I-------I-------
     2 I    42  I    76  I    56  I     174
       I  51.8I  70.9I  51.3I
-------I-------I-------I-------I-------
     3 I    10  I    23  I    37  I      70
       I  20.8I  28.5I  20.6I
-------I-------I-------I-------I-------
TOTALS I   119 I   163 I   118 I     400

TOTAL CHI SQUARE =

       9.13 +    .00 +  9.60 +
       1.84 +    .37 +   .42 +
       5.63 +   1.07 + 12.95 +

          =   41.01

DEGREES OF FREEDOM = ( 3-1) X ( 3-1) =    4
```

FIGURE 13.2 *Computer printout for analysis of r × c table.*

EXAMPLE The following table shows the results of a study in which independent random samples of workers in three parts of the country were asked whether they feel that unemployment or inflation is a more serious problem:

	Northeast	*Midwest*	*Southwest*
Unemployment	87	73	66
Inflation	113	77	84
	200	150	150

Test at the 0.05 level of significance whether the differences among the sample proportions of workers who chose unemployment,

$$\frac{87}{200} = 0.435, \qquad \frac{73}{150} = 0.487, \qquad \text{and} \qquad \frac{66}{150} = 0.440$$

are significant.

Solution   **1.** H_0: $\;p_1 = p_2 = p_3$
   H_A: $\;p_1, p_2,$ and p_3 are not all equal.

2. $\alpha = 0.05$

3. Reject the null hypothesis if $\chi^2 \geq 5.991$, where

$$\chi^2 = \sum \frac{(o - e)^2}{e}$$

and 5.991 is the value of $\chi^2_{0.05}$ for $(2 - 1)(3 - 1) = 2$ degrees of freedom; otherwise, state that the differences among the three sample proportions are not significant.

4. Since the total for the first row is $87 + 73 + 66 = 226$, we get expected frequencies of

$$\frac{226 \cdot 200}{500} = 90.4 \quad \text{and} \quad \frac{226 \cdot 150}{500} = 67.8$$

for the first two cells of the first row of the table. Then, by subtraction, we get

$$226 - 90.4 - 67.8 = 67.8$$

for the third cell of the first row, and

$$200 - 90.4 = 109.6$$
$$150 - 67.8 = 82.2$$

and

$$150 - 67.8 = 82.2$$

for the three cells of the second row. Finally, substituting all these values together with the observed frequencies into the formula for χ^2, we get

$$\chi^2 = \frac{(87 - 90.4)^2}{90.4} + \frac{(73 - 67.8)^2}{67.8} + \frac{(66 - 67.8)^2}{67.8}$$

$$+ \frac{(113 - 109.6)^2}{109.6} + \frac{(77 - 82.2)^2}{82.2} + \frac{(84 - 82.2)^2}{82.2}$$

$$= 1.048$$

5. Since $\chi^2 = 1.048$ does not exceed 5.991, the null hypothesis cannot be rejected; in other words, the differences among the three sample proportions are not significant.

Note that in an application like this the number of degrees of freedom is simply $c - 1$. Also, for $c = 2$ we have here an alternative method for testing the significance of the difference between two sample proportions (as in Section 13.4), but it can be used only when the alternative hypothesis is $p_1 \neq p_2$.

Let us also remind the reader that all these tests are approximate large-sample tests, which should not be used when one (or more) of the expected frequencies is less than 5. However, if this is the case when we compare c sample proportions, we can sometimes combine samples in such a way that none of the expected frequencies is less than 5.

EXERCISES

13.56 Use the alternative formula on page 373 to recalculate the value of χ^2 obtained on page 374 for the 3×3 contingency table.

13.57 Use the alternative formula on page 373 to recalculate the value of χ^2 obtained on page 376 for the comparison of the three sample proportions.

13.58 Suppose that we ask 200 automotive dealers selling American cars, 150 automotive dealers selling European cars, and 200 automotive dealers selling Japanese cars whether they expect sales to go down, remain the same, go up, or whether they can't tell. What hypotheses shall we want to test, if we are going to perform a chi-square analysis of the resulting 4×3 table?

13.59 Suppose that we take a random sample of 400 persons living in federal housing projects and classify each one according to whether he or she has part-time employment, full-time employment, or no employment, and also according to whether he or she has 0, 1, 2, 3, or 4 or more children. What hypotheses shall we want to test, if we are going to perform a chi-square analysis of the resulting 3×5 table?

13.60 Suppose that we take random samples of voters in Memphis, Nashville, Knoxville, and Chattanooga, and ask them whether they favor an increase in the state tax on cigarettes. What hypotheses shall we want to test, if we are going to perform a chi-square analysis of the resulting 2×4 table?

13.61 In a study to determine whether there is a relationship between bank employees' standard of dress and their professional advancement, a random sample of size $n = 300$ yielded the results shown in the following table:

	Speed of advancement		
	Slow	Average	Fast
Very well dressed	32	56	32
Well dressed	28	69	22
Poorly dressed	15	33	13
	75	158	67

Test at the 0.05 level of significance whether there is a real relationship between standard of dress and speed of professional advancement.

13.62 A sample survey, designed to show how students attending a large university get to their classes, yielded the following results:

	Freshman	Sophomore	Junior	Senior
Walk	104	87	89	72
Automobile	22	29	35	43
Bicycle	46	34	37	32
Other	28	50	39	53
	200	200	200	200

Use the 0.05 level of significance to test the null hypothesis that the same proportions of freshmen, sophomores, juniors, and seniors use these means of transportation.

13.63 The following sample data pertain to shipments received by a large firm from three different vendors:

	Vendor A	Vendor B	Vendor C
Rejected	12	8	20
Not perfect but acceptable	23	12	30
Perfect	85	60	110

Use the 0.01 level of significance to test whether the three vendors ship products of equal quality.

13.64 A market research organization wants to determine, on the basis of the following information, whether there is a relationship between the size of a tube of toothpaste which a shopper buys and the number of persons in the shopper's household:

		Number of persons			
		1–2	3–4	5–6	7 or more
Size of tube bought	Giant	23	116	78	43
	Large	54	25	16	11
	Small	31	68	39	8

At the 0.01 level of significance, is there a relationship?

13.65 Use the alternative formula on page 373 to recalculate the value of χ^2 for the preceding exercise.

13.66 Of a group of 200 persons suffering anxiety disorders, 100 received psychotherapy and 100 received psychological counseling. A panel of psychiatrists determined after six months whether their condition had deteriorated, remained unchanged, or improved.

Based on the results shown in the following table, test at the 0.05 level of significance whether the two treatment methods are equally effective:

	Psychotherapy	Psychological counseling
Deteriorated	8	11
Unchanged	58	62
Improved	34	27

13.67 In a study of students' parents' feelings about a required course in sex education, a random sample of 360 parents are classified according to whether they have one, two, or three or more children in the school system, and also whether they feel that the course is poor, adequate, or good. Supposing that the results are as shown in the following table, test at the 0.05 level of significance whether there is a relationship between parents' reaction to the course and the number of children they have in school:

	Number of children		
	1	2	3 or more
Poor	48	40	12
Adequate	55	53	29
Good	57	46	20

★ **13.68** If the analysis of a contingency table shows that there is a relationship between the two variables under consideration, the strength of the relationship may be measured by the **contingency coefficient**

Contingency coefficient

$$C = \sqrt{\frac{\chi^2}{\chi^2 + n}}$$

where χ^2 is the value of the chi-square statistic obtained for the table and n is the grand total of all the frequencies. This coefficient takes on values between 0 (corresponding to independence) and a maximum value less than 1 depending on the size of the table; for instance, it can be shown that for a 3×3 table the maximum value of C is $\sqrt{2/3} = 0.82$.

(a) Calculate the value of C for the example on page 372, which concerns the relationship between qualification test score and on-the-job performance.

(b) Find C for the contingency table of Exercise 13.64.

13.69 A market research study shows that among 100 men, 100 women, and 200 children interviewed, there were 44, 53, and 133 who did not like the flavor of a new toothpaste. Use the 0.05 level of significance to decide whether or not the differences among the three sample proportions, $\frac{44}{100} = 0.44$, $\frac{53}{100} = 0.53$, and $\frac{133}{200} = 0.665$, are significant.

13.70 On page 367 we referred to a study in which 24 of 200 brand A tires, 21 of 200 brand B tires, 18 of 200 brand C tires, and 33 of 200 brand D tires failed to last 20,000 miles. Based on these figures and at the 0.05 level of significance, is there a difference in the durability of the four kinds of tires?

13.71 The following table shows the results of a survey in which random samples of the parents of 12th graders in three school districts were asked whether they are for or against the removal of certain controversial books from a required reading list:

	District 1	District 2	District 3
For the removal of the books	8	13	12
Against the removal of the books	52	67	48

At the 0.05 level of significance, are the difference among $\frac{8}{60} = 0.133$, $\frac{13}{80} = 0.162$, and $\frac{12}{60} = 0.200$, the sample proportions obtained for the three school districts, significant?

13.72 Tests are made on the ability of three materials to withstand extreme temperature changes. If 43 of 120 specimens of material A, 28 of 80 specimens of material B, and 23 of 100 specimens of material C crumbled, test at the 0.05 level of significance whether the probability of crumbling is the same for the three kinds of materials.

13.73 The following table shows the results of a study in which random samples of the members of five large unions were asked whether they are for or against a certain political candidate:

	Union 1	Union 2	Union 3	Union 4	Union 5
For the candidate	74	81	69	75	91
Against the candidate	26	19	31	25	9

At the 0.01 level of significance, can we conclude that the differences among the five sample proportions are significant?

★ 13.74 The z test for comparing two proportions, introduced on page 366, considers x_1 out of n_1 successes in one group and x_2 out of n_2 successes in a second group. The information can be displayed in this table:

	Success	Failure
Group 1	x_1	$n_1 - x_1$
Group 2	x_2	$n_2 - x_2$

Suppose that we relabel the entries so that $x_1 = A$, $n_1 - x_1 = B$, $x_2 = C$, and $n_2 - x_2 = D$. Then the table is:

	Success	Failure
Group 1	A	B
Group 2	C	D

Let $N = A + B + C + D = n_1 + n_2$ be the total number of items. Show that the z test for comparing two proportions can be expressed as

$$z = \frac{\sqrt{N}(AD - BC)}{\sqrt{(A + B)(C + D)(A + C)(B + D)}}$$

★ 13.75 For the data in a 2×2 table relabeled as in the previous exercise, show that

$$\chi^2 = \frac{N(AD - BC)^2}{(A + B)(C + D)(A + C)(B + D)}$$

and hence that χ^2 is the square of the value of z.

13.76 Rework Exercise 13.49 on page 368 by analyzing the data like a 2×2 table, and verify that the value obtained here for χ^2 equals the square of the value obtained originally for z.

13.77 Suppose that all the values in an $r \times c$ table are doubled. Show that this will double the value of χ^2.

★ 13.78 Use an argument similar to that on page 371 to show that the rule for calculating the expected frequencies for an $r \times c$ table (multiplying the row total by the column total and then dividing by the grand total) applies also when the column totals are fixed sample sizes.

★ 13.79 Verify that if the expected frequencies for an $r \times c$ table are calculated with the rule on page 371, the sum of the expected frequencies for any row equals the sum of the corresponding observed frequencies.

★ 13.80 Verify that if the expected frequencies for an $r \times c$ table are calculated with the rule on page 371, the sum of the expected frequencies for any column equals the sum of the corresponding observed frequencies.

13.6

GOODNESS OF FIT

In this section we shall consider another application of the chi-square criterion, in which we compare an observed frequency distribution with a distribution we might expect according to theory or assumptions. We refer to such a comparison as a test of **goodness of fit**.

To illustrate, let us consider Table XI, the table of random numbers, which is supposed to have been constructed in such a way that each digit is a value of a random variable which takes on the values 0, 1, 2, 3, 4, 5, 6, 7, 8, and 9 with equal probabilities of 0.10. To see whether it is reasonable to maintain that this is, indeed, the case, we might count how many times each digit appears in the table or part of the table; specifically, let us take the 250 digits in five randomly-chosen columns of Table XI (the 14th column on page 543, the 8th and 29th columns on page 544, the 33rd column on page 545, and the 17th column on page 546). This yields the values shown in the "observed frequency" column of the following table:

Digit	Probability	Observed frequency o	Expected frequency e
0	0.10	21	25
1	0.10	28	25
2	0.10	24	25
3	0.10	33	25
4	0.10	23	25
5	0.10	21	25
6	0.10	23	25
7	0.10	23	25
8	0.10	21	25
9	0.10	33	25
		250	250

The expected frequencies in the right-hand column were obtained by multiplying each of the probabilities of 0.10 by 250, the total number of digits counted.

To test whether the discrepancies between the observed and expected frequencies can be attributed to chance, we use the same chi-square statistic as in Section 13.5:

Statistic for test of goodness of fit

$$\chi^2 = \sum \frac{(o - e)^2}{e}$$

calculating $\frac{(o - e)^2}{e}$ separately for each class of the distribution. Then, if the value we get for χ^2 exceeds χ^2_α, we reject the null hypothesis on which the expected frequencies are based at the level of significance α. The number of degrees of freedom is $k - m - 1$, where k is the number of terms $\frac{(o - e)^2}{e}$ added in the formula for χ^2, and m is the number of parameters of the probability distribution which have to be estimated from the sample data. The role of m will be made clear in the examples that follow.

EXAMPLE Based on the observed frequencies in the table given above, test at the 0.05 level of significance whether there is any indication that the digits in Table XI may not be regarded as random.

Solution **1.** H_0: The probability of each digit is 0.10.
H_A: The probabilities are not all 0.10.

2. $\alpha = 0.05$

3. Reject the null hypothesis if $\chi^2 \geq 16.919$, where

$$\chi^2 = \sum \frac{(o - e)^2}{e}$$

and 16.919 is the value of $\chi^2_{0.05}$ for $k - m - 1 = 10 - 0 - 1 = 9$ degrees of freedom; otherwise, state that there is no indication that the digits in Table XI may not be regarded as random. (Here $m = 0$ since none of the parameters of the probability distribution had to be estimated from the sample data.)

4. Substituting the observed and expected frequencies from the table on page 381 into the formula for χ^2, we get

$$\chi^2 = \frac{(21 - 25)^2}{25} + \frac{(28 - 25)^2}{25} + \cdots + \frac{(33 - 25)^2}{25}$$

$$= 7.92$$

5. Since $\chi^2 = 7.92$ does not exceed 16.919, the null hypothesis cannot be rejected; in other words, there is no indication that the digits in Table XI may not be regarded as random.

To give an example where we must estimate a parameter of the probability distribution from the observed data, suppose that a person takes a bus to and from work, and the following table shows how many times, Monday through Friday, the buses were late arriving at the respective stops in 100 weeks:

Number of times buses were late	Number of weeks
0	0
1	2
2	8
3	19
4	29
5	26
6	13
7	3
8	0
9	0
10	0

Based on this information, we want to test, using $\alpha = 0.05$, whether the data may be regarded as values of a random variable having a binomial distribution.

We know that $n = 10$, but since p is unknown, it will have to be estimated from the given data. We do this by calculating the mean of the observed data, putting it

```
MTB > BINOMIAL N=1Ø P=.42

   BINOMIAL PROBABILITIES FOR N =    10    AND  P =    .42ØØØØ

       K              P ( X = K)              P (X LESS OR = K)
       Ø                 .ØØ43                     .ØØ43
       1                 .Ø312                     .Ø355
       2                 .1Ø17                     .1372
       3                 .1963                     .3335
       4                 .2488                     .5822
       5                 .2162                     .7984
       6                 .13Ø4                     .9288
       7                 .Ø54Ø                     .9828
       8                 .Ø147                     .9975
       9                 .ØØ24                     .9998
      1Ø                 .ØØØ2                    1.ØØØØ
```

FIGURE 13.3 *Computer printout of the binomial distribution with n = 10 and p = 0.42.*

equal to *np* (the expression for the mean of the binomial distribution), and solving for *p*. Thus, we get

$$\bar{x} = \frac{0 \cdot 0 + 1 \cdot 2 + 2 \cdot 8 + \cdots + 10 \cdot 0}{100}$$

$$= 4.2$$

and, hence, $np = 10p = 4.2$, and $p = 0.42$. Since the binomial probabilities for $n = 10$ and $p = 0.42$ cannot be obtained from Table V at the end of the book, we shall have to refer to a more detailed table (say, the National Bureau of Standards table referred to on page 217) or use a computer as in the example on page 185. Copying the probabilities in the column $P(X = K)$ from Figure 13.3, we get

Number of times buses were late	Binomial probability	Observed frequency *o*	Expected frequency *e*
0	0.0043	0	0.4
1	0.0312	2	3.1
2	0.1017	8	10.2
3	0.1963	19	19.6
4	0.2488	29	24.9
5	0.2162	26	21.6
6	0.1304	13	13.0
7	0.0540	3	5.4
8	0.0147	0	1.5
9	0.0024	0	0.2
10	0.0002	0	0.0

where the expected frequencies in the right-hand column were obtained by multiplying each of the binomial probabilities by 100 and then rounding to one decimal.

Since the sampling distribution of the χ^2 statistic which we use to compare the o's and the e's is only approximately a chi-square distribution, we must combine the first three classes, and also the last four, so that none of the expected frequencies is less than 5. This leaves us with

Number of times buses were late	Observed frequency o	Expected frequency e
2 or less	10	13.7
3	19	19.6
4	29	24.9
5	26	21.6
6	13	13.0
7 or more	3	7.1

where we added the frequencies of the respective classes. Now we proceed as before.

EXAMPLE Test at the 0.05 level of significance whether the data on page 382 may be regarded as values of a random variable having a binomial distribution.

Solution
1. H_0: Random variable has a binomial distribution.
   H_A: Random variable does not have a binomial distribution.
2. $\alpha = 0.05$
3. Reject the null hypothesis if $\chi^2 \geq 9.488$, where

$$\chi^2 = \sum \frac{(o-e)^2}{e}$$

and 9.488 is the value of $\chi^2_{0.05}$ for $6 - 1 - 1 = 4$ degrees of freedom; otherwise, accept the null hypothesis or reserve judgment. (Here $k = 6$ since six terms must be added to obtain χ^2 and $m = 1$ since p had to be estimated from the observed data.)
4. Substituting the observed and expected frequencies from the table given above into the formula for χ^2, we get

$$\chi^2 = \frac{(10 - 13.7)^2}{13.7} + \frac{(19 - 19.6)^2}{19.6} + \frac{(29 - 24.9)^2}{24.9}$$

$$+ \frac{(26 - 21.6)^2}{21.6} + \frac{(13 - 13.0)^2}{13.0} + \frac{(3 - 7.1)^2}{7.1}$$

$$= 5.0$$

rounded to one decimal.

5. Since $\chi^2 = 5.0$ does not exceed 9.488, the null hypothesis cannot be rejected; in other words, there is no real evidence that the random variable (the number of buses that are late per week) does not have a binomial distribution.

The method we have illustrated here is used quite generally to test how well distributions, expected on the basis of theory or assumptions, fit, or describe, observed data. Thus, in the exercises which follow we shall test also whether observed distributions have (at least approximately) the shape of Poisson and normal distributions.

EXERCISES

13.81 To see whether a die is balanced, it is rolled 720 times and the following results are obtained: 1 showed 129 times, 2 showed 107 times, 3 showed 98 times, 4 showed 132 times, 5 showed 136 times, and 6 showed 118 times. At the 0.05 level of significance, do these results support the hypothesis that the die is balanced?

13.82 Ten years' data show that in a given city there were no bank robberies in 57 months, one bank robbery in 36 months, two bank robberies in 15 months, and three or more bank robberies in 12 months. At the 0.05 level of significance, does this substantiate the claim that the probabilities of 0, 1, 2, or 3 or more bank robberies in any one month are 0.40, 0.30, 0.20, and 0.10?

13.83 Four coins are tossed 320 times and 0, 1, 2, 3, and 4 heads showed 14, 68, 108, 104, and 26 times. Using the probabilities given on page 181, namely, $\frac{1}{16}, \frac{4}{16}, \frac{6}{16}, \frac{4}{16}$, and $\frac{1}{16}$, test at the 0.05 level of significance whether it is reasonable to suppose that the coins are balanced and randomly tossed.

13.84 A quality control engineer takes daily samples of four tractors coming off an assembly line and on 200 consecutive working days he obtains the data summarized in the following table:

Number of tractors requiring adjustments	Number of days
0	102
1	78
2	19
3	1
4	0

Use the 0.01 level of significance to test the null hypothesis that the data may be looked upon as random samples from a binomial population with $n = 4$ and $p = 0.10$.

13.85 In 400 five-minute intervals the air-traffic control of an airport received 0, 1, 2, 3, 4, 5, 6, 7, and 8 or more radio messages with respective frequencies of 2, 12, 54, 76, 68, 76, 45, 41, and 26. Use the 0.05 level of significance to test whether the number of radio messages which they receive during a five-minute interval is a random variable having the Poisson distribution with $\lambda = 4.6$, namely, that the probabilities for 0, 1, 2, 3, 4, 5, 6, 7, and 8 or more calls are 0.010, 0.046, 0.107, 0.163, 0.187, 0.173, 0.132, 0.087, and 0.095.

13.86 For 300 consecutive working days, a baker prepares three large chocolate cakes, and those not sold on the same day are donated to a charitable food bank. Given the data shown in the following table, test at the 0.05 level of significance whether they may be looked upon as values of a random variable having a binomial distribution:

Number of cakes sold	Number of days
0	2
1	14
2	46
3	238

Use the formula for the binomial distribution to calculate the required probabilities.

13.87 The following is the distribution of the hourly number of car drivers requesting gasoline at a rural gas station:

Car drivers requesting gasoline per hour	Frequency
0	33
1	68
2	96
3	98
4	106
5	65
6	25
7	8
8	1

Find the mean of this distribution, and using it as an estimate of the parameter λ, fit a Poisson distribution. That is, determine the corresponding Poisson probabilities by using the formula for the Poisson distribution, possibly using a calculator or suitable statistical software. Multiply each of the probabilities by the total frequency 500 to determine expected frequencies. Test at the 0.01 level of significance whether the hourly number of drivers requesting gasoline may be looked upon as a random variable having a Poisson distribution.

13.88 The following is the distribution of the number of minutes it took 80 persons to complete a certain tax form:

Time required to complete form (minutes)	Frequency
10–14	8
15–19	28
20–24	27
25–29	12
30–34	4
35–39	1
	80

As can easily be verified, the mean of this distribution is $\bar{x} = 20.7$ and its standard deviation is $s = 5.4$. To test the null hypothesis that these data constitute a random sample from a normal population, proceed with the following steps:

(a) Find the probabilities that a random variable having a normal distribution with $\mu = 20.7$ and $\sigma = 5.4$ will take on a value less than 14.5, between 14.5 and 19.5, between 19.5 and 24.5, between 24.5 and 29.5, between 29.5 and 34.5, and greater than 34.5.

(b) Changing the first and last classes of the distribution to "14 or less" and "35 or more," find the expected normal curve frequencies corresponding to the six classes of the distribution by multiplying the probabilities obtained in part (a) by the total frequency of 80.

(c) Test at the 0.05 level of significance whether the given data may be looked upon as a random sample from a normal population.

13.89 The following are the scores of 200 persons on a screening test required of applicants by a firm specializing in executive placement:

Score	Number of persons
20–24	3
25–29	12
30–34	50
35–39	75
40–44	40
45–49	15
50–54	4
55–59	1

As can easily be verified, the mean of this distribution is $\bar{x} = 37.10$ and its standard deviation is $s = 5.85$. To test the null hypothesis that these data constitute a random sample from a normal population, proceed with the following steps:

(a) Find the probabilities that a random variable having a normal distribution with $\mu = 37.10$ and $\sigma = 5.85$ will take on a value less than 24.5, between 24.5 and 29.5, between 29.5 and 34.5, between 34.5 and 39.5, between 39.5 and 44.5, between 44.5 and 49.5, between 49.5 and 54.5, and greater than 54.5.

(b) Changing the first and last classes of the distribution to "24 or less" and "55 or more," find the expected normal curve frequencies corresponding to the eight classes of the distribution by multiplying the probabilities obtained in part (a) by the total frequency of 200.

(c) Test at the 0.05 level of significance whether the given data may be looked upon as a random sample from a normal population.

13.7

CHECKLIST OF KEY TERMS

(with page references to their definitions)

Cell, 371
Chi-square statistic, 372
★ *Contingency coefficient*, 378
Contingency table, 370
Count data, 351
Expected cell frequencies, 372
Goodness of fit, 380
Grand total, 370
Multinomial population, 370

Observed cell frequencies, 372
Pooling, 365
★ Posterior probability, 356
★ Prior probability, 356
$r \times c$ table, 369
Sample proportion, 351
Standard error of a proportion, 352
Standard error of the difference between two proportions, 365

13.8

REVIEW EXERCISES

13.90 An airline has four flights between two cities. If 0, 1, 2, 3, or all 4 of the flights arrived on time 1, 13, 40, 137, and 59 times on 250 days, test at the 0.05 level of significance whether these data may be looked upon as a random sample from a binomial population.

13.91 In a random sample of 90 persons shopping at a department store, 63 made at least one purchase. Construct a 95 percent confidence interval for the probability that any one person shopping at the department store will make at least one purchase.

13.92 The following table shows how samples of the residents of three federally financed housing projects replied to the question whether they would continue to live there if they had the choice:

	Project 1	Project 2	Project 3
Yes	63	84	69
No	37	16	31

Test at the 0.01 level of significance whether the differences among the three sample proportions (of "yes" answers) may be attributed to chance.

13.93 Based on the result of $n = 15$ trials, we want to test the null hypothesis $p = 0.40$ against the alternative hypothesis $p > 0.40$. If we reject the null hypothesis when the number of successes is 8 or more (and accept the null hypothesis otherwise), find:
(a) the probability of a Type I error;
(b) the probability of a Type II error when $p = 0.50$;
(c) the probability of a Type II error when $p = 0.60$.

13.94 In a study conducted at a large airport, 81 of 300 persons who had just gotten off a plane and 32 of 200 persons who were about to board a plane admitted that they were afraid of flying. Use the z statistic to test at the 0.01 level of significance whether the difference between the corresponding sample proportions is significant.

13.95 Use the χ^2 statistic to rework the preceding exercise, and verify that the value obtained for χ^2 equals the square of the value obtained for z.

13.96 In a random sample of 200 retired persons, 137 stated that they prefer living in an apartment to living in a one-family home. At the 0.05 level of significance, does this refute the claim that 60 percent of all retired persons prefer living in an apartment to living in a one-family home?

13.97 In accordance with the rule that np and $n(1 - p)$ must both be greater than 5, for what values of p can we use the normal approximation to the binomial distribution when $n = 400$?

13.98 A sample check reveals that 186 of 200 of a professional football team's season-ticket holders intend to renew their tickets for the next season. Construct a 95 percent confidence interval for the true proportion of the team's season-ticket holders who intend to renew their tickets for the next season.

13.99 An undercover police officer wishes to determine what percentage of vendors at flea markets collect state sales tax. How large a sample will he need if he wants to assert with 95 confidence that the error of his estimate is at most 0.06 if:
    (a) he has no idea about the true value;
    (b) he is certain that the true value is in the range 10 percent to 30 percent?

13.100 Among 120 trucks stopped at a road block, 42 were overloaded. If we use the sample proportion, $\frac{42}{120} = 0.35$ to estimate the corresponding true proportion, what can we say with 98 percent confidence about the maximum error?

13.101 A geneticist found that in independent random samples of 100 men and 100 women there were 31 men and 24 women with a certain minor blood disorder. Can he conclude at the 0.01 level of significance that the corresponding true proportion for men is significantly greater than that for women?
    (a) Comment on the formulation of this question.
    (b) Restate the question as it should have been asked, and answer it by performing the appropriate test.

13.102 In $n = 11$ trials, a certain weed killer was effective nine times. What is the tail probability if we want to test the null hypothesis that the weed killer is effective 50 percent of the time against the alternative hypothesis that it is effective more than 50 percent of the time?

13.103 With reference to the preceding exercise, can the null hypothesis be rejected at the 0.04 level of significance?

13.104 Interviews of samples of 200 men and 200 women with minor emotional problems yielded the results shown in the following table:

	Men	Women
Never been in therapy	125	144
Therapy for six months or less	54	42
Therapy for more than six months	21	14

What hypothesis will we test, if we are going to perform a chi-square analysis of this 3×2 table?

13.105 Analyse the data of the preceding exercise by performing an appropriate chi-square test at the 0.05 level of significance.

13.106 In an experiment, an interviewer of job applicants is asked to write down his first impression (favorable or unfavorable) after two minutes and his final impression at the end of the interview. Use the following data and the 0.01 level of significance to test the interviewer's claim that his first and final impressions are the same 90 percent of the time:

		First impression	
		Favorable	Unfavorable
Final impression	Favorable	186	33
	Unfavorable	54	127

13.107 To see whether a newly discovered manuscript can be attributed to a famous nineteenth-century historian, his literary style was analyzed statistically, and it was found, among other things, that in many samples of 1,000 words of text material he used the word "from" 0 to 4 times 43 percent of the time, 5 to 7 times 32 percent of the time, and 8 or more times 25 percent of the time. Now, if in 20 samples of 1,000 words of text material from the newly discovered manuscript, the word "from" is used 0 to 4 times, 5 to 7 times, and 8 or more times in 15, 3, and 2 of the samples, is this evidence at the 0.05 level of significance that the manuscript is not the work of this historian?

13.108 Fifty chronic headache sufferers participated in an experiment in which they used products A and B on two separate headaches. The data from these subjects is summarized in this table:

		Relief with product B	
		Yes	No
Relief with Product A	Yes	22	8
	No	6	14

What hypothesis will we test if we are going to perform a chi-square analysis of this 2×2 table?

13.109 The following is the distribution of the sulfur oxides emission data obtained on page 23.

Tons of sulfur oxides	Frequency
5.0– 8.9	3
9.0–12.9	10
13.0–16.9	14
17.0–20.9	25
21.0–24.9	17
25.0–28.9	9
29.0–32.9	2
	80

We showed on page 61 that $\bar{x} = 18.85$, and it can be verified that $s = 5.55$. Test at the 0.05 level of significance whether the data may be looked upon as a random sample from a normal population.

★ **13.110** The purchasing agent of a firm feels that the probability is 0.80 that any one of several shipments of steel recently received will meet specifications. The head of the firm's quality control department feels that this probability is 0.90, and the chief engineer feels (somewhat more pessimistically) that it is 0.60.
   (a) If the managing director of the firm feels that in this matter the purchasing agent is 10 times as reliable as the chief engineer while the head of the quality control department is 14 times as reliable as the chief engineer, what prior probabilities would he assign to their claims?

   (b) If five of the shipments are inspected and only two meet specifications, what posterior probabilities should the managing director of the firm assign to the respective claims?

★ **13.111** Among the 500 children at a summer camp, a random sample of 120 are interviewed and only 27 of them prefer being at camp to vacationing with their parents. Use the formula of Exercise 13.16 on page 359 to construct a 95 percent confidence interval for the actual proportion of all the children at that camp, who prefer being there to traveling with their parents.

13.112 Tests of the fidelity and the selectivity of 190 radios produced the results shown in the following table:

		Fidelity		
		Low	Average	High
Selectivity	Low	7	12	31
	Average	35	59	18
	High	15	13	0

Use the 0.01 level of significance to test the null hypothesis that fidelity is independent of selectivity.

★ **13.113** With reference to the preceding exercise, calculate the value of the contingency coefficient defined in Exercise 13.68 on page 378.

13.114 Tests are made on the proportion of defective castings produced by two molds. If in a random sample of 100 castings from mold A there were 12 defectives and in a random sample of 200 castings from mold B there were 39 defectives, test at the 0.05 level of significance whether the corresponding true proportion for mold B exceeds that for mold A by more than 0.04.

13.115 In a random sample of 500 voters in a large city, 285 favor a new issue of municipal bonds. If we use $\frac{285}{500} = 0.57$ as an estimate of the true proportion of voters in this city who favor the bonds, what can we say with 99 percent confidence about the maximum error of this estimate?

13.116 A political leader hires a private poll to estimate what percentage of residents of his precinct believe a rumor about him spread by his opponent. How large a sample will the poll have to take to be able to assert with a probability of at least 0.90 that the sample percentage will be within 5 percent of the true value?

13.117 Suppose that we want to test the null hypothesis $p = 0.30$ on the basis of the number of successes, x, in 14 trials. For what values of x would we reject the null hypothesis if the level of significance and the alternative hypothesis are, respectively,

    (a) $\alpha = 0.05$ and $p \neq 0.30$;
    (b) $\alpha = 0.05$ and $p > 0.30$;
    (c) $\alpha = 0.05$ and $p < 0.30$;
    (d) $\alpha = 0.10$ and $p \neq 0.30$?

13.118 Among 210 persons with alcohol problems admitted to the psychiatric emergency room of a hospital, 26 were admitted on a Monday, 23 on a Tuesday, 19 on a Wednesday, 25 on a Thursday, 33 on a Friday, 44 on a Saturday, and 40 on a Sunday. Use the 0.05 level of significance to test the null hypothesis that the psychiatric emergency room can expect equally many persons with alcohol problems on each day of the week.

13.119 When asked to identify Jean Jacques Rousseau, 28 of 120 persons said that he is a deep-sea diver. If the probability of a Type I error is to be at most 0.05, does this substantiate the claim that at least 30 percent of all persons make this mistake?

13.120 A large manufacturer hires many handicapped workers. To see whether their handicaps affect their performance, the personnel manager obtained the following sample data:

		Deaf	Blind	Other handicap	No handicap
	Above average	11	3	14	36
Performance	Average	24	11	39	134
	Below average	5	6	7	30
		40	20	60	200

Explain why a "standard" chi-square analysis with $(3 - 1)(4 - 1) = 6$ degrees of freedom cannot be performed.

13.121 There are various ways in which the data of the preceding exercise can be salvaged. Use one of them and perform a suitable chi-square analysis at the 0.05 level of significance.

★ **13.122** The method of Section 13.2 can also be used to find the posterior distribution of the parameter λ of a Poisson distribution. Suppose for instance, that several brokers are discussing the sale of a large estate, and that one of them feels that a newspaper ad should produce three serious inquiries, a second feels that it should produce five serious inquiries, and a third feels that it should produce six.

    (a) If in the past the second broker has been twice as reliable as the first and the first has been three times as reliable as the third, what prior probabilities should we assign to their claims?

    (b) How would these probabilities be affected if the ad actually produced only one serious inquiry and it can be assumed that the number of serious inquiries is a random variable having the Poisson distribution with either $\lambda = 3$, $\lambda = 5$, or $\lambda = 6$ corresponding to the three claims?

13.123 Four coins are tossed 240 times and 0, 1, 2, 3, or 4 heads showed 32, 96, 62, 42, and 8 times. Use the 0.01 level of significance to test whether it is reasonable to suppose that the coins are balanced and randomly tossed.

★ 13.124 It costs more to test a certain type of ammunition than to manufacture it, and, hence, only two rounds are tested from each large lot. If lots are accepted without further inspection when both rounds function according to specifications, find the actual proportion of defectives for which this test procedure will cause a lot to be subjected to further inspection with a probability of 0.19.

13.9
REFERENCES

Tables for confidence limits for proportions, including those for small samples, were first published in Vol. 26 (1934) of *Biometrika*, a statistical journal. They have been reproduced in many textbooks, including the first five editions of this book, and among current texts they may be found, for example, in

MAXWELL, E. A., *Introduction to Statistical Thinking*. Englewood Cliffs, N.J.: Prentice-Hall, Inc., 1983.

The theory which underlies the various tests of this chapter is discussed in most textbooks on mathematical statistics; for instance, in Chapters 8 and 12 of the book by Freund and Walpole listed below. Details about contingency tables may be found in

EVERITT, B. S., *The Analysis of Contingency Tables*. New York: John Wiley & Sons, Inc., 1977.

In recent years, research has been done on the analysis of $r \times c$ tables with ordinal categorical variables. A readable introduction to this topic can be found in

FREUND, J. E., and SIMON, G. A., *Statistics: A First Course*, 5th ed. Englewood Cliffs, N.J.: Prentice-Hall, Inc., 1991.

More advanced material may be found in

AGRESTI, A., *Analysis of Ordinal Categorical Data*. New York: John Wiley & Sons, Inc., 1984.
GOODMAN, L. A., *The Analysis of Cross-Classified Data Having Ordered Categories*. Cambridge, Mass.: Harvard University Press, 1984.

An extension of the Bayesian method of Section 13.2 to the case where p can take on any value on the continuous interval from 0 to 1, and its prior distribution is a continuous curve, is discussed in Chapter 10 of

FREUND, J. E., and WALPOLE, E. R., *Mathematical Statistics*, *4th ed.* Englewood Cliffs, N.J.: Prentice-Hall, Inc., 1987.

14 ANALYSIS OF VARIANCE

To begin with, we shall extend the work of Sections 11.8 and 11.9 and consider problems in which we must decide whether observed differences among more than two sample means can be attributed to chance, or whether there are real differences among the means of the populations sampled. For instance, we may want to decide on the basis of sample data whether there really is a difference in the effectiveness of three methods of teaching a foreign language, we may want to compare the average yield per acre of several varieties of wheat, we may want to see whether there really is a difference in the average mileage obtained with four kinds of gasoline, we may want to judge whether there really is a difference in the durability of five kinds of carpet, and so on. The method we shall introduce for this purpose is a powerful statistical tool called **analysis of variance**, ANOVA for short.

However, this is not all. With reference to the first of the four examples above, we might go one step further and ask whether an observed difference in the effectiveness of the three teaching methods is really due to these methods and not due to the teachers who use them, to the quality of the texts that are being used, or, say, to the intelligence of the students who are being taught. Also, with regard to the different varieties of wheat, we might ask whether the differences we observe in their yield are really due to their quality and not due to the use of different fertilizers, to differences in the quality of the soil or, say, to differences in the amount of irrigation that is applied to the soil. As the saying goes, questions like these open a whole new can of worms, as they lead to the vast subject of **experimental design**; namely, to the problem of planning experiments in such a way that meaningful questions can be asked and put to a test.

Following an introductory example in Section 14.1, we shall present the one-way analysis of variance in Section 14.3, the two-way analysis of variance in Section 14.5, and a generalization in Section 14.7. Various related topics of experimental design are discussed in Sections 14.2, 14.4, 14.6, and 14.8.

14.1

DIFFERENCES AMONG k MEANS: AN EXAMPLE

Let us introduce the kind of problem we face here by means of an example. Suppose, for instance, that we want to compare the cleansing action of three detergents on the basis of the following whiteness readings made on fifteen swatches of white cloth, which were first soiled with India ink and then washed in an agitator-type washing machine with the respective detergents:

Detergent A:	77	81	71	76	80
Detergent B:	72	58	74	66	70
Detergent C:	76	85	82	80	77

The means of these three samples are 77, 68, and 80, and we would like to know whether the differences among them are significant or whether they can be attributed to chance.

In many problems like this, if $\mu_1, \mu_2, \ldots,$ and μ_k are the means of k populations sampled, we want to test the null hypothesis

$$\mu_1 = \mu_2 = \cdots = \mu_k$$

against the alternative hypothesis that these μ's are not all equal.[†] This null hypothesis would be supported if the differences among the sample means are small, and the alternative hypothesis would be supported if at least some of the differences among the sample means are large. Thus, we need a precise measure of the discrepancies among the $\bar{x}$'s, and with it a rule which tells us when the discrepancies are so large that the null hypothesis should be rejected.

Possible choices for such a measure are the standard deviation of the $\bar{x}$'s or their variance. To determine the latter for our example dealing with the whiteness readings obtained for the swatches washed with the three detergents, we first calculate the mean of the $\bar{x}$'s, getting

$$\frac{77 + 68 + 80}{3} = 75$$

Then, substituting into the formula which defines the sample variance, we get

$$s_{\bar{x}}^2 = \frac{(77 - 75)^2 + (68 - 75)^2 + (80 - 75)^2}{3 - 1}$$

$$= 39$$

where the subscript $\bar{x}$ is used to show that this is the variance of the sample means.

Next let us make two assumptions which are critical to the method by which we shall analyze our problem:

1. **The populations we are sampling can be approximated closely with normal distributions.**
2. **These populations all have the same standard deviation σ.**

With reference to our example, this means that we shall assume that the whiteness readings, for each detergent, are values of a random variable having at least approximately a normal distribution, and that these three random variables have the same standard deviation σ.

_____

[†] In connection with work later in this chapter, it is desirable to write these means as

$$\mu_1 = \mu + \alpha_1, \quad \mu_2 = \mu + \alpha_2, \ldots, \quad \text{and} \quad \mu_k = \mu + \alpha_k$$

Here

$$\mu = \frac{\mu_1 + \mu_2 + \cdots + \mu_k}{k}$$

is called the **grand mean** and the α's, whose sum is zero (see Exercise 14.18 on page 407), are called the **treatment effects**. In this notation, we test the null hypothesis $\alpha_1 = \alpha_2 = \cdots = \alpha_k = 0$ against the alternative that the α's are not all zero.

With these assumptions, and if the null hypothesis $\mu_1 = \mu_2 = \mu_3$ is true, we can look upon the three samples as if they came from one and the same (at least approximately normal) population and, hence, upon $s_{\bar{x}}^2$, the variance of their means, as an estimate of $\sigma_{\bar{x}}^2$, the square of the standard error of the mean. Now, since $\sigma_{\bar{x}} = \dfrac{\sigma}{\sqrt{n}}$ for random samples of size n from infinite populations, we can look upon $s_{\bar{x}}^2$ as an estimate of

$$\sigma_{\bar{x}}^2 = \left(\frac{\sigma}{\sqrt{n}}\right)^2 = \frac{\sigma^2}{n}$$

and, therefore, upon $n \cdot s_{\bar{x}}^2$ as an estimate of σ^2. For our example, we thus have

$$n \cdot s_{\bar{x}}^2 = 5 \cdot 39 = 195$$

as an estimate of σ^2, the common variance of the three populations sampled.

Note that this estimate is based on the variation among the three sample means. If σ^2 were known, we could compare $n \cdot s_{\bar{x}}^2$ with σ^2 and reject the null hypothesis that the population means are all equal if this value is much larger than σ^2. However, in most practical problems σ^2 is not known and we have no choice but to estimate it on the basis of the sample data. Having assumed under the null hypothesis that the three samples do, in fact, come from identical populations, we could use any one of their variances, s_1^2, s_2^2, or s_3^2, as a second estimate of σ^2, and we could also use their mean. Thus averaging, or **pooling**, the three sample variances in our example, we get

$$\frac{s_1^2 + s_2^2 + s_3^2}{3}$$

$$= \frac{1}{3}\left[\frac{(77-77)^2 + (81-77)^2 + (71-77)^2 + (76-77)^2 + (80-77)^2}{5-1}\right.$$

$$+ \frac{(72-68)^2 + (58-68)^2 + (74-68)^2 + (66-68)^2 + (70-68)^2}{5-1}$$

$$\left. + \frac{(76-80)^2 + (85-80)^2 + (82-80)^2 + (80-80)^2 + (77-80)^2}{5-1}\right]$$

$$= 23$$

Now we have two estimates of σ^2,

$$n \cdot s_{\bar{x}}^2 = 195 \quad \text{and} \quad \frac{s_1^2 + s_2^2 + s_3^2}{3} = 23$$

and it should be observed that the first estimate, which is based on the variation among the sample means, is much larger than the second, which is based on the variation within the samples and, hence, measures variation due to chance. This suggests that the three population means are probably not all equal; namely, that the null hypothesis ought to be rejected. To put this comparison on a rigorous basis, we use the statistic

$$F = \frac{\text{estimate of } \sigma^2 \text{ based on the variation among the } \bar{x}\text{'s}}{\text{estimate of } \sigma^2 \text{ based on the variation within the samples}}$$

which is referred to as an **F statistic** or as a **variance ratio**.

If the null hypothesis is true and the assumptions we made are valid, the sampling distribution of this statistic is the F distribution, which we introduced in Chapter 12. Since the null hypothesis will be rejected only when F is large (that is, when the variation among the $\bar{x}$'s is too great to be attributed to chance), we base our decision on the criterion of Figure 14.1. For $\alpha = 0.05$ or 0.01, the values of F_α may be looked up in Table IV at the end of the book, and if we compare the means of k random samples of size n, we have $k - 1$ degrees of freedom for the numerator and $k(n - 1)$ degrees of freedom for the denominator.

Returning to our numerical example, we find that $F = \frac{195}{23} = 8.48$, and since this exceeds 6.93, which is the value of $F_{0.01}$ for $k - 1 = 3 - 1 = 2$ and $k(n - 1) = 3(5 - 1) = 12$ degrees of freedom, the null hypothesis must be rejected at the 0.01 level of significance. In other words, the differences among the three sample means are too large to be entirely attributed to chance.

The technique we have just described is the simplest form of an **analysis of variance**. Although we could go ahead and perform F tests for differences among k means without further discussion, it will be instructive to look at the problem from an analysis-of-variance point of view, and we shall do so in Section 14.3.

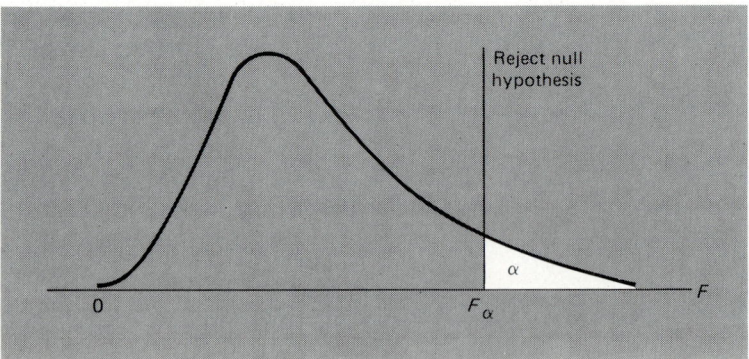

FIGURE 14.1 *Test criterion.*

EXERCISES

14.1 Samples of peanut butter produced by three different manufacturers are tested for aflatoxin content (ppb), with the following results:

Brand 1:	0.5	6.3	1.1	2.7	5.5	4.3
Brand 2:	2.5	1.8	3.6	5.2	1.2	0.7
Brand 3:	3.3	1.5	0.4	4.8	2.2	1.0

(a) Calculate $n \cdot s_{\bar{x}}^2$ for these data, and also the mean of the variances of the three samples and the value of F.

(b) Test at the 0.05 level of significance whether the differences among the means of the three samples can be attributed to chance.

14.2 An agronomist planted three test plots each with four varieties of wheat and obtained the following yields (in pounds per plot):

Variety A:	57	62	61
Variety B:	52	53	60
Variety C:	53	56	56
Variety D:	56	59	59

(a) Calculate $n \cdot s_{\bar{x}}^2$ for these data, and also the mean of the variances of the four samples and the value of F.

(b) Test at the 0.05 level of significance whether the differences among the means of the four samples can be attributed to chance.

14.3 The following are the mileages which a test driver got with four gallons each of five brands of gasoline:

Brand A:	30	25	27	26
Brand B:	29	26	29	28
Brand C:	32	32	35	37
Brand D:	29	34	32	33
Brand E:	32	26	31	27

(a) Calculate $n \cdot s_{\bar{x}}^2$ for these data, and also the mean of the variances of the five samples and the value of F.

(b) Test at the 0.01 level of significance whether the differences among the means of the five samples can be attributed to chance.

14.4 The following are the fourth grade reading comprehension scores on a standardized test for random samples of students from three large schools:

School 1:	81	83	77	72	86
	92	83	78	80	75
School 2:	73	112	66	104	95
	81	62	76	129	90
School 3:	84	89	81	76	79
	83	85	74	80	78

Explain why the method described in Section 14.1 should probably not be used to test for significant differences among the three sample means.

14.5 The following are the weights (pound) of random samples of the athletes playing basketball in three different conferences:

Conference X:	186	195	212	169	188	195
	175	201				
Conference Y:	205	114	127	186	179	117
	191	124				
Conference Z:	173	195	203	212	195	189
	190	169				

Explain why the method described in Section 14.1 should probably not be used to test for significant differences among the three sample means.

14.2

THE DESIGN OF EXPERIMENTS: RANDOMIZATION

In the example of the preceding section it may have seemed perfectly natural to conclude that the three detergents are not equally effective; and yet, a moment's reflection will show that this conclusion is not so "natural" at all. For all we know, the swatches cleaned with detergent B may have been more soiled than the others, the washing times may have been longer for detergent C, there may have been differences in water hardness or water temperature, and even the instruments used to make the whiteness readings may have gone out of adjustment after the readings for detergents A and C were made.

It is entirely possible, of course, that the differences among the three sample means, 77, 68, and 80, are due largely to differences in the quality of the three detergents, but we have just listed several other factors which could be responsible. It is important to remember that a significance test may show that differences among sample means are too large to be attributed to chance, but it cannot say why the differences occurred.

In general, if we want to show that one factor (among various others) can be considered the cause of an observed phenomenon, we must somehow make sure that none of the other factors can reasonably be held responsible. There are various ways in which this can be done; for instance, we can conduct a rigorously **controlled experiment** in which all variables except the one of concern are held fixed. To do this in the example dealing with the three detergents, we might soil the swatches with exactly equal amounts of India ink, always use exactly the same washing time, water of exactly the same temperature and hardness, and inspect the measuring instruments after each use. Under these rigid conditions, significant differences among the sample means cannot be due to differently soiled swatches, or differences in washing time, water temperature, water hardness, or measuring instruments. On the positive side, the differences show that the detergents are not all equally effective if they are used in this narrowly restricted way. Of course, we cannot say whether the same differences would exist if the washing time is longer or shorter, if the water has a different temperature or hardness, and so on.

In most cases, "overcontrolled" experiments like the one just described do not really provide us with the kind of information we want. Also, such experiments are rarely possible in actual practice; for example, it would have been difficult in our illustration to be sure that the instruments really were measuring identically on repeated washings or that some other factor, not thought of or properly controlled, was not responsible for the observed differences in whiteness. So, we look for alternatives. At the other extreme we can conduct an experiment in which none of the exraneous factors is controlled, but in which we protect ourselves against their effects by **randomization**. That is, we design, or plan, the experiment in such a way that the variations caused by these extraneous factors can all be combined under the general heading of "chance."

In our example dealing with the three detergents, we could accomplish this by numbering the swatches (which generally will not be exactly equally soiled) from 1 to 15, specifying the random order in which they are to be washed and measured, and randomly selecting the five swatches which are to be washed with each of the three detergents. When all the variations due to uncontrolled extraneous factors can thus be included under the heading of chance variation, we refer to the design of the experiment as a **completely randomized design**.

As should be apparent, randomization protects against the effects of the extraneous factors only in a probabilistic sort of way. For instance, in our example it is possible, though very unlikely, that detergent A will be randomly assigned to the five swatches which happen to be the least soiled, or that the water happens to be coldest when we wash the five swatches with detergent B. It is partly for this reason that we often try to control some of the factors and randomize the others, and thus use designs that are somewhere between the two extremes which we have described.

Randomization protects against the effects of factors that cannot be controlled perfectly. The experimenter is not relieved of the responsibility of designing the experiment carefully simply because randomization will be used. In our example, an honest effort should be made to prepare the swatches as equally soiled as possible.

Finally, we must make this very important point: *Randomization should be used even if the experimenter believes that all effects are carefully controlled.* In our example, an experimenter who has taken steps to control the soil level on the swatches, the wash temperature, the water hardness, and so on, must still use randomization to assign the swatches to the detergents.

The use of randomization is a cornerstone of the experimental sciences. An experiment in which randomization was possible, but not used, will never be taken completely seriously.

14.3

ONE-WAY ANALYSIS OF VARIANCE

In an analysis of variance, the basic idea is to express a measure of the total variation of a set of data as a sum of terms, which can be attributed to specific sources, or causes, of variation; in its simplest form, it applies to experiments which are planned as completely randomized designs. In the example of Section 14.1, two such sources of variation are (1) actual differences in the cleansing action of the three detergents, and (2) chance, which in problems of this kind is usually called the **experimental error**.

As a measure of the total variation of kn observations consisting of k samples of size n, we shall use the **total sum of squares**[†]

$$SST = \sum_{i=1}^{k} \sum_{j=1}^{n} (x_{ij} - \bar{x}..)^2$$

where x_{ij} is the jth observation of the ith sample ($i = 1, 2, \ldots, k$ and $j = 1, 2, \ldots, n$), and $\bar{x}..$ is the **grand mean**, the mean of all the kn measurements or observations. Note that if we divide the total sum of squares SST by $kn - 1$, we get the variance of the combined data.

If we let $\bar{x}_i.$ denote the mean of the ith sample (for $i = 1, 2, \ldots, k$), we can now write the following identity, which forms the basis of a **one-way analysis of**

---

† The use of double subscripts and double summations is treated briefly in Section 3.8.

variance:[†]

Identity for one-way analysis of variance with n observations in each of k groups.

$$SST = n \cdot \sum_{i=1}^{k} (\bar{x}_{i.} - \bar{x}..)^2 + \sum_{i=1}^{k} \sum_{j=1}^{n} (x_{ij} - \bar{x}_{i.})^2$$

It is customary to refer to the first term on the right, the quantity which measures the variation among the sample means, as the **treatment sum of squares** $SS(Tr)$, and to the second term, which measures the variation within the individual samples, as the **error sum of squares** SSE. The choice of the word "treatment" is explained by the origin of many analysis-of-variance techniques in agricultural experiments where different fertilizers, for example, are regarded as different **treatments** applied to the soil. So, we shall refer to the three detergents in our example as three different treatments, and in other problems we may refer to four nationalities as four different treatments, five different advertising campaigns as five different treatments, and so on. The word "error" in "error sum of squares" pertains to the experimental error, or chance.

In the notation of the preceding paragraph, the identity for a one-way analysis of variance reads

$$SST = SS(Tr) + SSE$$

and since its proof requires quite a bit of algebraic manipulation, let us merely verify it numerically for the three-detergents example of Section 14.1. Substituting the original whiteness readings, the three sample means, and the grand mean (see pages 393 and 394) into the formulas for the three sums of squares, we get

$$SST = (77 - 75)^2 + (81 - 75)^2 + (71 - 75)^2$$
$$+ (76 - 75)^2 + (80 - 75)^2 + (72 - 75)^2$$
$$+ (58 - 75)^2 + (74 - 75)^2 + (66 - 75)^2$$
$$+ (70 - 75)^2 + (76 - 75)^2 + (85 - 75)^2$$
$$+ (82 - 75)^2 + (80 - 75)^2 + (77 - 75)^2$$
$$= 666$$

[†] This identity may be derived by writing the total sum of squares as

$$SST = \sum_{i=1}^{k} \sum_{j=1}^{n} (x_{ij} - \bar{x}..)^2$$

$$= \sum_{i=1}^{k} \sum_{j=1}^{n} [(\bar{x}_{i.} - \bar{x}..) + (x_{ij} - \bar{x}_{i.})]^2$$

and then expanding the squares $[(\bar{x}_{i.} - \bar{x}..) + (x_{ij} - \bar{x}_{i.})]^2$ by means of the binomial theorem and simplifying algebraically.

$$SS(Tr) = 5[(77 - 75)^2 + (68 - 75)^2 + (80 - 75)^2]$$

$$= 390$$

$$SSE = (77 - 77)^2 + (81 - 77)^2 + (71 - 77)^2 + (76 - 77)^2$$

$$+ (80 - 77)^2 + (72 - 68)^2 + (58 - 68)^2 + (74 - 68)^2$$

$$+ (66 - 68)^2 + (70 - 68)^2 + (76 - 80)^2 + (85 - 80)^2$$

$$+ (82 - 80)^2 + (80 - 80)^2 + (77 - 80)^2$$

$$= 276$$

and it can be seen that

$$SS(Tr) + SSE = 390 + 276 = 666 = SST$$

Although this may not be immediately apparent, what we have done here is very similar to what we did in Section 14.1. Indeed, $SS(Tr)$ divided by $k - 1$ equals the quantity which we denoted by $n \cdot s_{\bar{x}}^2$ and put into the numerator of the F statistic on page 396. Called the **treatment mean square**, it measures the variation among the sample means and it is denoted by $MS(Tr)$. Thus,

$$MS(Tr) = \frac{SS(Tr)}{k - 1}$$

and for the three-detergents example we get $MS(Tr) = \frac{390}{2} = 195$. This equals the value we got for $n \cdot s_{\bar{x}}^2$ on page 395 (see also Exercise 14.21).

Similarly, SSE divided by $k(n - 1)$ equals the mean of the k sample variances, $\frac{1}{3}(s_1^2 + s_2^2 + s_3^2)$ in our example, which we put into the denominator of the F statistic on page 396. Called the **error mean square**, it measures the variation within the samples and it is denoted by MSE. Thus,

$$MSE = \frac{SSE}{k(n - 1)}$$

and for the three-detergents example we get $MSE = \dfrac{276}{3(5 - 1)} = \dfrac{276}{12} = 23$. This equals the value we got for $\dfrac{1}{3}(s_1^2 + s_2^2 + s_3^2)$ on page 395 (see also Exercise 14.21).

Since F was defined on page 396 as the ratio of these two measures of the variation among the sample means and within the samples, we can now write

Statistic for test concerning differences among means

$$F = \frac{MS(Tr)}{MSE}$$

In practice, we display the work required for the determination of F in the following kind of table, called an **analysis of variance table**:

Source of variation	Degrees of freedom	Sum of squares	Mean square	F
Treatments	$k - 1$	SS(Tr)	MS(Tr)	$\dfrac{MS(Tr)}{MSE}$
Error	$k(n - 1)$	SSE	MSE	
Total	$kn - 1$	SST		

The degrees of freedom for treatments and error are the numerator and denominator degrees of freedom referred to on page 396. Note that they equal the quantities we divide into the sums of squares to obtain the corresponding mean squares.

After we have calculated F, we proceed as in Section 14.1. Assuming again that the populations sampled can be approximated closely with normal distributions and that they have equal standard deviations, we reject the null hypothesis

$$\mu_1 = \mu_2 = \cdots = \mu_k$$

against the alternative hypotheses that these μ's are not all equal (or, in the notation introduced in the footnote to page 394, the null hypothesis

$$\alpha_1 = \alpha_2 = \cdots = \alpha_k = 0$$

against the alternative hypothesis that these treatment effects are not all zero), if the value obtained for F is greater than or equal to F_α for $k - 1$ and $k(n - 1)$ degrees of freedom.

EXAMPLE   Use the sums of squares on pages 400 and 401 to construct an analysis of variance table for the three-detergents example, and test at the 0.01 level of significance whether or not the three detergents are equally effective.

Solution   Since $k = 3$, $n = 5$, $SST = 666$, $SS(Tr) = 390$, and $SSE = 276$, we get $k - 1 = 2$, $k(n - 1) = 12$, $MS(Tr) = \frac{390}{2} = 195$, $MSE = \frac{276}{12} = 23$, and $F = \frac{195}{23} = 8.48$. All these results are summarized in the following table:

Source of variation	Degrees of freedom	Sum of squares	Mean square	F
Treatments	2	390	195	8.48
Error	12	276	23	
Total	14	666		

Note that the total degrees of freedom is simply the sum of the degrees of freedom for treatments and error (see also Exercise 14.19 on page 407).

Finally, since $F = 8.48$ exceeds 6.93, the value of $F_{0.01}$ for 2 and 12 degrees of freedom, the null hypothesis that the three detergents are equally effective must be rejected.

The numbers which we used in our illustration were intentionally chosen so that the calculations would be easy. In actual practice, the calculation of the sums of squares can be quite tedious unless we use the following computing formulas, in which $T_{i.}$ denotes the sum of the values for the ith treatment (that is, the sum of the values in the ith sample), and $T_{..}$ denotes the **grand total** of all the data:

Computing formulas for sums of squares with n observations in each of k groups

$$SST = \sum_{i=1}^{k} \sum_{j=1}^{n} x_{ij}^2 - \frac{1}{kn} \cdot T_{..}^2$$

$$SS(Tr) = \frac{1}{n} \cdot \sum_{i=1}^{k} T_{i.}^2 - \frac{1}{kn} \cdot T_{..}^2$$

and by subtraction

$$SSE = SST - SS(Tr)$$

EXAMPLE Use these computing formulas to verify the sums of squares on pages 400 and 401 for the three-detergents example.

Solution Substituting $k = 3$, $n = 5$, $T_{1.} = 385$, $T_{2.} = 340$, $T_{3.} = 400$, $T_{..} = 1,125$, and $\sum\sum x^2 = 85,041$ into the formulas, we get

$$SST = 85,041 - \tfrac{1}{15}(1,125)^2$$

$$= 85,041 - 84,375$$

$$= 666$$

$$SS(Tr) = \tfrac{1}{5}[385^2 + 340^2 + 400^2] - 84,375$$

$$= 390$$

and

$$SSE = 666 - 390 = 276$$

As can be seen, these results are identical with the ones obtained before.

A computer printout of the analysis of variance of this section is shown in Figure 14.2. Besides the degrees of freedom, the sums of squares, the mean squares, and the value of F, it provides information which permits further comparisons

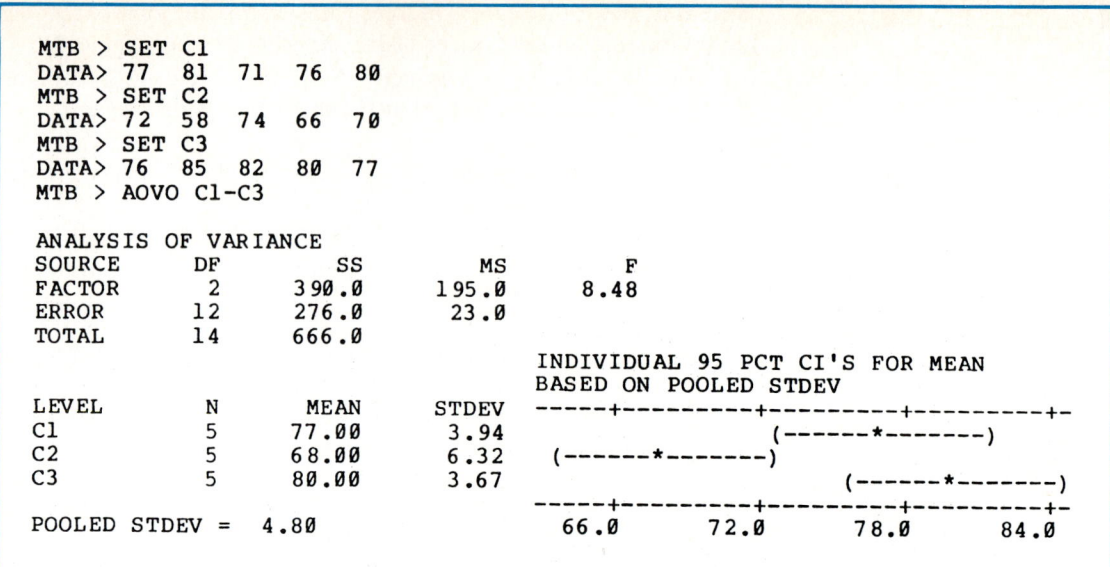

```
MTB > SET C1
DATA> 77   81   71   76   80
MTB > SET C2
DATA> 72   58   74   66   70
MTB > SET C3
DATA> 76   85   82   80   77
MTB > AOVO C1-C3

ANALYSIS OF VARIANCE
SOURCE      DF        SS        MS         F
FACTOR       2      390.0     195.0      8.48
ERROR       12      276.0      23.0
TOTAL       14      666.0
                                   INDIVIDUAL 95 PCT CI'S FOR MEAN
                                   BASED ON POOLED STDEV
LEVEL       N       MEAN      STDEV  -----+---------+---------+---------+-
C1           5      77.00      3.94                 (------*-------)
C2           5      68.00      6.32     (------*-------)
C3           5      80.00      3.67                      (------*-------)
                                       -----+---------+---------+---------+-
POOLED STDEV =     4.80               66.0      72.0      78.0      84.0
```

FIGURE 14.2 *Computer printout for one-way analysis of variance.*

among the population means. We shall not go into this here. Some statistical software also gives the tail probability, namely, the probability of getting a value greater than or equal to the observed value of F when the null hypothesis is true. For our example it is about 0.005.

The method we have discussed here applies only when the sample sizes are all equal, but minor modifications make it applicable also when the sample sizes are not all equal. If the ith sample is of size n_i, the computing formulas for the sums of squares become

Computing formulas for sums of squares (unequal sample sizes)

$$SST = \sum_{i=1}^{k} \sum_{j=1}^{n_i} x_{ij}^2 - \frac{1}{N} \cdot T_{..}^2$$

$$SS(Tr) = \sum_{i=1}^{k} \frac{T_{i\cdot}^2}{n_i} - \frac{1}{N} \cdot T_{..}^2$$

$$SSE = SST - SS(Tr)$$

where $N = n_1 + n_2 + \cdots + n_k$. The only other change is that the total number of degrees of freedom is $N - 1$, and the degrees of freedom for treatments and error are $k - 1$ and $N - k$.

EXAMPLE   A laboratory technician wants to compare the breaking strength of three kinds of thread and originally he had planned to repeat each determination six times. Not having enough time, however, he has to base his analysis on the following results (in ounces):

Thread 1:	18.0	16.4	15.7	19.6	16.5	18.2
Thread 2:	21.1	17.8	18.6	20.8	17.9	19.0
Thread 3:	16.5	17.8	16.1			

Perform an analysis of variance to test at the 0.05 level of significance whether the differences among the sample means are significant.

Solution

1. H_0: $\mu_1 = \mu_2 = \mu_3$
   H_A: The μ's are not all equal.
2. $\alpha = 0.05$
3. Reject the null hypothesis if $F \geq 3.89$, where F is to be determined by an analysis of variance and 3.89 is the value of $F_{0.05}$ for $k - 1 = 3 - 1 = 2$ and $N - k = 15 - 3 = 12$ degrees of freedom; otherwise, accept the null hypothesis or reserve judgment.
4. Substituting $n_1 = 6$, $n_2 = 6$, $n_3 = 3$, $N = 15$, $T_1. = 104.4$, $T_2. = 115.2$, $T_3. = 50.4$, $T.. = 270.0$, and $\sum\sum x^2 = 4{,}897.46$ into the computing formulas for the sums of squares, we get

$$SST = 4{,}897.46 - \frac{1}{15}(270.0)^2 = 37.46$$

$$SS(Tr) = \frac{104.4^2}{6} + \frac{115.2^2}{6} + \frac{50.4^2}{3} - \frac{1}{15}(270.0)^2$$

$$= 15.12$$

and

$$SSE = 37.46 - 15.12 = 22.34$$

Since the degrees of freedom are $k - 1 = 3 - 1 = 2$, $N - k = 15 - 3 = 12$, and $N - 1 = 14$, we then get

$$MS(Tr) = \frac{15.12}{2} = 7.56 \quad MSE = \frac{22.34}{12} = 1.86 \quad \text{and} \quad F = \frac{7.56}{1.86} = 4.06$$

and all these results are summarized in the following analysis-of-variance table:

Source of variation	Degrees of freedom	Sum of squares	Mean square	F
Treatments	2	15.12	7.56	4.06
Error	12	22.34	1.86	
Total	14	37.46		

5. Since $F = 4.06$ exceeds 3.89, the null hypothesis must be rejected; in other words, we conclude that there is a difference in the strength of the three kinds of thread.

This example could also utilize the p-value. The final three steps would be these:

3'. The statistic is

$$F = \frac{MS(Tr)}{MSE}$$

4'. Find

$$F = \frac{7.56}{1.86} = 4.06$$

Since $F = 4.06$ falls between 3.89 and 6.93, the values of $F_{0.05}$ and $F_{0.01}$ for 2 and 12 degrees of freedom, we find that $0.01 < p < 0.05$.

5'. Since the level of significance was specified as $\alpha = 0.05$, we conclude, as before, that the null hypothesis must be rejected. Had the level of significance not been specified, we might simply have stated that $0.01 < p < 0.05$.

EXERCISES

14.6 An experiment is performed to determine which of three golf ball designs, A, B, or C, will give the greatest distance when driven from a tee. Criticize the experiment if:
- (a) one golf pro hits all the design A balls, another hits all the design B balls, and a third hits all the design C balls;
- (b) all the design A balls are hit first, the design B balls next, and the design C balls last.

14.7 A botanist wants to compare three kinds of tulips having, respectively, red, white, and yellow flowers. She has four bulbs of each kind and plants them in a flower bed in the following pattern, where R, W, and Y denote the three colors:

R	R	R	R
W	W	W	W
Y	Y	Y	Y

When the plants reach maturity, she measures their height and performs an analysis of variance. Criticise this experiment and indicate how it might be improved.

★ 14.8 To compare three weight-reducing diets, 5 of 15 persons are randomly assigned to each of the diets. After they have been on these diets for two weeks, a one-way analysis of variance is performed on their weight losses to test the null hypothesis that the three diets are equally effective. It has been claimed that this procedure cannot yield a valid conclusion because the 5 persons who originally weighed the most might be assigned to the same diet. Verify that the probability of this happening by chance is about 0.001.

★ 14.9 With reference to the preceding exercise, suppose that 5 of the 15 persons are randomly assigned to each of the three diets, but it is discovered subsequently that the 5 persons who originally weighed the most are all assigned to the same diet. Should the one-way analysis of variance still be performed?

14.10 Rework part (b) of Exercise 14.1 on page 397 by performing an analysis of variance, using the computing formulas to obtain the necessary sums of squares. Compare the values of F obtained here and in part (a) of Exercise 14.1.

14.11 Rework part (b) of Exercise 14.2 on page 397 by performing an analysis of variance, using the computing formulas to obtain the necessary sums of squares. Compare the values of F obtained here and in part (a) of Exercise 14.2.

14.12 Rework part (b) of Exercise 14.3 on page 397 by performing an analysis of variance, using the computing formulas to obtain the necessary sums of squares. Compare the values of F obtained here and in part (a) of Exercise 14.3.

14.13 The following are the numbers of mistakes made on five occasions by four compositors setting the type for a technical report:

Compositor 1:	10	13	9	11	12
Compositor 2:	11	13	8	16	12
Compositor 3:	10	15	13	11	15
Compositor 4:	15	7	11	12	9

At the 0.05 level of significance, can we conclude that there is a real difference in the numbers of mistakes made in general by the four compositors?

14.14 The following data show the yields of soybeans (in bushels per acre) planted two inches apart on essentially similar plots with the rows 20, 24, 28, and 32 inches apart:

20 in.	24 in.	28 in.	32 in.
23.1	21.7	21.9	19.8
22.8	23.0	21.3	20.4
23.2	22.4	21.6	19.3
23.4	21.1	20.2	18.5
23.6	21.9	21.6	19.1
21.7	23.4	23.8	21.9

Test at the 0.05 level of significance whether the differences among the four sample means can be attributed to chance.

14.15 A large marketing firm owns many photocopy machines, several of each of four different models. Over the last six months, the office manager has tabulated for each machine the average number of minutes per week that it is out of service due to repairs, resulting in the following data:

Model G:	56	61	68	42	82	70
Model H:	74	77	92	63	54	
Model K:	25	36	29	56	44	48	38
Model M:	78	105	89	112	61	

Test at the 0.01 level of significance whether the differences among the four sample means can be attributed to chance.

14.16 The following are the weight losses of certain machine parts due to friction (in milligrams) when used with three different lubricants:

Lubricant X:	10 13 12 10 14 8 12 13
Lubricant Y:	9 8 12 9 8 11 7 6 8 11 9
Lubricant Z:	6 7 7 5 9 8 4 10

Test at the 0.01 level of significance whether the differences among the three sample means are significant.

14.17 To study its performance, a newly designed motorboat was timed over a marked course under various wind and water conditions. Use the following data (in minutes) to test, at the 0.05 level of significance, whether the differences among the three sample means are significant:

Calm conditions:	26 19 16 22
Moderate conditions:	25 27 25 20 18 23
Choppy conditions:	23 25 28 31 26

14.18 With reference to the footnote to page 394, verify that the sum of the treatment effects, the α's, is equal to zero.

14.19 Verify that $kn - 1$, the expression given for the total degrees of freedom in the table on page 402, equals the sum of $k - 1$ and $k(n - 1)$, namely, the sum of the degrees of freedom for treatments and error.

14.20 Verify that the quantity referred to as the POOLED STDEV (the pooled standard deviation) in Figure 14.2 is the square root of the value obtained for $\frac{1}{3}(s_1^2 + s_2^2 + s_3^2)$ on page 395.

★ **14.21** Verify symbolically that for a one-way analysis of variance

(a) $\dfrac{SS(Tr)}{k - 1} = n \cdot s_{\bar{x}}^2;$

(b) $\dfrac{SSE}{k(n - 1)} = \dfrac{1}{k} \cdot \sum_{i=1}^{k} s_i^2$, where s_i^2 is the variance of the ith sample.

14.22 When $k = 2$ and we are testing the null hypothesis $\mu_1 = \mu_2$ against the alternative hypothesis $\mu_1 \neq \mu_2$, the one-way analysis of variance is equivalent to the two-sample t test of Section 11.9. Indeed, it can be shown that F must equal t^2, where F and t are the statistics on which the respective tests are based. Perform an analysis of variance for the data of Exercise 11.106 on page 328 and show that the value obtained for F equals the square of the value obtained for t in Exercise 11.106.

14.4

THE DESIGN OF EXPERIMENTS: BLOCKING

To introduce another important concept in the design of experiments, suppose that a reading comprehension test is given to random samples of eighth graders from each of four schools, and that the results are

School A:	87	70	92
School B:	43	75	56
School C:	70	66	50
School D:	67	85	70

The means of these four samples are 83, 58, 62, and 77, and since the differences among them are very large, it would seem reasonable to conclude that there are some real differences in the average reading comprehension of eighth graders in the four schools. This does not follow, however, from a one-way analysis of variance. We get

Source of variation	Degrees of freedom	Sum of squares	Mean square	F
Treatments	3	1,278	426	2.90
Error	8	1,176	147	
Total	11	2,454		

and since $F = 2.90$ is less than 4.07, the value of $F_{0.05}$ for 3 and 8 degrees of freedom, the null hypothesis (that the population means are all equal) cannot be rejected at the 0.05 level of significance.

The reason for this is that there are not only considerable differences among the four means, but also very large differences among the values within the samples. In the first sample they range from 70 to 92, in the second sample from 43 to 75, in the third sample from 50 to 70, and in the fourth sample from 67 to 85. Giving this some thought, it would seem reasonable to conclude that these differences within the samples may well be due to differences in intelligence, an extraneous factor (we might call it a "nuisance" factor) which was randomized by taking a random sample of eighth graders from each school. Thus, variations due to differences in intelligence were included in the experimental error; this "inflated" the error sum of squares which went into the denominator of the F statistic, and the results were not significant.

To avoid this kind of situation, we could hold the extraneous factor fixed, but this will seldom give us the information we want. In our example, we could limit the study to eighth graders with a scholastic grade-point average (GPA) of 90 or above, but then the results would apply only to eighth graders with a GPA of 90 or above. Another possibility is to vary the known source of variability (the extraneous factor) deliberately over as wide a range as necessary, and to do it in such a way that the variability it causes can be measured and, hence, eliminated from the experimental error. This means that we should plan the experiment in such a way that we can perform a **two-way analysis of variance**, in which the total variability of the data is partitioned into three components attributed, respectively, to treatments (in our example, the four schools), the extraneous factor, and experimental error.

As we shall see later, this can be accomplished in our example by randomly selecting from each school one eighth grader with a low GPA, one eighth grader with a typical GPA, and one eighth grader with a high GPA, where "low," "typical," and "high" are presumably defined in a rigorous way. Suppose, then, that we proceed in this way and get the results shown in the following table:

	Low GPA	Typical GPA	High GPA
School A	71	92	89
School B	44	51	85
School C	50	64	72
School D	67	81	86

What we have done here is called **blocking**, and the three levels of GPA are called **blocks**. In general, blocks are the levels at which we hold an extraneous factor fixed, so that we can measure its contribution to the total variability of the data by means of a two-way analysis of variance. In the scheme we chose for our example, we are dealing with **complete blocks**—they are complete in the sense that each treatment appears the same number of times in each block. There is one eighth grader from each school in each block.

Suppose, furthermore, that the order in which the students are tested may have some effect on the results. If the order is randomized within each block (that is, for each level of GPA), we refer to the design of the experiment as a **randomized block design**.

14.5

TWO-WAY ANALYSIS OF VARIANCE

In this section we shall present the theory of a **two-way analysis of variance** in connection with experiments where blocking is used to reduce the error sum of squares, and, hence, we refer to the two variables under consideration as "treatments" and "blocks." As we shall indicate later, the same kind of analysis applies also to **two-factor experiments**, where both variables are of material concern.

Before we go into any details, let us point out that there are essentially two ways of analyzing such two-variable experiments, and they depend on whether the two variables are independent, or whether there is an **interaction**. Suppose, for instance, that a tire manufacturer is experimenting with different kinds of treads, and that he finds that one kind is especially good for use on dirt roads while another kind is especially good for use on hard pavement. If this is the case, we say that there is an interaction between road conditions and tread design. On the other hand, if each of the treads is affected equally by the different road conditions, we would say that there is no interaction and that the two variables (road conditions and tread design) are independent. In this book we shall study only the case where there is no interaction.

To formulate the hypotheses to be tested in the two-variable case, let us write μ_{ij} for the population mean which corresponds to the ith treatment and the jth block (in our numerical example, the average reading comprehension score of eighth graders with the grade point average level j in the ith school) and express it as

$$\mu_{ij} = \mu + \alpha_i + \beta_j$$

As in the footnote to page 394, μ is the grand mean (the average of all the population means μ_{ij}) and the α_i are the treatment effects (whose sum is zero). Correspondingly, we refer to the β_j as the **block effects** (whose sum is also zero), and write the two null hypotheses we want to test as

$$\alpha_1 = \alpha_2 = \cdots = \alpha_k = 0$$

and

$$\beta_1 = \beta_2 = \cdots = \beta_n = 0$$

The alternative to the first null hypothesis (which in our illustration amounts to the hypothesis that the average reading comprehension of eighth graders is the same in all four schools) is that the treatment effects α_i are not all zero; the alternative to the second null hypothesis (which in our illustration amounts to the hypothesis that the average reading comprehension of eighth graders is the same for all three levels of GPA) is that the block effects β_j are not all zero.

To test the second of the null hypotheses, we need a quantity, similar to the treatment sum of squares, which measures the variation among the block means (58, 72, and 83 for the data on page 409). So, if we let $T_{.j}$ denote the total of all the values in the jth block, substitute it for $T_{i.}$ in the computing formula for $SS(Tr)$ on page 403, sum on j instead of i, and interchange n and k, we obtain, analogous to $SS(Tr)$ the **block sum of squares**

Computing formula for block sum of squares

$$SSB = \frac{1}{k} \cdot \sum_{j=1}^{n} T_{\cdot j}^2 - \frac{1}{kn} \cdot T_{\cdot\cdot}^2$$

In a two-way analysis of variance (with no interaction) we compute SST and $SS(Tr)$ according to the formulas on page 403, SSB according to the formula immediately above, and then we get SSE by subtraction. Since

$$SST = SS(Tr) + SSB + SSE$$

we have

Error sum of squares (two-way analysis of variance)

$$SSE = SST - [SS(Tr) + SSB]$$

Observe that the error sum of squares for a two-way analysis of variance does not equal the error sum of squares for a one-way analysis of variance performed on the same data, even though we denote both with the symbol SSE. In fact, we are now partitioning the error sum of squares for the one-way analysis of variance into two terms: the block sum of squares, SSB, and the remainder which is the new error sum of squares, SSE.

We can now construct the following analysis-of-variance table for a two-way analysis of variance (with no interaction):

Source of variation	Degrees of freedom	Sum of squares	Mean square	F
Treatments	$k - 1$	$SS(Tr)$	$MS(Tr) = \dfrac{SS(Tr)}{k - 1}$	$\dfrac{MS(Tr)}{MSE}$
Blocks	$n - 1$	SSB	$MSB = \dfrac{SSB}{n - 1}$	$\dfrac{MSB}{MSE}$
Error	$(k - 1)(n - 1)$	SSE	$MSE = \dfrac{SSE}{(k - 1)(n - 1)}$	
Total	$kn - 1$	SST		

The mean squares are again the sums of squares divided by their respective degrees of freedom, and the two F values are the mean squares for treatments and blocks divided by the mean square for error. Also, the degrees of freedom for blocks are

$n - 1$ (like those for treatments with n substituted for k), and the degrees of freedom for error are found by subtracting the degrees of freedom for treatments and blocks from $kn - 1$, the total number of degrees of freedom:

$$(kn - 1) - (k - 1) - (n - 1) = kn - k - n + 1$$
$$= (k - 1)(n - 1)$$

Thus, in the significance test for treatments the numerator and denominator degrees of freedom for F are $k - 1$ and $(k - 1)(n - 1)$, and in the significance test for blocks the numerator and denominator degrees of freedom for F are $n - 1$ and $(k - 1)(n - 1)$.

EXAMPLE   With reference to the data on page 409, namely,

	Low GPA	Typical GPA	High GPA
School A	71	92	89
School B	44	51	85
School C	50	64	72
School D	67	81	86

test at the 0.05 level of significance whether the differences among the means obtained for the four schools (treatments) are significant, and also whether the differences among the means obtained for the three levels of GPA (blocks) are significant.

Solution   1. H_0's: $\alpha_1 = \alpha_2 = \alpha_3 = \alpha_4 = 0$
$\beta_1 = \beta_2 = \beta_3 = 0$
H_A's: The treatment effects are not all equal to zero; the block effects are not all equal to zero.
2. $\alpha = 0.05$ for both tests.
3. For treatments, reject the null hypothesis if $F \geq 4.76$, where F is to be determined by a two-way analysis of variance and 4.76 is the value of $F_{0.05}$ for $k - 1 = 4 - 1 = 3$ and $(k - 1)(n - 1) = (4 - 1)(3 - 1) = 6$ degrees of freedom. For blocks, reject the null hypothesis if $F \geq 5.14$, where F is to be determined by a two-way analysis of variance and 5.14 is the value of $F_{0.05}$ for $n - 1 = 3 - 1 = 2$ and $(k - 1)(n - 1) = (4 - 1)(3 - 1) = 6$ degrees of freedom. If either null hypothesis cannot be rejected, accept it or reserve judgment.
4. Substituting $k = 4$, $n = 3$, $T_{1.} = 252$, $T_{2.} = 180$, $T_{3.} = 186$, $T_{4.} = 234$, $T_{.1} = 232$, $T_{.2} = 288$, $T_{.3} = 332$, $T_{..} = 852$, and $\sum\sum x^2 = 63{,}414$ into the computing formulas for the sums of squares, we get

$$SST = 63{,}414 - \tfrac{1}{12}(852)^2$$
$$= 63{,}414 - 60{,}492$$
$$= 2{,}922$$

$$SS(Tr) = \tfrac{1}{3}[252^2 + 180^2 + 186^2 + 234^2] - 60,492$$

$$= 1,260$$

$$SSB = \tfrac{1}{4}[232^2 + 288^2 + 332^2] - 60,492$$

$$= 1,256$$

and

$$SSE = 2,922 - [1,260 + 1,256]$$

$$= 406$$

Since the degrees of freedom are $k - 1 = 4 - 1 = 3$, $n - 1 = 3 - 1 = 2$, $(k - 1)(n - 1) = (4 - 1)(3 - 1) = 6$, and $kn - 1 = 4 \cdot 3 - 1 = 11$, we then get $MS(Tr) = \dfrac{1,260}{3} = 420$, $MSB = \dfrac{1,256}{2} = 628$, $MSE = \dfrac{406}{6} = 67.67$, $F = \dfrac{420}{67.67} = 6.21$ for treatments, and $F = \dfrac{628}{67.67} = 9.28$ for blocks, and all these results are summarized in the following analysis-of-variance table:

Source of variation	Degrees of freedom	Sum of squares	Mean square	F
Treatments	3	1,260	420	6.21
Blocks	2	1,256	628	9.28
Error	6	406	67.67	
Total	11	2,922		

5. For treatments, since $F = 6.21$ exceeds 4.76, the null hypothesis must be rejected; for blocks, since $F = 9.28$ exceeds 5.14, the null hypothesis must be rejected. In other words, we conclude that the average reading comprehension of eighth graders is not the same for the four schools, and also that it is not the same for the three levels of GPA.

Observe that by blocking we were able to show that the differences among the means obtained for the four schools are significant, whereas without blocking, in the experiment described on page 408, the differences among the means obtained for the four schools were not significant.

Needless to say, perhaps, there exists extensive software for a two-way analysis of variance. Many of the programs, but not the one used for the printout shown in Figure 14.3, provide a considerable amount of additional information.

```
MTB > READ COMP SCORES IN Cl SCHOOL IN C2 IQ IN C3
DATA> 71  1  1
DATA> 44  2  1
DATA> 50  3  1
DATA> 67  4  1
DATA> 92  1  2
DATA> 51  2  2
DATA> 64  3  2
DATA> 81  4  2
DATA> 89  1  3
DATA> 85  2  3
DATA> 72  3  3
DATA> 86  4  3
MTB > TWOW Cl C2 C3

ANALYSIS OF VARIANCE ON Cl

SOURCE          DF          SS          MS
C2               3      1260.0       420.0
C3               2      1256.0       628.0
ERROR            6       406.0        67.7
TOTAL           11      2922.0
```

FIGURE 14.3 *Computer printout for two-way analysis of variance of the reading comprehension data.*

As we already pointed out on page 409, a two-way analysis of variance can also be used in connection with **two-factor experiments**, where both variables (factors) are of material concern. It could be used, for example, in the analysis of the following data collected in an experiment designed to test whether the range of a missile flight (in miles) is affected by differences among launchers and also by differences among fuels (see Exercise 14.29 on page 416):

	Fuel 1	Fuel 2	Fuel 3	Fuel 4
Launcher X	45.9	57.6	52.2	41.7
Launcher Y	46.0	51.0	50.1	38.8
Launcher Z	45.7	56.9	55.3	48.1

Note that we used a different format for this table to distinguish between two-factor experiments (where uncontrolled extraneous factors are usually randomized over the entire experiment) and experiments where we deal with treatments and blocks.

Also, when a two-way analysis of variance is used in this way, we usually call the two variables **factor A** and **factor B** (instead of treatments and blocks) and write SSA instead of $SS(Tr)$; we still write SSB, but now B stands for the factor B instead of blocks.

Throughout this section we have assumed exactly one observation in each treatment-by-block combination, so that the experiment consists of exactly kn data values. Experiments with more than one observation in each treatment-by-block combination are done frequently, but we shall not discuss them in this book.

EXERCISES

14.23 To compare the amounts of time that three television stations allot to commercials, a research worker measures the time devoted to commercials in random samples of 15 shows on each station. To her dismay, she discovers that there is so much variation within the samples—for one station the figures vary from 6 to 35 minutes—that it is virtually impossible to get significant results. Is there a way in which she might overcome this obstacle?

14.24 To compare five word processors, A, B, C, D, and E, four persons, 1, 2, 3, and 4, were timed in preparing a certain kind of report on each of the machines. The results (in minutes) are shown in the following table:

	1	2	3	4
A	49.1	48.2	52.3	57.0
B	47.5	40.9	44.6	49.5
C	76.2	46.8	50.1	55.3
D	50.7	43.4	47.0	52.6
E	55.8	48.3	82.6	57.8

Explain why these data should not be analyzed by the method of Section 14.5.

14.25 The following are the cholesterol contents (in milligrams per package) which four laboratories obtained for 6-ounce packages of three very similar diet foods:

	Laboratory 1	Laboratory 2	Laboratory 3	Laboratory 4
Diet food A	3.7	2.8	3.1	3.4
Diet food B	3.1	2.6	2.7	3.0
Diet food C	3.5	3.4	3.0	3.3

Perform a two-way analysis of variance, using the 0.05 level of significance for both tests.

14.26 The calorie content of six different brands of orange juice were determined by three different machines. The numbers below are the determination in calories per 6 fluid ounces:

	Orange juice brand					
	A	B	C	D	E	F
Machine 1	89	97	92	105	100	91
Machine 2	92	101	94	110	100	95
Machine 3	90	98	94	109	99	94

Perform a two-way analysis of variance, using the 0.01 level of significance for both tests.

14.27 A laboratory technician measured the breaking strength of each of five kinds of linen threads by using four different measuring instruments, I_1, I_2, I_3, and I_4, and obtained the following results (in ounces):

	I_1	I_2	I_3	I_4
Thread 1	20.9	20.4	19.9	21.9
Thread 2	25.0	26.2	27.0	24.8
Thread 3	25.5	23.1	21.5	24.4
Thread 4	24.8	21.2	23.5	25.7
Thread 5	19.6	21.2	22.1	21.1

Perform a two-way analysis of variance, using the 0.05 level of significance for both tests.

14.28 The following are eight consecutive weeks' earnings (in dollars) of three salespersons who sell cosmetics door-to-door:

	Week number							
	1	*2*	*3*	*4*	*5*	*6*	*7*	*8*
Salesperson 1	186	222	198	216	210	194	203	219
Salesperson 2	197	203	194	208	220	209	190	205
Salesperson 3	174	213	190	197	233	206	199	221

(a) Use a one-way analysis of variance and the 0.05 level of significance to test the null hypothesis that on the average the three salespersons' weekly earnings are the same.

(b) Retest the null hypothesis of part (a), regarding the successive weeks as blocks.

14.29 With reference to the missile-range data on page 414, perform a two-way analysis of variance, using the 0.05 level of significance for both tests.

14.30 The following are the numbers of defectives produced by four workmen operating, in turn, three different machines:

		Workman			
		B_1	B_2	B_3	B_4
Machine	A_1	35	38	41	32
	A_2	31	40	38	31
	A_3	36	35	43	25

Perform a two-way analysis of variance, using the 0.05 level of significance for both tests.

14.6

THE DESIGN OF EXPERIMENTS: REPLICATION

In Section 14.4 we showed how we can increase the amount of information to be gained from an experiment by blocking, that is, by eliminating the effect of an extraneous factor. Another way to increase the amount of information to be gained from an experiment is to increase the volume of the data. For instance, in the example on page 408 we might increase the size of the samples and give the reading comprehension test to twenty eighth graders from each school instead of three. For more complicated designs, the same thing can be accomplished by executing the entire experiment more than once, and this is called **replication**. With reference to the example on page 409, we might conduct the experiment (select and test twelve eighth graders) in one week, and then replicate (repeat) the entire experiment in the next week.

Conceptually, replication does not present any difficulties, but computationally it does, and that is why we shall not go into this any further. Indeed, if an experiment requiring a two-way analysis of variance is replicated, it will then require a three-way analysis of variance, since replication, itself, may be a source of variation in the data. For instance, this might be the case in our example if it got very hot and humid during the second week, making it difficult for the students to concentrate.

14.7

LATIN SQUARES ★

In Section 14.5 we saw how blocking can be used to eliminate the variability due to one extraneous factor from the experimental error, and, in principle, two or more extraneous sources of variation can be handled in the same way. The only real problem is that this may inflate the size of an experiment beyond practical bounds. Suppose, for instance, that in the example dealing with the reading comprehension of eighth graders we would also like to eliminate whatever variability there may be due to differences in age (12, 13, or 14) and in sex. Allowing for all possible combinations of GPA, age, and sex, we will have to use $3 \cdot 3 \cdot 2 = 18$ different blocks, and if there is to be one eighth grader from each school in each block, we will have to select and test $18 \cdot 4 = 72$ eighth graders in all. If we also wanted to eliminate whatever variability there may be due to ethnic background, for which we might consider five categories, this would raise the required number of eighth graders to $72 \cdot 5 = 360$.

In this section we will show how problems like this can sometimes be resolved, at least in part, by planning experiments as **Latin squares**. At the same time, we also hope to impress upon the reader that it is through proper design that experiments can be made to yield a wealth of information. To give an example, suppose that a market research organization wants to compare four ways of packaging a breakfast food, but it is concerned about possible regional differences in the popularity of the breakfast food, and also about the effects of promoting the breakfast food in different ways. So, it decides to test market the different kinds of packaging in the northeastern, southeastern, northwestern, and southwestern parts of the United States and to promote them with discounts, lotteries, coupons, and two-for-one sales. Thus, there are $4 \cdot 4 = 16$ blocks (combinations of regions and methods of promotion) and it would take $16 \cdot 4 = 64$ market areas (cities) to promote each kind of packaging once within each block within each region of the country. Such a marketing project would be incredibly expensive. Moreover, the test markets must be separated from each other so that the promotion methods do not interfere with each other, and the United States simply does not have 64 sufficiently widely separated test markets. It is of interest to note, however, that with proper planning 16 market areas (cities) will suffice. To illustrate, let us consider the following arrangement, called a **Latin square**, in which the letters A, B, C, and D represent the four kinds of packaging:

	Discounts	Lotteries	Coupons	Two-for-one sales
Northeast	A	B	C	D
Southeast	B	C	D	A
Northwest	C	D	A	B
Southwest	D	A	B	C

In general, a Latin square is a square array of the letters A, B, C, D, ..., of the English (Latin) alphabet, which is such that each letter occurs once and only once in each row and in each column.

The above Latin square, looked upon as an experimental design, requires that discounts be used with packaging A in a city in the Northeast, with packaging B in a city in the Southeast, with packaging C in a city in the Northwest, and with packaging D in a city in the Southwest; that lotteries be used with packaging B in a city in the Northeast, with packaging C in a city in the Southeast, with packaging D in a city in the Northwest, and with packaging A in a city in the Southwest; and so on. Note that each kind of promotion is used once in each region and once with each kind of packaging; each kind of packaging is used once in each region and once with each kind of promotion; and each region is used once with each kind of packaging and once with each kind of promotion. As we shall see, this will enable us to perform an analysis of variance leading to significance tests for all three variables.

The analysis of an $r \times r$ Latin square is very similar to a two-way analysis of variance. The total sum of squares and the sums of squares for rows and columns are calculated in the same way in which we previously calculated SST, $SS(Tr)$, and

SSB, but we must find an extra sum of squares which measures the variability due to the variable represented by the letters $A, B, C, D, \ldots$, namely, a new treatment sum of squares. The formula for this sum of squares is

$$SS(Tr) = \frac{1}{r} \cdot (T_A^2 + T_B^2 + T_C^2 + \cdots) - \frac{1}{r^2} \cdot T_{..}^2$$

where T_A is the total of the observations corresponding to treatment A, T_B is the total of the observations corresponding to treatment B, and so forth. Finally, the error sum of squares is again obtained by subtraction:

$$SSE = SST - [SSR + SSC + SS(Tr)]$$

where *SSR* and *SSC* are the sums of squares for rows and columns.

We can now construct the following analysis-of-variance table for the analysis of an $r \times r$ Latin square:

Source of variation	Degrees of freedom	Sum of squares	Mean square	F
Rows	$r - 1$	SSR	$MSR = \dfrac{SSR}{r-1}$	$\dfrac{MSR}{MSE}$
Columns	$r - 1$	SSC	$MSC = \dfrac{SSC}{r-1}$	$\dfrac{MSC}{MSE}$
Treatments	$r - 1$	SS(Tr)	$MS(Tr) = \dfrac{SS(Tr)}{r-1}$	$\dfrac{MS(Tr)}{MSE}$
Error	$(r-1)(r-2)$	SSE	$MSE = \dfrac{SSE}{(r-1)(r-2)}$	
Total	$r^2 - 1$	SST		

The mean squares are again the sums of squares divided by their respective degrees of freedom, and the three F-values are the mean squares for rows, columns, and treatments divided by the mean square for error. The degrees of freedom for rows, columns, and treatments are all $r - 1$, and, by subtraction, the degree of freedom for error is

$$(r^2 - 1) - (r - 1) - (r - 1) - (r - 1) = r^2 - 3r + 2$$
$$= (r - 1)(r - 2)$$

Thus, for each of the three significance tests the numerator and denominator degrees of freedom for F are $r - 1$ and $(r - 1)(r - 2)$.

EXAMPLE Suppose that in the breakfast-food study referred to on page 418, the market research organization actually gets the data shown in the following table, where the figures are a week's sales in \$10 thousand:

	Discounts	Lotteries	Coupons	Two-for-one Sales
Northeast	A 48	B 38	C 42	D 53
Southeast	B 39	C 43	D 50	A 54
Northwest	C 42	D 50	A 47	B 44
Southwest	D 46	A 48	B 46	C 52

Analyze this Latin square, using the 0.05 level of significance for each test.

Solution   1. H_0's: The row, column, and treatment effects (defined as in the footnote to page 394 and on page 410) are all equal to zero.
H_A's: The respective effects are not all equal to zero.
2. $\alpha = 0.05$ for each test.
3. For rows, columns, or treatments, reject the null hypothesis if $F \geq 4.76$, where the F's are obtained by means of an analysis of variance 4.76 is the value of $F_{0.05}$ for $r - 1 = 4 - 1 = 3$ and $(r - 1)(r - 2) = (4 - 1)(4 - 2) = 6$ degrees of freedom.
4. Substituting $r = 4$, $T_1. = 181$, $T_2. = 186$, $T_3. = 183$, $T_4. = 192$, $T._1 = 175$, $T._2 = 179$, $T._3 = 185$, $T._4 = 203$, $T_A = 197$, $T_B = 167$, $T_C = 179$, $T_D = 199$, $T.. = 742$, and $\sum\sum x^2 = 34{,}756$ into the computing formulas for the sums of squares, we get

$$SST = 34{,}756 - \tfrac{1}{16}(742)^2 = 34{,}756 - 34{,}410.25$$

$$= 345.75$$

$$SSR = \tfrac{1}{4}[181^2 + 186^2 + 183^2 + 192^2] - 34{,}410.25$$

$$= 17.25$$

$$SSC = \tfrac{1}{4}[175^2 + 179^2 + 185^2 + 203^2] - 34{,}410.25$$

$$= 114.75$$

$$SS(Tr) = \frac{1}{4}[197^2 + 167^2 + 179^2 + 199^2] - 34{,}410.25$$

$$= 174.75$$

and

$$SSE = 345.75 - [17.25 + 114.75 + 174.75]$$

$$= 39.00$$

The remainder of the work is shown in the following analysis-of-variance table:

Source of variation	Degrees of freedom	Sum of squares	Mean square	F
Rows (regions)	3	17.25	$\frac{17.25}{3} = 5.75$	$\frac{5.75}{6.5} = 0.9$
Columns (promotion methods)	3	114.75	$\frac{114.75}{3} = 38.25$	$\frac{38.25}{6.5} = 5.9$
Treatments (packaging)	3	174.75	$\frac{174.75}{3} = 58.25$	$\frac{58.25}{6.5} = 9.0$
Error	6	39.00	$\frac{39.00}{6} = 6.5$	
Total	15	345.75		

5. For rows, since $F = 0.9$ is less than 4.76, the null hypothesis cannot be rejected; for columns, since $F = 5.9$ exceeds 4.76, the null hypothesis must be rejected; for treatments, since $F = 9.0$ exceeds 4.76, the null hypothesis must be rejected. In other words, we conclude that differences in promotion and packaging, but not the different regions, affect the breakfast food's sales.

14.8

THE DESIGN OF EXPERIMENTS: SOME FURTHER CONSIDERATIONS ★

There are many other experimental designs besides the ones we have discussed, and they serve a great variety of special purposes. Widely used, for example, are the **incomplete block designs**, which apply when it is impossible to have each treatment in each block.

The need for such a design arises, for instance, when we want to compare 13 kinds of tires but cannot put them all on a test car at the same time. Numbering the tires from 1 to 13, we might use the following experimental design

Test run	Kinds of tires			
1	1	2	4	10
2	2	3	5	11
3	3	4	6	12
4	4	5	7	13
5	5	6	8	1
6	6	7	9	2
7	7	8	10	3
8	8	9	11	4
9	9	10	12	5
10	10	11	13	6
11	11	12	1	7
12	12	13	2	8
13	13	1	3	9

Here there are 13 test runs, or blocks, and since each kind of tire appears together with each other kind of tire once within the same block, the design is referred to as a **balanced incomplete block design**. The fact that each kind of tire appears together with each other kind of tire once within the same block is important; it facilitates the statistical analysis because it assures that we have the same amount of information for comparing each pair of tires. In general, the analysis of incomplete block designs is fairly complicated, and we shall not go into it here, as it has been our purpose only to demonstrate what can be accomplished by the careful design of an experiment.

EXERCISES

⋆ **14.31** Suppose that we want to compare the number of defective pieces produced by four machine operators working on five different machine parts (1, 2, 3, 4, and 5) in three different shifts (I, II, and III).

    (a) If the machine operators are to be regarded as the "treatments," list the blocks (combinations of machine parts and shifts) that would be required if each part is to be produced in each shift.

    (b) How many observations would be required if each machine operator is to work once on each machine part during each shift?

⋆ **14.32** An agronomist wants to compare the yield of 16 varieties of corn, and at the same time study the effect of five different fertilizers and two methods of irrigation. How many test plots must he plant if each variety of corn is to be grown once with each possible combination of fertilizers and methods of irrigation?

⋆ **14.33** A manufacturer of pharmaceuticals wants to market a new cold remedy which is actually a combination of four medications, and he wants to experiment first with two dosages for each medication. If A_L and A_H denote the low and high dosage of medication A, B_L and B_H the low and high dosage of medication B, C_L and C_H the low and high dosage of medication C, and D_L and D_H the low and high dosage of medication D, list the 16 preparations he has to test if each dosage of each medication is to be used once in combination with each dosage of each of the other medications.

⋆ **14.34** Making use of the fact that each of the letters must occur once and only once in each row and each column, complete the following Latin squares:

(a)

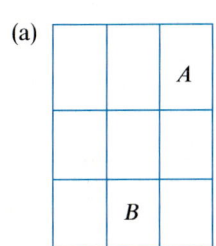

(b)

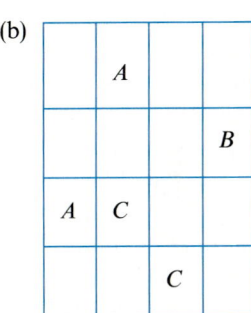

(c)

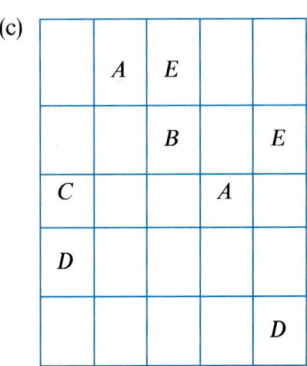

★ 14.35 The sample data in the following 3×3 Latin square are the scores in an American history test obtained by nine college students of various ethnic backgrounds and of various professional interests, who were taught by instructors A, B, and C:

| | Ethnic background | | |
	Mexican	German	Polish
Law	A 75	B 86	C 69
Medicine	B 95	C 79	A 86
Engineering	C 70	A 83	B 93

Analyze this Latin square, using the 0.05 level of significance for each test.

★ 14.36 To compare four different golf-ball designs, A, B, C, and D, each kind was driven by each of four golf pros, P_1, P_2, P_3, and P_4, using once each the four different drivers, D_1, D_2, D_3, and D_4. The distances from the tee to the points where the balls came to rest (in yards) were as shown in the following table:

	D_1	D_2	D_3	D_4
P_1	D 231	B 215	A 261	C 199
P_2	C 234	A 300	B 280	D 266
P_3	A 301	C 208	D 247	B 255
P_4	B 253	D 258	C 210	A 290

Analyze this Latin square, using the 0.05 level of significance for each test.

★ 14.37 Among the nine persons interviewed in a poll, three are Easterners, three are Southerners, and three are Westerners. By profession, three of them are teachers, three are lawyers, and three are doctors, and no two of the same profession come from the same part of the United States. Also, three are Democrats, three are Republicans, and three are Independents, and no two of the same political affiliation are of the same profession or come from the same part of the United States. If one of the teachers is an Easterner and an Independent, another teacher is a Southerner and a Republican, and one of the lawyers is a Southerner and a Democrat, what is the political affiliation of the doctor who is a Westerner? (*Hint*: Construct a 3×3 Latin square; this exercise is a simplified version of a famous problem posed by R. A. Fisher in his classical work *The Design of Experiments*.)

★ 14.38 To test their ability to make decisions under pressure, the nine senior executives of a company are to be interviewed by each of four psychologists. As it takes a psychologist a full day to interview three of the executives, the schedule for the interviews is arranged as follows, where the nine executives are denoted by A, B, C, D, E, F, G, H, and I:

Day	Psychologist	Executives		
March 2	I	B	C	?
March 3	I	E	F	G
March 4	I	H	I	A
March 5	II	C	?	H
March 6	II	B	F	A
March 9	II	D	E	?
March 10	III	D	G	A
March 11	III	C	F	?
March 12	III	B	E	H
March 13	IV	B	?	I
March 16	IV	C	?	A
March 17	IV	D	F	H

Replace the six question marks with the appropriate letters, given that each of the nine executives is to be interviewed together with each of the other executives once and only once on the same day. Note that this will make the arrangement a balanced incomplete block design, which may be important because each executive is tested together with each other executive once under identical conditions.

★ **14.39** A newspaper regularly prints the columns of seven writers but has room for only three in each edition. Complete the following schedule, in which the writers are numbered 1–7, so that each writer's column appears three times per week, and a column of each writer appears together with a column of each other writer once per week:

Day	Writers		
Monday	1	2	3
Tuesday	4		
Wednesday	1	4	5
Thursday	2		
Friday	1	6	7
Saturday	5		
Sunday	2	4	6

14.9
CHECKLIST OF KEY TERMS
(with page references to their definitions)

14.10
REVIEW EXERCISES

14.40 With reference to the golf-ball experiment of Exercise 14.6 on page 406, criticize the experiment if all the design *A* balls are hit from the No. 1 tee, all the design *B* balls are hit from the No. 2 tee, and all the design *C* balls are hit from the No. 3 tee.

14.41 To find the best arrangement of instruments on a control panel of an airplane, three different arrangements were tested by simulating emergency conditions and observing the reaction time required to correct the condition. The reaction times (in tenths of a second) of twelve pilots (randomly assigned to the different arrangements) were as follows:

Arrangement 1:	8	15	10	11
Arrangement 2:	16	11	14	19
Arrangement 3:	12	7	13	8

(a) Calculate $n \cdot s_{\bar{x}}^2$ for these data, and also the mean of the variances of the three samples and the value of F.

(b) Test at the 0.01 level of significance whether the differences among the three sample means can be attributed to chance.

14.42 Rework part (b) of the preceding exercise, using the computing formulas to obtain the necessary sums of squares. Compare the values of F obtained here and in part (a) of the preceding exercise.

★ 14.43 A school has seven department heads who are assigned to seven different committees as shown in the following table:

Committee	Department heads
Textbooks	Dodge, Fleming, Griffith, Anderson
Athletics	Bowman, Evans, Griffith, Anderson
Band	Bowman, Carlson, Fleming, Anderson
Dramatics	Bowman, Carlson, Dodge, Griffith
Tenure	Carlson, Evans, Fleming, Griffith
Salaries	Bowman, Dodge, Evans, Fleming
Discipline	Carlson, Dodge, Evans, Anderson

(a) Verify that this arrangement is a balanced incomplete block design.

(b) If Dodge, Bowman, and Carlson are (in that order) appointed chairpersons of the first three committees, how will the chairpersons of the other four committees have to be chosen so that each of the department heads is chairperson of one of the committees?

14.44 Give the degrees of freedom for treatments, blocks, and errors in a two-way analysis of variance with:
(a) $k = 6$ and $n = 7$;
(b) $k = 6$ and $n = 10$;
(c) $k = 5$ and $n = 4$;
(d) $k = 8$ and $n = 7$.

14.45 The sample data in the following table are the grades in a statistics test obtained by nine college students from three majors who were taught by three different instructors:

	Instructor A	Instructor B	Instructor C
Marketing	77	88	71
Finance	88	97	81
Insurance	85	95	72

Analyze this two-factor experiment, using the 0.05 level of significance.

★ 14.46 The figures in the following 5 × 5 Latin square are the numbers of minutes engines E_1, E_2, E_3, E_4, and E_5, tuned up by mechanics M_1, M_2, M_3, M_4, and M_5, ran with a gallon of fuel A, B, C, D, or E:

	E_1	E_2	E_3	E_4	E_5
M_1	A 31	B 24	C 20	D 20	E 18
M_2	B 21	C 27	D 23	E 25	A 31
M_3	C 21	D 27	E 25	A 29	B 21
M_4	D 21	E 25	A 33	B 25	C 22
M_5	E 21	A 37	B 24	C 24	D 20

Analyze this Latin square, using the 0.01 level of significance for each of the tests of significance.

14.47 The manager of a restaurant wants to determine whether the sales of chicken dinners depend on how this entree is described on the menu. He has three kinds of menus printed, listing chicken dinners among the other entrees, or featuring them as "Chef's Special," or as "Gourmet's Delight," and he intends to use each kind of menu on six different Sundays. Actually, the manager collects only the following data, showing the number of chicken dinners sold on twelve Sundays:

Listed among other entrees	76 94 85 77
Featured as Chef's Special	109 117 102 92 115
Featured as Gourmet's Delight	100 83 102

Perform a one-way analysis of variance at the 0.05 level of significance.

★ 14.48 A food chemist is considering the use of four spices in a recipe for chili with beans. Let A, B, C, and D denote that the four spices are used, while a, b, c, and d denote that they are not used. Then $AbcD$, for example, denotes that the first and fourth spices are used and the second and third are not used.
  (a) List the 16 possible ways in which the food chemist can use or not use these four spices.
  (b) In how many of the 16 ways are the first two spices both used?

★ 14.49 To study the earnings of faculty members in certain areas from speeches, writing, and consulting, a research worker interviews six groups of faculty members:
  (1) four male assistant professors of economics;
  (2) four male professors of economics;
  (3) four female professors of statistics;
  (4) four male associate professors of economics;
  (5) four female assistant professors of statistics;
  (6) four female associate professors of statistics.
If he combines groups 1 and 5, groups 2 and 3, and groups 4 and 6, and then performs an analysis of variance with $k = 3$ and $n = 8$, and gets a significant value of F, to what sources of variation (sex, rank, or subject) can this be attributed?

★ 14.50 With reference to the preceding exercise, explain why there is no way in which the research worker can use the data to test whether there is a significant difference that can be attributed to sex.

14.51 The following data are the amounts of time (in minutes) that it took a person to drive to work, Monday through Friday, along four different routes:

	Monday	Tuesday	Wednesday	Thursday	Friday
Route 1	27	26	26	32	23
Route 2	28	29	27	30	26
Route 3	30	34	31	34	27
Route 4	29	28	31	31	27

Perform a two-way analysis of variance, using the 0.05 level of significance for both tests.

14.11
REFERENCES

The following are some of the many books that have been written on the subject of analysis of variance:

GUENTHER, W. C., *Analysis of Variance.* Englewood Cliffs, N.J.: Prentice-Hall, Inc., 1964.
SNEDECOR, G. W., and COCHRAN, W. G., *Statistical Methods, 6th ed.* Ames, Iowa: Iowa State University Press, 1973.

Problems relating to the design of experiments are also treated in the above books and in

Anderson, V. L., and McLean, R. A., *Design of Experiments: A Realistic Approach.* New York: Marcel Dekker, Inc., 1974.

Box, G. E. P., Hunter, W. G., and Hunter, J. S., *Statistics for Experimenters.* New York: John Wiley & Sons, Inc., 1978.

Cochran, W. G., and Cox, G. M., *Experimental Design, 2nd ed.* New York: John Wiley & Sons, Inc., 1957.

Finney, D. J., *An Introduction to the Theory of Experimental Design.* Chicago: The University of Chicago Press, 1960.

Fleiss, J., *The Design and Analysis of Clinical Experiments,* New York. John Wiley & Sons, Inc., 1986.

Hicks, C. R., *Fundamental Concepts in the Design of Experiments, 2nd ed.* New York: Holt, Rinehart and Winston, 1973.

Romano, A., *Applied Statistics for Science and Industry.* Boston: Allyn and Bacon, Inc., 1977.

A table of Latin squares for $r = 3, 4, 5, \ldots,$ and 12 may be found in the above-mentioned book by W. G. Cochran and G. M. Cox.

Informally, some questions of experimental design are discussed in Chapters 18 and 19 of

Brook, R. J., Arnold, G. C., Hassard, T. H., and Pringle, R. M., eds., *The Fascination of Statistics.* New York: Marcel Dekker, Inc., 1986.

15 REGRESSION

In many statistical investigations, the main goal is to establish relationships which make it possible to predict one or more variables in terms of others. For instance, studies are made to predict the future sales of a product in terms of its price, a person's weight loss in terms of the number of weeks he or she has been on an 800-calories-per-day diet, family expenditures on medical care in terms of family income, the per capita consumption of certain food items in terms of their nutritional value and the amount of money spent advertising them on television, and so forth.

Of course, it would be ideal if we could predict one quantity exactly in terms of another, but this is seldom possible. In most instances we must be satisfied with predicting averages or expected values. For instance, we cannot predict exactly how much money a specific college graduate will earn ten years after graduation, but given suitable data we can predict the average earnings of all college graduates ten years after graduation. Similarly, we can predict the average yield of a variety of wheat in terms of the total rainfall in July, and we can predict the expected grade-point average of a student starting law school in terms of his or her IQ. This problem of predicting the average value of one variable in terms of the known value of another variable (or the known values of other variables) is called the problem of **regression**. This term dates back to Francis Galton (1822–1911), who used it first in connection with a study of the relationship between the heights of fathers and sons.

After a general introduction to curve fitting and the method of least squares in Sections 15.1 and 15.2, questions concerning inferences based on least-squares lines are discussed in Section 15.3, and problems where predictions are based on more than one variable and problems where the relationship is not linear are treated in the two optional sections, Sections 15.4 and 15.5.

15.1
CURVE FITTING

Whenever possible, we try to express, or approximate, relationships between known quantities and quantities that are to be predicted in terms of mathematical equations. This has been very successful in the natural sciences, where it is known, for instance, that at a constant temperature the relationship between the volume, y, and the pressure, x, of a gas is given by the formula

$$y = \frac{k}{x}$$

where k is a numerical constant. Also, it has been shown that the relationship between the size of a culture of bacteria, y, and the length of time, x, it has been exposed to certain environmental conditions is given by the formula

$$y = a \cdot b^x$$

where a and b are numerical constants. More recently, equations like these have also been used to describe relationships in the behavioral sciences, the social sciences, and other fields. For instance, the first of the equations above is often used in economics to describe the relationship between price and quantity demanded, and the second has been used to describe the growth of one's vocabulary or the accumulation of wealth.

Whenever we use observed data to arrive at a mathematical equation which describes the relationship between two variables—a procedure known as **curve fitting**—we must face three kinds of problems:

1. **We must decide what kind of curve, and hence what kind of "predicting" equation we want to use.**
2. **We must find the particular equation which is "best" in some sense.**
3. **We must investigate certain questions regarding the merits of the particular equation, and of predictions made from it.**

The second of these problems will be discussed in some detail in Section 15.2, and the third in Section 15.3.

The first kind of problem is usually decided by direct inspection of the data. We plot the data on ordinary (arithmetic) graph paper, sometimes on special graph paper with special scales (see Section 15.5), and we decide by visual inspection upon the kind of curve (a straight line, a parabola,...) which best describes the overall pattern of the data. There are methods by which this can be done more objectively, but they are fairly advanced and they will not be discussed in this book.

So far as our work here is concerned, we shall concentrate mainly on **linear equations** in two unknowns. They are of the form

$$y = a + bx$$

where a is the **y-intercept** (the value of y for $x = 0$) and b is the **slope** of the line (namely, the change in y which accompanies an increase of one unit in x).[†] Linear equations are useful and important not only because many relationships are actually of this form, but also because they often provide close approximations to relationships which would otherwise be difficult to describe in mathematical terms.

The term "linear equation" arises from the fact that the graph of $y = a + bx$ is a straight line. That is, all pairs of values of x and y which satisfy an equation of the form $y = a + bx$ constitute points which fall on a straight line. In practice, the values of a and b are usually estimated from observed data, and once they have been determined, we can substitute values of x into the equation and calculate the corresponding predicted values of y.

To illustrate, suppose that we are given data on a Midwestern county's yield of wheat (in bushels per acre), y, and its annual rainfall (in inches measured from

---

[†] In other branches of mathematics, linear equations in two unknowns are often written as $y = mx + b$. Of course, such choices of notation are arbitrary, but the one used widely in statistics, $y = a + bx$, has the advantage that it lends itself readily to generalizations. For instance, the equation of a parabola can be written as $y = a + bx + cx^2$.

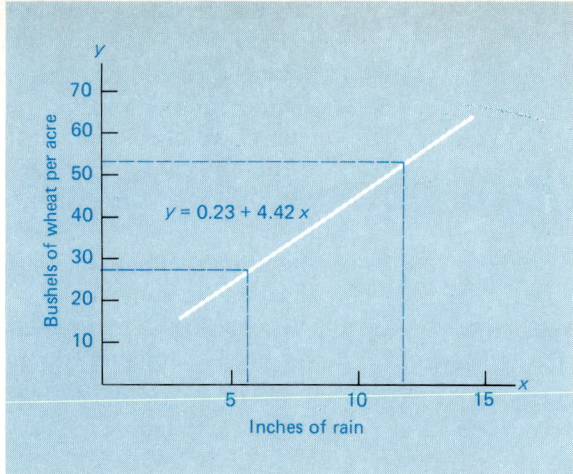

FIGURE 15.1 *Graph of linear equation.*

September through August), *x*, and that by the method which will be discussed in Section 15.2 we obtained the predicting equation

$$y = 0.23 + 4.42x$$

(see Exercise 15.7). The corresponding graph is shown in Figure 15.1, and it should be observed that for any pair of values of *x* and *y* which are such that $y = 0.23 + 4.42x$, we get a point (*x*, *y*) which falls on the line. Substituting *x* = 6, for instance, we find that when there is an annual rainfall of 6 inches, we can expect a yield of $y = 0.23 + 4.42 \cdot 6 = 26.75$ bushels per acre; similarly, substituting *x* = 12, we find that when there is an annual rainfall of 12 inches, we can expect a yield of $y = 0.23 + 4.42 \cdot 12 = 53.27$ bushels per acre. The points (6, 26.75) and (12, 53.27) lie on the straight line of Figure 15.1, and this is true for any other points obtained in the same way.

15.2
THE METHOD OF LEAST SQUARES

Once we have decided to fit a straight line to a given set of data, we face the second kind of problem, namely, that of finding the equation of the particular line which in some sense provides the best possible fit. To show what is involved, let us consider the following sample data obtained in a study of the relationship between the number of years that applicants for certain foreign service jobs studied German in high school or college and the scores which they received in a proficiency test in that language:

Number of years x	Score in test y
3	57
4	78
4	72
2	58
5	89
3	63
4	73
5	84
3	75
2	48

If we plot the points which correspond to these ten pairs of values as in Figure 15.2, we observe that, even though the points do not all fall on a straight line, the overall pattern of the relationship is reasonably well described by the white line. There is

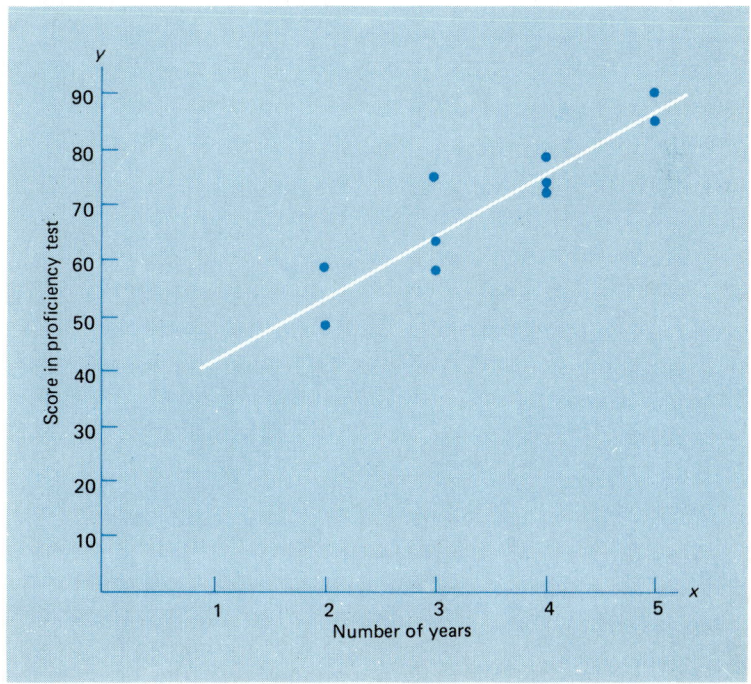

FIGURE 15.2 *Data on number of years studied German and score on test.*

no noticeable departure from linearity in the scatter of the points, so we feel justified in deciding that a straight line is a suitable description of the underlying relationship.

We now face the problem of finding the equation of the line which in some sense provides the best fit to the data and which, it is hoped, will later yield the best possible predictions of y from x. Logically speaking, there is no limit to the number of straight lines which can be drawn on a piece of graph paper. Some of these lines would fit the data so poorly that we could not consider them seriously, but many others would seem to provide more or less "good" fits, and the problem is to find the one line which fits the data "best" in some well-defined sense. If all the points actually fall on a straight line there is no problem, but this is an extreme case which we rarely encounter in practice. In general, we have to be satisfied with a line having certain desirable properties, short of perfection.

The criterion which, today, is used almost exclusively for defining a "best" fit dates back to the early part of the nineteenth century and the work of the French mathematician Adrien Legendre; it is known as the **method of least squares**. As it will be used here, this method requires that the line which we fit to our data be such that the sum of the squares of the vertical distances from the points to the line is a minimum.

To explain why this is done, let us refer to the following data which might represent the numbers of correct answers, x and y, that four students got on two parts of a multiple-choice test:

x	y
4	6
9	10
1	2
6	2

In Figure 15.3 we plotted the corresponding data points, and through them we drew two lines to describe the overall pattern.

If we use the horizontal line in the diagram on the left to "predict" y for the given values of x, we would get $y = 5$ in each case, and the errors of these "predictions" are $6 - 5 = 1$, $10 - 5 = 5$, $2 - 5 = -3$, and $2 - 5 = -3$. In Figure 15.3, they are the vertical distances from the points to the line.

The sum of these errors is $1 + 5 + (-3) + (-3) = 0$, but this is no indication of their size and we find ourselves in a position similar to that on page 71, which led to the definition of the standard deviation. Squaring the errors as we squared the deviations from the mean on page 75, we find that the sum of the squares of the errors is $1^2 + 5^2 + (-3)^2 + (-3)^2 = 44$.

Now let us consider the line in the diagram on the right, which was drawn so that it passes through the points $(1, 2)$ and $(9, 10)$; as can easily be verified, its equation is $y = 1 + x$. Judging by eye, this line seems to provide a much better fit than the horizontal line in the diagram on the left, and if we use it to "predict" y for

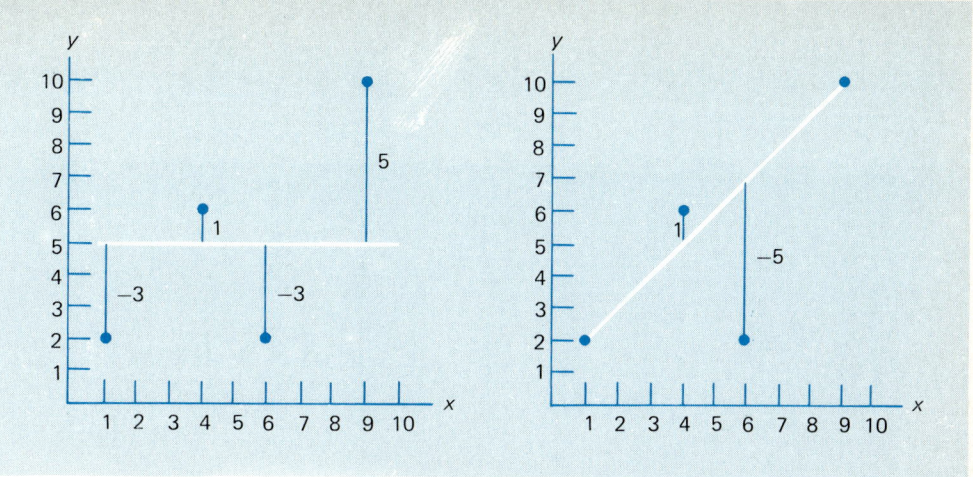

FIGURE 15.3 *Two lines fit to four pairs of data.*

the given values of x, we would get $1 + 4 = 5$, $1 + 9 = 10$, $1 + 1 = 2$, and $1 + 6 = 7$. The errors of these "predictions," which in the figure on the right are the vertical distances from the points to the line, are $6 - 5 = 1$, $10 - 10 = 0$, $2 - 2 = 0$, and $2 - 7 = -5$.

The sum of these errors is $1 + 0 + 0 + (-5) = -4$, which is numerically greater than the sum we obtained for the errors made with the other line of Figure 15.3, but this is of no consequence. The sum of the squares of the errors is now $1^2 + 0^2 + 0^2 + (-5)^2 = 26$, and this is much less than the 44 which we obtained before. In this sense, the line on the right provides a much better fit to the data than the horizontal line on the left.

We can go even one step further, and ask for the equation of the line for which the sum of the squares of the errors is least, and in Exercise 15.12 on page 442 the reader will be asked to verify that its equation is $y = \frac{15}{17} + \frac{14}{17}x$ for our example. We refer to this line as a **least-squares line**.

To show how a least-squares line is actually fit to a set of **data points**, that is, to a set of paired data, let us consider n pairs of numbers (x_1, y_1), $(x_2, y_2), \ldots,$ and (x_n, y_n), which might represent such things as the thrust and the speed of n rockets, the height and weight of n persons, the reading rate and reading comprehension of n students, or the number of persons unemployed in two cities in n years. If we write the equation of the line as $\hat{y} = a + bx$, where the symbol $\hat{y}$ (y-hat) is used to distinguish between an observed value of y and the corresponding value $\hat{y}$ on the line, the least-squares criterion requires that we minimize the sum of the squares of the differences between the y's and the $\hat{y}$'s (see Figure 15.4). This means that we must find the numerical values of the constants a and b appearing in the equation $\hat{y} = a + bx$ for which

$$\sum (y - \hat{y})^2 = \sum [y - (a + bx)]^2$$

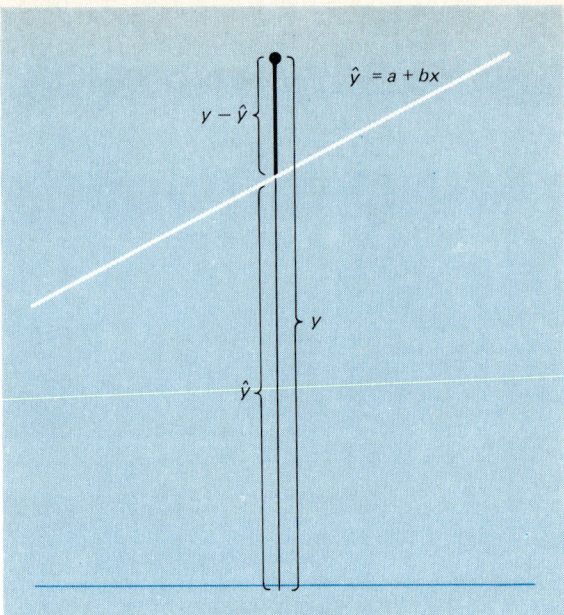

FIGURE 15.4 Least-squares line.

is as small as possible. As it takes calculus or fairly tedious algebra to find the expressions for a and b which minimize $\sum (y - \hat{y})^2$, let us merely state the result that they are given by the solutions for a and b of the following system of two linear equations:

Normal equations

$$\sum y = na + b(\sum x)$$

$$\sum xy = a(\sum x) + b(\sum x^2)$$

In these equations, called the **normal equations**, n is the number of pairs of observations, $\sum x$ and $\sum y$ are the sums of the observed x's and y's, $\sum x^2$ is the sum of the squares of the x's, and $\sum xy$ is the sum of the products obtained by multiplying each x by the corresponding y.

EXAMPLE Fit a least-squares line to the data on page 433, which pertain to the numbers of years that certain applicants for foreign service jobs have studied German in high school or college and the scores which they received in a proficiency test in that language.

Solution The sums needed for substitution into the normal equations are obtained by performing the calculations shown in the following table:

Number of years x	Test score y	x^2	xy
3	57	9	171
4	78	16	312
4	72	16	288
2	58	4	116
5	89	25	445
3	63	9	189
4	73	16	292
5	84	25	420
3	75	9	225
2	48	4	96
35	697	133	2,554

(There are many desk calculators, or hand-held calculators, on which the various sums can be accumulated directly, so that there is no need to fill in all the details. Indeed, on some calculators the values of a and b can be obtained directly by recording the data and then pressing the appropriate buttons.) Substituting $\sum x = 35$, $\sum y = 697$, $\sum x^2 = 133$, $\sum xy = 2{,}554$, and $n = 10$ into the normal equations, we get

$$697 = 10a + 35b$$

$$2{,}554 = 35a + 133b$$

and we must now solve these two simultaneous linear equations for a and b. There are several ways in which this can be done; from elementary algebra we can use either the method of elimination or the method of determinants. Using the first, we get $a = 31.55$ and $b = 10.90$.

As an alternative to this procedure we will develop a number of formulas which will be extremely useful in later calculations that we will make in connection with least-squares problems.

Useful formulas for least-squares calculations

$$\bar{x} = \frac{\sum x}{n} \qquad \bar{y} = \frac{\sum y}{n}$$

$$S_{xx} = \sum x^2 - \frac{(\sum x)^2}{n} \qquad S_{yy} = \sum y^2 - \frac{(\sum y)^2}{n}$$

$$S_{xy} = \sum xy - \frac{(\sum x) \cdot (\sum y)}{n}$$

The formulas for $\bar{x}$ and $\bar{y}$ are familiar. The new formulas S_{xx}, S_{xy}, and S_{yy} will appear in many of the formulas that follow. They give us an easy numerical technique for solving the normal equations.

Solutions of normal equations

$$b = \frac{S_{xy}}{S_{xx}} \qquad a = \bar{y} - b\bar{x}$$

The computational process for finding the least-squares estimates (the solutions of the normal equations) involves several steps. First find the values of the five sums $\sum x$, $\sum y$, $\sum x^2$, $\sum y^2$, and $\sum xy$. Then find the five quantities $\bar{x}$, $\bar{y}$, S_{xx}, S_{xy}, S_{yy}. Finally, use the formulas above to find b and a. Note that you must find b first since a is calculated in terms of b. The quantity S_{yy} has not been used yet, but it will appear in Section 15.3.

EXAMPLE Use these formulas to rework the preceding example.

Solution First substituting $n = 10$ and the various sums obtained in the table on page 437 into the formulas for S_{xx} and S_{xy}, we get

$$S_{xx} = 133 - \frac{1}{10}(35)^2 = 10.5$$

and

$$S_{xy} = 2{,}554 - \frac{1}{10}(35)(697) = 114.5$$

Thus,

$$b = \frac{114.5}{10.5} = 10.90 \quad \text{and} \quad a = \frac{697 - 10.90(35)}{10} = 31.55$$

As the reader may well suspect, there exists computer software for fitting least-squares lines. A printout obtained by means of such a program is shown in Figure 15.5. The values of a and b are shown to be 31.533 and 10.905 in the column headed "COEFFICIENT," and the differences between these values and those obtained above are due to rounding. The bottom part of the printout is not needed here, but we shall refer to it in Chapter 16.

Once we have determined the equation of a least-squares line, we can use it to make predictions.

EXAMPLE Use the least-squares line $\hat{y} = 31.55 + 10.90x$ to predict the proficiency score of an applicant who has studied German in high school or college for two years.

Solution Substituting $x = 2$ into the equation, we get

$$\hat{y} = 31.55 + 10.90(2) = 53.35$$

and this is the best prediction we can make in the least-squares sense.

```
MTB > SET Cl
DATA> 3  4  4  2  5  3  4  5  3  2
MTB > SET C2
DATA> 57  78  72  58  89  63  73  84  75  48
MTB > REGR C2 1 Cl

THE REGRESSION EQUATION IS
C2 = 31.5 + 10.9 Cl


                                   ST. DEV.    T-RATIO =
COLUMN       COEFFICIENT           OF COEF.    COEF/S.D.
                31.533               6.360        4.96
Cl              10.905               1.744        6.25

S = 5.651

R-SQUARED = 83.0 PERCENT
R-SQUARED = 80.9 PERCENT, ADJUSTED FOR D.F.

ANALYSIS OF VARIANCE

  DUE TO      DF           SS         MS=SS/DF
REGRESSION    1          1248.6        1248.6
RESIDUAL      8           255.5          31.9
TOTAL         9          1504.1

                       Y     PRED. Y    ST.DEV.
  ROW        Cl        C2     VALUE      PRED. Y    RESIDUAL    ST.RES.
   9        3.00      75.00   64.25        1.99      10.75      2.03R

R DENOTES AN OBS. WITH A LARGE ST. RES.

DURBIN-WATSON STATISTIC = 2.52
```

FIGURE 15.5 *Computer printout for least-squares line.*

When we make a prediction like this, we cannot really expect that we will always hit the answer right on the nose; in fact, we cannot possibly be right when the answer has to be a whole number, as in our illustration, and our prediction is 53.35. With reference to this example, it would be very unreasonable to expect that every applicant who has studied German for a given number of years will get the same score in the proficiency test; indeed, the data on page 433 show that this is not the case. Thus, to make meaningful predictions based on least-squares lines, we must look upon the values of $\hat{y}$ obtained by substituting given values of x as averages, or expected values. Interpreted in this way, we refer to least-squares lines as **regression lines**, or better as **estimated regression lines**, since the values of a and b are estimated on the basis of sample data. Questions relating to the goodness of these estimates will be discussed in Section 15.3.

In the discussion of this section we have considered only the problem of fitting a straight line to paired data. More generally, the method of least squares can also be used to fit other kinds of curves and to derive predicting equations in more than two unknowns. The problem of fitting some curves other than straight lines by the method of least squares will be discussed briefly in Section 15.5, and a simple example of a predicting equation in more than two unknowns will be given in Section 15.4.

EXERCISES

15.1 A dog which had six hours of obedience training made five mistakes at a dog show, a dog which had twelve hours of obedience training made six mistakes, and a dog which had eighteen hours of obedience training made only one mistake. If we let x denote the number of hours of obedience training and y the number of mistakes, which of the two lines, $y = 10 - \frac{1}{2}x$ or $y = 8 - \frac{1}{3}x$, provides a better fit to the three data points in the sense of least squares?

15.2 With reference to the preceding exercise, is the line which provides the better fit a least-squares line?

15.3 The following table shows how many weeks six persons worked at an automobile inspection station and the number of cars each one inspected between noon and 2 p.m. on a given day:

Number of weeks employed x	Number of cars inspected y
2	13
7	21
9	23
1	14
5	15
12	21

(a) Find the equation of the least-squares line which will enable us to predict y in terms of x.
(b) Use part (a) to estimate how many cars someone who has been working at the inspection station for eight weeks can be expected to inspect during the given two-hour period.
(c) Explain why you would not use the results of part (a) to predict the number of cars that would be inspected during the given two-hour period by someone who has been working at the inspection station for 120 weeks.

15.4 The following table shows the number of hours a runner has run during each of eight consecutive weeks and the corresponding times in which she ran a mile at the end of the week:

Number of hours run	Times of mile (minutes)
13	5.2
15	5.1
18	4.9
20	4.6
19	4.7
17	4.8
21	4.6
16	4.9

(a) Find the equation of the least-squares line which will allow us to predict the runner's time for the mile from the number of hours she ran that week.
(b) Predict how fast the runner would run a mile at the end of a week in which she ran for 18 hours.
(c) Explain why your solution to part (b) is different from the 4.9 minutes which she actually achieved at the end of the third week (during which she ran for 18 hours).

15.5 Market research shows that if a new toy is priced at $1.50 it will yield a profit of $72,000, if it is priced at $2.00 it will yield a profit of $40,000, and if it is priced at $2.40 it will yield a profit of $28,800. Discerning a fairly linear pattern, the toy manufacturer fits a least-squares line and gets $\hat{y} = 142,623 - 48,656x$, where x is the price in dollars and y is the profit in dollars. Can the toy manufacturer conclude that if he literally gives the toys away at ten cents apiece, he can expect a profit of $\hat{y} = 142,623 - 48,656(0.10) = \$137,757$?

15.6 The following data pertain to the chlorine residue in a swimming pool at various times after it has been treated with chemicals:

Number of hours	Chlorine residue (parts per million)
0	2.2
2	1.8
4	1.5
6	1.4
8	1.1
10	1.1
12	0.9

The reading at 0 hour was taken immediately after the chemical treatment was completed.

(a) Fit a least-squares line from which we can predict the chlorine residue in terms of the number of hours since the pool has been treated with chemicals.

(b) Use the equation of the least-squares line to estimate the chlorine residue in the pool 5 hours after it has been treated with chemicals.

(c) Use the equation of the least-squares line to estimate the chlorine residue in the pool 8 hours after it has been treated with chemicals. Why is your answer somewhat different from the 1.1 parts per million actually observed at 8 hours?

(d) Suppose that you learned that the data values given above were all taken over a 12-hour period on the same day. Why might the results parts (a), (b), and (c) be misleading?

15.7 Verify that the equation of the example on pages 431 and 432 can be obtained by fitting a least-squares line to the following data:

Rainfall (inches)	Yield of wheat (bushels per acre)
12.9	62.5
7.2	28.7
11.3	52.2
18.6	80.6
8.8	41.6
10.3	44.5
15.9	71.3
13.1	54.4

15.8 The following are the midterm and final examination grades of eight students in a course in European history:

Midterm	Final examination
75	81
66	57
92	95
86	77
65	71
44	62
60	63
79	84

(a) Find the equation of the least-squares line which will enable us to predict final examination grades in this course from midterm grades.

(b) Use the least-squares equation obtained in part (a) to predict the final examination grade of a student who got a 68 on the midterm test.

15.9 1983 government statistics show that for couples with 0, 1, 2, or 3 children, the relationship between the number of children, x, and family income in dollars, y, is fairly well described by the least-squares line $\hat{y} = 22,000 + 2,500x$. If a childless couple has twins, will this increase their income by $2,500(2) = \$5,000$?

15.10 The following show the improvement (gain in reading speed) of eight students in a speed-reading program, and the number of weeks they have been in the program:

Number of weeks x	Speed gain (words per minute) y
3	86
5	118
2	49
8	193
6	164
9	232
3	73
4	109

(a) Plot the eight data points to verify that it is reasonable to assume that the relationship between average speed gain and time is linear.

(b) Find the equation of the least-squares line which will enable us to predict speed gain from the number of weeks that a student has been in the program.

(c) Use the results of part (b) to predict the speed gain of a student after he or she has been in the program for seven weeks.

15.11 A sociologist exploring the relationship between family size and food bills selected at random six female customers at a supermarket. Each selected customer was asked how many children under the age of 18 lived with her, and she was also asked the number of quarts of milk consumed weekly by her household, on average. The data resulting from this inquiry are these:

Number of children under 18	Weekly milk consumption (quarts)
2	14
2	20
2	9
2	25
2	16
2	14

The sociologist wanted to find the least-squares line which would enable her to predict the milk consumption from the number of children. What computational problems will be encountered?

15.12 Verify for the example on page 434 that the equation of the least-squares line is $y = \frac{15}{17} + \frac{14}{17}x$, and calculate the corresponding sum of the squares of the vertical deviations from the points to the line. How does this sum of squares compare with those obtained for the other two lines on pages 434 and 435?

15.13 A study of the relationship between the IQ's of husbands and wives yielded the least-squares equation $\hat{y} = 48 + 0.5x$, where x is the IQ of the husband and y is the IQ of the wife. Given that this equation is based on the following data:

x	y
90	90
114	102
102	

where one of the y-values has been misplaced, find the missing value of y.

15.14 Raw material used in the production of a synthetic fiber is stored in a place which has no humidity control. Measurements of the relative humidity in the storage place and the moisture content of a sample of the raw material (both in percentages) on 12 days yielded the following results:

Humidity	Moisture content
46	12
53	14
37	11
42	13
34	10
29	8
60	17
44	12
41	10
48	15
33	9
40	13

(a) Fit a least-squares line from which we can predict the moisture content in terms of the humidity.

(b) Use the result of part (a) to estimate the moisture content when the relative humidity is 38 percent.

★ **15.15** Suppose that in the preceding exercise we had wanted to estimate what humidity will yield a moisture content of 10 percent. We could substitute $\hat{y} = 10$ into the equation obtained in part (a) of the preceding exercise and solve for x, but this would not provide an estimate in the least-squares sense. To make the best possible least-squares predictions, or estimates, of humidity in terms of moisture content, we denote the moisture contents by x, the humidity readings by y, and then fit a least-squares line to these data. Find the equation of this line and use it to estimate the humidity which will yield a moisture content of 10 percent.

15.16 When the x's are equally spaced (that is, when the differences between successive values of x are all equal), the calculation of a and b can be simplified by coding the x's by assigning them the values $\ldots, -3,$ $-2, -1, 0, 1, 2, 3, \ldots$, when n is odd, or the values $\ldots, -5, -3, -1, 1, 3, 5, \ldots$, when n is even. With this coding, the sum of the x's is zero, and the formulas for a and b on page 438 become

$$a = \frac{\sum y}{n} \quad \text{and} \quad b = \frac{\sum xy}{\sum x^2}$$

Of course, the equation of the resulting least-squares line expresses y in terms of the coded x's, and we have to account for this when we use the equation to make predictions.

    (a) During its first five years of operation, a company's gross income from sales was 1.4, 2.1, 2.6, 3.5, and 3.7 million dollars. Fit a least-squares line and, assuming that the trend continues, predict the company's gross income from sales during its sixth year of operation.

    (b) At the end of eight successive years, a manufacturing company had 1.0, 1.7, 2.3, 3.1, 3.5, 3.4, 3.9, and 4.7 million dollars invested in plants and equipment. Fit a least-squares line and, assuming that the trend continues, predict the company's investment in plants and equipment at the end of the tenth year.

★ **15.17** Ten bright students from a certain high school were ranked among themselves according to their performance on the PSAT test and then again according to their performance on the SAT test, for which they had extensive tutoring. The results were as follows:

Rank on PSAT test x	Rank on SAT test	Change in rank y
1	4	−3
2	7	−5
3	5	−2
4	3	1
5	8	−3
6	10	−4
7	1	6
8	2	6
9	9	0
10	6	4

    (a) Verify that the equation of the least-squares line fit to these data is $\hat{y} = -5.07 + 0.92x$.

    (b) Substitute $x = 1$ and $x = 10$ into the equation obtained in part (a), and thus determine the expected change in rank of the students who ranked first and tenth on the PSAT test.

    (c) Can we conclude that the extensive tutoring had a positive effect on the lower-scoring students and an adverse effect on the higher-scoring students?

15.18 The following are alternative formulas for the symbolic solution of the two normal equations:

$$a = \frac{(\sum y)(\sum x^2) - (\sum x)(\sum xy)}{n(\sum x^2) - (\sum x)^2}$$

$$b = \frac{n(\sum xy) - (\sum x)(\sum y)}{n(\sum x^2) - (\sum x)^2}$$

    (a) Use these formulas to recalculate the values of a and b obtained for the example on page 437.

    (b) Use these formulas to rework Exercise 15.8.

15.3
REGRESSION ANALYSIS

In the preceding section we used a least-squares line to predict that someone who has studied German in high school or college for two years will score 53.35 in the proficiency test, but even if we interpret the line correctly as a regression line (that is, treat predictions made from it as averages or expected values), there are questions that remain to be answered. For instance,

How good are the values we obtained for _a_ and _b_ in the least-squares equation $\hat{y} = 31.55 + 10.90x$?

How good an estimate is $\hat{y} = 53.35$ of the average score of persons who have had two years of German in high school or college?

After all, $a = 31.55$, $b = 10.90$, and $\hat{y} = 53.35$ for $x = 2$ are only estimates based on sample data, and if we base our work on ten other applicants for such foreign service jobs, the method of least squares would probably yield different values for a and b, and a different value for $\hat{y}$ for $x = 2$. Also, for making predictions, we might ask

Can we give an interval for which we can assert with some degree of confidence that it will contain the score of a person who has studied German in high school or college for two years?

With regard to the first of these questions, we said that $a = 31.55$ and $b = 10.90$ are "only estimates based on sample data," and this implies the existence of corresponding true values, usually denoted by α and β and referred to as the **regression coefficients**. Accordingly, there is also a true regression line $\mu_{y|x} = \alpha + \beta x$, where $\mu_{y|x}$ is the true mean of y for a given value of x. To distinguish between a and α and b and β, we refer to a and b as the **estimated regression coefficients**.

To clarify the idea of a true regression line, let us consider Figure 15.6, where we have drawn the distributions of y for several values of x. With reference to our numerical example, these curves are the distributions of the proficiency scores of persons who have had one, two, or three years of German in high school or college, and to complete the picture we can visualize similar curves for all other values of x within the range of values under consideration. Note that the means of all the distributions of Figure 15.6 lie on the true regression line $\mu_{y|x} = \alpha + \beta x$.

In **linear regression analysis** we assume that the x's are constants, not values of random variables, and that for each value of x the variable to be predicted, y, has a certain distribution (as pictured in Figure 15.6) whose mean is $\alpha + \beta x$. In **normal regression analysis** we assume, furthermore, that these distributions are all normal distributions with the same standard deviation σ.

Based on these assumptions, it can be shown that the estimated regression coefficients a and b, obtained by the method of least squares, are values of random variables having normal distributions with the means α and β and the standard deviations

$$\sigma\sqrt{\frac{1}{n} + \frac{\bar{x}^2}{S_{xx}}} \quad \text{and} \quad \frac{\sigma}{\sqrt{S_{xx}}}$$

The estimated regression coefficients a and b are, however, not statistically independent. Note that both of these standard error formulas require that we estimate σ, the common standard deviation of the normal distributions pictured in Figure 15.6. Otherwise, the x's are assumed to be constants, so there is no problem in determining $\bar{x}$ and S_{xx}. The estimate of σ we shall use is called the **standard error of**

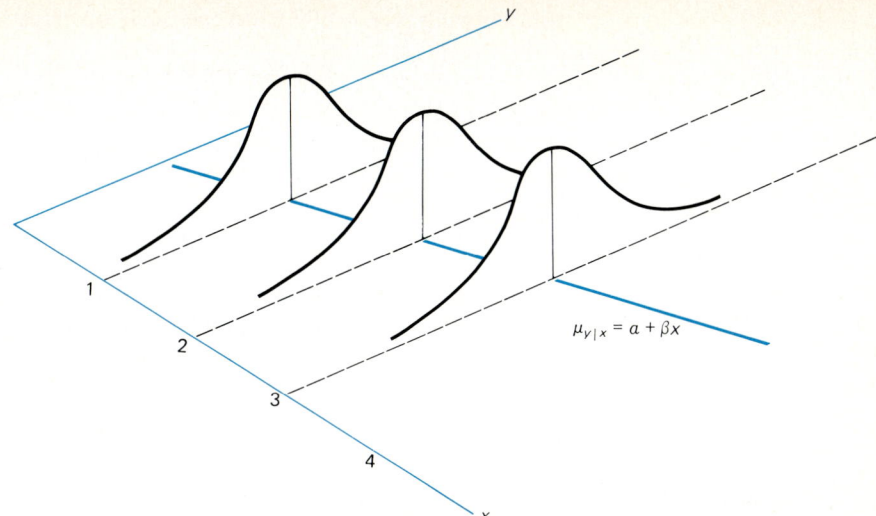

FIGURE 15.6 *Distributions of y for given values of x.*

estimate and it is denoted by s_e. Its formula is

$$s_e = \sqrt{\frac{\sum (y - \hat{y})^2}{n - 2}}$$

where, again, the y's are the observed values of y and the $\hat{y}$'s are the corresponding values on the least-squares line. Observe that s_e^2 is the sum of the squares of the vertical deviations from the points to the line (namely, the quantity which we minimized by the method of least squares) divided by $n - 2$.

The formula above defines s_e, but in practice we calculate its value by means of the computing formula

Standard error of estimate (computing formula)

$$s_e = \sqrt{\frac{S_{yy} - \dfrac{(S_{xy})^2}{S_{xx}}}{n - 2}}$$

The symbol S_{yy} was defined on page 437.

EXAMPLE Find s_e for the least-squares line which we fit to the data pertaining to the numbers of years that certain applicants for foreign service jobs have studied German in high school or college and the scores which they received in a proficiency test in that language.

Solution Since $n = 10$ and we have already shown on page 437 that $b = 10.90$ and $S_{xy} = 114.5$, the only other quantity needed is S_{yy}. Calculating

$$\sum y^2 = 57^2 + 78^2 + \cdots + 48^2$$

$$= 50{,}085$$

and copying $\sum y = 697$ from the table on page 437, we get

$$S_{yy} = 50{,}085 - \tfrac{1}{10}(697)^2 = 1{,}504.1$$

and hence

$$s_e = \sqrt{\frac{1{,}504.1 - (114.5)^2/10.5}{10 - 2}} = 5.651$$

The result we obtained here can also be read from the printout of Figure 15.5.

If we make all the assumptions of normal regression analysis (see page 444), inferences about the regression coefficients α and β can be based on the statistics

Statistics for inferences about regression coefficients

$$t = \frac{a - \alpha}{s_e\sqrt{\dfrac{1}{n} + \dfrac{\bar{x}^2}{S_{xx}}}}$$

$$t = \frac{b - \beta}{s_e/\sqrt{S_{xx}}}$$

whose sampling distributions are t distributions with $n - 2$ degrees of freedom. Note that the quantities in the denominators are estimates of the corresponding standard errors with s_e substituted for σ.

The following example illustrates how we test hypotheses about either of the regression coefficients α and β.

EXAMPLE Suppose that it has been claimed that $\beta = 12.5$ in the example dealing with the applicants for the foreign service jobs and their proficiency in German, and that the whole study was actually made to test this claim at the 0.05 level of significance.

Solution 1. H_0: $\beta = 12.5$
 H_A: $\beta \neq 12.5$
2. $\alpha = 0.05$
3. Reject the null hypothesis if $t \leq -2.306$ or $t \geq 2.306$, where

$$t = \frac{b - \beta}{s_e/\sqrt{S_{xx}}}$$

and 2.306 is the value of $t_{0.025}$ for $10 - 2 = 8$ degrees of freedom; otherwise, accept the null hypothesis or reserved judgment.
4. Since we know from page 438 and the above that $b = 10.90$, $S_{xx} = 10.5$, and $s_e = 5.651$, substitution of these values together with $\beta = 12.5$ yields

$$t = \frac{10.90 - 12.5}{5.651/\sqrt{10.5}} = -0.92$$

5. Since $t = -0.92$ falls on the interval from -2.306 and 2.306, the null hypothesis cannot be rejected; that is, there is no real evidence that the slope is not $\beta = 12.5$.

We could have saved ourselves some work in the preceding example by referring to the printout of Figure 15.5 on page 439. In the column headed "ST. DEV. OF COEF." it shows that the estimated standard error, the quantity which goes into the denominator of the t statistic, is 1.744, so we can write directly

$$t = \frac{10.90 - 12.5}{1.744} = -0.92$$

Tests concerning the regression coefficient α are performed in the same way, except that we use the first, instead of the second, of the two t statistics. In most practical applications, however, the regression coefficient α is not of much interest—it is just the y-intercept, namely, the value of y which corresponds to $x = 0$. In many cases it has no real meaning.

To construct confidence intervals for the regression coefficients α and β, we substitute for the middle term of $-t_{\alpha/2} < t < t_{\alpha/2}$ the appropriate t statistic from page 446. Then, simple algebra leads to the formulas

Confidence limits for regression coefficients

$$a \pm t_{\alpha/2} \cdot s_e \sqrt{\frac{1}{n} + \frac{\bar{x}^2}{S_{xx}}}$$

and

$$b \pm t_{\alpha/2} \cdot \frac{s_e}{\sqrt{S_{xx}}}$$

where the degree of confidence is $(1 - \alpha)100$ percent and $t_{\alpha/2}$ is the entry in Table II for $n - 2$ degrees of freedom.

EXAMPLE   The following data show the average numbers of hours which six students spent on homework per week and their grade-point indexes for the courses they took in a given semester:

Hours spent on homework x	Grade-point index y
15	2.0
28	2.7
13	1.3
20	1.9
4	0.9
10	1.7

Construct a 95 percent confidence interval for β, the amount by which a student in the population sampled can expect to raise his or her grade-point index by studying an extra hour per week.

Solution   First calculating the necessary sums we get $\sum x = 90$, $\sum y = 10.5$, $\sum x^2 = 1{,}694$, $\sum xy = 181.1$, and $\sum y^2 = 20.29$, so that

$$S_{xx} = 1{,}694 - \tfrac{1}{6}(90)^2 = 344$$

$$S_{yy} = 20.29 - \tfrac{1}{6}(10.5)^2 = 1.915$$

and

$$S_{xy} = 181.1 - \tfrac{1}{6}(90)(10.5) = 23.6$$

Then we get

$$b = \frac{23.6}{344} = 0.0686$$

$$s_e = \sqrt{\frac{1.915 - \dfrac{(23.6)^2}{344}}{6 - 2}} = 0.272$$

and since $t_{0.025} = 2.776$ for $6 - 2 = 4$ degrees of freedom, we arrive at the confidence limits

$$0.0686 \pm 2.776 \cdot \frac{0.272}{\sqrt{344}}$$

or 0.0686 ± 0.0407. Rounding to three decimals, we can write the 95 percent confidence interval for β as

$$0.028 < \beta < 0.109$$

This confidence interval is rather wide, and this is due to two things—the very small size of the sample and the variation measured by s_e, namely, the variation among the grade-point indexes of students doing the same amount of homework.

The work we have done here can be simplified greatly by using a printout like that of Figure 15.5 on page 439. For our new example we get the printout shown in Figure 15.7, and we find that $b = 0.0686$ and that the estimate of the standard error, by which we have to multiply $t_{\alpha/2}$, is 0.01467. Using again $t_{0.025} = 2.776$, we thus get the confidence limits

$$0.0686 \pm 2.776(0.01467)$$

or 0.0686 ± 0.0407. This is precisely what we had before.

To answer the second question asked on page 444, the one concerning the estimation, or prediction, of the average value of y for a given value of x, we use a method that is very similar to the one just discussed. Basing our argument on

```
MTB > SET C1
DATA> 15  28  13  20  4  10
MTB > SET C2
DATA> 2.0  2.7  1.3  1.9  0.9  1.7
MTB > REGR C2 1 C1

THE REGRESSION EQUATION IS
C2 = 0.721 + 0.0686 C1

                                        ST. DEV.      T-RATIO =
   COLUMN         COEFFICIENT           OF COEF.      COEF/S.D.
                     0.7209              0.2464         2.93
   C1                0.06860             0.01467        4.68

   S = 0.2720

   R-SQUARED = 84.5 PERCENT
   R-SQUARED = 80.7 PERCENT, ADJUSTED FOR D.F.

   ANALYSIS OF VARIANCE

     DUE TO        DF            SS         MS=SS/DF
   REGRESSION      1           1.6191        1.6191
   RESIDUAL        4           0.2959        0.0740
   TOTAL           5           1.9150


   DURBIN-WATSON STATISTIC = 1.18
```

FIGURE 15.7 *Computer printout for regression example.*

another t statistic, we arrive at the following $(1 - \alpha)100$ percent confidence limits for $\mu_{y|x_0}$, the mean of y when $x = x_0$:

Confidence limits for mean of y when $x = x_0$

$$(a + bx_0) \pm t_{\alpha/2} \cdot s_e \sqrt{\frac{1}{n} + \frac{(x_0 - \bar{x})^2}{S_{xx}}}$$

As before, the number of degrees of freedom is $n - 2$ and the corresponding value of $t_{\alpha/2}$ may be read from Table II.

EXAMPLE Referring again to the data on page 433, suppose that we want to estimate the mean proficiency score of applicants who have had two years of German in high school or college. Construct a 99 percent confidence interval for this mean.

Solution Copying $\sum x = 35$, $S_{xx} = 10.5$, $a + bx_0 = 31.55 + 10.90(2) = 53.35$, and $s_e = 5.651$ from pages 438 and 446, and substituting these values together with $n = 10$. $\bar{x} = \frac{35}{10} = 3.5$, and $t_{0.005} = 3.355$ (for $10 - 2 = 8$ degrees of freedom) into the confidence interval formula, we get

$$53.35 \pm (3.355)(5.651) \cdot \sqrt{\frac{1}{10} + \frac{(2 - 3.5)^2}{10.5}}$$

or 53.35 ± 10.65. Hence, we can write the 99 percent confidence interval for the mean proficiency scores of applicants with two years of German in high school or college as

$$42.70 < \mu_{y|x} < 64.00$$

The third question asked on page 444 differs from the other two in that it does not concern the estimation of a population parameter, but the prediction of a single future observation. The endpoints of an interval for which we can assert with a given degree of confidence that it will contain such an observation are called **limits of prediction**, and the calculation of such limits will answer the third kind of question. Basing our argument on yet another t statistic, we arrive at the following $(1 - \alpha)100$ percent limits of prediction for a value of y when $x = x_0$:

Limits of prediction

$$(a + bx_0) \pm t_{\alpha/2} \cdot s_e \sqrt{1 + \frac{1}{n} + \frac{(x_0 - \bar{x})^2}{S_{xx}}}$$

Again, the number of degrees of freedom is $n - 2$ and the corresponding value of $t_{\alpha/2}$ may be read from Table II.

EXAMPLE Referring again to the example on page 433, find 99 percent limits of prediction for the proficiency score of an applicant who has studied German in high school or college for two years.

Solution Noting that the only difference between the limits above and the confidence limits for $\mu_{y|x_0}$ is that we add 1 to the quantity under the square-root sign, we can immediately write the limits of prediction as

$$53.35 \pm (3.355)(5.66) \cdot \sqrt{1 + \frac{1}{10} + \frac{(2 - 3.5)^2}{10.5}}$$

or

$$53.35 \pm 21.77$$

Thus, the 99 percent limits of prediction are 31.58 and 75.12.

The interval for $\mu_{y|x}$, the mean score of all students who have had two years of German, is considerably shorter than the interval of the previous example, which applies to a prediction for one person.

Let us remind the reader that all these methods are based on the very stringent assumptions of normal regression analysis. Furthermore, if we base more than one inference on the same data, we will run into problems with regard to the levels of significance and/or degrees of confidence. The random variables on which the various procedures are based are clearly not independent.

EXERCISES

15.19 With reference to Exercise 15.3 on page 440, test the null hypothesis $\beta = 1.2$ (namely, the hypothesis that each additional week on the job adds 1.2 to the number of cars a person can be expected to inspect in the given period of time) against the alternative hypothesis $\beta < 1.2$ at the 0.05 level of significance.

15.20 With reference to Exercise 15.6 on page 441, test the null hypothesis $\beta = -0.12$ against the alternative hypothesis $\beta > -0.12$ at the 0.05 level of significance. Also, state in words what hypothesis is being tested.

15.21 The following table shows the assessed values and the selling prices of eight houses, constituting a random sample of all the houses sold recently in a suburban area:

Assessed value (thousands of dollars)	Selling price (thousands of dollars)
4.03	163.4
7.20	218.3
3.25	155.2
4.48	174.0
2.79	148.8
5.16	181.1
8.04	223.2
5.80	192.5

Assessed values are used for local tax purposes and do not necessarily represent actual value.

(a) Fit a least-squares line which will enable us to predict the selling price of a house in that area in terms of its assessed value.
(b) Test the null hypothesis $\beta = 10.3$ against the alternative hypothesis $\beta > 10.3$ at the 0.05 level of significance.

15.22 The following data show the advertising expenses (expressed as a percentage of total expenses) and the net operating profits (expressed as a percentage of

total sales) in a random sample of six drugstores:

Advertising expenses	Net operating profits
1.5	3.6
1.0	2.8
2.8	5.4
0.4	1.9
1.3	2.9
2.0	4.3

(a) Fit a least-squares line which will enable us to predict net operating profits in terms of advertising expenses.
(b) Test the null hypothesis $\beta = 1.60$ against the alternative hypothesis $\beta \neq 1.60$ at the 0.01 level of significance.

15.23 With reference to the Exercise 15.6 on page 441, test the null hypothesis $\alpha = 2.1$ against the alternative hypothesis $\alpha < 2.1$ at the 0.05 level of significance. Also state in words what hypothesis is being tested.

★ **15.24** With reference to the example on page 437 test the null hypothesis $\alpha = 0.85$ against the alternative hypothesis $\alpha < 0.85$ at the 0.05 level of significance, getting the value of a and the estimate of the standard error directly from Figure 15.7.

15.25 The following sample data show the demand for a product (in thousands of units) and its price (in cents) charged in six different market areas:

Price	18	10	14	11	16	13
Demand	9	125	57	90	22	79

(a) Fit a least-squares line from which we can predict the demand for the product in terms of its price.

(b) Construct a 95 percent confidence interval for β and explain in words what quantity is thus being estimated.

15.26 With reference to Exercise 15.14 on page 442, construct a 99 percent confidence interval for β.

15.27 With reference to Exercise 15.22, construct a 98 percent confidence interval for β and state in words what quantity is thus being estimated.

15.28 With reference to Exercise 15.6 on page 441, construct a 95 percent confidence interval for α.

15.29 With reference to Exercise 15.22, construct a 99 percent confidence interval for α.

★ **15.30** With reference to the example on page 433, the one dealing with the language proficiency of applicants to certain foreign service jobs, construct a 99 percent confidence interval for α, getting the value of a and the estimate of the standard error directly from Figure 15.5.

15.31 With reference to Exercise 15.3 on page 440, construct a 95 percent confidence interval for the average number of cars a person will inspect in the given time period if he or she has been working at the inspection station for eight weeks. (See also Exercise 15.34.)

15.32 With reference to Exercise 15.21, construct a 95 percent confidence interval for the average selling price of a house in the given suburban area which has an assessed value of $4,500. (See also Exercise 15.35.)

15.33 With reference to Exercise 15.22, construct a 95 percent confidence interval for the mean net operating profits (expressed as a percentage of total sales) when the advertising expenses are 2.0 percent of total expenses.

15.34 With reference to Exercise 15.3 on page 440, find 95 percent limits of prediction for the number of cars a person will inspect in the given time period if he or she has been working at the inspection station for eight weeks.

15.35 With reference to Exercise 15.21, find 95 percent limits of prediction for the selling price of a house in the given suburban area which has an assessed value of $4,500.

15.36 True or false?
  (a) A confidence interval for the mean of y for a given value of x can be made as narrow as we want by making n sufficiently large.
  (b) The difference between upper and lower limits of prediction can be made as small as we want by making n sufficiently large.

15.37 Verify that the following formula for s_e is equivalent to the one given on page 445.

$$s_e = \sqrt{\frac{S_{xx} \cdot S_{yy} - (S_{xy})^2}{(n-2)S_{xx}}}$$

Also use this alternative formula to recalculate the value of s_e given on page 445.

15.4

MULTIPLE REGRESSION ★[†]

Although there are many problems in which one variable can be predicted quite accurately in terms of another, it stands to reason that predictions should improve if one considers additional relevant information. For instance, we should be able to make better predictions of the performance of newly hired teachers if we consider not only their education, but also their years of experience and their

---

[†] This section is marked optional because the calculations in most of the exercises would be very tedious, in some instances prohibitive, unless suitable computer software is available.

personality. Also, we should be able to make better predictions of a new textbook's success if we consider not only the quality of the work, but also the potential demand and the competition.

Many mathematical formulas can serve to express relationships among more than two variables, but most commonly used in statistics (partly for reasons of convenience) are linear equations of the form

$$y = b_0 + b_1 x_1 + b_2 x_2 + \cdots + b_k x_k$$

Here y is the variable which is to be predicted, $x_1, x_2, \ldots$, and x_k are the k known variables on which predictions are to be based, and $b_0, b_1, b_2, \ldots$, and b_k are numerical constants which must be determined from observed data.

To illustrate, consider the following equation which was obtained in a study of the demand for different meats:

$$\hat{y} = 3.489 - 0.090 x_1 + 0.064 x_2 + 0.019 x_3$$

Here y denotes the total consumption of federally inspected beef and veal in millions of pounds, x_1 denotes a composite retail price of beef in cents per pound, x_2 denotes a composite retail price of pork in cents per pound, and x_3 denotes income as measured by a certain payroll index. With this equation, we can predict the total consumption of federally inspected beef and veal on the basis of specified values of x_1, x_2, and x_3.

The problem of determining a linear equation in more than two variables which best describes a given set of data is that of finding numerical values for $b_0, b_1, b_2, \ldots$, and b_k. This is usually done by the method of least squares; that is, we minimize the sum of squares $\sum (y - \hat{y})^2$, where as before the y's are the observed values and the $\hat{y}$'s are the values calculated by means of the linear equation. In principle, the problem of determining the values of $b_0, b_1, b_2, \ldots$, and b_k is the same as it is in the two-variable case, but manual solutions may be very tedious because the method of least squares requires that we solve as many normal equations as there are unknown constants $b_0, b_1, b_2, \ldots$, and b_k. For instance, when there are two independent variables x_1 and x_2, and we want to fit the equation $y = b_0 + b_1 x_1 + b_2 x_2$, we must solve the three normal equations

Normal equations (two independent variables)

$$\sum y = n \cdot b_0 + b_1(\sum x_1) + b_2(\sum x_2)$$
$$\sum x_1 y = b_0(\sum x_1) + b_1(\sum x_1^2) + b_2(\sum x_1 x_2)$$
$$\sum x_2 y = b_0(\sum x_2) + b_1(\sum x_1 x_2) + b_2(\sum x_2^2)$$

Here $\sum x_1 y$ is the sum of the products obtained by multiplying each given value of x_1 by the corresponding value of y, $\sum x_1 x_2$ is the sum of the products obtained by multiplying each given value of x_1 by the corresponding value of x_2, and so on.

The following data show the number of bedrooms, the number of baths, and the prices at which eight one-family houses sold recently in a certain community:

Number of bedrooms x_1	Number of baths x_2	Price (dollars) y
3	2	78,800
2	1	74,300
4	3	83,800
2	1	74,200
3	2	79,700
2	2	74,900
5	3	88,400
4	2	82,900

Find a linear equation which will enable us to predict the average sales price of a one-family house in the given community in terms of the number of bedrooms and the number of baths.

Solution The quantities needed for substitution into the three normal equations are $n = 8$, $\sum x_1 = 25$, $\sum x_2 = 16$, $\sum y = 637{,}000$, $\sum x_1^2 = 87$, $\sum x_1 x_2 = 55$, $\sum x_2^2 = 36$, $\sum x_1 y = 2{,}031{,}100$, and $\sum x_2 y = 1{,}297{,}700$, and we get

$$637{,}000 = 8b_0 + 25b_1 + 16b_2$$

$$2{,}031{,}100 = 25b_0 + 87b_1 + 55b_2$$

$$1{,}297{,}700 = 16b_0 + 55b_1 + 36b_2$$

We could solve these equations by the method of elimination or by using determinants, but in view of the rather tedious calculations, such work is nowadays left to computers. So, let us refer to the printout of Figure 15.8. Here we find in the column headed "COEFFICIENT" that $b_0 = 65{,}191.7$, $b_1 = 4{,}133.3$, and $b_2 = 758.3$, and in the line immediately above the coefficients that, after rounding, the least-squares equation becomes

$$\hat{y} = 65{,}192 + 4{,}133x_1 + 758x_2$$

This equations tells us that (in the given community and at the time the study was made) each extra bedroom adds on the average $4,133, and each bath $758, to the sales price of a house.

EXAMPLE Based on the result of the preceding example, predict the sales price of a three-bedroom house with two baths.

Solution Substituting $x_1 = 3$ and $x_2 = 2$ into the equation obtained above, we get

$$\hat{y} = 65{,}192 + 4{,}133(3) + 758(2)$$

$$= \$79{,}107$$

or approximately $79,100.

```
MTB > SET C1
DATA> 3   2   4   2   3   2   5   4
MTB > SET C2
DATA> 2   1   3   1   2   2   3   2
MTB > SET C3
DATA> 78800   74300   83800   74200   79700   74900   88400   82900
MTB > REGR C3 2 C1 C2


THE REGRESSION EQUATION IS
C3 = 65192 + 4133 C1 + 758 C2


                                           ST. DEV.      T-RATIO =
          COLUMN         COEFFICIENT        OF COEF.     COEF/S.D.
                          65191.7            418.0        155.96
          C1              4133.3             228.6         18.08
          C2               758.3             340.5          2.23


          S = 370.4


          R-SQUARED = 99.6 PERCENT
          R-SQUARED = 99.5 PERCENT, ADJUSTED FOR D.F.
```

FIGURE 15.8 *Computer printout for multiple regression problem.*

Although we shall not go into this here, let us point out that most printouts like that of Figure 15.8 also provide information which makes it easy to test hypotheses about the true regression coefficients (the quantities estimated by b_0, b_1, b_2, ...) or to construct confidence intervals.

EXERCISES

★ 15.38 The following are data on the ages and incomes of a random sample of five executives working for a large multinational company, and the number of years each went to college:

Age x_1	Years college x_2	Income (dollars) y
38	4	81,700
46	0	73,300
39	5	89,500
43	2	79,800
32	4	69,900

(a) Fit an equation of the form $y = b_0 + b_1x_1 + b_2x_2$ to the given data.
(b) Use the equation obtained in part (a) to estimate the average income of 39-year-old executives with four years of college who work for this company.

★ 15.39 The following are sample data provided by a moving company on the weights of six shipments, the distances they were moved, and the damage that was incurred:

Weight (thousands of pounds) x_1	Distance (thousands of miles) x_2	Damage (dollars) y
4.0	1.5	160
3.0	2.2	112
1.6	1.0	69
1.2	2.0	90
3.4	0.8	123
4.8	1.6	186

Hardness (Rockwell 30-T) y	Copper content (percent) x_1	Annealing temperature (degrees F) x_2
78.9	0.02	1,000
55.2	0.02	1,200
80.9	0.10	1,000
57.4	0.10	1,200
85.3	0.18	1,000
60.7	0.18	1,200

(a) Fit an equation of the form $y = b_0 + b_1 x_1 + b_2 x_2$ to the given data.

(b) Use the equation obtained in part (a) to predict the damage when a shipment weighing 2,400 pounds is moved 1,200 miles.

(c) Repeat part (b) for a shipment weighing 2,400 pounds and moving 1,500 miles.

(d) Repeat part (b) for a shipment weighing 3,600 pounds and moving 1,200 miles.

(e) Repeat part (b) for a shipment weighing 3,600 pounds and moving 1,500 miles.

★ 15.40 The following data were collected to determine the relationship between two processing variables and the hardness of a certain kind of steel:

(a) Fit an equation of the form $y = b_0 + b_1 x_1 + b_2 x_2$ to the given data.

(b) Use the equation obtained in part (a) to predict the hardness of the steel when its copper content is 0.14 percent and the annealing temperature is 1,100 degrees F.

★ 15.41 When the x_1's and/or the x_2's are equally spaced, the calculation of $b_0, b_1, b_2, \ldots,$ can usually be simplified by using coding like that of Exercise 15.16 on page 443. Rework the preceding exercise after coding the three x_1-values -1, 0, and 1, and the two x_2-values -1 and 1.

15.5

NONLINEAR REGRESSION ★

When data depart more or less widely from linearity, we must consider fitting some curve other than a straight line. In this section we shall first give two cases where the relationship is not linear but the method of Section 15.2 can nevertheless be employed; then we shall give an example of **polynomial curve fitting** by fitting a **parabola** whose equation is

$$y = a + bx + cx^2$$

It is common practice to plot paired data on various kinds of graph paper, to see whether there are scales for which the points fall close to a straight line. Of course, when this is the case for ordinary graph paper, we proceed as in Section 15.2. If it is the case when we use **semilog paper** (with equal subdivisions for

x and a logarithmic scale for y, as shown in Figure 15.10), this indicates that an **exponential curve** will provide a good fit. The equation of such a curve is

$$y = a \cdot b^x$$

or in logarithmic form

$$\log y = \log a + x(\log b)$$

which is a linear equation in x and $\log y$, where "log" stands for logarithm to the base 10. Observe that if we write A, B, and Y for $\log a$, $\log b$, and $\log y$, the equation becomes $Y = A + Bx$, which is the usual equation of a straight line.

For finding an exponential curve, we can simply apply the procedure on pages 437 and 438 to the problem expressed as $Y = A + Bx$.

EXAMPLE The following are data on a company's net profits during the first six years that it has been in business:

Year	Net profit (thousands of dollars)
1	112
2	149
3	238
4	354
5	580
6	867

Figure 15.9 on page 458, where these data are plotted on ordinary graph paper shows that the relationship is not linear, but Figure 15.10, where a logarithmic scale is used for the y's, shows that the overall pattern is remarkably "straightened out." Thus, fit an exponential curve to the given data.

Solution Obtaining the logarithms of the y's with a calculator, or perhaps looking them up in Table XIV, we find that:

x	y	$Y = \log y$
1	112	2.0492
2	149	2.1732
3	238	2.3766
4	354	2.5490
5	580	2.7634
6	867	2.9380

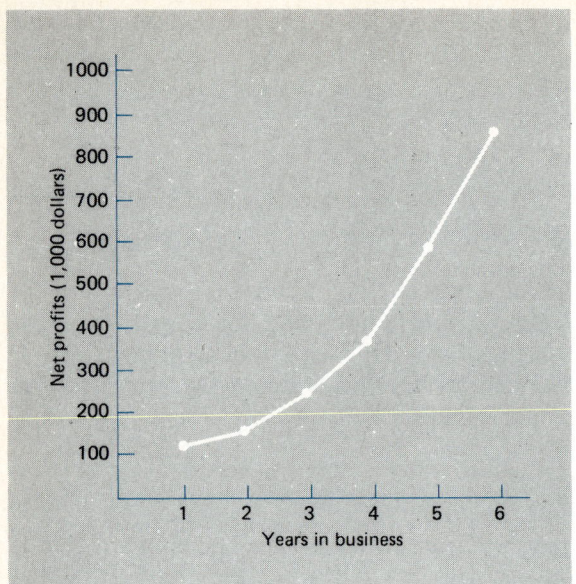

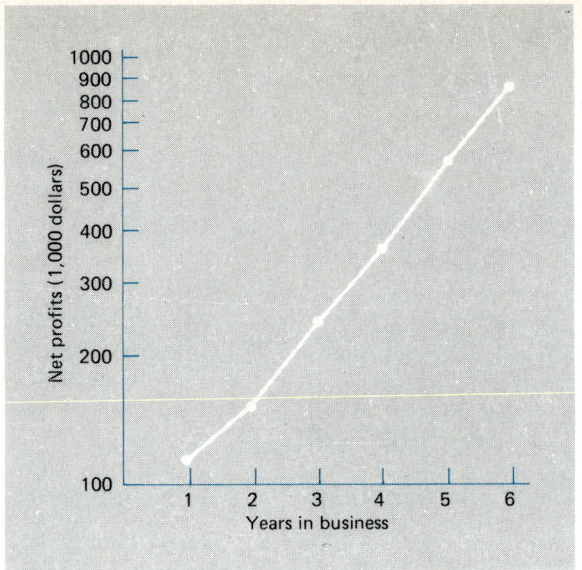

FIGURE 15.9 *Net profits plotted on ordinary graph paper.*   FIGURE 15.10 *Net profits plotted on semi-log paper.*

Then we note that $n = 6$, and we find $\sum x = 21$, $\sum x^2 = 91$, $\sum Y = 14.8494$, and $\sum xY = 55.1664$. These calculations are found using $Y = \log y$. This enables us to find

$$\bar{x} = 3.5, \qquad \bar{Y} = 2.4749,$$

$$S_{xx} = 91 - \frac{(21)^2}{6} = 17.5, \quad \text{and}$$

$$S_{xY} = 55.1664 - \frac{(21)(14.8494)}{6} = 3.1935$$

Then
$$B = \frac{S_{xY}}{S_{xx}} = 0.1825$$

and
$$A = \bar{Y} - B\bar{x} = 2.4749 - (0.1825)(3.5) = 1.8362$$

The equation which describes the relationship is then

$$\hat{Y} = \log \hat{y} = 1.8362 + 0.1825x$$

Since 1.8362 and 0.1825 are the estimates corresponding to $\log a$ and $\log b$, we find by taking antilogarithms that $a = 68.58$ and $b = 1.52$. Thus, the equation of the exponential curve which best describes the relationship between the company's net profit and the number of years it has been in business is given by

$$\hat{y} = 68.58(1.52)^x$$

where $\hat{y}$ is in thousands of dollars.

Once we have fit an exponential curve to a set of paired data, we can predict a future value of y by substituting into its equation the corresponding value of x, but it is usually more convenient to substitute x into the logarithmic form of the equation, namely, $\log \hat{y} = \log a + x(\log b)$.

EXAMPLE With reference to the preceding example, predict the company's net profit for the eighth year that it will have been in business.

Solution Substituting $x = 8$ into the logarithmic form of the exponential curve, we get

$$\log \hat{y} = 1.8362 + 8(0.1825)$$

$$= 3.2962$$

and hence $\hat{y} = 1,980$ or $\$1,980,000$.

If points representing paired data fall close to a straight line when plotted on **log-log paper** (with logarithmic scales for both x and y), this indicates that an equation of the form

$$y = a \cdot x^b$$

will provide a good fit. In the logarithmic form, the equation of such a **power function** is

$$\log y = \log a + b(\log x)$$

which is a linear equation in $\log x$ and $\log y$. (Writing A, X, and Y for $\log a$, $\log x$, and $\log y$, the equation becomes $Y = A + bX$, which is the usual equation of a straight line.) For finding a power curve, we can simply apply the procedure on pages 437 and 438 to the problem expressed as $Y = A + bX$. The work in fitting a power function is similar to that required to fit an exponential curve, and we shall not illustrate it here. In Exercises 15.45 and 15.46 on page 462 the reader will find data to which the method can be applied.

When the values of y first increase and then decrease, or first decrease and then increase, a **parabola** having the equation

$$y = a + bx + cx^2$$

will often provide a good fit. This equation can also be written as

$$y = b_0 + b_1 x + b_2 x^2$$

to conform with the notation of Section 15.4, and we can thus look upon parabolas as linear equations in the two unknowns $x_1 = x$ and $x_2 = x^2$. Thus, fitting a parabola to a set of paired data is nothing new—we simply use the method of Section 15.4.

EXAMPLE   The following are data on the drying time of a varnish and the amount of a certain chemical additive:

Amount of additive (grams) x	Drying time (hours) y
1	7.2
2	6.7
3	4.7
4	3.7
5	4.7
6	4.2
7	5.2
8	5.7

(a) Fit a parabola which, as would appear from Figure 15.11, should be the right kind of curve to fit to the given data.

(b) Use the result of part (a) to predict the drying time of the varnish when 6.5 grams of the chemical are added.

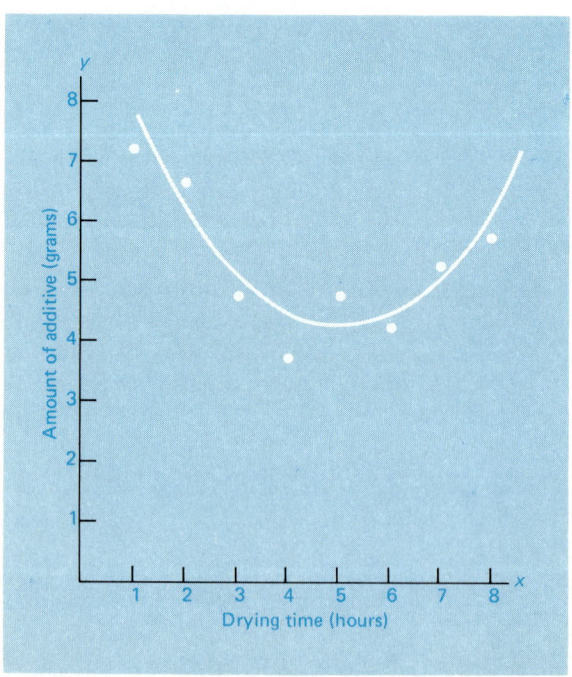

FIGURE 15.11 *Parabola fitted to varnish-drying-time data.*

Solution (a) Using a computer to fit a linear equation of the form $\hat{y} = b_0 + b_1 x_1 + b_2 x_2$ to the following data

$x_1 = x$	$x_2 = x^2$	y
1	1	7.2
2	4	6.7
3	9	4.7
4	16	3.7
5	25	4.7
6	36	4.2
7	49	5.2
8	64	5.7

we get the printout shown in Figure 15.12. Here, the coefficients are $b_0 = 9.2446$, $b_1 = -2.0149$, and $b_2 = 0.19940$, and if we round to two decimals we can write the equation of the parabola as

$$\hat{y} = 9.24 - 2.01x + 0.20x^2$$

(b) Substituting $x = 6.5$ into the equation, we get

$$\hat{y} = 9.24 - 2.01(6.5) + 0.20(6.5)^2$$

$$= 4.62 \text{ hours}$$

```
MTB > SET C1
DATA> 1  2  3  4  5  6  7  8
MTB > LET C2=C1**2
MTB > SET C3
DATA> 7.2  6.7  4.7  3.7  4.7  4.2  5.2  5.7
MTB > REGR C3 2 C1 C2

THE REGRESSION EQUATION IS
C3 = 9.24 - 2.01 C1 + 0.199 C2

                                     ST. DEV.      T-RATIO =
COLUMN        COEFFICIENT            OF COEF.      COEF/S.D.
                 9.2446             0.7645          12.09
C1              -2.0149             0.3898          -5.17
C2               0.19940            0.04228          4.72

S = 0.5480

R-SQUARED = 85.3 PERCENT
R-SQUARED = 79.4 PERCENT, ADJUSTED FOR D.F.
```

FIGURE 15.12 *Computer printout for fitting parabola.*

On page 459 we introduced parabolas as curves which bend once—that is, their values first increase and then decrease, or first decrease and then increase. For patterns which bend more than once, **polynomial equations** of higher degree, such as

$$y = a + bx + cx^2 + dx^3 \qquad \text{or} \qquad y = a + bx + cx^2 + dx^3 + ex^4$$

can be fit by the same technique. In practice, we often use sections of such curves, especially parts of parabolas, when there is only a slight curvature in the pattern we want to describe (see Exercise 15.50 below).

EXERCISES

★ **15.42** Fit an exponential curve to the following data on the percentage of the radial tires made by a certain manufacturer that are still usable after having been driven the given numbers of miles:

Miles driven (thousands) x	Percentage usable y
1	97.2
2	91.8
5	82.5
10	64.4
20	41.0
30	29.9
40	17.6
50	11.3

Also estimate the percentage of the tires we can expect to be usable after they have been driven for 25,000 miles.

★ **15.43** The following data pertain to the growth of a colony of bacteria in a culture medium:

Days since inoculation x	Bacteria count (thousands) y
2	112
4	148
6	241
8	363
10	585

Fit an exponential curve and use it to estimate the bacteria count at the end of the fifth day.

★ **15.44** When the x's are equally spaced, the calculations needed to fit an exponential curve can be simplified appreciably by using coding like that of Exercise 15.16 on page 443.
  (a) Coding the years -5, -3, -1, 1, 3, and 5, rework the illustration on page 457.
  (b) What coded value of x do we have to substitute into the least-squares equation obtained in part (a), if we want to predict the company's profit for the eighth year?

★ **15.45** The following data pertain to the demand for a product (in thousands of units) and its price (in cents) charged in five different market areas:

Price	20	16	10	11	14
Demand	22	41	120	89	56

Fit a power function and use it to estimate the demand when the price of the product is 12 cents.

★ **15.46** The following data pertain to the volume of a gas (in cubic inches) and its pressure (in pounds per square inch), when the gas is compressed at a constant temperature:

Volume x	Pressure y
50	16.0
30	40.1
20	78.0
10	190.5
5	532.2

Fit a power function and use it to estimate the pressure of this gas when it is compressed to a volume of 15 cubic inches.

★ **15.47** The following are data on the amount of fertilizer applied to the soil, x in pounds per square foot, and the yield of a certain food crop, y in pounds per square yard:

x	y
0.5	32.0
1.1	34.3
2.2	15.7
0.2	20.8
1.6	33.5
2.0	21.5

(a) Plot these data to verify that it is reasonable to fit a parabola.
(b) Fit a parabola by the method of least squares.
(c) Use the equation obtained in part (a) to predict the yield when 1.5 pounds of the fertilizer are applied per square foot.

★ **15.48** In the years 1978–1983 the production of uranium in the United States was 18,490, 18,730, 21,850, 19,240, 13,430, and 10,600 short tons (according to the 1985 *Statistical Abstract of the United States*). Code the years -5, -3, -1, 1, 3, and 5, and fit a parabola by the method of least squares.

★ **15.49** In the years 1977–1983 there were 3.1, 3.2, 3.2, 3.0, 3.2, 3.8, and 4.1 million new passports issued in the United States (according to the 1985 *Statistical Abstract of the United States*). Code the years -3, -2, -1, 0, 1, 2, and 3, and fit a parabola by the method of least squares.

★ **15.50** Fit a parabola to the data of Exercise 15.45 and use it to estimate the demand when the price of the product is 12 cents.

★ **15.51** Market research shows that weekly sales of a new candy bar will be related to its price as follows:

Price (cents)	Weekly sales (number of bars)
50	232,000
55	194,000
60	169,000
65	157,000

Finding that the parabola

$$\hat{y} = 1,130,000 - 28,000x + 200x^2$$

provides an excellent fit, the person conducting the study substitutes $x = 85$; gets $y = 195,000$, and then predicts that weekly sales will total 195,000 if the candy is priced at 85 cents. Comment on this argument.

15.6

CHECKLIST OF KEY TERMS
(with page references to their definitions)

15.7
REVIEW EXERCISES

15.52 The following are the numbers of hours which ten persons (interviewed as part of a sample survey) spent watching television, x, and reading books or magazines, y, per week:

x	y
18	7
25	5
19	1
12	5
12	10
27	2
15	3
9	9
12	8
18	4

(a) Fit a least-squares line which will enable us to predict y in terms of x.
(b) If a person spends 20 hours watching television per week, predict how many hours he or she will spend reading books or magazines.

15.53 With reference to the preceding exercise, construct a 95 percent confidence interval for the regression coefficient β.

15.54 With reference to Exercise 15.52, construct a 99 percent confidence interval for the number of hours a person who watches 20 hours of television per week can be expected to read books or magazines per week.

★ 15.55 For the last five months, the ice cream sales, in thousands of gallons, for a certain chain of supermarkets were 46, 55, 57, 52, and 44. Use coding like that of Exercise 15.16 on page 443 to fit a parabola for these data.

★ 15.56 The following are data on the average weekly profit (in \$1,000) of five restaurants, their seating capacities, and the average daily traffic (in thousands of cars) which passes their locations:

Seating capacity x_1	Traffic count x_2	Weekly net profit y
120	19	23.8
200	8	24.2
150	12	22.0
180	15	26.2
240	16	33.5

(a) Use the method of least squares to fit an equation of the form

$$y = b_0 + b_1 x_1 + b_2 x_2$$

to these data.
(b) Use the equation obtained in part (a) to predict the average weekly net profit of a restaurant with a seating capacity of 210 at a location where the daily traffic count averages 14,000 cars.

★ 15.57 The following data pertain to the cosmic-ray doses measured at various altitudes:

Altitude (hundreds of feet) x	Dose rate (mrem/year) y
0.5	28
4.5	30
7.8	32
12.0	36
48.0	58
53.0	69

Fit an exponential curve and use it to estimate the cosmic radiation at 6,000 feet.

15.58 The following are the high school averages, x, and first-year college grade-point indexes, y, of ten students:

x	y
3.0	2.6
2.7	2.4
3.8	3.9
2.6	2.1
3.2	2.6
3.4	3.3
2.8	2.2
3.1	3.2
3.5	2.8
3.3	2.5

(a) Fit a least-squares line which will enable us to predict first-year college grade-point indexes in terms of high school averages.
(b) Use the equation obtained in part (a) to predict the first-year college grade-point index of a student with a high school average of 2.5.

15.59 With reference to the preceding exercise, test the null hypothesis $\beta = 0.50$ against the alternative hypothesis $\beta > 0.50$ at the 0.05 level of significance.

15.60 The following are the processing times (minutes), x, and hardness readings, y, of certain machine parts:

x	y
20	282
34	275
19	171
10	142
24	145
31	340
25	282
13	105
29	233

(a) Fit a least-squares line which will enable us to predict hardness from processing time.

(b) Use the equation obtained in part (a) to predict the hardness reading of a machine part which was processed for 25 minutes.

15.61 With reference to the preceding exercise, test the null hypothesis $\alpha = 55.0$ against the alternative hypothesis $\alpha \neq 55.0$ at the 0.01 level of significance.

15.62 With reference to Exercise 15.60, find 95 percent limits of prediction for the hardness reading of a machine part which has been processed for 25 minutes.

★ **15.63** The following data show the scores which ten students got in an examination, their total SAT (Scholastic Aptitude Test) scores, and the number of hours that they studied for the examination:

Number of hours studied x_1	SAT score x_2	Score in examination y
9	1,160	70
3	1,013	38
16	1,022	90
19	995	99
6	1,180	59
11	1,117	75
14	1,062	77
12	1,207	84
6	1,014	49
9	1,008	61

(a) Use the method of least squares to fit an equation of the form $y = b_0 + b_1 x_1 + b_2 x_2$.
(b) Use the equation obtained in part (a) to predict the scores of students with these combinations of SAT scores and study times:

Number of hours studied	SAT score
12	1,000
18	1,000
12	1,200
18	1,200

15.8
REFERENCES

Methods of deciding which kind of curve to fit to a given set of paired data may be found in books on numerical analysis and in more advanced texts in statistics. Further information about the material of this chapter may be found in

CHATTERJEE, S., and PRICE, B., *Regression Analysis by Example.* New York: John Wiley & Sons, Inc., 1977.

DANIEL, C., and WOOD, F., *Fitting Equations to Data, 2nd ed.* New York: John Wiley & Sons, Inc., 1980.

DRAPER, N. R., and SMITH, H., *Applied Regression Analysis, 2nd ed.* New York: John Wiley & Sons, Inc., 1981.

EZEKIEL, M., and FOX, K. A., *Methods of Correlation and Regression Analysis, 3rd ed.* New York: John Wiley & Sons, Inc., 1959.

GUNST, R. F., and MASON, R. L., *Regression Analysis and Its Applications: A Data-Oriented Approach.* New York: Marcel Dekker, Inc., 1980.

HARRIS, R. J., *A Primer of Multivariate Statistics.* New York: Academic Press, Inc., 1971.

MOSTELLER, F., and TUKEY, J. W., *Data Analysis and Regression.* Reading, Mass.: Addison-Wesley Publishing Company, Inc., 1977.

WEISBERG, S., *Applied Linear Regression, 2nd ed.* New York: John Wiley & Sons, Inc., 1985.

WONNACOTT, T. H., and WONNACOTT, R. J., *Regression: A Second Course in Statistics.* New York: John Wiley & Sons, Inc., 1981.

CORRELATION

Having learned how to fit a least-squares line to paired data, we turn now to the problem of determining how well such a line actually fits the data. Of course, we can get some idea of this by inspecting a diagram which shows the line together with the data, but to demonstrate how we can be more objective, let us refer back to the original data of the example dealing with the foreign service job applicants' proficiency in German, namely,

Years studied German x	Score in test y
3	57
4	78
4	72
2	58
5	89
3	63
4	73
5	84
3	75
2	48

As can be seen from this table, there are substantial differences among the scores, the smallest being 48 and the largest being 89. However, we also see that the 48 was obtained by a person who had studied German for two years, while the 89 was obtained by a person who had studied German for five years, and this suggests that the differences among the scores may well be due, at least in part, to differences in the length of time that the applicants had studied German. This observation raises the following question, which we shall answer in this chapter: Of the total variation among the y's, how much can be attributed to the relationship between the two variables x and y (that is, to the fact that the observed y's correspond to different values of x), and how much can be attributed to chance.

In Section 16.1 we shall introduce the coefficient of correlation as a measure of the strength of the linear relationship between two variables, in Section 16.2 we shall learn how to interpret it, and in Section 16.3 we shall study related problems of inference; the problems of multiple and partial correlation will be touched upon lightly in Section 16.5.

16.1
THE COEFFICIENT OF CORRELATION

With regard to the question raised in the chapter opening given above, we are faced here with an analysis of variance. Figure 16.1 shows what we mean. As can be seen from the diagram, the deviation of an observed value of y from the mean of all the y's, $y - \bar{y}$, can be written as a sum of two parts. The first part is the deviation of $\hat{y}$ (the value on the line corresponding to an observed value of x) from the mean of all the y's, $\hat{y} - \bar{y}$; the second part is the deviation of the observed value of y from the

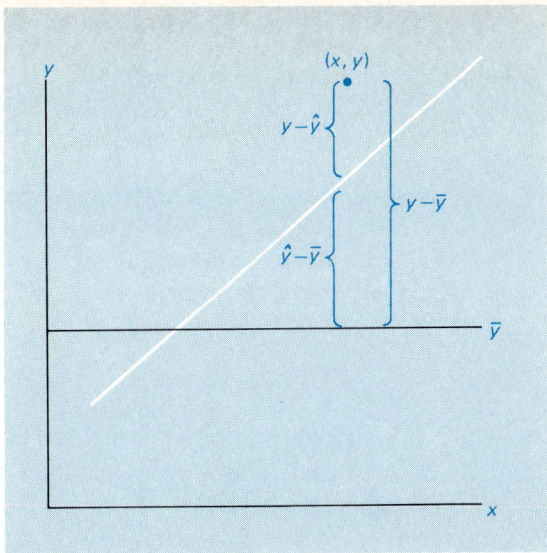

FIGURE 16.1 *Illustration to show that* $y - \bar{y} = (\hat{y} - \bar{y}) + (y - \hat{y})$.

corresponding value on the line, $y - \hat{y}$. Symbolically, we write

$$y - \bar{y} = (\hat{y} - \bar{y}) + (y - \hat{y})$$

for any observed value y, and if we square the expressions on both sides of this identity and sum over all n values of y, we find that algebraic simplifications lead to

$$\sum (y - \bar{y})^2 = \sum (\hat{y} - \bar{y})^2 + \sum (y - \hat{y})^2$$

The quantity on the left above measures the total variation of the y's and we call it the **total sum of squares**; note that $\sum (y - \bar{y})^2$ is just the variance of the y's multiplied by $n - 1$. The first of the two sums on the right, $\sum (\hat{y} - \bar{y})^2$, is called the **regression sum of squares** and it measures that part of the total variation of the y's which can be ascribed to the relationship between the two variables x and y; indeed, if all the points lie on the least-squares line, then $y = \hat{y}$ and the regression sum of squares equals the total sum of squares. In practice, this is hardly, if ever, the case, and the fact that all the points do not lie on a least-squares line is an indication that there are other factors than differences among the x's which affect the values of y. It is customary to combine all these other factors under the general heading of "chance." Chance variation is thus measured by the amounts by which the points deviate from the line; specifically, it is measured by $\sum (y - \hat{y})^2$, called the **residual sum of squares**, which is the second of the two components into which we partitioned the total sum of squares.

To determine these sums of squares for the example dealing with the language proficiency of certain foreign service job applicants, we could substitute the values of y given on page 468, $\bar{y}$, and the values of $\hat{y}$ obtained by substituting the x's into $\hat{y} = 31.55 + 10.90x$ (see page 438), but there are simplifications. First, for $\sum (y - \bar{y})^2$ we have the computing formula

$$S_{yy} = \sum y^2 - \frac{1}{n} \left(\sum y \right)^2$$

and on page 446 we showed that it equals 1,504.1 for our example. Second, $\sum (y - \hat{y})^2$ is the quantity which we minimized by the method of least squares, and which appears in the numerator of the formula for s_e. Copying the numerator from its computing formula on page 445, we get

$$\sum (\hat{y} - \bar{y})^2 = S_{yy} - \frac{(S_{xy})^2}{S_{xx}}$$

and $1,504.1 - (114.5)^2/10.5 = 255.50$ for our example. (The values of S_{xx} and S_{xy} were determined on page 438.) Finally, by subtraction, the regression sum of squares is given by

$$\sum (\hat{y} - \bar{y})^2 = \sum (y - \bar{y})^2 - \sum (y - \hat{y})^2$$
$$= S_{yy} - \left(S_{yy} - \frac{(S_{xy})^2}{S_{xx}} \right)$$
$$= \frac{(S_{xy})^2}{S_{xx}}$$

and for our example we get $(114.5)^2/10.5 = 1,248.59$.

It is of interest to note that all the quantities we have calculated here could have been obtained directly from the computer printout of Figure 15.5 on page 439. Under ANALYSIS OF VARIANCE, in the column headed SS, we find that the total sum of squares is 1,504.1, the residual sum of squares is 255.5, and the regression sum of squares is 1,248.6. The differences between the values shown here and above are, of course, due to rounding.

We are now ready to examine the sums of squares, and comparing the regression sum of squares with the total sum of squares, we find that

$$\frac{\sum (\hat{y} - \bar{y})^2}{\sum (y - \bar{y})^2} = \frac{1,248.59}{1,504.1} = 0.83$$

is the proportion of the total variation of the scores which can be attributed to the relationship with x, namely, to the differences in the number of years which the applicants had studied German in high school or college. This quantity is referred to as the **coefficient of determination** and it is denoted by r^2. Note that the coefficient of determination is also given in the printout of Figure 15.5 on page 439; near the middle it says "R-SQUARED = 83.0 PERCENT."

If we take the square root of the coefficient of determination, we get the **coefficient of correlation**, which is denoted by the letter r. Its sign is chosen so that it is like that of the estimated regression coefficient b, and for our example, where b

is positive, we get

$$r = \sqrt{0.83} = 0.91$$

rounded to two decimals.

It follows that the correlation coefficient is positive when the least-squares line has an upward slope, namely, when the relationship between x and y is such that small values of y tend to go with small values of x and large values of y tend to go with large values of x. Also, the correlation coefficient is negative when the least-squares line has a downward slope, namely, when large values of y tend to go with small values of x and small values of y tend to go with large values of x. Examples of **positive** and **negative correlations** are shown in the first two diagrams below.

Since part of the variation of the y's cannot exceed their total variation, $\sum (y - \hat{y})^2$ cannot exceed $\sum (y - \bar{y})^2$, and it follows from the formula defining r that correlation coefficients must lie on the interval from -1 to $+1$. If all the points actually fall on a straight line, the residual sum of squares, $\sum (y - \hat{y})^2$, is zero, $\sum (\hat{y} - \bar{y})^2 = \sum (y - \bar{y})^2$, and the resulting value of r, -1 or $+1$, is indicative of a perfect fit. If, however, the scatter of the points is such that the least-squares line is a horizontal line coincident with $\bar{y}$ (that is, a line with slope 0 which intersects the y-axis at $a = \bar{y}$), then $\sum (y - \hat{y})^2$ equals $\sum (y - \bar{y})^2$ and $r = 0$. In that case none of the variation of the y's can be attributed to their relationship with x, and the fit is so poor that knowledge of x is of no help in predicting y. The predicted value of y is $\bar{y}$ regardless of x. An example of this is shown in the third diagram of Figure 16.2.

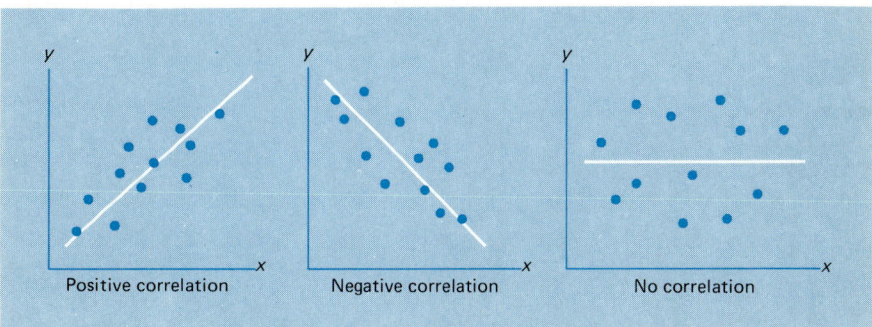

FIGURE 16.2 *Types of correlation.*

The formula which defines r shows clearly the nature, or essence, of the coefficient of correlation, but in actual practice it is seldom used to determine its value. To derive a computing formula for r, we first substitute

$$\sum (y - \bar{y})^2 = S_{yy} \quad \text{and} \quad \sum (\hat{y} - \bar{y})^2 = \frac{(S_{xy})^2}{S_{xx}}$$

from page 470 into the formula for r^2, getting

$$r^2 = \frac{S_{xy}^2}{S_{xx} \cdot S_{yy}}$$

and hence that

Computing formula for coefficient of correlation

$$r = \frac{S_{xy}}{\sqrt{S_{xx} \cdot S_{yy}}}$$

For easy reference, let us remind the reader that

$$S_{xx} = \sum x^2 - \frac{1}{n}\left(\sum x\right)^2$$

$$S_{yy} = \sum y^2 - \frac{1}{n}\left(\sum y\right)^2$$

and

$$S_{xy} = \sum xy - \frac{1}{n}\left(\sum x\right)\left(\sum y\right)$$

EXAMPLE The following are the scores which 12 students received in final examinations in economics and anthropology:

Economics	Anthropology
51	74
68	70
72	88
97	93
55	67
73	73
95	99
74	73
20	33
91	91
74	80
80	86

Calculate r.

Solution First calculating the necessary sums, we get $\sum x = 850$, $\sum x^2 = 65{,}230$, $\sum y = 927$, $\sum y^2 = 74{,}883$, and $\sum xy = 69{,}453$. Then, substituting these values together with

$n = 12$ into the formulas for S_{xx}, S_{yy}, S_{xy}, and finally r, we find that

$$S_{xx} = 65,230 - \tfrac{1}{12}(850)^2 = 5,021.67$$

$$S_{yy} = 74,883 - \tfrac{1}{12}(927)^2 = 3,272.25$$

$$S_{xy} = 69,453 - \tfrac{1}{12}(850)(927) = 3,790.5$$

and

$$r = \frac{3,790.5}{\sqrt{(5,021.67)(3,272.25)}} = 0.935$$

Observe that if S_{xy} is positive, then both b and r will be positive; negative S_{xy} leads to negative b and r.

The quantity S_{xy} in the numerator of the formula for r is actually a computing formula for $\sum (x - \bar{x})(y - \bar{y})$, which, divided by n, is called the first **product moment**. For this reason, r is also referred to at times as the **product-moment coefficient of correlation**. Note that in $\sum (x - \bar{x})(y - \bar{y})$ we add the products obtained by multiplying the deviation of each x from $\bar{x}$ by the deviation of the corresponding y from $\bar{y}$. In this way we literally measure how the x's and y's vary together. If their relationship is such that large values of x tend to go with large values of y, and small values of x with small values of y, the deviations $x - \bar{x}$ and $y - \bar{y}$ tend to be both positive or both negative, and most of the products $(x - \bar{x})(y - \bar{y})$ will be positive. On the other hand, if the relationship is such that large values of x tend to go with small values of y, and small values of x with large values of y, the deviations $x - \bar{x}$ and $y - \bar{y}$ tend to be of opposite sign, and most of the products $(x - \bar{x})(y - \bar{y})$ will be negative. For this reason, $\sum (x - \bar{x})(y - \bar{y})$, divided by $n - 1$, is called the **sample covariance**.

16.2

THE INTERPRETATION OF r

When r equals $+1$, -1, or 0, there is no problem about the interpretation of the coefficient of correlation. As we have already indicated, it is $+1$ or -1 when all the points actually fall on a straight line, and it is zero when the fit of the least-squares line is so poor that knowledge of x does not help in the prediction of y. In general, the definition of r tells us that $100r^2$ is the percentage of the total variation of the y's which is explained by, or is due to, their relationship with x. This itself is an important measure of the relationship between two variables; beyond this, it permits valid comparisons of the strength of several relationships. For instance, if $r = 0.80$ in one study and $r = 0.40$ in another study, it would be very misleading to report that the 0.80 correlation is "twice as good" or "twice as strong" as the 0.40 correlation. When $r = 0.80$, then $100(0.80)^2 = 64$ percent of the variation of the y's is accounted for by the relationship with x, and when $r = 0.40$, only $100(0.40)^2 = 16$ percent of the variation of the y's is accounted for by the relationship with

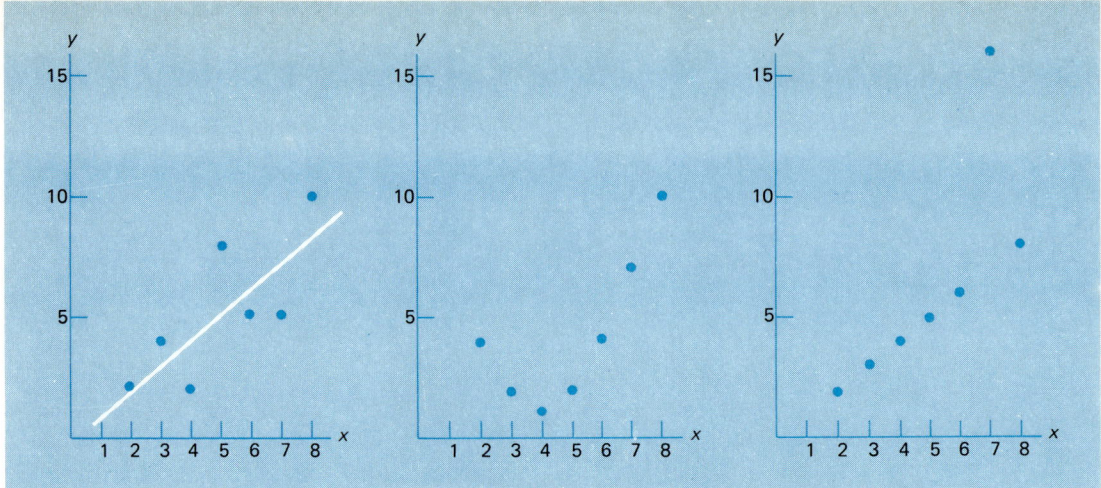

FIGURE 16.3 *Three sets of paired data for which r = 0.75.*

x. Thus, in the sense of "percentage of variation accounted for" we can say that the 0.80 correlation is four times as strong as the 0.40 correlation. In the same way, we say that a relationship for which $r = 0.60$ is nine times as strong as a relationship for which $r = 0.20$.

There are several pitfalls in the interpretation of the coefficient of correlation. First, it is often overlooked that r measures only the strength of linear relationships; second, it should be remembered that a strong correlation (a value of r close to $+1$ or -1) does not necessarily imply a cause–effect relationship.

If r is calculated indiscriminately, for instance, for the three sets of data of Figure 16.3, we get $r = 0.75$ in each case, but it is a meaningful measure of the strength of the relationship only in the first case. In the second case there is a very strong curvilinear relationship between the two variables, and in the third case six of the seven points actually fall on a straight line, but the seventh point is so far off that it suggests the possibility of a gross error of measurement or an error in recording the data. Thus, before we calculate r, we should always plot the data to see whether there is reason to believe that the relationship is, in fact, linear.

The fallacy of interpreting a high value of r (that is, a value close to $+1$ or -1) as an indication of a cause–effect relationship, is best explained with a few examples. Frequently used as an illustration, is the high positive correlation between the annual sales of chewing gum and the incidence of crime in the United States. Obviously, one cannot conclude that crime might be reduced by prohibiting the sale of chewing gum; both variables depend upon the size of the population, and it is this mutual relationship with a third variable (population size) which produces the positive correlation. Another example is the strong positive correlation which was observed between the number of storks seen nesting in English villages and the number of children born in the same villages. We leave it to the reader's ingenuity to explain why there might be a strong correlation in this case in the absence of any cause–effect relationship.

EXERCISES

16.1 Use the values of S_{xx}, S_{yy}, and S_{xy} obtained on page 448 to calculate r for the data dealing with the grade-point indexes of six students and the average amount of time they spent on homework per week.

16.2 The following are the numbers of minutes it took 12 mechanics to assemble a piece of machinery in the morning, x, and in the late afternoon, y:

x	y
12	14
11	11
9	14
13	11
10	12
11	15
12	12
14	13
10	16
9	10
11	10
12	14

Calculate r.

16.3 With reference to the preceding exercise, what percentage of the variation of the y's is due to differences among the x's?

16.4 After a student calculated r for a large set of paired data, she discovered to her dismay that the variable which should have been labeled x was labeled y and the variable which should have been labeled y was labeled x. Is there any reason for being dismayed?

16.5 The following table shows the percentages of the vote predicted by a poll for eight candidates for the U.S. Senate in different states, x, and the percentages of the vote which they actually received, y:

x	y
43	50
46	42
51	57
59	55
41	46
53	48
52	53
62	56

Calculate r.

16.6 The calculation of r can often be simplified by adding a suitable positive or negative number to each x, each y, or both. Rework Exercise 16.5 after subtracting 41 from each x and 42 from each y.

16.7 After a student computed the correlation between height and weight for a large number of people, he realized that the weights were given in kilograms and the heights were given in centimeters. He wanted to obtain the correlation between height in inches and weight in pounds. If there are 0.3937 inch per centimeter and 2.2 pounds per kilogram, how should he correct his calculations?

16.8 If we calculate r for each of the following sets of data, should we be surprised if we get $r = 1$ and $r = -1$? Explain your answers.

(a)

x	y
6	9
14	11

(b)

x	y
12	5
8	15

16.9 The following data were obtained in a study of the relationship between the resistance (ohms) and the failure time (minutes) of certain overloaded resistors:

Resistance	Failure time
48	45
28	25
33	39
40	45
36	36
39	35
46	36
40	45
30	34
42	39
44	51
48	41
39	38
34	32
47	45

Calculate r and determine what percentage of the variation in failure time is due to differences in resistance.

16.10 A student computed the correlation between height and weight for a large group of third grade school children and obtained a value of $r = 0.32$. She was unable to decide whether she should conclude that being tall causes a child to put on more weight or that having excess weight enables a child to grow taller. Help her solve this dilemma.

16.11 State in each case whether you would expect a positive correlation, a negative correlation, or no correlation:
  (a) the ages of husbands and wives;
  (b) the amount of rubber on tires and the number of miles they have been driven;
  (c) the number of hours that golfers practice and their scores;
  (d) shoe size and IQ;
  (e) the weight of the load of trucks and their gasoline consumption.

16.12 State in each case whether you would expect a positive correlation, a negative correlation, or no correlation:
  (a) pollen count and the sale of anti-allergy drugs;
  (b) income and education;
  (c) the number of sunny days in August in Detroit and the attendance at the Detroit Zoo;
  (d) shirt size and sense of humor;
  (e) number of persons getting flu shots and number of persons catching the flu.

16.13 With reference to Exercise 15.40 on page 456,
  (a) calculate r for the copper content and the hardness of the given specimens of steel;
  (b) rework part (a) after subtracting 0.10 from each copper content and then dividing by 0.08.

16.14 With reference to Exercise 15.38 on page 455, calculate r for each of the following pairs of variables:
  (a) age and income;
  (b) age and years college;
  (c) years college and income.

16.15 If $r = 0.41$ for one set of paired data and $r = 0.29$ for another set of paired data, compare the strengths of the two relationships.

16.16 In a medical study it was found that $r = 0.70$ for the weight of babies six months old and their weight at birth, and that $r = 0.60$ for the weight of babies six months old and their average daily intake of food. Give a counterexample to show that it is not valid to conclude that weight at birth and food intake to-

gether account for
$$100(0.70)^2 + 100(0.60)^2 = 85 \text{ percent}$$
of the variation of babies' weight when they are six months old. Can you explain why the conclusion is not valid?

16.17 Working with various socioeconomic data for the years 1975–1982, a researcher got $r = 0.9225$ for the number of foreign language degrees offered by U.S. colleges and universities and the mileage of railroad track owned by U.S. railroads. Can we conclude that $100(0.9225)^2 = 85.1$ percent of the variation in the foreign language degrees is accounted for by differences in the ownership of railroad track?

★ **16.18** Correlation methods can be used to study internal relationships within a series of data collected at regular intervals. Such data are called **time series** and are commonly encountered in economics, where data are reported annually, monthly, weekly, and so on. The following numbers are total gross sales (in thousands of dollars) for 15 consecutive weeks at a wholesale beverage store:

60	58	62	56	47
44	48	49	47	42
41	42	46	44	50

We can see how the sales may be related to sales in previous weeks by computing the **autocorrelation coefficients** corresponding to various time lags. For example, the autocorrelation coefficient of lag 1 compares y_t, the sales in week t, with y_{t-1}, the sales in week $t - 1$. This requires the computation of the correlation coefficient for these pairs of values:

y_t: 60 58 62 56 47 44 48 49 47 42 41 42 46 44
y_{t-1}: 58 62 56 47 44 48 49 47 42 41 42 46 44 50

Observe that there are only 14 pairs in this list.
  Similarly, the autocorrelation coefficient of lag 2 requires the computation of the correlation coefficient for these 13 pairs of values:

y_t: 60 58 62 56 47 44 48 49 47 42 41 42 46
y_{t-2}: 62 56 47 44 48 49 47 42 41 42 46 44 50

(a) Find the value of the autocorrelation coefficient of lag 1.

(b) Find the value of the autocorrelation coefficient of lag 2.

★ 16.19 Correlation methods are sometimes used to study the relationship between two (time) series of data which are recorded annually, monthly, weekly, daily, and so on. Suppose, for instance, that in the years 1973–1986 a large textile manufacturer spent 0.8, 0.5, 0.8, 1.0, 1.0, 0.9, 0.8, 1.2, 1.0, 0.9, 0.8, 1.0, 1.0, and 0.8 million dollars on research and development, and that in these years its share of the market was 20.4, 18.6, 19.1, 18.0, 18.2, 19.6, 20.0, 20.4, 19.2, 20.5, 20.8, 18.9, 19.0, and 19.8 percent. To see whether and how the company's share of the market in a given year may be related to its expenditure on research and development in prior years, let x_t denote the company's research and development expenditures and y_t its market share in the year t, and calculate

(a) the correlation coefficient for y_t and x_{t-1};

(b) the correlation coefficient for y_t and x_{t-2};

(c) the correlation coefficient for y_t and x_{t-3};

(d) the correlation coefficient for y_t and x_{t-4}.

For instance, in part (a) calculate r after pairing the 1974 percentage share of the market with the 1973 expenditures on research and development, the 1975 market share with the 1974 expenditures, and so on, ..., and in part (d) calculate r after pairing the 1977 percentage share of the market with the 1973 expenditures on research and development, the 1978 market share with the 1974 expenditures, and so on. These time-lag correlations are called **cross correlations**. To continue,

(e) discuss the apparent duration of the effect of expenditures on research and development on the company's share of the market.

16.3
CORRELATION ANALYSIS

When r is calculated on the basis of sample data, we may get a strong positive or negative correlation purely by chance, even though there is actually no relationship whatever between the two variables under consideration.

Suppose, for instance, that we take a pair of dice, one red and one green, roll them five times, and get the following results:

Red die x	Green die y
4	5
2	2
4	6
2	1
6	4

Presumably, there is no relationship between x and y, the numbers of points showing on the two dice. It is hard to see why large values of x should go with large values of y and small values of x with small values of y, but calculating r, we get the surprisingly high value $r = 0.66$. This raises the question of whether something is wrong with the assumption that there is no relationship between x and y, and to answer it we shall have to see whether the high value of r may be attributed to chance.

When a correlation coefficient is calculated from sample data, as in the above example, the value we get for r is only an estimate of a corresponding parameter, the **population correlation coefficient**, which we denote by ρ (the Greek letter *rho*). What r measures for a sample, ρ measures for a population.

To make inferences about ρ on the basis of r, we shall have to make several assumptions about the distributions of the random variables whose values we observe. In **normal correlation analysis** we make the same assumptions as in normal regression analysis (see page 444), except that the x's are not constants, but values of a random variable having a normal distribution.

Since the sampling distribution of r is rather complicated under these assumptions, it is common practice to base inferences about ρ on the **Fisher Z transformation**, a change of scale from r to Z, which is given by

$$Z = \frac{1}{2} \cdot \ln \frac{1 + r}{1 - r}$$

Here ln denotes "natural logarithm," that is, logarithm to the base e, where $e = 2.71828...$. This transformation is named after R. A. Fisher, a prominent statistician, who showed that under the assumptions of normal correlation analysis and for any value of ρ, the distribution of Z is approximately normal with

$$\mu_Z = \frac{1}{2} \cdot \ln \frac{1 + \rho}{1 - \rho} \quad \text{and} \quad \sigma_Z = \frac{1}{\sqrt{n - 3}}$$

Hence,

Statistic for inferences about ρ

$$z = \frac{Z - \mu_Z}{\sigma_Z} = (Z - \mu_Z)\sqrt{n - 3}$$

has approximately the standard normal distribution. The application of this theory is greatly facilitated by the use of Table IX at the end of the book, which gives the values of Z corresponding to $r = 0.00, 0.01, 0.02, 0.03, \ldots$, and 0.99. Observe that only positive values are given in this table; if r is negative, we simply look up $-r$ and take the negative of the corresponding Z. Note also that the formula for μ_Z is like that for Z with r replaced by ρ; therefore, Table IX can be used to look up values of μ_Z corresponding to given values of ρ.

EXAMPLE Use the 0.05 level of significance to test the null hypothesis of no correlation (that is, the null hypothesis $\rho = 0$) for the example where we rolled a pair of dice five times and got $r = 0.66$.

Solution
1. H_0: $\rho = 0$
   H_A: $\rho \neq 0$
2. $\alpha = 0.05$
3. Since $\mu_Z = 0$ for $\rho = 0$, reject the null hypothesis if $z \leq -1.96$ or $z \geq 1.96$, where

$$z = Z \cdot \sqrt{n - 3}$$

Otherwise, state that the value of r is not significant.

4. Substituting $n = 5$ and $Z = 0.793$, the value of Z corresponding to $r = 0.66$ according to Table IX, we get

$$z = 0.793\sqrt{5 - 3}$$

$$= 1.12$$

5. Since $z = 1.12$ falls on the interval from -1.96 to 1.96, the null hypothesis cannot be rejected; in other words, the value obtained for r is not significant, as should have been expected.

An alternative way of handling this kind of problem (namely, testing the null hypothesis $\rho = 0$) is given in Exercise 16.24 on page 481.

EXAMPLE With reference to the example dealing with the proficiency scores of certain applicants for foreign service jobs, where we had $n = 10$ and $r = 0.91$, test the null hypothesis $\rho = 0.70$ against the alternative hypothesis $\rho > 0.70$ at the 0.05 level of significance.

Solution

1. H_0: $\rho = 0.70$
   H_A: $\rho > 0.70$
2. $\alpha = 0.05$
3. Reject the null hypothesis if $z \geq 1.645$, where

$$z = (Z - \mu_Z)\sqrt{n - 3}$$

and otherwise, accept the null hypothesis or reserve judgment.
4. Substituting $n = 10$, $Z = 1.528$ corresponding to $r = 0.91$, and $\mu_Z = 0.867$ corresponding to $\rho = 0.70$ according to Table IX, we get

$$z = (1.528 - 0.867)\sqrt{10 - 3}$$

$$= 1.75$$

5. Since $z = 1.75$ exceeds 1.645, the null hypothesis must be rejected; we conclude that $\rho > 0.70$ for the proficiency scores of applicants to the foreign service jobs and the number of years that they have studied German in high school or college.

To construct confidence intervals for ρ, we first construct confidence intervals for μ_Z, and then convert to r and ρ by means of Table IX. A confidence interval formula for μ_Z may be obtained by substituting

$$z = (Z - \mu_Z)\sqrt{n - 3}$$

for the middle term of the double inequality $-z_{\alpha/2} < z < z_{\alpha/2}$, and then manipulating the terms algebraically so that the middle term is μ_Z. This leads to the following $(1 - \alpha)100$ percent confidence interval for μ_Z:

Confidence interval for μ_Z

$$Z - \frac{z_{\alpha/2}}{\sqrt{n - 3}} < \mu_Z < Z + \frac{z_{\alpha/2}}{\sqrt{n - 3}}$$

EXAMPLE If $r = 0.62$ for the cost estimates of two mechanics for a random sample of 30 repair jobs, construct a 95 percent confidence interval for the population correlation coefficient.

Solution Getting $Z = 0.725$, corresponding to $r = 0.62$, from Table IX, and substituting it together with $n = 30$ and $z_{0.025} = 1.96$ into the above confidence interval formula for μ_Z, we find that

$$0.725 - \frac{1.96}{\sqrt{27}} < \mu_Z < 0.725 + \frac{1.96}{\sqrt{27}}$$

or

$$0.348 < \mu_Z < 1.102$$

Finally, looking up the values of r which come closest to $Z = 0.348$ and $Z = 1.102$ in Table IX, we get the 95 percent confidence interval

$$0.33 < \rho < 0.80$$

for the true strength of the linear relationship between cost estimates made by the two mechanics.

EXAMPLE If $r = 0.20$ for a random sample of $n = 40$ paired data, construct a 95 percent confidence interval for ρ.

Solution Getting $Z = 0.203$, corresponding to $r = 0.20$, from Table IX, and substituting it together with $n = 40$ and $z_{0.025} = 1.96$ into the above confidence interval formula for μ_Z, we find that

$$0.203 - \frac{1.96}{\sqrt{37}} < \mu_Z < 0.203 + \frac{1.96}{\sqrt{37}}$$

or

$$-0.119 < \mu_Z < 0.525$$

Then, looking up in Table IX the values of r which come closest to $Z = 0.119$ and $Z = 0.525$, we get the 95 percent confidence interval

$$-0.12 < \rho < 0.48$$

for the population correlation coefficient.

EXERCISES

16.20 With reference to Exercise 16.2 on page 475, test the null hypothesis of no correlation at the 0.05 level of significance.

16.21 With reference to Exercise 16.9 on page 475, test the null hypothesis $\rho = 0$ against the alternative hypothesis $\rho \neq 0$ at the 0.05 level of significance.

16.22 With reference to Exercise 15.14 on page 442,
    (a) calculate r for the given data;
    (b) test at the 0.05 level of significance whether the value of r obtained in part (a) is significant.

16.23 With reference to Exercise 15.7 on page 441,
    (a) determine what percentage of the variation in the yield can be attributed to difference in the rainfall;
    (b) test the null hypothesis of no correlation (between rainfall and yield) at the 0.01 level of significance.

16.24 Under the assumptions of normal correlation analysis, the test of the null hypothesis $\rho = 0$ may also be based on the statistic

$$t = \frac{r\sqrt{n-2}}{\sqrt{1-r^2}}$$

which has the t distribution with $n - 2$ degrees of freedom. Use this statistic to test in each case whether the value of r is significant at the 0.05 level of significance:
    (a) $n = 12$ and $r = 0.50$;
    (b) $n = 20$ and $r = 0.62$.

16.25 Use the t statistic of the preceding exercise to test in each case whether the value of r is significant at the 0.01 level of significance:
    (a) $n = 14$ and $r = 0.78$;
    (b) $n = 16$ and $r = 0.51$.

16.26 Use the t statistic of Exercise 16.24 to rework Exercise 16.20.

16.27 With reference to Exercise 15.7 on page 441,
    (a) use the t statistic of Exercise 16.24 to test the null hypothesis of no correlation;

    (b) find the t statistic which tests the null hypothesis $\beta = 0$;
    (c) compare the results of (a) and (b).

16.28 The t statistic given in Exercise 16.24 is numerically identical to the t statistic which tests $\beta = 0$ in the regression of Y on X. Show that this is so by expressing both t statistics in terms of S_{xx}, S_{xy}, S_{yy}, and n.

16.29 In a study of the relationship between the death rate from lung cancer and the per capita consumption of cigarettes twenty years earlier, data for $n = 9$ countries yielded $r = 0.73$. At the 0.05 level of significance, test the null hypothesis $\rho = 0.50$ against the alternative hypothesis $\rho > 0.50$.

16.30 In a study of the relationship between the available heat (per cord) of green wood and air-dried wood, data for $n = 13$ kinds of wood yielded $r = 0.94$. Use the 0.01 level of significance to test the null hypothesis $\rho = 0.75$ against the alternative hypothesis $\rho \neq 0.75$.

16.31 If $n = 18$ and $r = -0.64$ for certain paired data, test the null hypothesis $\rho = -0.30$ against the alternative hypothesis $\rho < -0.30$ at the 0.05 level of significance.

16.32 Assuming that the conditions underlying normal correlation analysis are met, use the Fisher Z transformation to construct approximate 95 percent confidence intervals for ρ when
    (a) $r = 0.80$ and $n = 15$;
    (b) $r = -0.22$ and $n = 30$;
    (c) $r = 0.64$ and $n = 100$.

16.33 Assuming that the conditions underlying normal correlation analysis are met, use the Fisher Z transformation to construct approximate 99 percent confidence intervals for ρ when
    (a) $r = -0.87$ and $n = 19$;
    (b) $r = 0.39$ and $n = 24$;
    (c) $r = 0.16$ and $n = 40$.

16.4

CORRELATIONS FOR COUNT DATA

Correlations can also be computed for cross-classified count data, provided that both categories of classifications are ordered. This is the analysis suggested in Section 13.5 when we first encountered such data, and the details will be completed here. The data in the table below give the numbers of broken eggs in each of 600 cartons received at a particular supermarket loading dock. The data have been displayed in this fashion to ask whether there is any relationship between egg size and the tendency to break.

						Total
Medium	52	26	12	7	3	100
Large	98	35	10	6	1	150
X-Large	188	40	16	4	2	250
Jumbo	80	16	3	1	0	100
	0	1	2	3	4	

Number of broken eggs

This table must be read carefully. The shipment involved 100 cartons of medium eggs; of these 100 cartons, 52 cartons had no broken eggs, 26 cartons had one broken egg, and so on.

A great many analyses are possible for data such as these. We could, for example, compute the χ^2 statistic to test whether the probabilities of the various numbers of broken eggs are the same for each size category. We could also utilize the fact that there are four different types of cartons, and we could do an analysis of variance. These methods are both described in Chapters 13 and 14, and they are both valid. The χ^2 statistic ignores the orderings on the sizes of the eggs and on the numbers of broken eggs. The analysis of variance utilizes the numbers of broken eggs as data values, but ignores ordering on the sizes of the eggs. These analyses are valid tests of the null hypothesis of no relation between size of egg and likelihood of breakage, but they are less likely to produce significant test statistics than the test based on the correlation coefficient.

We can calculate the correlation coefficient most easily by assigning integer scores to the numbers of broken eggs and to the egg sizes. The numbers of broken eggs are already given as integers 0, 1, 2, 3, and 4, and we should not alter these scores. The egg sizes can be encoded in many ways, but it is simplest to assign the values as 1, 2, 3, and 4. The choice of scores is of course subjective. The row totals will be needed in the calculations, and these are already provided in the table. The column totals can be found to be 418, 117, 41, 18, and 6. We had 418 cartons with no broken eggs, 117 cartons with one broken egg, 41 with two broken eggs, 18 with three broken, and 6 with four broken. There were 600 cartons in this problem, and indeed $418 + 117 + 41 + 18 + 6 = 600$.

Let x correspond to the egg size. The possible values for x are 1, 2, 3, and 4, so that $\sum x = 100 \cdot 1 + 150 \cdot 2 + 250 \cdot 3 + 100 \cdot 4 = 1{,}550$. In similar fashion we can find $\sum x^2 = 100 \cdot 1^2 + 150 \cdot 2^2 + 250 \cdot 3^2 + 100 \cdot 4^2 = 4{,}550$.

Let y correspond to the number of broken eggs. The possible values for y are 0, 1, 2, 3, and 4. Then $\sum y = 418 \cdot 0 + 117 \cdot 1 + 41 \cdot 2 + 18 \cdot 3 + 6 \cdot 4 = 277$. We will then find $\sum y^2 = 418 \cdot 0^2 + 117 \cdot 1^2 + 41 \cdot 2^2 + 18 \cdot 3^2 + 6 \cdot 4^2 = 539$.

The sum $\sum xy$ is also needed. This will use the individual entries of the table, and it is calculated as

$$
\begin{aligned}
\sum xy = \quad & 52 \cdot 0 + 26 \cdot 1 + 12 \cdot 2 + 7 \cdot 3 + 3 \cdot 4 \\
+ \; & 98 \cdot 0 + 35 \cdot 2 + 10 \cdot 4 + 6 \cdot 6 + 1 \cdot 8 \\
+ \; & 188 \cdot 0 + 40 \cdot 3 + 16 \cdot 6 + 4 \cdot 9 + 2 \cdot 12 \\
+ \; & 80 \cdot 0 + 16 \cdot 4 + 3 \cdot 8 + 1 \cdot 12 + 0 \cdot 16 \\
= \; & 613
\end{aligned}
$$

We can now find

$$
S_{xx} = 4{,}550 - \frac{1550^2}{600} = 545.8333
$$

$$
S_{yy} = 539 - \frac{277^2}{600} = 411.1183
$$

and

$$
S_{xy} = 613 - \frac{1{,}550 \cdot 277}{600} = -102.5833
$$

We substitute these into the formula for r to get

$$
r = \frac{-102.5833}{\sqrt{(545.8333)(411.1183)}} = -0.2166
$$

The numeric value of r conveys much more information about the data than would the chi-squared test or the analysis of variance. There are many steps in the calculations for r, and use of a computer is recommended.

16.5

MULTIPLE AND PARTIAL CORRELATION ★

In Section 16.1 we introduced the coefficient of correlation as a measure of the goodness of the fit of a least-squares line to a set of paired data. If predictions are to be made with an equation of the form

$$
\hat{y} = b_0 + b_1 x_1 + b_2 x_2 + \cdots + b_k x_k
$$

obtained by the method of least squares as in Section 15.4, we define the **multiple correlation coefficient** in the same way in which we originally defined r. We take the square root of the quantity

$$
\frac{\sum (\hat{y} - \bar{y})^2}{\sum (y - \bar{y})^2}
$$

which is the proportion of the total variation of the y's that can be attributed to the relationship with the x's. The only difference is that we now calculate $\hat{y}$ by means of the multiple regression equation instead of the equation $\hat{y} = a + bx$.

EXAMPLE In the example on page 454, where we derived the multiple regression equation

$$\hat{y} = 65{,}192 + 4{,}133x_1 + 758x_2$$

it can be shown that

$$\sum (\hat{y} - \bar{y})^2 = 185{,}269{,}168 \quad \text{and} \quad \sum (y - \bar{y})^2 = 185{,}955{,}008.$$

What is the value of the multiple correlation coefficient? (Actually, these values were obtained with a continuation of the computer program which yielded the printout shown in Figure 15.8; indeed, the square of the multiple correlation coefficient is shown in the lower left-hand corner of Figure 15.8.)

Solution Since

$$\frac{185{,}269{,}168}{185{,}955{,}008} = 0.9963$$

it follows that the multiple correlation coefficient is $\sqrt{0.9963} = 0.998$.

This example also serves to illustrate that adding more independent variables in a correlation study may not be sufficiently productive to justify the extra work. As it can be shown that $r = 0.996$ for y and x_1 alone, it is apparent that very little is gained by considering also x_2. In other words, predictions based on the number of bedrooms alone are virtually as good as predictions which account also for the number of baths. The situation is quite different, though, in Exercise 16.35, where two independent variables together account for a much higher proportion of the total variation in y than does either x_1 or x_2 alone.

When we discussed the problem of correlation and causation, we showed that a strong correlation between two variables may be due entirely to their dependence on a third variable. We illustrated this with the examples of chewing gum sales and the crime rate, and child births and the number of storks. To give another example, let us consider the two variables x_1, the weekly amount of hot chocolate sold by a refreshment stand at a summer resort, and x_2, the weekly number of visitors to the resort. If, on the basis of suitable data, we get $r = -0.30$ for these variables, this should come as a surprise—after all, we would expect more sales of hot chocolate when there are more visitors, and vice versa, and hence a positive correlation.

However, if we think for a moment, we may surmise that the negative correlation of -0.30 may well be due to the fact that the variables x_1 and x_2 are both related to a third variable x_3, the average weekly temperature at the resort. If the temperature is high, there will be more visitors, but they will prefer cold drinks to hot chocolate; if the temperature is low, there will be fewer visitors, but they will prefer hot chocolate to cold drinks. So, let us suppose that further data yield $r = -0.70$ for x_1 and x_3, and $r = 0.80$ for x_2 and x_3. These values seem reasonable since low sales of hot chocolate should go with high temperatures and vice versa, while the number of visitors should be high when the temperature is high, and low when the temperature is low.

In the preceding example, we should really have investigated the relationship between x_1 and x_2 (hot chocolate sales and the number of visitors to the resort) when all other factors, primarily temperature, are held fixed. As it is seldom possible to control matters to such an extent, it has been found that a statistic called the **partial correlation coefficient** does a fair job of eliminating the effects of other variables. If we write the ordinary correlation coefficients for x_1 and x_2, x_1 and x_3, and x_2 and x_3, as r_{12}, r_{13}, and r_{23}, the partial correlation coefficient for x_1 and x_2 with x_3 fixed is given by

Partial correlation coefficient

$$r_{12.3} = \frac{r_{12} - r_{13} \cdot r_{23}}{\sqrt{1 - r_{13}^2}\sqrt{1 - r_{23}^2}}$$

EXAMPLE   Calculate $r_{12.3}$ for the example above.

Solution   Substituting $r_{12} = -0.30$, $r_{13} = -0.70$, and $r_{23} = 0.80$ into the formula for $r_{12.3}$, we get

$$r_{12.3} = \frac{(-0.30) - (-0.70)(0.80)}{\sqrt{1 - (-0.70)^2}\sqrt{1 - (0.80)^2}} = 0.61$$

This shows that, as we expected, there is a positive relationship between the sales of hot chocolate and the number of visitors to the resort when the effect of differences in temperature is eliminated.

We have given this example mainly to illustrate what we mean by partial correlation, but it also serves as a reminder that correlation coefficients can be very misleading unless they are interpreted with care. Partial correlation coefficients do not solve the problem of interpreting ordinary correlation coefficients, so they should also be used with considerable caution.

EXERCISES

★ 16.34 The data in the table shown at the end of this Exercise summarize the results of a survey question given to 122 students at a particular college. The question, "Do you favor the retention of the college's current humanities requirements?" had seven response categories. These categories ranged from "strongly disagree" to "strongly agree." Compute the correlation coefficient to assess the relationship between opinion and year in college. Assign the scores 1, 2, 3, and 4 to the rows of the table. To the columns of the table assign scores -3, -2, -1, 0, 1, 2, and 3.

Responses of 122 college students to a question on the humanities requirement.

	Very strongly disagree	Strongly disagree	Disagree	Neutral	Agree	Strongly Agree	Very strongly Agree	Total
Freshman	2	1	2	14	8	3	1	31
Sophomore	0	2	2	9	8	4	2	27
Junior	1	0	0	10	13	2	2	28
Senior	1	0	1	8	17	5	4	36
Total	4	3	5	41	46	14	9	122

16.35 Exercise 13.61 on page 377 examined the relationship between bank employees' standard of dress and their professional advancement. Compute a correlation coefficient between these two variables, using evenly spaced scores. (*Hint*: The arithmetic will be relatively simple with scores -1, 0, and 1 for each of the variables.)

★ **16.36** Residents of a long-term health-care facility were asked to rate the quality of a new chicken salad recipe, choosing responses from the list *poor, acceptable, good,* and *delicious.* In fact, the residents were given, at random, one of three different versions of the chicken salad recipe, differing only in the amount of salt. Here is a summary of their responses:

	Poor	Acceptable	Good	Delicious
100 mg salt	6	4	1	0
200 mg salt	0	2	7	3
400 mg salt	0	1	4	5

The salt amounts are specified in milligrams of sodium per 4 ounces of chicken salad.

(a) Compute the correlation coefficient between the amount of salt and the ratings, using evenly spaced scores.
(b) Compute the correlation coefficient between these variables, using evenly spaced scores for the ratings and scores 1, 2, and 4 for the salt amounts.
(c) If you repeated part (b), using scores -1, 0, and 2 for the salt amounts, would you get the same value for the correlation coefficient?

★ **16.37** In a multiple regression problem, the regression sum of squares is $\sum (\hat{y} - \bar{y})^2 = 45{,}225$, and the total sum of squares is $\sum (y - \bar{y})^2 = 136{,}210$. Find the value of the multiple correlation coefficient.

★ **16.38** Use the least-squares equation obtained in Exercise 15.38 on page 455 to calculate $\hat{y}$ for each of the five executives. Then determine the values of $\sum (\hat{y} - \bar{y})^2$ and $\sum (y - \bar{y})^2$, and calculate the multiple correlation coefficient. Also compare the result with the values obtained for r in parts (a) and (c) of Exercise 16.14 on page 476.

★ **16.39** Use the least-squares equation obtained in Exercise 15.39 on page 455 to calculate $\hat{y}$ for each of the six shipments. Then calculate the regression sum of squares, the total sum of squares, and the multiple correlation coefficient.

★ 16.40 An experiment was conducted to determine if the height of certain rose bushes can be predicted on the basis of the amount of rose fertilizer and the amount of irrigation that is applied to the soil. For predicting the height on the basis of both variables, they obtained a multiple correlation coefficient of 0.58; for predicting the height on the basis of the amount of rose fertilizer alone they got $r = 0.66$. Comment on these results.

★ 16.41 Using the illustration on pages 484 and 485, find the partial correlation coefficient for x_1 (sales of hot chocolate) and x_3 (temperature) when x_2 (number of visitors) remains fixed.

★ 16.42 Use the results of Exercise 16.14 on page 476 to find:
(a) the partial correlation coefficient for age and income when years of college remains fixed;
(b) the partial correlation coefficient for years of college and income when age remains fixed.

★ 16.43 An experiment yielded the following results: $r_{12} = 0.80$, $r_{13} = -0.70$, and $r_{23} = 0.90$. Explain why these figures cannot all be correct.

16.6

CHECKLIST OF KEY TERMS
(with page references to their definitions)

Auto correlation, 476
Coefficient of correlation, 470
Coefficient of determination, 470
Cross correlation, 477
Fisher *Z* transformation, 478
★ Multiple correlation coefficient, 483
Negative correlation, 471
Normal correlation analysis, 478
★ Partial correlation coefficient, 485

Population correlation coefficient, 478
Positive correlation, 471
Product moment, 473
Product-moment coefficient of correlation, 473
Regression sum of squares, 469
Residual sum of squares, 469
Sample covariance, 473
Time series, 476
Total sum of squares, 469

16.7

REVIEW EXERCISES

16.44 If $r = 0.56$ for one set of paired data and $r = -0.97$ for another, compare the strengths of the two relationships.

★ 16.45 Use the least-squares equation obtained in Exercise 15.56 on page 464 to calculate $\hat{y}$ for each of the five restaurants. Then calculate the regression and total sums of squares, and the multiple correlation coefficient.

16.46 If $r = 0.56$ for one set of paired data and $r = -0.97$ for another, compare the strengths of the two relationships.

16.47 Working a homework assignment, a student got $S_{xx} = 145.22$, $S_{xy} = -210.58$, and $S_{yy} = 287.45$ for a given set of paired data. Explain why there must be an error in her calculations.

16.48 A set of $n = 22$ paired observations yielded $r = 0.36$. Test the null hypothesis of no correlation at the 0.05 level of significance:
  (a) using the test based on the Fisher Z transformation;
  (b) using the t statistic of Exercise 16.24 on page 481.

16.49 The following data pertain to a study of the effects of environmental pollution on wildlife; in particular, the relationship between DDT and the thickness of the eggshells of certain birds:

DDT residue in yolk lipids (parts per million)	Thickness of eggshells (millimeters)
117	0.49
65	0.52
303	0.37
98	0.53
122	0.49
150	0.42

Calculate r for these data.

16.50 With reference to the preceding exercise, test at the 0.05 level of significance whether the value obtained for r is significant.

16.51 If $r = 0.25$ for the ages of a group of college students and their knowledge of foreign affairs, what percentage of the variation of their knowledge of foreign affairs can be attributed to differences in age?

16.52 Assuming that the conditions underlying normal correlation analysis are met, use the Fisher Z transformation to construct approximate 95 percent confidence intervals for ρ when:
  (a) $r = 0.44$ and $n = 20$;
  (b) $r = -0.32$ and $n = 38$.

16.53 State in each case whether you would expect a positive correlation, a negative correlation, or no correlation:
  (a) the number of hours that sprinters practice and the time in which they run 100 yards;
  (b) family expenditures on food and family expenditures on clothing;
  (c) daily high humidity and the Dow Jones index;
  (d) malpractice insurance claims and malpractice insurance rates.

★ **16.54** In a multiple regression problem, the residual sum of squares is $\sum (y - \hat{y})^2 = 926$ and the total sum of squares is $\sum (y - \bar{y})^2 = 1,702$. Find the value of the multiple correlation coefficient.

16.55 Calculate r for the data of Exercise 15.52 on page 464 and test the null hypothesis of no correlation at the 0.01 level of significance.

16.56 The following are the numbers of inquiries which a real estate agency received in eight weeks about houses for rent, x, and houses for sale, y:

x	y
60	82
72	85
47	62
38	53
17	29
45	50
33	69
57	88

Calculate r.

16.57 With reference to the preceding exercise, test the null hypothesis $\rho = 0.70$ against the alternative hypothesis $\rho > 0.70$ at the 0.05 level of significance.

16.58 With reference to Exercise 16.56, recalculate r after subtracting 40 from each x and 60 from each y.

★ **16.59** In connection with the study referred to in Exercise 16.17 on page 476, the researcher coded the years -3, -2, -1, 0, 1, 2, 3, and 4, and got $r = -0.9725$ for the coded years and the number of foreign language degrees conferred, and $r = -0.9821$ for the coded years and the ownership of railroad track. Find the partial correlation coefficient for the number of foreign language degrees conferred and the ownership of railroad track when time is held fixed.

16.8
REFERENCES

More detailed information about multiple and partial correlation may be found in

EZEKIEL, M., and FOX, K. A., *Methods of Correlation and Regression Analysis, 3rd ed.* New York: John Wiley & Sons, Inc., 1959.

HARRIS, R. J., *A Primer of Multivariate Statistics.* New York: Academic Press, Inc., 1975.

and an advanced theoretical treatment is given in Volume 2 of

KENDALL, M. G., and STUART, A., *The Advanced Theory of Statistics, 3rd ed.* New York: Hafner Press, 1973.

Volume 1 of this book provides the theoretical foundation of significance tests for r.

NONPARAMETRIC TESTS

Most of the tests discussed in Chapters 11 through 16 require specific assumptions about the population, or populations, sampled. In many cases we must assume that the populations have roughly the shape of normal distributions, that their variances are known or are known to be equal, or that the samples are independent. Since there are many situations where it is doubtful whether all the necessary assumptions can be met, statisticians have developed alternative procedures based on less stringent assumptions, which have become known as **nonparametric tests**.

Aside from the fact that nonparametric tests can be used under more general conditions than the standard tests which they replace, they are often easier to explain and understand; moreover, in many nonparametric tests the computational burden is so light that they come under the heading of "quick and easy" or "shortcut" techniques. For all these reasons, nonparametric tests have become quite popular, and extensive literature is devoted to their theory and application.

In Sections 17.1 and 17.2 we present the sign test as a nonparametric alternative to tests concerning means and tests concerning differences between means based on paired data; another nonparametric test which serves the same purpose but is less wasteful of information, is given in Sections 17.3 and 17.4. In Sections 17.5 through 17.7 we present a nonparametric alternative to tests concerning differences between means based on independent samples and a somewhat similar nonparametric alternative to the one-way analysis of variance. In Sections 17.8 through 17.10 we shall learn how to test the randomness of a sample after the data have actually been obtained; and in Section 17.11 we present a nonparametric test of the significance of a relationship between paired data. In Section 17.12 we mention some of the weaknesses of nonparametric methods, and in Section 17.13 we give a table which lists the various nonparametric tests and the "standard" tests which they replace.

17.1
THE SIGN TEST

Except for the large-sample tests, all the tests concerning means which we studied in Chapter 11 were based on the assumption that the populations sampled have roughly the shape of normal distributions. When this assumption is untenable in practice, these standard tests can be replaced by any one of several nonparametric alternatives, among them the **sign test** which we shall study in this section and in Section 17.2.

Small samples usually cannot tell us whether the population sampled follows a normal distribution. A common indicator of a departure from normality is the occurrence of one or more values which straggle far away from the other sample values. If this is the case, most statisticians would be unwilling to assume that they are sampling a normal population. Without a lot of data, decisions about the appropriateness of normal distributions are quite subjective. The nonparametric procedures used in this chapter are valid under fairly general conditions.

They should certainly be used when there is any doubt about the appropriateness of assuming normal distributions, and they may even be used when the assumption of normal distributions would be reasonable.

The **one-sample sign test** concerns the median $\tilde{\mu}$ of a continuous population. The probability of getting a sample value less than the median and the probability of getting a sample value greater than the median are both $\frac{1}{2}$. If the population values are symmetrically distributed about the median, then the median $\tilde{\mu}$ and the mean μ are equal. Only occasionally are we in a situation which allows us to assume the symmetry of the population. The procedures that follow will usually be described in terms of the median $\tilde{\mu}$, but we should remember that in certain situations the median and the mean are equal.

Then, to test the null hypothesis $\tilde{\mu} = \tilde{\mu}_0$ against an appropriate alternative on the basis of a random sample of size n, we replace each sample value greater than $\tilde{\mu}_0$ with a plus sign and each sample value less than μ_0 with a minus sign. Then we test the null hypothesis that the total number of plus signs is the value of a random variable having the binomial distribution with $p = \frac{1}{2}$. If a sample value equals $\tilde{\mu}_0$ exactly, which is certainly possible with rounded data, it is discarded.

EXAMPLE   The following data constitute a random sample of 15 measurements of the octane rating of a certain kind of gasoline:

$$
\begin{array}{ccccc}
97.5 & 95.2 & 97.3 & 96.0 & 96.8 \\
100.3 & 97.4 & 95.3 & 93.2 & 99.1 \\
96.1 & 97.6 & 98.2 & 98.5 & 94.9
\end{array}
$$

Use the one-sample sign test to test the null hypothesis $\tilde{\mu} = 98.5$ against the alternative hypothesis $\tilde{\mu} < 98.5$ at the 0.05 level of significance.

Solution   Since one of the values is exactly equal to 98.5, the sample size for the one-sample sign test is only $n = 14$.

1. H_0: $\tilde{\mu} = 98.5$
   H_A: $\tilde{\mu} < 98.5$
2. $\alpha = 0.05$
3. The criterion may be based on the number of plus signs or the number of minus signs. Using the number of plus signs, denoted by x, the null hypothesis should be rejected if $x \leq 3$. Table I for $n = 14$ and $p = 0.50$, indicates that $P[x \leq 3] = 0.0289$, which is less than 0.05. $P[x \leq 4]$ exceeds 0.05.
4. Replace each value greater than 98.5 with a plus sign and each value less than 98.5 with a minus sign. The next-to-last value is 98.5 exactly and will be discarded. The sample values then yield

$$
\begin{array}{ccccc}
- & - & - & - & - \\
+ & - & - & - & + \\
- & - & - & & -
\end{array}
$$

The calculation is x itself; here $x = 2$.

5. Since $x = 2 < 3$, the null hypothesis must be rejected. We conclude that the median octane rating of this kind of gasoline is less than 98.5.

This hypothesis test can also be done with p-values. Replace steps 3, 4, and 5 with $3'$, $4'$, and $5'$.

$3'$. The test statistic is x itself.
$4'$. Here $x = 2$. Table I shows that for $n = 14$ and $p = 0.50$ the probability that $x \leq 2$ is $0.001 + 0.006 = 0.007$. The p-value is therefore 0.007.
$5'$. Since 0.007 is less than 0.05, the null hypothesis must be rejected.

Observe that the procedure based on the p-value is a bit simpler because of the structure of Table V. Accordingly, the hypothesis tests in this chapter which use Table V will be discussed only through the p-value.

The sign test can also be used when we deal with paired data as in Section 11.10. In such problems, each pair of sample values is replaced with a plus sign if the first value is greater than the second, with a minus sign if the first value is smaller than the second, and it is discarded when the two values are equal. Then we proceed as before.

The sign test for the paired-sample situation asks whether the population of within-pairs differences has a median which is equal to zero. In most practical situations, the median is zero only when the two populations have the same distribution. The procedure that follows will therefore be described in terms of comparing the means of the two populations.

EXAMPLE To determine the effectiveness of a new traffic control system, the number of accidents that occurred at ten dangerous intersections during four weeks before and four weeks after the installation of the new system was observed and the following data were obtained:

3 and 1,	4 and 2,	2 and 3,	5 and 2,	3 and 3,
2 and 0,	3 and 2,	6 and 3,	1 and 2,	1 and 0

Use the sign test to test the null hypothesis that the new traffic control system is only as effective as the old system at the 0.05 level of significance.

Solution Since one of the pairs, 3 and 3, has to be discarded, the sample size for the sign test is only $n = 9$.

1. H_0: $\mu_1 = \mu_2$, where μ_1 and μ_2 are the mean numbers of accidents in four weeks at a dangerous intersection with the old and the new control systems.
   H_A: $\mu_1 > \mu_2$
2. $\alpha = 0.05$
$3'$. The statistic is x, the number of plus signs.

4'. Replacing each pair of values with a plus sign if the first value is greater than the second or with a minus sign if the first value is smaller than the second, the nine unequal sample pairs yield

$$
\begin{array}{ccccc}
+ & + & - & + & \\
+ & + & + & - & +
\end{array}
$$

Thus $x = 7$. Using Table V for $n = 9$ and $p = 0.50$, the probability of $x \geq 7$ is $0.070 + 0.018 + 0.002 = 0.090$. The p-value is thus 0.090.

5'. Since 0.090 exceeds 0.05, the null hypothesis cannot be rejected. In other words, we cannot conclude on the basis of this test that the new control system is more effective than the old system.

The test we have described here is only one of several nonparametric ways of analyzing such paired sample data. Another popular test used for this purpose is the Wilcoxon signed-rank test, which may be found in the books listed in the Bibliography at the end of the chapter.

17.2

THE SIGN TEST (Large Samples)

When np and $n(1 - p)$ are both greater than 5, so that we can use the normal approximation to the binomial distribution, the sign test may be based on the large-sample test of Section 13.3, namely, on the statistic

$$
z = \frac{x - np_0}{\sqrt{np_0(1 - p_0)}}
$$

with $p_0 = 0.50$, which has approximately the standard normal distribution.

EXAMPLE The following are measurements of the ocean depth in a certain location (in fathoms): 46.4, 48.3, 51.9, 38.8, 46.5, 45.6, 52.1, 41.0, 54.2, 44.9, 52.3, 43.6, 48.7, 42.2, and 44.9. Use the sign test at the level of significance $\alpha = 0.05$ to test the null hypothesis $\tilde{\mu} = 43.0$ (the previously recorded ocean depth in that location) against the alternative hypothesis $\tilde{\mu} \neq 43.0$.

Solution
1. H_0: $\tilde{\mu} = 43.0$
   H_A: $\tilde{\mu} \neq 43.0$
2. $\alpha = 0.05$
3. Reject the null hypothesis if $z \leq -1.96$ or $z \geq 1.96$, where

$$
z = \frac{x - np_0}{\sqrt{np_0(1 - p_0)}}
$$

with $p_0 = 0.50$; otherwise, accept it or reserve judgment.

4. Replacing each value greater than 43.0 with a plus sign and each value less than 43.0 with a minus sign, we get

$$+ \; + \; + \; - \; + \; + \; + \; - \; + \; + \; + \; + \; + \; - \; +$$

Thus, $x = 12$ and substitution into the formula yields

$$z = \frac{12 - 15(0.50)}{\sqrt{15(0.50)(0.50)}} = 2.32$$

5. Since $z = 2.32$ exceeds 1.96, the null hypothesis must be rejected; in other words, the median ocean depth at the given location is not 43.0 fathoms, as had been previously reported.

The steps which use the p-value are these.

3′. The test statistic is

$$z = \frac{x - np_0}{\sqrt{np_0(1 - p_0)}}$$

4′. Find, as above, that $x = 12$ and substitute into the formula to obtain $z = 2.32$. Table V indicates that $P[|z| \geq 2.32] = 2(0.5000 - 0.4898) = 0.0204$, which is the p-value.

5′. Since $0.0204 \leq 0.05$, the null hypothesis must be rejected.

EXERCISES

17.1 On 12 occasions, Ms. Brown had to wait 3, 6, 7, 6, 4, 8, 6, 2, 8, 6, 1, and 9 minutes for the bus that takes her to work. Use the sign test based on Table I and the 0.05 level of significance to test the null hypothesis $\tilde{\mu} = 5$ against the alternative hypothesis $\tilde{\mu} \neq 5$.

17.2 Nine women buying new eyeglasses tried on, respectively, 12, 11, 14, 15, 10, 14, 11, 8, and 12 different frames. Use the sign test at the 0.05 level of significance to test the null hypothesis $\tilde{\mu} = 10$ against the alternative hypothesis $\tilde{\mu} > 10$.

17.3 In six rounds of golf at the Paradise Valley Country Club, a professional scored 71, 69, 72, 74, 71, and 72. Use the sign test at the 0.05 level of significance to test the null hypothesis $\tilde{\mu} = 70$ against the alternative hypothesis $\tilde{\mu} > 70$.

17.4 The following data are the weights (in grams) of 14 packages of a certain kind of candy: 101.0, 99.8, 100.9, 103.6, 97.1, 100.0, 102.5, 100.5, 101.0, 98.2, 100.3, 102.6, 100.0, and 100.8. Use the sign test based on Table I and the level of significance $\alpha = 0.05$ to test the null hypothesis $\tilde{\mu} = 100.0$ against the alternative hypothesis $\tilde{\mu} \neq 100.0$.

17.5 The following are the numbers of passengers carried on flights 136 and 137 between Chicago and Phoenix on 12 days: 232 and 189, 265 and 230, 249 and 236, 250 and 261, 255 and 249, 236 and 218, 270 and 258, 266 and 253, 249 and 251, 240 and 233, 257 and 254, and 239 and 249. Use the sign test based on Table I and the 0.05 level of significance to test the null hypothesis $\mu_1 = \mu_2$ against the alternative hypothesis $\mu_1 > \mu_2$.

17.6 The following are the grades which 15 students received on the midterm and final examinations in a course in European history: 66 and 73, 88 and 91, 75 and 78, 90 and 86, 63 and 69, 58 and 67, 75 and 75, 82 and 80, 73 and 76, 84 and 89, 85 and 81, 93 and 96, 70 and 76, 85 and 82, and 90 and 97. Use the sign test based on Table I and the 0.05 level of significance to test the null hypothesis $\mu_1 = \mu_2$ against the alternative hypothesis $\mu_1 < \mu_2$.

17.7 The following are the numbers of speeding tickets issued by two police officers on 20 days: 6 and 9, 11 and 13, 12 and 12, 10 and 17, 15 and 13, 7 and 11, 9 and 13, 7 and 12, 14 and 15, 11 and 13, 14 and 10,

6 and 12, 9 and 9, 12 and 14, 8 and 13, 16 and 11, 10 and 15, 12 and 14, 15 and 15, and 12 and 18. Use the sign test based on the normal approximation to the binomial distribution and the level of significance $\alpha = 0.01$ to test the null hypothesis that on the average the two police officers issue equally many speeding tickets.

17.8 The following are data on the daily sulfur oxides emission of an industrial plant (in tons):

17	15	20	29	19	18	22	25	27	9
24	20	17	6	24	14	15	23	24	26
19	23	28	19	16	22	24	17	20	13
19	10	23	18	31	13	20	17	24	14

Use the sign test and the 0.01 level of significance to test the null hypothesis $\tilde{\mu} = 23$ against the alternative hypothesis $\tilde{\mu} < 23$.

17.9 Rework Exercise 17.4 using the sign test based on the normal approximation to the binomial distribution.

17.10 The following are the miles per gallon obtained with 40 tankfuls of a certain kind of gas

24.1	24.8	24.2	24.2	24.5
24.5	24.0	23.8	24.5	24.0
25.2	24.7	24.6	24.1	24.2
24.8	25.6	25.1	24.3	24.7
25.0	24.3	25.3	23.6	24.4
23.2	23.8	25.3	24.6	25.2
24.4	24.1	24.9	25.8	24.2
24.1	24.5	24.6	25.2	23.3

Use the sign test at the 0.01 level of significance to test the null hypothesis $\tilde{\mu} = 24.2$ against the alternative hypothesis $\tilde{\mu} > 24.2$.

17.11 The following are the numbers of artifacts dug up by two archaeologists at an ancient cliff dwelling on 30 days: 1 and 0, 0 and 0, 2 and 1, 3 and 0, 1 and 2, 0 and 0, 2 and 0, 2 and 1, 3 and 1, 0 and 2, 1 and 0, 1 and 1, 4 and 2, 1 and 1, 2 and 1, 1 and 0, 3 and 2, 5 and 2, 2 and 6, 1 and 0, 3 and 2, 2 and 3, 4 and 0, 1 and 2, 3 and 1, 2 and 0, 0 and 1, 2 and 0, 4 and 1, and 2 and 0. Use the sign test at the level of significance $\alpha = 0.01$ to test the null hypothesis that the two archaeologists are equally good at finding artifacts against the alternative hypothesis that the first one is better.

17.12 Use the large-sample sign test to rework Exercise 11.110 on page 329.

17.13 The following are the numbers of employees absent from two departments of a large firm on 25 days: 4 and 3, 2 and 5, 6 and 6, 3 and 6, 1 and 4, 2 and 4, 5 and 2, 1 and 4, 3 and 4, 6 and 5, 2 and 5, 7 and 1, 4 and 6, 1 and 3, 2 and 5, 0 and 3, 6 and 5, 4 and 6, 1 and 2, 4 and 1, 2 and 4, 0 and 1, 5 and 3, 2 and 3, and 2 and 4. Use the sign test at the 0.05 level of significance to test the null hypothesis $\mu_1 = \mu_2$ against the alternative hypothesis $\mu_1 < \mu_2$.

17.3

RANK SUMS: THE *U* TEST

In this section we shall present a nonparametric alternative to the small-sample *t* test concerning the difference between two means. It is called the **U test**, the **Wilcoxon test**, or the **Mann–Whitney test**, named after the statisticians who contributed to its development. The three different names refer to different methods of organizing the calculations, but the procedures are logically equivalent. We will be able to test the null hypothesis that the two samples come from identical populations without having to assume that the populations sampled have normal distributions; in fact, the test requires only that the populations sampled are continuous to avoid ties, and in practice it does not even matter whether this assumption is satisfied.

Suppose that you randomly select samples from each of the two populations being compared. The U test, as originally defined, is based on the number of times values of the sample from one population are exceeded by values from the other population. This is a somewhat complicated notion, and our work here will be described in terms of the ranks of the jointly ranked data from the two populations.

To illustrate how the U test is thus performed, suppose that we want to compare the grain size of sand obtained from two different locations on the moon on the basis of the following diameters (in millimeters):

Location 1: 0.37, 0.70, 0.75, 0.30, 0.45, 0.16, 0.62, 0.73, 0.33
Location 2: 0.86, 0.55, 0.80, 0.42, 0.97, 0.84, 0.24, 0.51, 0.92, 0.69

The means of these two samples are 0.49 and 0.68, and their difference is large, but it remains to be seen whether it is significant.

To perform the U test, we first arrange the data jointly, as if they comprise one sample, in an increasing order of magnitude. For our data, we get

0.16	0.24	0.30	0.33	0.37	0.42	0.45	0.51	0.55	0.62
1	2	1	1	1	2	1	2	2	1

0.69	0.70	0.73	0.75	0.80	0.84	0.86	0.92	0.97
2	1	1	1	2	2	2	2	2

where we indicated for each value whether it came from location 1 or location 2. Assigning the data, in this order, the ranks 1, 2, 3, . . . , and 19, we find that the values of the first sample (location 1) occupy ranks 1, 3, 4, 5, 7, 10, 12, 13, and 14, while those of the second sample (location 2) occupy ranks 2, 6, 8, 9, 11, 15, 16, 17, 18, and 19. There are no ties here, but if there were, we would assign each of the tied observations the mean of the ranks which they jointly occupy. For instance, if the third and fourth values were the same, we would assign each the rank $\frac{3+4}{2} = 3.5$, and if the ninth, tenth, and eleventh values were the same, we would assign each the rank $\frac{9+10+11}{3} = 10$.

Now, if there is an appreciable difference between the means of the two populations, most of the lower ranks are likely to go to the values of one sample while most of the higher ranks are likely to go to the values of the other sample. The test of the null hypothesis that the two samples come from identical populations may thus be based on W_1, the sum of the ranks of the values of the first sample, or on W_2, the sum of the ranks of the values of the second sample. In practice, it does not matter which sample we refer to as sample 1 and which sample we refer to as sample 2, and whether we base the test on W_1 or W_2.[†] If the sample sizes

---

[†] When the sample sizes are unequal, it is common practice to let n_1 be the smaller of the two; this is not required, however, for the work in this book.

are n_1 and n_2, the sum of W_1 and W_2 is simply the sum of the first $n_1 + n_2$ positive integers, which is known to equal

$$\frac{(n_1 + n_2)(n_1 + n_2 + 1)}{2}$$

This formula enables us to find W_2 if we know W_1, and vice versa. For our illustration we get

$$W_1 = 1 + 3 + 4 + 5 + 7 + 10 + 12 + 13 + 14$$
$$= 69$$

and since the sum of the first 19 positive integers is $\dfrac{19 \cdot 20}{2} = 190$, it follows that $W_2 = 190 - 69 = 121$. (This value may be checked by actually adding 2, 6, 8, 9, 11, 15, 16, 17, 18, and 19.)

When the use of **rank sums** was first proposed as a nonparametric alternative to the two-sample t test, the decision was based on W_1 or W_2. Nowadays, the decision is based on either of the related statistics

U_1 and U_2 statistics

$$U_1 = W_1 - \frac{n_1(n_1 + 1)}{2}$$

or

$$U_2 = W_2 - \frac{n_2(n_2 + 1)}{2}$$

or on the statistic U, which always equals the smaller of the two. The resulting tests are equivalent to those based on W_1 or W_2, but they have the advantage that they lend themselves more readily to the construction of tables of critical values. Not only do U_1 and U_2 take on values on the interval from 0 to $n_1 n_2$—indeed, their sum is always equal to $n_1 n_2$—but their sampling distributions are symmetrical about $\dfrac{n_1 n_2}{2}$. The use of U, which always equals the smaller of the values of U_1 and U_2, has the added advantage that the resulting test is one-tailed, and hence easier to tabulate.

Accordingly, we test the null hypothesis that the two samples come from identical populations against the alternative hypothesis that the two populations have unequal means with the following criterion:

Reject the null hypothesis if

$$U \le U'_\alpha$$

where U'_α is given in Table VII for $n_1 \le 15$, $n_2 \le 15$, and $\alpha = 0.05$ or $\alpha = 0.01$.

In the construction of Table VII, U'_α is the largest value of U for which the probability of $U \le U'_\alpha$ is less than or equal to α, and the blank spaces indicate that the null hypothesis cannot be rejected at the given level of significance regardless of the value which we obtain for U. More extensive tables may be found in handbooks of statistical tables, but when n_1 and n_2 are both greater than 8, it is generally considered reasonable to use the large-sample test described in Section 17.4.

Observe that the null hypothesis is rejected for small values of U. In every case, either $U = U_1$ or $U = U_2$. When U actually equals U_1, then it is sample 1 which is producing the small values. Similarly, when U is equal to U_2, then sample 2 is producing the small values.

EXAMPLE   With reference to the grain-size data on page 497, use the U test at the 0.05 level of significance to test the null hypothesis that the two samples come from identical populations against the alternative hypothesis that the two populations have unequal means.

Solution   1. H_0:   Populations are identical.
  H_A: $\mu_1 \neq \mu_2$
2. $\alpha = 0.05$
3. Reject the null hypothesis if $U \leq 20$, which is the value of U'_α for $n_1 = 9$, $n_2 = 10$, and $\alpha = 0.05$; otherwise, accept it or reserve judgment.
4. Having shown that $W_1 = 69$ and $W_2 = 121$, we get

$$U_1 = 69 - \frac{9 \cdot 10}{2} = 24$$

$$U_2 = 121 - \frac{10 \cdot 11}{2} = 66$$

and, hence, $U = 24$. Note that $U_1 + U_2 = 24 + 66 = 90$, which equals $n_1 n_2 = 9 \cdot 10$.
5. Since $U = 24$ exceeds 20, the null hypothesis cannot be rejected; in other words, we cannot conclude that there is a real difference in the mean grain size of sand from the two locations on the moon. The organization of Table VII does not facilitate the use of p-values. For these data we are able to state only that $p > 0.05$.

The test which we have described here can also be used when the alternative is $\mu_1 < \mu_2$ or $\mu_1 > \mu_2$. However, since the procedure is more complicated in that case—we will have to use U_1 or U_2 instead of U—we shall discuss it only for large samples in Section 17.4.

17.4
RANK SUMS: THE U TEST (Large Samples)

The large-sample U test may be based on either U_1 or U_2, as given on page 498. The resulting tests will be equivalent, so that it does not matter which sample we identify as sample 1. In the description that follows, we shall use the statistic U_1.

Under the null hypothesis that the two samples come from identical populations, it can be shown that the mean and the standard deviation of the sampling distribution of U_1 are[†]

Mean and standard deviation of U_1 statistic

$$\mu_{U_1} = \frac{n_1 n_2}{2}$$

and

$$\sigma_{U_1} = \sqrt{\frac{n_1 n_2 (n_1 + n_2 + 1)}{12}}$$

Furthermore, if n_1 and n_2 are both greater than 8, the sampling distribution of U_1 can be approximated closely with a normal curve. Thus, we base the test of the null hypothesis that the two samples come from identical populations on the statistic

Statistic for large-sample U test

$$z = \frac{U_1 - \mu_{U_1}}{\sigma_{U_1}}$$

which has approximately the standard normal distribution. If the alternative hypothesis is $\mu_1 \neq \mu_2$, we reject the null hypothesis for $z \leq -z_{\alpha/2}$ or $z \geq z_{\alpha/2}$; if the alternative hypothesis is $\mu_1 < \mu_2$, we reject the null hypothesis for $z \leq -z_\alpha$ since small values of U_1 correspond to small values of the rank sum W_1. If the alternative hypothesis is $\mu_1 > \mu_2$, we reject the null hypothesis for $z \geq z_\alpha$ since large values of U_1 correspond to large values of the rank sum W_1.

Finding ranks when n_1 and n_2 are large can be a surprisingly messy task. However, you can do this easily by first constructing a stem-and-leaf display for the values of the first sample. Then, using a different-color pencil, continue the stem-and-leaf display for the values of the second sample.

EXAMPLE   The following are the weight gains (in pounds) of young turkeys, which are fed two different diets but are otherwise kept under identical conditions:

Diet 1:   16.3, 10.1, 10.7, 13.5, 14.9, 11.8, 14.3, 10.2, 12.0, 14.7, 23.6, 15.1, 14.5, 18.4, 13.2, 14.0

Diet 2:   21.3, 23.8, 15.4, 19.6, 12.0, 13.9, 18.8, 19.2, 15.3, 20.1, 14.8, 18.9, 20.7, 21.1, 15.8, 16.2

Use the large-sample U test at the 0.01 level of significance to test the null hypothesis that the two populations sampled are identical against the alternative hypothesis that on the average the second diet produces a greater gain in weight.

[†] If there are ties in rank, these formulas provide only approximations, but if the number of ties is small, there is no need to make a correction.

500   CHAP. 17 / NONPARAMETRIC TESTS

Solution

1. H_0: Populations are identical.
   H_A: $\mu_1 < \mu_2$
2. $\alpha = 0.01$
3. Reject the null hypothesis if $z \leq -2.33$, where

$$z = \frac{U_1 - \mu_{U_1}}{\sigma_{U_1}}$$

Otherwise, accept it or reserve judgment.

4. Arranging the data according to size, we get

10.1	10.2	10.7	11.8	12.0	12.0	13.2	13.5	13.9	14.0
1	1	1	1	1	2	1	1	2	1

14.3	14.5	14.7	14.8	14.9	15.1	15.3	15.4	15.8	16.2
1	1	1	2	1	1	2	2	2	2

16.3	18.4	18.8	18.9	19.2	19.6	20.1	20.7	21.1	21.3
1	1	2	2	2	2	2	2	2	2

23.6	23.8
1	2

Below the weight gains are listed the samples from which they came. Observe that there is tie at 12.0 pounds, involving one turkey from each diet. These are the fifth and sixth positions in the list, so each is assigned the rank $\frac{5+6}{2} = 5.5$. The sum of the rank positions occupied by the turkeys of the first sample (diet 1) is

$$\begin{aligned}
W_1 = \quad & 1 + \ 2 + \ 3 + 4 + 5.5 \qquad + 7 + 8 \qquad + 10 \\
& + 11 + 12 + 13 \qquad + 15 + 16 \\
& + 21 + 22 \\
& + 31 \\
= \ & 181.5
\end{aligned}$$

and

$$U_1 = 181.5 - \frac{16 \cdot 17}{2} = 45.5$$

Since

$$\mu_{U_1} = \frac{16 \cdot 16}{2} = 128 \quad \text{and} \quad \sigma_{U_1} = \sqrt{\frac{16 \cdot 16 \cdot 33}{12}} = 26.53,$$

it follows that

$$z = \frac{45.5 - 128}{26.53} = -3.11$$

5. Since $z = -3.11$ is less than -2.33, the null hypothesis must be rejected; in other words, we conclude that on the average the second diet produces a greater gain in weight.

The procedure using p-values replaces steps 3, 4, and 5 with these.

3′. The test statistic is

$$z = \frac{U_1 - \mu_{U_1}}{\sigma_{U_1}}$$

4′. Find, as shown above, that $U_1 = 45.5$, $\mu_{U_1} = 128$, and $\sigma_{U_1} = 26.53$, leading to $z = -3.11$. Table II shows that $P[z \leq -3.11]$ is less than 0.0010. Therefore $p < 0.0010$.

5′. Certainly $p < 0.01$, so that the null hypothesis must be rejected.

EXERCISES

17.14 The following are figures on the number of assaults committed in a city in six weeks in the spring and in six weeks in the fall:

> Spring: 46, 37, 42, 48, 38, 45
> Fall: 35, 30, 25, 39, 28, 32

Use the U test at the level of significance 0.01 to check the claim that on the average there are equally many assaults per week in the given city in the spring and in the fall.

17.15 The following are the Rockwell hardness numbers obtained for six aluminum die castings randomly selected from production lot A and for eight aluminum die castings randomly selected from production lot B:

> *Production lot A:*   75, 56, 63, 70, 58, 74
> *Production lot B:*   63, 85, 77, 80, 86, 76, 72, 82

Use the U test at the 0.05 level of significance to check the claim that the average hardness of die castings from the two production lots is the same.

17.16 Tests made on two kinds of 9-volt batteries showed the following lifetimes (in hours) of continuous use:

> *Brand A:*   11.7, 12.0, 10.8, 11.1, 11.9, 12.9, 12.4
> *Brand B:*   11.5, 12.8, 13.5, 13.6, 11.1, 12.4, 13.3

Use the U test at the 0.05 level of significance to test whether the difference between the two sample means, 11.8 and 12.6, can be attributed to chance.

17.17 The following are the numbers of misprints counted on pages selected at random from two Sunday editions of a newspaper:

> *May 10:*   12, 6, 11, 11, 15, 7
> *May 24:*   10, 3, 6, 8, 7, 5

Use the U test at the level of significance $\alpha = 0.05$ to test the null hypothesis that the two samples come from identical populations against the alternative hypothesis that the two populations have unequal means.

17.18 The following are the numbers of minutes it took a sample of 15 men and 12 women to complete a short screening test given to job applicants at a large bank:

> *Men:*   8.8, 7.8, 6.6, 10.7, 8.9, 8.4, 6.9, 6.4, 6.3,
>   8.0, 8.6, 8.1, 9.1, 9.7, 9.9
>
> *Women:* 7.5, 8.7, 8.3, 6.2, 6.5, 7.7, 9.8, 9.6, 9.2,
>   10.4, 8.2, 8.5

Use the U test based on Table VII and the 0.01 level of significance to test the null hypothesis that the two samples come from identical populations against the alternative hypothesis that the two populations have unequal means.

17.19 The following are the scores which samples of students from two minority groups obtained on a current events test:

> *Minority group 1:* 70, 62, 91, 55, 72, 94, 80, 96, 73, 44, 87, 78
>
> *Minority group 2:* 81, 23, 71, 30, 71, 54, 64, 93, 58, 41, 47, 56

Use the large-sample U test at the 0.05 level of significance to test whether students from the two minority groups can be expected to score equally well on this test.

17.20 Use the normal approximation to the sampling distribution of U to rework the example on page 497, which dealt with the grain size of sand from two locations on the moon.

17.21 Comparing two kinds of emergency flares, a consumer testing service obtained the following burning times (rounded to the nearest tenth of a minute):

> *Brand X:* 17.2, 18.1, 21.2, 19.3, 14.4, 21.1, 14.6, 19.1, 18.8, 15.2, 20.3, 17.5

> *Brand Y:* 13.6, 13.7, 11.8, 14.6, 15.2, 14.3, 22.5, 12.3, 13.5, 10.9, 14.4, 8.0

Use the large-sample U test at the 0.01 level of significance to see whether it is reasonable to say that in general brand X flares are better (last longer) than brand Y flares.

17.22 Use the large-sample U test to rework Exercise 17.18.

17.23 The following are the weekly food expenditures (in dollars) of families with two children chosen at random from two suburbs of a large city:

Suburb A:	278.60	267.89	286.45	271.15
	270.78	275.38	264.19	
	270.50	272.00	267.95	

Suburb B:	262.63	275.91	278.19	260.78
	263.12	255.35	263.76	
	275.16	266.51	271.72	

Use the large-sample U test at the 0.05 level of significance to check the claim that on the average such weekly food expenditures are higher in suburb A than in suburb B.

17.5

RANK SUMS: THE H TEST

The **H test**, or **Kruskal–Wallis test**, is a rank-sum test which serves to test the null hypothesis that k independent random samples come from identical populations against the alternative hypothesis that the means of these populations are not all equal. Unlike the standard test which it replaces, the one-way analysis of variance of Section 14.3, it does not require the assumption that the populations sampled have, at least approximately, normal distributions.

As in the U test, the data are ranked jointly from low to high as though they constitute a single sample. Then, if R_i is the sum of the ranks assigned to the n_i values of the ith sample and $n = n_1 + n_2 + \cdots + n_k$, the H test is based on the statistic

Statistic for H test

$$H = \frac{12}{n(n + 1)} \sum_{i=1}^{k} \frac{R_i^2}{n_i} - 3(n + 1)$$

If the null hypothesis is true and each sample has at least five observations, the sampling distribution of H can be approximated closely with a chi-square distribution with $k - 1$ degrees of freedom. Consequently, we reject the null hypothesis that the populations sampled are identical and accept the alternative hypothesis that the means of these populations are not all equal, if the value we get for H is greater than or equal to χ_α^2 for $k - 1$ degrees of freedom.

EXAMPLE The following are the final examination scores of samples of students who are taught German by three different methods (classroom instruction and language laboratory, only classroom instruction, and only self-study in language laboratory):

Method 1:	94, 87, 91, 74, 87, 97
Method 2:	85, 82, 79, 84, 61, 72, 80
Method 3:	89, 67, 72, 76, 69

Use the H test at the 0.05 level of significance to test the null hypothesis that the three populations sampled are identical against the alternative hypothesis that their means are not all equal.

Solution 1. H_0: Populations are identical.
    H_A: The population means are not all equal.
  2. $\alpha = 0.05$
  3. Reject the null hypothesis if $H \geq 5.991$, the value of $\chi_{0.05}^2$ for $3 - 1 = 2$ degrees of freedom, where H is calculated in accordance with the formula above. Otherwise, accept it or reserve judgment.
  4. Arranging the data jointly according to size, we get 61, 67, 69, 72, 72, 74, 76, 79, 80, 82, 84, 85, 87, 87, 89, 91, 94, and 97. Assigning the data, in this order, the ranks 1, 2, 3, . . . , and 18, we find that the values of the first sample occupy ranks 6, 13, 14, 16, 17, and 18, while those of the second sample occupy ranks 1, 4.5, 8, 9, 10, 11, and 12, and those of the third sample occupy ranks 2, 3, 4.5, 7, and 15. (Since the two 87's belong to the same sample, we simply assign them ranks 13 and 14.) Thus,

$$R_1 = 6 + 13 + 14 + 16 + 17 + 18 = 84$$

$$R_2 = 1 + 4.5 + 8 + 9 + 10 + 11 + 12 = 55.5$$

$$R_3 = 2 + 3 + 4.5 + 7 + 15 = 31.5$$

and it follows that

$$H = \frac{12}{18 \cdot 19} \left(\frac{84^2}{6} + \frac{55.5^2}{7} + \frac{31.5^2}{5} \right) - 3 \cdot 19$$

$$= 6.67$$

  5. Since $H = 6.67$ exceeds 5.991, the null hypothesis must be rejected; in other words, we conclude that the three methods of teaching German are not all equally effective.

The structure of the chi-square table, Table III, does not allow us to determine the p-value exactly. However, we are able to determine that $0.01 < p < 0.05$.

EXERCISES

17.24 The following are the miles per gallon which a test driver got for six tankfuls each of three kinds of gasoline:

Gasoline 1:	28, 23, 26, 31, 14, 29
Gasoline 2:	21, 31, 32, 19, 27, 16
Gasoline 3:	24, 17, 21, 31, 22, 18

Use the H test at the level of significance $\alpha = 0.05$ to check the claim that there is no difference in the true average mileage yield of the three kinds of gasoline.

17.25 To compare four bowling balls, a professional bowler bowled five games with each ball and got the following results:

Bowling ball D:	221, 232, 207, 198, 212
Bowling ball E:	202, 225, 252, 218, 226
Bowling ball F:	210, 205, 189, 196, 216
Bowling ball G:	229, 192, 247, 220, 208

Use the H test at the 0.05 level of significance to test the null hypothesis that the bowler performs equally well with the four bowling balls.

17.26 Three groups of guinea pigs were injected, respectively, with 0.5 mg, 1.0 mg, and 1.5 mg of a tranquilizer, and the following are the numbers of seconds it took them to fall asleep:

0.5-mg dose:	8.2, 10.0, 10.2, 13.7, 14.0, 7.8, 12.7, 10.9
1.0-mg dose:	9.7, 13.1, 11.0, 7.5, 13.3, 12.5, 8.8, 12.9, 7.9, 10.5
1.5-mg dose:	12.0, 7.2, 8.0, 9.4, 11.3, 9.0, 11.5, 8.5

Use the H test at the level of significance $\alpha = 0.01$ to test the null hypothesis that the differences in dosage have no effect on the length of time it takes guinea pigs to fall asleep.

17.27 Use the H test to rework Exercise 14.28 on page 416.

17.6
TESTS OF RANDOMNESS: RUNS

All the methods of inference which we have discussed are based on the assumption that the samples are random; yet, there are many applications where it is difficult to decide whether this assumption is justifiable. This is true, particularly, when we have little or no control over the selection of the data, as is the case, for example, when we rely on whatever records are available to make long-range predictions of the weather, when we use whatever data are available to estimate the mortality rate of a disease, or when we use sales records for past months to make predictions of a department store's sales. None of this information constitutes a random sample in the strict sense.

There are several methods of judging the randomness of a sample on the basis of the order in which the observations are obtained; they enable us to decide, after the data have been collected, whether patterns that look suspiciously nonrandom may be attributed to chance. The technique we shall describe here and in the next two sections is based on the **theory of runs**.

A **run** is a succession of identical letters (or other kinds of symbols) which is followed and preceded by different letters or no letters at all. To illustrate, consider the following arrangement of healthy, H, and diseased, D, elm trees that were planted many years ago along a country road:

$$\underline{H\ H\ H\ H}\ \underline{D\ D\ D}\ \underline{H\ H\ H\ H\ H\ H\ H}\ \underline{D\ D}\ \underline{H\ H}\ \underline{D\ D\ D\ D}$$

Using underlines to combine the letters which constitute the runs, we find that there is first a run of four H's, then a run of three D's, then a run of seven H's, then a run of two D's, then a run of two H's, and finally a run of four D's.

The **total number of runs** appearing in an arrangement of this kind is often a good indication of a possible lack of randomness. If there are too few runs we might suspect a definite grouping or clustering, or perhaps a trend; if there are too many runs, we might suspect some sort of repeated alternating, or cyclical, pattern. In the example above there seems to be a definite clustering—the diseased trees seem to come in groups—but it remains to be seen whether this is significant or whether it can be attributed to chance.

If there are n_1 letters of one kind, n_2 letters of another kind, and u runs, we base this kind of decision on the following criterion:

Reject the null hypothesis of randomness if

$$u \leq u'_{\alpha/2} \qquad \text{or} \qquad u \geq u_{\alpha/2}$$

where $u'_{\alpha/2}$ and $u_{\alpha/2}$ are given in Table VIII for $n_1 \leq 15$, $n_2 \leq 15$, and $\alpha = 0.05$ or $\alpha = 0.01$.

In the construction of Table VIII, $u'_{\alpha/2}$ is the largest value of u for which the probability of $u \leq u'_{\alpha/2}$ is less than or equal to $\alpha/2$, $u_{\alpha/2}$ is the smallest value of u for which the probability of $u \geq u_{\alpha/2}$ is less than or equal to $\alpha/2$, and the blank spaces indicate that the null hypothesis of randomness cannot be rejected for values in that tail of the sampling distribution of u regardless of the value which we obtain for u. More extensive tables for the **u test** may be found in handbooks of statistical tables, but when n_1 and n_2 are both at least 10, it is generally considered reasonable to use the large-sample test described in Section 17.7.

EXAMPLE With reference to the arrangement of healthy and diseased elm trees given above use the u test at the 0.05 level of significance to test the null hypothesis of randomness against the alternative hypothesis that the arrangement is not random.

Solution

1. H_0: Arrangement is random.
   H_A: Arrangement is not random.
2. $\alpha = 0.05$
3. Since $n_1 = 13$, $n_2 = 9$, and $\alpha = 0.05$, we get $u'_{0.025} = 6$ and $u_{0.025} = 17$ from Table VIII; thus, the null hypothesis must be rejected if $u \leq 6$ or $u \geq 17$. Otherwise, accept it or reserve judgment.
4. $u = 6$, as can be seen from the data.
5. Since $u = 6$ is less than or equal to 6, the null hypothesis must be rejected; in other words, we conclude that the arrangement of healthy and diseased elm trees is not random. Indeed, it seems that the diseased trees come in clusters.

The structure of Table VIII does not allow us to determine the p-value exactly. However, we are able to determine that $0.01 < p \leq 0.05$.

17.7

TESTS OF RANDOMNESS: RUNS (Large Samples)

Under the null hypothesis that n_1 letters of one kind and n_2 letters of another kind are arranged at random, it can be shown that the mean and the standard deviation of u, the total number of runs, are

Mean and standard deviation of u

$$\mu_u = \frac{2n_1 n_2}{n_1 + n_2} + 1$$

and

$$\sigma_u = \sqrt{\frac{2n_1 n_2 (2n_1 n_2 - n_1 - n_2)}{(n_1 + n_2)^2 (n_1 + n_2 - 1)}}$$

Furthermore, if neither n_1 nor n_2 is less than 10, the sampling distribution of u can be approximated closely with a normal curve. Thus, we base the test of the null hypothesis of randomness on the statistic

Statistic for large-sample u test

$$z = \frac{u - \mu_u}{\sigma_u}$$

which has approximately the standard normal distribution. If the alternative hypothesis is that the arrangement is not random, we reject the null hypothesis for $z \leq -z_{\alpha/2}$ or $z \geq z_{\alpha/2}$; if the alternative hypothesis is that there is a clustering or a trend, we reject the null hypothesis for $z \leq -z_\alpha$; and if the alternative hypothesis is that there is an alternating, or cyclical, pattern, we reject the null hypothesis for $z \geq z_\alpha$.

EXAMPLE The following is an arrangement of men, M, and women, W, lined up to purchase tickets for a rock concert:

$$M\ W\ M\ M\ W\ M\ M\ M\ W\ M\ W\ M\ M\ M\ W\ W\ M$$
(cont.) $\quad M\ M\ M\ W\ W\ M\ W\ M\ M\ M\ W\ M\ M\ M\ W\ W$
(cont.) $\quad W\ M\ W\ M\ M\ M\ W\ M\ W\ M\ M\ M\ M\ W\ W\ M$

Test for randomness at the 0.05 level of significance.

Solution
1. H_0: Arrangement is random.
   H_A: Arrangement is not random.
2. $\alpha = 0.05$
3. Reject the null hypothesis if $z \leq -1.96$ or $z \geq 1.96$, where

$$z = \frac{u - \mu_u}{\sigma_u}$$

Otherwise, accept it or reserve judgment.
4. Since $n_1 = 30$, $n_2 = 18$, and $u = 27$, we get

$$\mu_u = \frac{2 \cdot 30 \cdot 18}{30 + 18} + 1 = 23.5$$

$$\sigma_u = \sqrt{\frac{2 \cdot 30 \cdot 18(2 \cdot 30 \cdot 18 - 30 - 18)}{(30 + 18)^2(30 + 18 - 1)}} = 3.21$$

and, hence,

$$z = \frac{27 - 23.5}{3.21} = 1.09$$

5. Since $z = 1.09$ falls between -1.96 and 1.96, the null hypothesis of randomness cannot be rejected: there is no real evidence of any lack of randomness.

To use the p-value replace steps 3, 4, and 5 with these.

3'. The test statistic is

$$z = \frac{\mu - \mu_u}{\sigma_u}$$

4'. Find, as shown above, that $u = 27$, $\mu_u = 23.5$, and $\sigma_u = 3.21$, leading to $z = 1.09$. Table I shows that $P[|z| \geq 1.09] = 0.2757$, which is the p-value.

5'. Since p is greater than $x = 0.05$, the null hypothesis must be accepted.

17.8

TESTS OF RANDOMNESS: RUNS ABOVE AND BELOW THE MEDIAN

The u test is not limited to tests of the randomness of sequences of attributes, such as the H's and D's, or M's and W's, of our examples. Any sample consisting of numerical measurements or observations can be treated similarly by using the letters a and b to denote, respectively, values falling above and below the median of the sample. Numbers equal to the median are omitted. The resulting sequence of a's and b's (representing the data in their original order) can then be tested for randomness on the basis of the total number of runs of a's and b's, namely, the total number of **runs above and below the median**. Depending on the size of n_1 and n_2, we use Table VIII or the large-sample test of Section 17.7.

EXAMPLE On 24 successive runs between two cities, a bus carried 24, 19, 32, 28, 21, 23, 26, 17, 20, 28, 30, 24, 13, 35, 26, 21, 19, 29, 27, 18, 26, 14, 21, and 23 passengers. Use the total number of runs above and below the median and the level of significance $\alpha = 0.01$, to decide whether it is reasonable to treat these data as if they constitute a random sample.

Solution Since the median is 23.5, as can easily be verified, we get the following arrangement of values above and below the median:

$$a\ b\ a\ a\ b\ b\ a\ b\ b\ a\ a\ a\ b\ a\ a\ b\ b\ a\ a\ b\ a\ b\ b\ b$$

1. H_0: Arrangement is random.
   H_A: Arrangement is not random.
2. $\alpha = 0.01$
3. Since $n_1 = 12$, $n_2 = 12$, and $\alpha = 0.01$, we get $u'_{0.005} = 6$ and $u_{0.005} = 20$ from Table VIII; thus, the null hypothesis must be rejected if $u \leq 6$ or $u \geq 20$; otherwise, accept it or reserve judgment.
4. As can be seen from the arrangements of a's and b's above, there are $u = 14$ runs.
5. Since $u = 14$ falls between 6 and 20, the null hypothesis cannot be rejected; in other words, there is no real evidence to suggest that the data cannot be treated as if they constitute a random sample.

The structure of Table VIII does not allow us to determine the p-value exactly. However, we can certainly say that $p > 0.05$.

EXERCISES

17.28 The following sequence of *C*'s and *A*'s shows the order in which 25 cars with California or Arizona license plates crossed the Colorado River at Blyth, California, to enter Arizona.

$$C\ A\ A\ C\ A\ C\ C\ A\ C\ C\ C\ A\ A\ C$$
(cont.) $A\ A\ A\ C\ A\ C\ C\ A\ C\ C$

Test for randomness at the level of significance $\alpha = 0.05$.

17.29 The following is the order in which red, *R*, and black, *B*, cards were dealt to a bridge player:

$$B\ B\ B\ R\ R\ R\ R\ B\ B\ R\ R\ R$$

Test for randomness at the 0.05 level of significance.

17.30 Test at the 0.01 level of significance whether the following arrangement of defective, *D*, and nondefective, *N*, pieces coming off an assembly line may be regarded as random:

$$N\ N\ N\ N\ N\ N\ N\ D\ D$$
(cont.) $D\ D\ N\ N\ N\ D\ D\ N\ N\ N$

17.31 A driver buys gasoline either at a Shell station, *S*, or at a Chevron station, *C*, and the following arrangement shows where he purchased gasoline (in the given order) over a certain period of time:

$$C\ C\ C\ S\ C\ S\ C\ S\ S\ C\ C$$
(cont.) $S\ C\ S\ C\ S\ C\ S\ S\ C\ S\ C$

Test for randomness at the 0.05 level of significance.

17.32 Representing each 0, 2, 4, 6, and 8 by the letter *E* and each 1, 3, 5, 7, and 9 by the letter *O*, check at the 0.05 level of significance whether the arrangements of the 50 digits in the first column of the random-number table on page 543, may be regarded as random.

17.33 The following arrangement shows whether 50 persons interviewed consecutively in the given order are for, *F*, or against, *A*, an increase in the city sales tax:

$$A\ A\ A\ A\ A\ F\ A\ A\ F\ F\ A\ A\ A\ A\ A\ A\ A\ F$$
(cont.) $A\ A\ A\ A\ F\ F\ F\ A\ A\ A\ F\ A\ A\ F\ A\ A\ A\ A$
(cont.) $F\ F\ A\ A\ A\ A\ A\ A\ A\ A\ F\ A\ A\ A$

Test for randomness at the level of significance $\alpha = 0.05$.

17.34 Use the large-sample *u* test to rework Exercise 17.28.

17.35 To test whether a radio signal contains a message or constitutes random noise, an interval of time is subdivided into a number of very short intervals and for each of these it is determined whether the signal strength exceeds, *E*, or does not exceed, *N*, a certain level of background noise. Test at the level of significance $\alpha = 0.05$ whether the following arrangement, thus obtained, may be regarded as random, and hence that the signal contains no message and may be regarded as random noise:

$$E\ N\ N\ N\ N\ E\ N\ E\ N\ N\ N\ E\ E\ N\ N\ N\ E$$
(cont.) $E\ N\ E\ N\ N\ N\ E\ E\ N\ N\ N\ N\ N\ E\ E\ N\ E$
(cont.) $N\ N\ E\ N\ N\ N\ E\ E\ N\ N\ N\ E\ N\ E\ N\ N$
(cont.) $N\ N\ N\ E\ N$

17.36 Mentally simulate fifty flips of a coin, and test at the 0.05 level of significance whether the resulting sequence of *H*'s and *T*'s (heads and tails) may be regarded as random.

17.37 The following are the numbers of students absent from a school on 24 consecutive days:

	38	31	32	27	28	30	26	33
(cont.)	36	30	28	35	33	31	29	35
(cont.)	31	33	31	28	30	28	25	29

Test for randomness at the 0.05 level of significance.

17.38 The following are the numbers of business lunches that an insurance agent had in 30 consecutive months: 6, 7, 5, 6, 8, 6, 8, 6, 6, 4, 3, 2, 4, 4, 3, 4, 7, 5, 6, 8, 6, 6, 3, 4, 2, 5, 4, 4, 3, and 7. Discarding the three values which equal the median, 5, test for randomness at the level of significance $\alpha = 0.01$.

17.39 The following are the examination grades of 40 students in the order in which they finished an examination:

	75	95	77	93	89	83	69	77	92	88
(cont.)	62	64	91	72	76	83	50	65	84	67
(cont.)	63	54	58	76	70	62	65	41	63	55
(cont.)	32	58	61	68	54	28	35	49	82	60

Test for randomness at the level of significance $\alpha = 0.05$.

17.40 The total number of retail stores opening for business and also quitting business within the calendar years 1948–1980 in a large city were 108, 103, 109, 107, 125, 142, 147, 122, 116, 153, 144, 162, 143, 126, 145, 129, 134, 137, 143, 150, 148, 152, 125, 106, 112, 139, 132, 122, 138, 148, 155, 146, and 158. Making use of the fact that the median is 138, test at the 0.05 level of significance whether there is a significant trend.

17.41 The following are the weights in pounds of tomatoes harvested from 50 consecutive tomato plants:

	5.1	5.0	5.2	5.8	6.6	4.9	5.2	6.1	7.1	5.8
(cont.)	6.4	6.1	5.2	7.0	5.4	5.6	6.9	5.0	5.9	5.3
(cont.)	5.9	6.1	7.2	5.5	6.2	6.4	6.0	7.4	6.7	6.0
(cont.)	5.3	6.5	6.7	5.5	6.9	6.5	6.8	5.8	6.4	7.4
(cont.)	7.2	7.6	7.8	6.6	7.8	5.9	6.7	7.0	6.8	7.2

Making use of the fact that the median is 6.3 pounds, test for randomness at the level of significance $\alpha = 0.05$ by using runs above and below the median.

17.42 A habitual coffee drinker consumed these numbers of cups on each of the thirty days of April:

	5	3	4	3	5	4	2	2	4	4
(cont.)	3	4	3	5	6	6	6	7	5	4
(cont.)	3	4	3	3	3	2	2	4	4	5

Note that she drank

2 cups of coffee on 4 days,
3 cups of coffee on 8 days,
4 cups of coffee on 9 days,
5 cups of coffee on 5 days,
6 cups of coffee on 3 days, and
7 cups of coffee on 1 day.

Test for randomness at the level of significance $\alpha = 0.05$ by using runs above and below the median. Observe that after removing the median, you will have $n_1 \neq n_2$.

17.9 RANK CORRELATION

Since the significance test for r of Section 16.3 is based on very stringent assumptions, we sometimes use a nonparametric alternative which can be applied under much more general conditions. This test of the null hypothesis of no correlation is based on the **rank-correlation coefficient**, often called **Spearman's rank-correlation coefficient** and denoted by r_S.

The calculation of the rank-correlation coefficient for a given set of n pairs of x's and y's requires several steps. We first rank the x's among themselves from low to high (or high to low). Then we rank the y's in the same way. Then we find the d's as the differences between the ranks and substitute into the formula

Rank-correlation coefficient

$$r_S = 1 - \frac{6(\sum d^2)}{n(n^2 - 1)}$$

When there are ties in rank, we proceed as before and assign to each of the tied observations the mean of the ranks which they jointly occupy.

EXAMPLE The following are the numbers of hours which ten students studied for an examination and the grades which they received:

Number of hours studied x	Grade in examination y
9	56
5	44
11	79
13	72
10	70
5	54
18	94
15	85
2	33
8	65

Calculate r_S.

Solution Ranking the x's among themselves from low to high, and also the y's, we get the ranks shown in the first two columns of the following table:

Rank of x	Rank of y	d	d^2
5	4	1.0	1.00
2.5	2	0.5	0.25
7	8	-1.0	1.00
8	7	1.0	1.00
6	6	0.0	0.00
2.5	3	-0.5	0.25
10	10	0.0	0.00
9	9	0.0	0.00
1	1	0.0	0.00
4	5	-1.0	1.00
			4.50

Then, determining the d's and their squares, and substituting $n = 10$ and $\sum d^2 = 4.50$ into the formula for r_S, we get

$$r_S = 1 - \frac{6(4.50)}{10(10^2 - 1)} = 0.97$$

As can be seen from this example, r_S is easy to compute manually, and this is why it is sometimes used instead of r when no calculator is available. When there are no ties, r_S actually equals the correlation coefficient r calculated for the two sets of ranks; when ties exist there may be a small (but usually negligible) difference. Of course, by using ranks instead of the original data we lose some information, but this is usually offset by the rank-correlation coefficient's computational ease.

When we use r_S to test the null hypothesis of no correlation between two variables x and y, we do not have to make any assumptions about the nature of the populations sampled. Under the null hypothesis of no correlation—indeed, the null hypothesis that the x's and y's are randomly matched—the sampling distribution of r_S has the mean 0 and the standard deviation

$$\sigma_{r_S} = \frac{1}{\sqrt{n-1}}$$

Since this sampling distribution can be approximated with a normal distribution even for relatively small values of n, we base the test of the null hypothesis of no correlation on the statistic

Statistic for testing significance of r_S

$$z = \frac{r_S - 0}{1/\sqrt{n-1}} = r_S\sqrt{n-1}$$

which has approximately the standard normal distribution.

EXAMPLE   With reference to the preceding example where we had $n = 10$ and $r_S = 0.97$, test the significance of this value of r_S at the 0.01 level of significance.

Solution   1. H_0: $\rho = 0$ (no correlation)
H_A: $\rho \neq 0$
2. $\alpha = 0.01$
3. Reject the null hypothesis if $z \leq -2.575$ or $z \geq 2.575$, where

$$z = r_S\sqrt{n-1}$$

Otherwise, accept it or reserve judgment.
4. For $n = 10$ and $r_S = 0.97$, we get

$$z = 0.97\sqrt{10-1} = 2.91$$

5. Since $z = 2.91$ exceeds 2.575, the null hypothesis must be rejected; in other words, we conclude that there is a relationship between study time and examination grades in the populations sampled.

The p-value method replaces steps 3, 4, and 5 with these.

3'. The test statistic is

$$z = r_S\sqrt{n-1}$$

4'. Find, as shown above, that for $n = 10$ and $r_S = 0.97$, we calculate $z = 2.91$. Table I shows that $P[|z| \geq 2.91] = 0.0036$, which is the p-value.
5'. Since p is smaller than $\alpha = 0.01$, the null hypothesis must be rejected.

EXERCISES

17.43 Calculate r_S for the following data representing the statistics grades, x, and psychology grades, y, of a sample of 15 students:

x	y
75	70
69	64
73	70
96	89
73	84
52	70
57	66
61	63
71	68
70	83
93	91
77	79
85	73
93	84
87	85

17.44 Test at the level of significance $\alpha = 0.05$ whether the value obtained for r_S in the preceding exercise is significant.

17.45 Calculate r_S for the following sample data representing the number of minutes it took 12 technicians to assemble an electric motor in the morning, x, and in the late afternoon, y:

x	y
10.8	15.1
16.6	16.8
11.1	10.9
10.3	14.2
12.0	13.8
15.1	21.5
13.7	13.2
18.5	21.1
17.3	16.4
14.2	19.3
14.8	17.4
15.3	19.0

17.46 Use the level of significance $\alpha = 0.05$ to test whether the value obtained for r_S in the preceding exercise is significant.

17.47 Ten weeks' sales of a downtown department store, x, and its suburban branch, y, are

x	y
71	49
64	31
58	24
80	68
63	30
69	40
76	62
60	22
66	35
55	16

where the units are \$10,000. Calculate r_S.

17.48 Assuming that the data of the preceding exercise may be looked upon as random samples of the two stores' sales, use the level of significance $\alpha = 0.01$ to test whether the value obtained for r_S is significant.

17.49 In Exercise 16.5 we gave the percentages of the vote predicted by a poll for eight candidates for the U.S. Senate in different states, x, and the corresponding percentages of the vote which they actually received, y, as

x	y
43	50
46	42
51	57
59	55
41	46
53	48
52	53
62	56

Calculate r_S and compare it with the value of r obtained for these data in Exercise 16.5 on page 475.

17.50 If a sample of $n = 20$ pairs of data yielded $r_S = 0.41$, is this rank-correlation coefficient significant at the 0.05 level of significance?

17.51 The following table shows how a panel of nutrition experts and a panel of heads of household ranked 15 breakfast foods on their palatability:

Breakfast food	Nutrition experts	Heads of household
I	7	5
II	3	4
III	11	8
IV	9	14
V	1	2
VI	4	6
VII	10	12
VIII	8	7
IX	5	1
X	13	9
XI	12	15
XII	2	3
XIII	15	10
XIV	6	11
XV	14	13

Calculate r_S as a measure of the consistency of the two rankings.

17.52 The following are the rankings which three judges gave to the work of ten artists:

Judge A: 5, 8, 4, 2, 3, 1, 10, 7, 9, 6
Judge B: 3, 10, 1, 4, 2, 5, 6, 7, 8, 9
Judge C: 8, 5, 6, 4, 10, 2, 3, 1, 7, 9

Calculate r_S for each pair of rankings and decide
(a) which two judges are most alike in their opinions about these artists;
(b) which two judges differ the most in their opinions about these artists.

17.10

SOME FURTHER CONSIDERATIONS

Although nonparametric tests have a great deal of intuitive appeal and are widely applicable, it should not be overlooked that they are usually **less efficient** than the standard tests which they replace. To illustrate what we mean here by "less efficient," let us refer to the example on page 273, where we showed that the mean of a random sample of size $n = 128$ is as reliable an estimate of the mean of a symmetrical population as the median of a random sample of size $n = 200$. Thus, the median requires a larger sample, and this is what we mean when we say that it is "less efficient."

To put it another way, nonparametric tests tend to be wasteful of information, and they should not be used indiscriminately when the assumptions underlying the corresponding standard tests are satisfied.

17.11

SUMMARY

The table which follows summarizes the various nonparametric tests we have discussed (except for the tests of randomness based on runs) and the corresponding standard tests which they replace. In each case we give the section or sections of the book where they are discussed.

Null hypothesis	Standard tests	Nonparametric tests
$\mu = \mu_0$	One-sample t test (Section 11.7) or corresponding large-sample test (Section 11.6)	One-sample sign test (Sections 17.1 and 17.2)
$\mu_1 = \mu_2$ (independent samples)	Two-sample t test (Section 11.9) or corresponding large-sample test (Section 11.8)	U test (Sections 17.3 and 17.4)
$\mu_1 = \mu_2$ (paired data)	Paired-sample t test (Section 11.10) or corresponding large-sample test (Section 11.10)	Paired-sample sign test (Sections 17.1 and 17.2)
$\mu_1 = \mu_2 = \cdots = \mu_k$	One-way analysis of variance (Section 14.3)	H test (Section 17.5)
$\rho = 0$	Test based on Fisher Z transformation (Section 16.3)	Test based on rank-correlation coefficient (Section 17.9)

17.12
CHECKLIST OF KEY TERMS
(with page references to their definitions)

H test, 503
Kruskal–Wallis test, 503
Mann–Whitney test, 496
Nonparametric tests, 491
One-sample sign test, 492
Paired-sample sign test, 493
Rank sums, 498
Rank-correlation coefficient, 511
Runs, 506
Runs above and below the median, 509

Sign test, 491
Signed-rank test, 494
Spearman's rank-correlation coefficient, 511
Theory of runs, 506
Total number of runs, 506
u test, 506
U test, 496
Wilcoxon signed-rank test, 494
Wilcoxon test, 496

17.13
REVIEW EXERCISES

17.53 The following are the prices (in dollars) charged for a certain camera in a random sample of 15 discount stores: 57.25, 58.14, 54.19, 56.17, 57.21, 55.38, 54.75, 57.29, 57.80, 54.50, 55.00, 56.35, 54.26, 60.23, and 53.99. Use the sign test based on Table V to test at the 0.05 level of significance whether or not the median price charged for the camera in the population sampled is $55.00.

17.54 Use the large-sample sign test to rework Exercise 17.53.

17.55 The following sequence shows whether a certain member was present, *P*, or absent, *A*, at twenty consecutive meetings of a fraternal organization:

$$P\ P\ P\ P\ P\ P\ P\ P\ A\ P\ P\ P\ P\ P\ P\ P\ A\ A\ A\ A$$

At the 0.05 level of significance, is there any real indication of a lack of randomness?

17.56 The following are the numbers of minutes that patients had to wait for their appointments with four doctors:

Doctor A:	18	26	29	22	16
Doctor B:	9	11	28	26	15
Doctor C:	20	13	22	25	10
Doctor D:	21	26	39	32	24

Use the *H* test at the 0.05 level of significance to test the null hypothesis that the means of the populations sampled are equal against the alternative hypothesis that these means are not all equal.

17.57 The following are data on the percentage kill of two kinds of insecticides used against mosquitos:

Insecticide X:	41.9	46.9	44.6	43.9	42.0	44.0
(cont.)	41.0	43.1	39.0	45.2	44.6	42.0
Insecticide Y:	45.7	39.8	42.8	41.2	45.0	40.2
(cont.)	40.2	41.7	37.4	38.8	41.7	38.7

Use the *U* test based on Table VII to test at the 0.05

level of significance whether or not the two insecticides are on the average equally effective.

17.58 Use the large-sample *U* test to rework the preceding exercise.

17.59 The following are the batting averages, *x*, and home runs hit, *y*, by a random sample of fifteen major league baseball players during the first half of the season:

x	y
0.252	12
0.305	6
0.299	4
0.303	15
0.285	2
0.191	2
0.283	16
0.272	6
0.310	8
0.266	10
0.215	0
0.211	3
0.272	14
0.244	6
0.320	7

Calculate the rank-correlation coefficient and test whether it is significant at the 0.01 level of significance.

17.60 The following are the numbers of persons who attended a "singles only" dance on twelve Saturdays: 172, 208, 169, 232, 123, 165, 197, 178, 221, 195, 209, and 182. Use the sign test based on Table V to test at the 0.05 level of significance whether or not the true median of the population sampled is $\tilde{\mu} = 169$.

17.61 Use the large-sample sign test to rework Exercise 17.60.

17.62 The following are the numbers of burglaries committed in two cities on twenty days: 83 and 80, 98 and 87, 115 and 86, 112 and 122, 77 and 102, 103 and 94, 116 and 81, 136 and 96, 156 and 158, 83 and 127, 105 and

104, 117 and 102, 86 and 100, 150 and 108, 119 and 124, 111 and 91, 137 and 103, 160 and 153, 121 and 140, and 143 and 105. Use the large-sample sign test at the 0.05 level of significance to test whether or not on the average there are as many burglaries per day in the two cities.

17.63 On what statistic do we base our decision and for what values of the statistic do we reject the null hypothesis $\mu_1 = \mu_2$, if we have random samples of size $n_1 = 8$ and $n_2 = 11$ and are using the U test based on Table VII at the 0.05 level of significance to test the null hypothesis against the alternative hypothesis

(a) $\mu_1 < \mu_2$;
(b) $\mu_1 > \mu_2$;
(c) $\mu_1 \neq \mu_2$?

17.64 The following sequence shows whether a television news program had at least 25 percent of a city's viewing audience, A, or less than 25 percent, L, on 36 consecutive weekday evenings:

$$L\ L\ L\ L\ A\ A\ L\ L\ L\ A\ L\ L\ L\ A\ A\ A\ A\ L$$
(cont.) $\quad A\ L\ L\ L\ A\ A\ L\ L\ L\ L\ L\ A\ L\ L\ L\ L\ A$

Test for randomness at the 0.05 level of significance.

17.65 If $r_s = 0.27$ for $n = 25$ paired data, is this rank-correlation coefficient significant at the 0.05 level of significance?

17.66 The following are the scores of sixteen golfers on the first two days of a tournament: 68 and 71, 73 and 76, 70 and 73, 74 and 71, 69 and 72, 72 and 74, 67 and 70, 72 and 68, 71 and 72, 73 and 74, 68 and 69, 70 and 72, 73 and 70, 71 and 75, 67 and 69, and 73 and 71. Use the sign test at the 0.05 level of significance to test whether on the average the hundreds of golfers parti-

cipating in the tournament scored equally well on the first two days or whether they tended to score lower on the first day.

17.67 The following are the closing prices of a commodity (in dollars) on twenty consecutive trading days: 378, 379, 379, 378, 377, 376, 374, 374, 373, 373, 374, 375, 376, 376, 376, 375, 374, 374, 373, and 374. Test for randomness at the 0.01 level of significance.

17.68 The following are the numbers of minutes it took two ambulance services to reach the scenes of accidents:

Ambulance service 1:	9.3	5.5	13.1	10.0	7.6
(cont.)	9.2	11.2	6.4	14.0	10.3

Ambulance service 2:	12.7	6.6	9.1	4.5	7.2
(cont.)	6.4	7.5			

Assuming that the data constitute random samples, use the U test at the 0.05 level of significance to test the null hypothesis that on the average the two ambulance services respond with equal speed against the alternative hypothesis that on the average the first ambulance service is slower.

17.69 On what statistic do we base our decision and for what values of the statistic do we reject the null hypothesis $\mu_1 = \mu_2$, if we have a random sample of $n = 11$ pairs of values and are using the signed-rank test at the 0.01 level of significance to test the null hypothesis against the alternative hypothesis

(a) $\mu_1 < \mu_2$;
(b) $\mu_1 > \mu_2$;
(c) $\mu_1 \neq \mu_2$?

17.70 Rework the preceding exercise with the level of significance changed to 0.05.

17.14

REFERENCES

Further information about the nonparametric tests discussed in this chapter and many others may be found in

CONOVER, W. J., *Practical Nonparametric Statistics.* New York: John Wiley & Sons, Inc., 1971.

DANIEL, W. W., *Applied Nonparametric Statistics.* Boston: Houghton Mifflin Company, 1978.

GIBBONS, J. D., *Nonparametric Statistical Inference.* New York: McGraw-Hill Book Company, 1971.

HOLLANDER, M., and WOLFE, D., *Nonparametric Statistical Methods.* New York: John Wiley & Sons, Inc., 1973.

LEHMANN, E. L., *Nonparametrics: Statistical Methods Based on Ranks.* San Francisco: Holden-Day, Inc., 1975.

MOSTELLER, F., and ROURKE, R. E. K., *Sturdy Statistics, Nonparametrics and Order Statistics.* Reading, Mass.: Addison-Wesley Publishing Company, Inc., 1973.

NOETHER, G. E., *Elements of Nonparametric Statistics, 2nd ed.* New York: John Wiley & Sons, Inc., 1976.

SIEGEL, S., *Nonparametric Statistics for the Behavioral Sciences.* New York: McGraw-Hill Book Company, 1956.

STATISTICAL TABLES

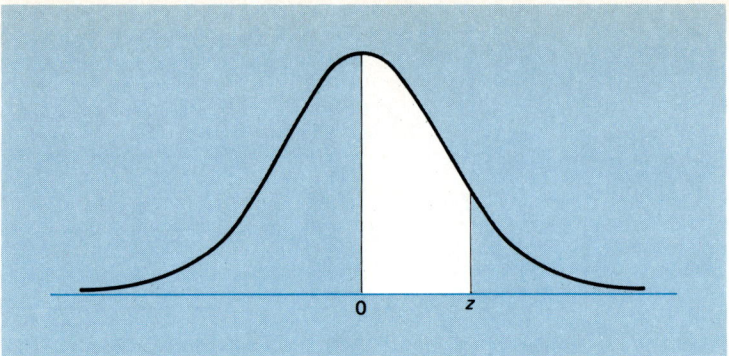

The entries in Table 1 are the probabilities that a random variable having the standard normal distribution will take on a value between 0 and z; they are given by the area of the white region under the curve in the figure shown above.

TABLE I Normal-Curve Areas

z	.00	.01	.02	.03	.04	.05	.06	.07	.08	.09
0.0	.0000	.0040	.0080	.0120	.0160	.0199	.0239	.0279	.0319	.0359
0.1	.0398	.0438	.0478	.0517	.0557	.0596	.0636	.0675	.0714	.0753
0.2	.0793	.0832	.0871	.0910	.0948	.0987	.1026	.1064	.1103	.1141
0.3	.1179	.1217	.1255	.1293	.1331	.1368	.1406	.1443	.1480	.1517
0.4	.1554	.1591	.1628	.1664	.1700	.1736	.1772	.1808	.1844	.1879
0.5	.1915	.1950	.1985	.2019	.2054	.2088	.2123	.2157	.2190	.2224
0.6	.2257	.2291	.2324	.2357	.2389	.2422	.2454	.2486	.2517	.2549
0.7	.2580	.2611	.2642	.2673	.2704	.2734	.2764	.2794	.2823	.2852
0.8	.2881	.2910	.2939	.2967	.2995	.3023	.3051	.3078	.3106	.3133
0.9	.3159	.3186	.3212	.3238	.3264	.3289	.3315	.3340	.3365	.3389
1.0	.3413	.3438	.3461	.3485	.3508	.3531	.3554	.3577	.3599	.3621
1.1	.3643	.3665	.3686	.3708	.3729	.3749	.3770	.3790	.3810	.3830
1.2	.3849	.3869	.3888	.3907	.3925	.3944	.3962	.3980	.3997	.4015
1.3	.4032	.4049	.4066	.4082	.4099	.4115	.4131	.4147	.4162	.4177
1.4	.4192	.4207	.4222	.4236	.4251	.4265	.4279	.4292	.4306	.4319
1.5	.4332	.4345	.4357	.4370	.4382	.4394	.4406	.4418	.4429	.4441
1.6	.4452	.4463	.4474	.4484	.4495	.4505	.4515	.4525	.4535	.4545
1.7	.4554	.4564	.4573	.4582	.4591	.4599	.4608	.4616	.4625	.4633
1.8	.4641	.4649	.4656	.4664	.4671	.4678	.4686	.4693	.4699	.4706
1.9	.4713	.4719	.4726	.4732	.4738	.4744	.4750	.4756	.4761	.4767
2.0	.4772	.4778	.4783	.4788	.4793	.4798	.4803	.4808	.4812	.4817
2.1	.4821	.4826	.4830	.4834	.4838	.4842	.4846	.4850	.4854	.4857
2.2	.4861	.4864	.4868	.4871	.4875	.4878	.4881	.4884	.4887	.4890
2.3	.4893	.4896	.4898	.4901	.4904	.4906	.4909	.4911	.4913	.4916
2.4	.4918	.4920	.4922	.4925	.4927	.4929	.4931	.4932	.4934	.4936
2.5	.4938	.4940	.4941	.4943	.4945	.4946	.4948	.4949	.4951	.4952
2.6	.4953	.4955	.4956	.4957	.4959	.4960	.4961	.4962	.4963	.4964
2.7	.4965	.4966	.4967	.4968	.4969	.4970	.4971	.4972	.4973	.4974
2.8	.4974	.4975	.4976	.4977	.4977	.4978	.4979	.4979	.4980	.4981
2.9	.4981	.4982	.4982	.4983	.4984	.4984	.4985	.4985	.4986	.4986
3.0	.4987	.4987	.4987	.4988	.4988	.4989	.4989	.4989	.4990	.4990

Also, for $z = 4.0$, 5.0, and 6.0, the areas are 0.49997, 0.4999997, and 0.499999999.

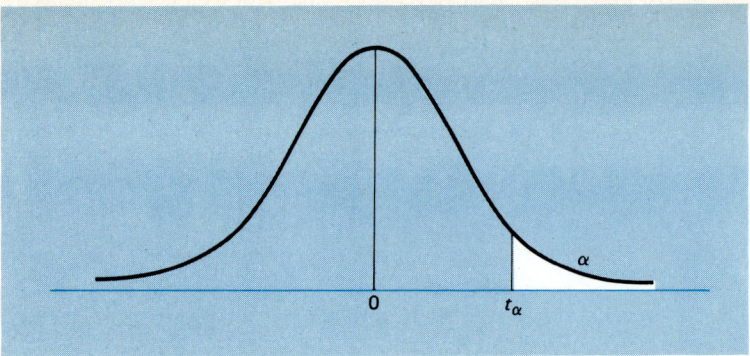

The entries in Table II are values for which the area to their right under the t distribution with given degrees of freedom (the area of the white region under the curve shown above) is equal to α.

TABLE II   Critical Values of t[†]

d.f.	$t_{.100}$	$t_{.050}$	$t_{.025}$	$t_{.010}$	$t_{.005}$	d.f.
1	3.078	6.314	12.706	31.821	63.657	1
2	1.886	2.920	4.303	6.965	9.925	2
3	1.638	2.353	3.182	4.541	5.841	3
4	1.533	2.132	2.776	3.747	4.604	4
5	1.476	2.015	2.571	3.365	4.032	5
6	1.440	1.943	2.447	3.143	3.707	6
7	1.415	1.895	2.365	2.998	3.499	7
8	1.397	1.860	2.306	2.896	3.355	8
9	1.383	1.833	2.262	2.821	3.250	9
10	1.372	1.812	2.228	2.764	3.169	10
11	1.363	1.796	2.201	2.718	3.106	11
12	1.356	1.782	2.179	2.681	3.055	12
13	1.350	1.771	2.160	2.650	3.012	13
14	1.345	1.761	2.145	2.624	2.977	14
15	1.341	1.753	2.131	2.602	2.947	15
16	1.337	1.746	2.120	2.583	2.921	16
17	1.333	1.740	2.110	2.567	2.898	17
18	1.330	1.734	2.101	2.552	2.878	18
19	1.328	1.729	2.093	2.539	2.861	19
20	1.325	1.725	2.086	2.528	2.845	20
21	1.323	1.721	2.080	2.518	2.831	21
22	1.321	1.717	2.074	2.508	2.819	22
23	1.319	1.714	2.069	2.500	2.807	23
24	1.318	1.711	2.064	2.492	2.797	24
25	1.316	1.708	2.060	2.485	2.787	25
26	1.315	1.706	2.056	2.479	2.779	26
27	1.314	1.703	2.052	2.473	2.771	27
28	1.313	1.701	2.048	2.467	2.763	28
29	1.311	1.699	2.045	2.462	2.756	29
inf.	1.282	1.645	1.960	2.326	2.576	inf.

[†] From Richard A. Johnson and Dean W. Wichern, *Applied Multivariate Statistical Analysis*, © 1982, p. 582. Adapted by permission of Prentice-Hall, Inc., Englewood Cliffs, N.J.

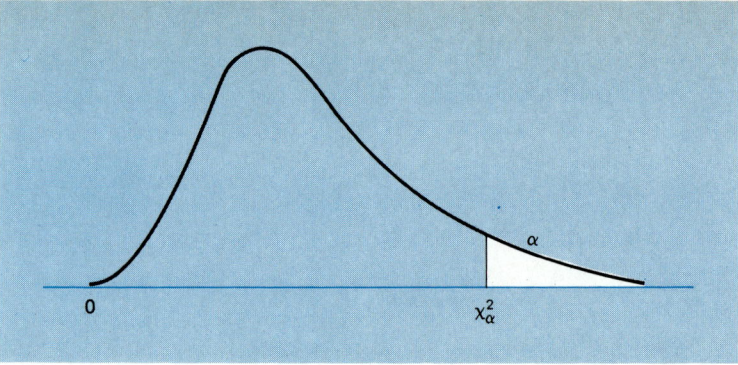

The entries in Table III are values for which the area to their right under the chi-square distribution with given degrees of freedom (the area of the white region under the curve shown above) is equal to α.

TABLE III Critical Values of χ^2 †

d.f.	$\chi^2_{.995}$	$\chi^2_{.99}$	$\chi^2_{.975}$	$\chi^2_{.95}$	$\chi^2_{.05}$	$\chi^2_{.025}$	$\chi^2_{.01}$	$\chi^2_{.005}$	d.f.
1	.0000393	.000157	.000982	.00393	3.841	5.024	6.635	7.879	1
2	.0100	.0201	.0506	.103	5.991	7.378	9.210	10.597	2
3	.0717	.115	.216	.352	7.815	9.348	11.345	12.838	3
4	.207	.297	.484	.711	9.488	11.143	13.277	14.860	4
5	.412	.554	.831	1.145	11.070	12.832	15.086	16.750	5
6	.676	.872	1.237	1.635	12.592	14.449	16.812	18.548	6
7	.989	1.239	1.690	2.167	14.067	16.013	18.475	20.278	7
8	1.344	1.646	2.180	2.733	15.507	17.535	20.090	21.955	8
9	1.735	2.088	2.700	3.325	16.919	19.023	21.666	23.589	9
10	2.156	2.558	3.247	3.940	18.307	20.483	23.209	25.188	10
11	2.603	3.053	3.816	4.575	19.675	21.920	24.725	26.757	11
12	3.074	3.571	4.404	5.226	21.026	23.337	26.217	28.300	12
13	3.565	4.107	5.009	5.892	22.362	24.736	27.688	29.819	13
14	4.075	4.660	5.629	6.571	23.685	26.119	29.141	31.319	14
15	4.601	5.229	6.262	7.261	24.996	27.488	30.578	32.801	15
16	5.142	5.812	6.908	7.962	26.296	28.845	32.000	34.267	16
17	5.697	6.408	7.564	8.672	27.587	30.191	33.409	35.718	17
18	6.265	7.015	8.231	9.390	28.869	31.526	34.805	37.156	18
19	6.844	7.633	8.907	10.117	30.144	32.852	36.191	38.582	19
20	7.434	8.260	9.591	10.851	31.410	34.170	37.566	39.997	20
21	8.034	8.897	10.283	11.591	32.671	35.479	38.932	41.401	21
22	8.643	9.542	10.982	12.338	33.924	36.781	40.289	42.796	22
23	9.260	10.196	11.689	13.091	35.172	38.076	41.638	44.181	23
24	9.886	10.856	12.401	13.848	36.415	39.364	42.980	45.558	24
25	10.520	11.524	13.120	14.611	37.652	40.646	44.314	46.928	25
26	11.160	12.198	13.844	15.379	38.885	41.923	45.642	48.290	26
27	11.808	12.879	14.573	16.151	40.113	43.194	46.963	49.645	27
28	12.461	13.565	15.308	16.928	41.337	44.461	48.278	50.993	28
29	13.121	14.256	16.047	17.708	42.557	45.722	49.588	52.336	29
30	13.787	14.953	16.791	18.493	43.773	46.979	50.892	53.672	30

† Based on Table 8 of *Biometrika Tables for Statisticians, Volume I* (Cambridge University Press, 1954), by permission of the *Biometrika* trustees.

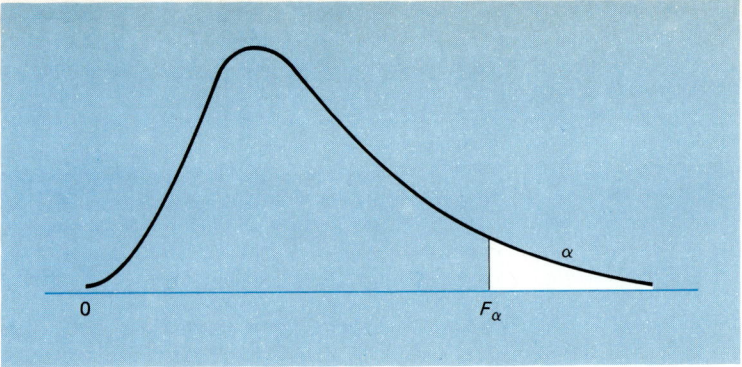

The entries in Table IV are values for which the area to their right under the F distribution with given degrees of freedom (the area of the white region under the curve shown above) is equal to α.

TABLE IV Critical Values of F[†]

Values of $F_{0.05}$

Degrees of freedom for numerator

Denom. df	1	2	3	4	5	6	7	8	9	10	12	15	20	24	30	40	60	120	∞
1	161	200	216	225	230	234	237	239	241	242	244	246	248	249	250	251	252	253	254
2	18.5	19.0	19.2	19.2	19.3	19.3	19.4	19.4	19.4	19.4	19.4	19.4	19.4	19.5	19.5	19.5	19.5	19.5	19.5
3	10.1	9.55	9.28	9.12	9.01	8.94	8.89	8.85	8.81	8.79	8.74	8.70	8.66	8.64	8.62	8.59	8.57	8.55	8.53
4	7.71	6.94	6.59	6.39	6.26	6.16	6.09	6.04	6.00	5.96	5.91	5.86	5.80	5.77	5.75	5.72	5.69	5.66	5.63
5	6.61	5.79	5.41	5.19	5.05	4.95	4.88	4.82	4.77	4.74	4.68	4.62	4.56	4.53	4.50	4.46	4.43	4.40	4.37
6	5.99	5.14	4.76	4.53	4.39	4.28	4.21	4.15	4.10	4.06	4.00	3.94	3.87	3.84	3.81	3.77	3.74	3.70	3.67
7	5.59	4.74	4.35	4.12	3.97	3.87	3.79	3.73	3.68	3.64	3.57	3.51	3.44	3.41	3.38	3.34	3.30	3.27	3.23
8	5.32	4.46	4.07	3.84	3.69	3.58	3.50	3.44	3.39	3.35	3.28	3.22	3.15	3.12	3.08	3.04	3.01	2.97	2.93
9	5.12	4.26	3.86	3.63	3.48	3.37	3.29	3.23	3.18	3.14	3.07	3.01	2.94	2.90	2.86	2.83	2.79	2.75	2.71
10	4.96	4.10	3.71	3.48	3.33	3.22	3.14	3.07	3.02	2.98	2.91	2.85	2.77	2.74	2.70	2.66	2.62	2.58	2.54
11	4.84	3.98	3.59	3.36	3.20	3.09	3.01	2.95	2.90	2.85	2.79	2.72	2.65	2.61	2.57	2.53	2.49	2.45	2.40
12	4.75	3.89	3.49	3.26	3.11	3.00	2.91	2.85	2.80	2.75	2.69	2.62	2.54	2.51	2.47	2.43	2.38	2.34	2.30
13	4.67	3.81	3.41	3.18	3.03	2.92	2.83	2.77	2.71	2.67	2.60	2.53	2.46	2.42	2.38	2.34	2.30	2.25	2.21
14	4.60	3.74	3.34	3.11	2.96	2.85	2.76	2.70	2.65	2.60	2.53	2.46	2.39	2.35	2.31	2.27	2.22	2.18	2.13
15	4.54	3.68	3.29	3.06	2.90	2.79	2.71	2.64	2.59	2.54	2.48	2.40	2.33	2.29	2.25	2.20	2.16	2.11	2.07
16	4.49	3.63	3.24	3.01	2.85	2.74	2.66	2.59	2.54	2.49	2.42	2.35	2.28	2.24	2.19	2.15	2.11	2.06	2.01
17	4.45	3.59	3.20	2.96	2.81	2.70	2.61	2.55	2.49	2.45	2.38	2.31	2.23	2.19	2.15	2.10	2.06	2.01	1.96
18	4.41	3.55	3.16	2.93	2.77	2.66	2.58	2.51	2.46	2.41	2.34	2.27	2.19	2.15	2.11	2.06	2.02	1.97	1.92
19	4.38	3.52	3.13	2.90	2.74	2.63	2.54	2.48	2.42	2.38	2.31	2.23	2.16	2.11	2.07	2.03	1.98	1.93	1.88
20	4.35	3.49	3.10	2.87	2.71	2.60	2.51	2.45	2.39	2.35	2.28	2.20	2.12	2.08	2.04	1.99	1.95	1.90	1.84
21	4.32	3.47	3.07	2.84	2.68	2.57	2.49	2.42	2.37	2.32	2.25	2.18	2.10	2.05	2.01	1.96	1.92	1.87	1.81
22	4.30	3.44	3.05	2.82	2.66	2.55	2.46	2.40	2.34	2.30	2.23	2.15	2.07	2.03	1.98	1.94	1.89	1.84	1.78
23	4.28	3.42	3.03	2.80	2.64	2.53	2.44	2.37	2.32	2.27	2.20	2.13	2.05	2.01	1.96	1.91	1.86	1.81	1.76
24	4.26	3.40	3.01	2.78	2.62	2.51	2.42	2.36	2.30	2.25	2.18	2.11	2.03	1.98	1.94	1.89	1.84	1.79	1.73
25	4.24	3.39	2.99	2.76	2.60	2.49	2.40	2.34	2.28	2.24	2.16	2.09	2.01	1.96	1.92	1.87	1.82	1.77	1.71
30	4.17	3.32	2.92	2.69	2.53	2.42	2.33	2.27	2.21	2.16	2.09	2.01	1.93	1.89	1.84	1.79	1.74	1.68	1.62
40	4.08	3.23	2.84	2.61	2.45	2.34	2.25	2.18	2.12	2.08	2.00	1.92	1.84	1.79	1.74	1.69	1.64	1.58	1.51
60	4.00	3.15	2.76	2.53	2.37	2.25	2.17	2.10	2.04	1.99	1.92	1.84	1.75	1.70	1.65	1.59	1.53	1.47	1.39
120	3.92	3.07	2.68	2.45	2.29	2.18	2.09	2.02	1.96	1.91	1.83	1.75	1.66	1.61	1.55	1.50	1.43	1.35	1.25
∞	3.84	3.00	2.60	2.37	2.21	2.10	2.01	1.94	1.88	1.83	1.75	1.67	1.57	1.52	1.46	1.39	1.32	1.22	1.00

Degrees of freedom for denominator

[†] Reproduced from M. Merrington and C. M. Thompson, "Tables of percentage points of the inverted beta (F) distribution," *Biometrika*, vol. 33 (1943), by permission of the *Biometrika* trustees.

TABLE IV Critical Values of F (Continued)

Values of $F_{0.01}$

Degrees of freedom for numerator

den\\num	1	2	3	4	5	6	7	8	9	10	12	15	20	24	30	40	60	120	∞
1	4,052	5,000	5,403	5,625	5,764	5,859	5,928	5,982	6,023	6,056	6,106	6,157	6,209	6,235	6,261	6,287	6,313	6,339	6,366
2	98.5	99.0	99.2	99.2	99.3	99.3	99.4	99.4	99.4	99.4	99.4	99.4	99.4	99.5	99.5	99.5	99.5	99.5	99.5
3	34.1	30.8	29.5	28.7	28.2	27.9	27.7	27.5	27.3	27.2	27.1	26.9	26.7	26.6	26.5	26.4	26.3	26.2	26.1
4	21.2	18.0	16.7	16.0	15.5	15.2	15.0	14.8	14.7	14.5	14.4	14.2	14.0	13.9	13.8	13.7	13.7	13.6	13.5
5	16.3	13.3	12.1	11.4	11.0	10.7	10.5	10.3	10.2	10.1	9.89	9.72	9.55	9.47	9.38	9.29	9.20	9.11	9.02
6	13.7	10.9	9.78	9.15	8.75	8.47	8.26	8.10	7.98	7.87	7.72	7.56	7.40	7.31	7.23	7.14	7.05	6.97	6.88
7	12.2	9.55	8.45	7.85	7.46	7.19	6.99	6.84	6.72	6.62	6.47	6.31	6.16	6.07	5.99	5.91	5.82	5.74	5.65
8	11.3	8.65	7.59	7.01	6.63	6.37	6.18	6.03	5.91	5.81	5.67	5.52	5.36	5.28	5.20	5.12	5.03	4.95	4.86
9	10.6	8.02	6.99	6.42	6.06	5.80	5.61	5.47	5.35	5.26	5.11	4.96	4.81	4.73	4.65	4.57	4.48	4.40	4.31
10	10.0	7.56	6.55	5.99	5.64	5.39	5.20	5.06	4.94	4.85	4.71	4.56	4.41	4.33	4.25	4.17	4.08	4.00	3.91
11	9.65	7.21	6.22	5.67	5.32	5.07	4.89	4.74	4.63	4.54	4.40	4.25	4.10	4.02	3.94	3.86	3.78	3.69	3.60
12	9.33	6.93	5.95	5.41	5.06	4.82	4.64	4.50	4.39	4.30	4.16	4.01	3.86	3.78	3.70	3.62	3.54	3.45	3.36
13	9.07	6.70	5.74	5.21	4.86	4.62	4.44	4.30	4.19	4.10	3.96	3.82	3.66	3.59	3.51	3.43	3.34	3.25	3.17
14	8.86	6.51	5.56	5.04	4.70	4.46	4.28	4.14	4.03	3.94	3.80	3.66	3.51	3.43	3.35	3.27	3.18	3.09	3.00
15	8.68	6.36	5.42	4.89	4.56	4.32	4.14	4.00	3.89	3.80	3.67	3.52	3.37	3.29	3.21	3.13	3.05	2.96	2.87
16	8.53	6.23	5.29	4.77	4.44	4.20	4.03	3.89	3.78	3.69	3.55	3.41	3.26	3.18	3.10	3.02	2.93	2.84	2.75
17	8.40	6.11	5.19	4.67	4.34	4.10	3.93	3.79	3.68	3.59	3.46	3.31	3.16	3.08	3.00	2.92	2.83	2.75	2.65
18	8.29	6.01	5.09	4.58	4.25	4.01	3.84	3.71	3.60	3.51	3.37	3.23	3.08	3.00	2.92	2.84	2.75	2.66	2.57
19	8.19	5.93	5.01	4.50	4.17	3.94	3.77	3.63	3.52	3.43	3.30	3.15	3.00	2.92	2.84	2.76	2.67	2.58	2.49
20	8.10	5.85	4.94	4.43	4.10	3.87	3.70	3.56	3.46	3.37	3.23	3.09	2.94	2.86	2.78	2.69	2.61	2.52	2.42
21	8.02	5.78	4.87	4.37	4.04	3.81	3.64	3.51	3.40	3.31	3.17	3.03	2.88	2.80	2.72	2.64	2.55	2.46	2.36
22	7.95	5.72	4.82	4.31	3.99	3.76	3.59	3.45	3.35	3.26	3.12	2.98	2.83	2.75	2.67	2.58	2.50	2.40	2.31
23	7.88	5.66	4.76	4.26	3.94	3.71	3.54	3.41	3.30	3.21	3.07	2.93	2.78	2.70	2.62	2.54	2.45	2.35	2.26
24	7.82	5.61	4.72	4.22	3.90	3.67	3.50	3.36	3.26	3.17	3.03	2.89	2.74	2.66	2.58	2.49	2.40	2.31	2.21
25	7.77	5.57	4.68	4.18	3.86	3.63	3.46	3.32	3.22	3.13	2.99	2.85	2.70	2.62	2.53	2.45	2.36	2.27	2.17
30	7.56	5.39	4.51	4.02	3.70	3.47	3.30	3.17	3.07	2.98	2.84	2.70	2.55	2.47	2.39	2.30	2.21	2.11	2.01
40	7.31	5.18	4.31	3.83	3.51	3.29	3.12	2.99	2.89	2.80	2.66	2.52	2.37	2.29	2.20	2.11	2.02	1.92	1.80
60	7.08	4.98	4.13	3.65	3.34	3.12	2.95	2.82	2.72	2.63	2.50	2.35	2.20	2.12	2.03	1.94	1.84	1.73	1.60
120	6.85	4.79	3.95	3.48	3.17	2.96	2.79	2.66	2.56	2.47	2.34	2.19	2.03	1.95	1.86	1.76	1.66	1.53	1.38
∞	6.63	4.61	3.78	3.32	3.02	2.80	2.64	2.51	2.41	2.32	2.18	2.04	1.88	1.79	1.70	1.59	1.47	1.32	1.00

Degrees of freedom for denominator

† Reproduced from M. Merrington and C. M. Thompson, "Tables of percentage points of the inverted beta (F) distribution," *Biometrika*, vol. 33 (1943), by permission of the *Biometrika* trustees.

TABLE V Binomial Probabilities

n	x	p 0.05	0.1	0.2	0.3	0.4	0.5	0.6	0.7	0.8	0.9	0.95
2	0	0.902	0.810	0.640	0.490	0.360	0.250	0.160	0.090	0.040	0.010	0.002
	1	0.095	0.180	0.320	0.420	0.480	0.500	0.480	0.420	0.320	0.180	0.095
	2	0.002	0.010	0.040	0.090	0.160	0.250	0.360	0.490	0.640	0.810	0.902
3	0	0.857	0.729	0.512	0.343	0.216	0.125	0.064	0.027	0.008	0.001	
	1	0.135	0.243	0.384	0.441	0.432	0.375	0.288	0.189	0.096	0.027	0.007
	2	0.007	0.027	0.096	0.189	0.288	0.375	0.432	0.441	0.384	0.243	0.135
	3		0.001	0.008	0.027	0.064	0.125	0.216	0.343	0.512	0.729	0.857
4	0	0.815	0.656	0.410	0.240	0.130	0.062	0.026	0.008	0.002		
	1	0.171	0.292	0.410	0.412	0.346	0.250	0.154	0.076	0.026	0.004	
	2	0.014	0.049	0.154	0.265	0.346	0.375	0.346	0.265	0.154	0.049	0.014
	3		0.004	0.026	0.076	0.154	0.250	0.346	0.412	0.410	0.292	0.171
	4			0.002	0.008	0.026	0.062	0.130	0.240	0.410	0.656	0.815
5	0	0.774	0.590	0.328	0.168	0.078	0.031	0.010	0.002			
	1	0.204	0.328	0.410	0.360	0.259	0.156	0.077	0.028	0.006		
	2	0.021	0.073	0.205	0.309	0.346	0.312	0.230	0.132	0.051	0.008	0.001
	3	0.001	0.008	0.051	0.132	0.230	0.312	0.346	0.309	0.205	0.073	0.021
	4			0.006	0.028	0.077	0.156	0.259	0.360	0.410	0.328	0.204
	5				0.002	0.010	0.031	0.078	0.168	0.328	0.590	0.774
6	0	0.735	0.531	0.262	0.118	0.047	0.016	0.004	0.001			
	1	0.232	0.354	0.393	0.303	0.187	0.094	0.037	0.010	0.002		
	2	0.031	0.098	0.246	0.324	0.311	0.234	0.138	0.060	0.015	0.001	
	3	0.002	0.015	0.082	0.185	0.276	0.312	0.276	0.185	0.082	0.015	0.002
	4		0.001	0.015	0.060	0.138	0.234	0.311	0.324	0.246	0.098	0.031
	5			0.002	0.010	0.037	0.094	0.187	0.303	0.393	0.354	0.232
	6				0.001	0.004	0.016	0.047	0.118	0.262	0.531	0.735
7	0	0.698	0.478	0.210	0.082	0.028	0.008	0.002				
	1	0.257	0.372	0.367	0.247	0.131	0.055	0.017	0.004			
	2	0.041	0.124	0.275	0.318	0.261	0.164	0.077	0.025	0.004		
	3	0.004	0.023	0.115	0.227	0.290	0.273	0.194	0.097	0.029	0.003	
	4		0.003	0.029	0.097	0.194	0.273	0.290	0.227	0.115	0.023	0.004
	5			0.004	0.025	0.077	0.164	0.261	0.318	0.275	0.124	0.041
	6				0.004	0.017	0.055	0.131	0.247	0.367	0.372	0.257
	7					0.002	0.008	0.028	0.082	0.210	0.478	0.698
8	0	0.663	0.430	0.168	0.058	0.017	0.004	0.001				
	1	0.279	0.383	0.336	0.198	0.090	0.031	0.008	0.001			
	2	0.051	0.149	0.294	0.296	0.209	0.109	0.041	0.010	0.001		
	3	0.005	0.033	0.147	0.254	0.279	0.219	0.124	0.047	0.009		
	4		0.005	0.046	0.136	0.232	0.273	0.232	0.136	0.046	0.005	
	5			0.009	0.047	0.124	0.219	0.279	0.254	0.147	0.033	0.005
	6			0.001	0.010	0.041	0.109	0.209	0.296	0.294	0.149	0.051
	7				0.001	0.008	0.031	0.090	0.198	0.336	0.383	0.279
	8					0.001	0.004	0.017	0.058	0.168	0.430	0.663

All values omitted in this table are 0.0005 or less.

TABLE V   Binomial Probabilities (*Continued*)

n	x	0.05	0.1	0.2	0.3	0.4	0.5	0.6	0.7	0.8	0.9	0.95
9	0	0.630	0.387	0.134	0.040	0.010	0.002					
	1	0.299	0.387	0.302	0.156	0.060	0.018	0.004				
	2	0.063	0.172	0.302	0.267	0.161	0.070	0.021	0.004			
	3	0.008	0.045	0.176	0.267	0.251	0.164	0.074	0.021	0.003		
	4	0.001	0.007	0.066	0.172	0.251	0.246	0.167	0.074	0.017	0.001	
	5		0.001	0.017	0.074	0.167	0.246	0.251	0.172	0.066	0.007	0.001
	6			0.003	0.021	0.074	0.164	0.251	0.267	0.176	0.045	0.008
	7				0.004	0.021	0.070	0.161	0.267	0.302	0.172	0.063
	8					0.004	0.018	0.060	0.156	0.302	0.387	0.299
	9						0.002	0.010	0.040	0.134	0.387	0.630
10	0	0.599	0.349	0.107	0.028	0.006	0.001					
	1	0.315	0.387	0.268	0.121	0.040	0.010	0.002				
	2	0.075	0.194	0.302	0.233	0.121	0.044	0.011	0.001			
	3	0.010	0.057	0.201	0.267	0.215	0.117	0.042	0.009	0.001		
	4	0.001	0.011	0.088	0.200	0.251	0.205	0.111	0.037	0.006		
	5		0.001	0.026	0.103	0.201	0.246	0.201	0.103	0.026	0.001	
	6			0.006	0.037	0.111	0.205	0.251	0.200	0.088	0.011	0.001
	7			0.001	0.009	0.042	0.117	0.215	0.267	0.201	0.057	0.010
	8				0.001	0.011	0.044	0.121	0.233	0.302	0.194	0.075
	9					0.002	0.010	0.040	0.121	0.268	0.387	0.315
	10						0.001	0.006	0.028	0.107	0.349	0.599
11	0	0.569	0.314	0.086	0.020	0.004						
	1	0.329	0.384	0.236	0.093	0.027	0.005	0.001				
	2	0.087	0.213	0.295	0.200	0.089	0.027	0.005	0.001			
	3	0.014	0.071	0.221	0.257	0.177	0.081	0.023	0.004			
	4	0.001	0.016	0.111	0.220	0.236	0.161	0.070	0.017	0.002		
	5		0.002	0.039	0.132	0.221	0.226	0.147	0.057	0.010		
	6			0.010	0.057	0.147	0.226	0.221	0.132	0.039	0.002	
	7			0.002	0.017	0.070	0.161	0.236	0.220	0.111	0.016	0.001
	8				0.004	0.023	0.081	0.177	0.257	0.221	0.071	0.014
	9				0.001	0.005	0.027	0.089	0.200	0.295	0.213	0.087
	10					0.001	0.005	0.027	0.093	0.236	0.384	0.329
	11						0.004	0.020	0.086	0.314	0.569	
12	0	0.540	0.282	0.069	0.014	0.002						
	1	0.341	0.377	0.206	0.071	0.017	0.003					
	2	0.099	0.230	0.283	0.168	0.064	0.016	0.002				
	3	0.017	0.085	0.236	0.240	0.142	0.054	0.012	0.001			
	4	0.002	0.021	0.133	0.231	0.213	0.121	0.042	0.008	0.001		
	5		0.004	0.053	0.158	0.227	0.193	0.101	0.029	0.003		
	6			0.016	0.079	0.177	0.226	0.177	0.079	0.016		
	7			0.003	0.029	0.101	0.193	0.227	0.158	0.053	0.004	
	8			0.001	0.008	0.042	0.121	0.213	0.231	0.133	0.021	0.002
	9				0.001	0.012	0.054	0.142	0.240	0.236	0.085	0.017
	10					0.002	0.016	0.064	0.168	0.283	0.230	0.099
	11						0.003	0.017	0.071	0.206	0 377	0.341
	12							0.002	0.014	0.069	0.282	0.540

TABLE V Binomial Probabilities (*Continued*)

n	x	0.05	0.1	0.2	0.3	0.4	0.5	0.6	0.7	0.8	0.9	0.95
13	0	0.513	0.254	0.055	0.010	0.001						
	1	0.351	0.367	0.179	0.054	0.011	0.002					
	2	0.111	0.245	0.268	0.139	0.045	0.010	0.001				
	3	0.021	0.100	0.246	0.218	0.111	0.035	0.006	0.001			
	4	0.003	0.028	0.154	0.234	0.184	0.087	0.024	0.003			
	5		0.006	0.069	0.180	0.221	0.157	0.066	0.014	0.001		
	6		0.001	0.023	0.103	0.197	0.209	0.131	0.044	0.006		
	7			0.006	0.044	0.131	0.209	0.197	0.103	0.023	0.001	
	8			0.001	0.014	0.066	0.157	0.221	0.180	0.069	0.006	
	9				0.003	0.024	0.087	0.184	0.234	0.154	0.028	0.003
	10				0.001	0.006	0.035	0.111	0.218	0.246	0.100	0.021
	11					0.001	0.010	0.045	0.139	0.268	0.245	0.111
	12						0.002	0.011	0.054	0.179	0.367	0.351
	13							0.001	0.010	0.055	0.254	0.513
14	0	0.488	0.229	0.044	0.007	0.001						
	1	0.359	0.356	0.154	0.041	0.007	0.001					
	2	0.123	0.257	0.250	0.113	0.032	0.006	0.001				
	3	0.026	0.114	0.250	0.194	0.085	0.022	0.003				
	4	0.004	0.035	0.172	0.229	0.155	0.061	0.014	0.001			
	5		0.008	0.086	0.196	0.207	0.122	0.041	0.007			
	6		0.001	0.032	0.126	0.207	0.183	0.092	0.023	0.002		
	7			0.009	0.062	0.157	0.209	0.157	0.062	0.009		
	8			0.002	0.023	0.092	0.183	0.207	0.126	0.032	0.001	
	9				0.007	0.041	0.122	0.207	0.196	0.086	0.008	
	10				0.001	0.014	0.061	0.155	0.229	0.172	0.035	0.004
	11					0.003	0.022	0.085	0.194	0.250	0.114	0.026
	12					0.001	0.006	0.032	0.113	0.250	0.257	0.123
	13						0.001	0.007	0.041	0.154	0.356	0.359
	14							0.001	0.007	0.044	0.229	0.488
15	0	0.463	0.206	0.035	0.005							
	1	0.366	0.343	0.132	0.031	0.005						
	2	0.135	0.267	0.231	0.092	0.022	0.003					
	3	0.031	0.129	0.250	0.170	0.063	0.014	0.002				
	4	0.005	0.043	0.188	0.219	0.127	0.042	0.007	0.001			
	5	0.001	0.010	0.103	0.206	0.186	0.092	0.024	0.003			
	6		0.002	0.043	0.147	0.207	0.153	0.061	0.012	0.001		
	7			0.014	0.081	0.177	0.196	0.118	0.035	0.003		
	8			0.003	0.035	0.118	0.196	0.177	0.081	0.014		
	9			0.001	0.012	0.061	0.153	0.207	0.147	0.043	0.002	
	10				0.003	0.024	0.092	0.186	0.206	0.103	0.010	0.001
	11				0.001	0.007	0.042	0.127	0.219	0.188	0.043	0.005
	12					0.002	0.014	0.063	0.170	0.250	0.129	0.031
	13						0.003	0.022	0.092	0.231	0.267	0.135
	14							0.005	0.031	0.132	0.343	0.366
	15								0.005	0.035	0.206	0.463

TABLE V Binomial Probabilities (*Continued*)

n	x	0.05	0.1	0.2	0.3	0.4	0.5	0.6	0.7	0.8	0.9	0.95
16	0	0.440	0.185	0.028	0.003							
	1	0.371	0.329	0.113	0.023	0.003						
	2	0.146	0.275	0.211	0.073	0.015	0.002					
	3	0.036	0.142	0.246	0.146	0.047	0.009	0.001				
	4	0.006	0.051	0.200	0.204	0.101	0.028	0.004				
	5	0.001	0.014	0.120	0.210	0.162	0.067	0.014	0.001			
	6		0.003	0.055	0.165	0.198	0.122	0.039	0.006			
	7			0.020	0.101	0.189	0.175	0.084	0.019	0.001		
	8			0.006	0.049	0.142	0.196	0.142	0.049	0.006		
	9			0.001	0.019	0.084	0.175	0.189	0.101	0.020		
	10				0.006	0.039	0.122	0.198	0.165	0.055	0.003	
	11				0.001	0.014	0.067	0.162	0.210	0.120	0.014	0.001
	12					0.004	0.028	0.101	0.204	0.200	0.051	0.006
	13					0.001	0.009	0.047	0.146	0.246	0.142	0.036
	14						0.002	0.015	0.073	0.211	0.275	0.146
	15							0.003	0.023	0.113	0.329	0.371
	16								0.003	0.028	0.185	0.440
17	0	0.418	0.167	0.023	0.002							
	1	0.374	0.315	0.096	0.017	0.002						
	2	0.158	0.280	0.191	0.058	0.010	0.001					
	3	0.041	0.156	0.239	0.125	0.034	0.005					
	4	0.008	0.060	0.209	0.187	0.080	0.018	0.002				
	5	0.001	0.017	0.136	0.208	0.138	0.047	0.008	0.001			
	6		0.004	0.068	0.178	0.184	0.094	0.024	0.003			
	7		0.001	0.027	0.120	0.193	0.148	0.057	0.009			
	8			0.008	0.064	0.161	0.185	0.107	0.028	0.002		
	9			0.002	0.028	0.107	0.185	0.161	0.064	0.008		
	10				0.009	0.057	0.148	0.193	0.120	0.027	0.001	
	11				0.003	0.024	0.094	0.184	0.178	0.068	0.004	
	12				0.001	0.008	0.047	0.138	0.208	0.136	0.017	0.001
	13					0.002	0.018	0.080	0.187	0.209	0.060	0.008
	14						0.005	0.034	0.125	0.239	0.156	0.041
	15						0.001	0.010	0.058	0.191	0.280	0.158
	16							0.002	0.017	0.096	0.315	0.374
	17								0.002	0.023	0.167	0.418
18	0	0.397	0.150	0.018	0.002							
	1	0.376	0.300	0.081	0.013	0.001						
	2	0.168	0.284	0.172	0.046	0.007	0.001					
	3	0.047	0.168	0.230	0.105	0.025	0.003					
	4	0.009	0.070	0.215	0.168	0.061	0.012	0.001				
	5	0.001	0.022	0.151	0.202	0.115	0.033	0.004				
	6		0.005	0.082	0.187	0.166	0.071	0.015	0.001			
	7		0.001	0.035	0.138	0.189	0.121	0.037	0.005			
	8			0.012	0.081	0.173	0.167	0.077	0.015	0.001		
	9			0.003	0.039	0.128	0.185	0.128	0.039	0.003		
	10			0.001	0.015	0.077	0.167	0.173	0.081	0.012		
	11				0.005	0.037	0.121	0.189	0.138	0.035	0.001	

TABLE V Binomial Probabilities (*Continued*)

n	x	0.05	0.1	0.2	0.3	0.4	0.5	0.6	0.7	0.8	0.9	0.95
	12				0.001	0.015	0.071	0.166	0.187	0.082	0.005	
	13					0.004	0.033	0.115	0.202	0.151	0.022	0.001
	14					0.001	0.012	0.061	0.168	0.215	0.070	0.009
	15						0.003	0.025	0.105	0.230	0.168	0.047
	16						0.001	0.007	0.046	0.172	0.284	0.168
	17							0.001	0.013	0.081	0.300	0.376
	18								0.002	0.018	0.150	0.397
19	0	0.377	0.135	0.014	0.001							
	1	0.377	0.285	0.068	0.009	0.001						
	2	0.179	0.285	0.154	0.036	0.005						
	3	0.053	0.180	0.218	0.087	0.017	0.002					
	4	0.011	0.080	0.218	0.149	0.047	0.007	0.001				
	5	0.002	0.027	0.164	0.192	0.093	0.022	0.002				
	6		0.007	0.095	0.192	0.145	0.052	0.008	0.001			
	7		0.001	0.044	0.153	0.180	0.096	0.024	0.002			
	8			0.017	0.098	0.180	0.144	0.053	0.008			
	9			0.005	0.051	0.146	0.176	0.098	0.022	0.001		
	10			0.001	0.022	0.098	0.176	0.146	0.051	0.005		
	11				0.008	0.053	0.144	0.180	0.098	0.017		
	12				0.002	0.024	0.096	0.180	0.153	0.044	0.001	
	13				0.001	0.008	0.052	0.145	0.192	0.095	0.007	
	14					0.002	0.022	0.093	0.192	0.164	0.027	0.002
	15					0.001	0.007	0.047	0.149	0.218	0.080	0.011
	16						0.002	0.017	0.087	0.218	0.180	0.053
	17							0.005	0.036	0.154	0.285	0.179
	18							0.001	0.009	0.068	0.285	0.377
	19								0.001	0.014	0.135	0.377
20	0	0.358	0.122	0.012	0.001							
	1	0.377	0.270	0.058	0.007							
	2	0.189	0.285	0.137	0.028	0.003						
	3	0.060	0.190	0.205	0.072	0.012	0.001					
	4	0.013	0.090	0.218	0.130	0.035	0.005					
	5	0.002	0.032	0.175	0.179	0.075	0.015	0.001				
	6		0.009	0.109	0.192	0.124	0.037	0.005				
	7		0.002	0.055	0.164	0.166	0.074	0.015	0.001			
	8			0.022	0.114	0.180	0.120	0.035	0.004			
	9			0.007	0.065	0.160	0.160	0.071	0.012			
	10			0.002	0.031	0.117	0.176	0.117	0.031	0.002		
	11				0.012	0.071	0.160	0.160	0.065	0.007		
	12				0.004	0.035	0.120	0.180	0.114	0.022		
	13				0.001	0.015	0.074	0.166	0.164	0.055	0.002	
	14					0.005	0.037	0.124	0.192	0.109	0.009	
	15					0.001	0.015	0.075	0.179	0.175	0.032	0.002
	16						0.005	0.035	0.130	0.218	0.090	0.013
	17						0.001	0.012	0.072	0.205	0.190	0.060
	18							0.003	0.028	0.137	0.285	0.189
	19								0.007	0.058	0.270	0.377
	20								0.001	0.012	0.122	0.358

TABLE VI   Critical Values of $T^{†}$

n	$T_{0.10}$	$T_{0.05}$	$T_{0.02}$	$T_{0.01}$
4				
5	1			
6	2	1		
7	4	2	0	
8	6	4	2	0
9	8	6	3	2
10	11	8	5	3
11	14	11	7	5
12	17	14	10	7
13	21	17	13	10
14	26	21	16	13
15	30	25	20	16
16	36	30	24	19
17	41	35	28	23
18	47	40	33	28
19	54	46	38	32
20	60	52	43	37
21	68	59	49	43
22	75	66	56	49
23	83	73	62	55
24	92	81	69	61
25	101	90	77	68

† From F. Wilcoxon and R. A. Wilcox, *Some Rapid Approximate Statistical Procedures*, American Cyanamid Company, Pearl River, N.Y., 1964. Reproduced with permission of American Cyanamid Company.

TABLE VII   Critical values of U[†]

Values of $U_{0.10}$

n_1 \ n_2	2	3	4	5	6	7	8	9	10	11	12	13	14	15
2				0	0	0	1	1	1	1	2	2	3	3
3		0	0	1	2	2	3	4	4	5	5	6	7	7
4		0	1	2	3	4	5	6	7	8	9	10	11	12
5	0	1	2	4	5	6	8	9	11	12	13	15	16	18
6	0	2	3	5	7	8	10	12	14	16	17	19	21	23
7	0	2	4	6	8	11	13	15	17	19	21	24	26	28
8	1	3	5	8	10	13	15	18	20	23	26	28	31	33
9	1	4	6	9	12	15	18	21	24	27	30	33	36	39
10	1	4	7	11	14	17	20	24	27	31	34	37	41	44
11	1	5	8	12	16	19	23	27	31	34	38	42	46	50
12	2	5	9	13	17	21	26	30	34	38	42	47	51	55
13	2	6	10	15	19	24	28	33	37	42	47	51	56	61
14	3	7	11	16	21	26	31	36	41	46	51	56	61	66
15	3	7	12	18	23	28	33	39	44	50	55	61	66	72

Values of $U_{0.05}$

n_1 \ n_2	2	3	4	5	6	7	8	9	10	11	12	13	14	15
2							0	0	0	0	1	1	1	1
3				0	1	1	2	2	3	3	4	4	5	5
4			0	1	2	3	4	4	5	6	7	8	9	10
5		0	1	2	3	5	6	7	8	9	11	12	13	14
6		1	2	3	5	6	8	10	11	13	14	16	17	19
7		1	3	5	6	8	10	12	14	16	18	20	22	24
8	0	2	4	6	8	10	13	15	17	19	22	24	26	29
9	0	2	4	7	10	12	15	17	20	23	26	28	31	34
10	0	3	5	8	11	14	17	20	23	26	29	33	36	39
11	0	3	6	9	13	16	19	23	26	30	33	37	40	44
12	1	4	7	11	14	18	22	26	29	33	37	41	45	49
13	1	4	8	12	16	20	24	28	33	37	41	45	50	54
14	1	5	9	13	17	22	26	31	36	40	45	50	55	59
15	1	5	10	14	19	24	29	34	39	44	49	54	59	64

[†] This table is based on Table 11.4 of D. B. Owen, *Handbook of Statistical Tables*, © 1962, U.S. Department of Energy. Published by Addison-Wesley Publishing Company, Inc., Reading Mass. Reprinted with permission of the publisher.

TABLE VII   Critical Values of U (*Continued*)

Values of $U_{0.02}$

n_1 \ n_2	2	3	4	5	6	7	8	9	10	11	12	13	14	15
2												0	0	0
3						0	0	1	1	1	2	2	2	3
4				0	1	1	2	3	3	4	5	5	6	7
5			0	1	2	3	4	5	6	7	8	9	10	11
6			1	2	3	4	6	7	8	9	11	12	13	15
7		0	1	3	4	6	7	9	11	12	14	16	17	19
8		0	2	4	6	7	9	11	13	15	17	20	22	24
9		1	3	5	7	9	11	14	16	18	21	23	26	28
10		1	3	6	8	11	13	16	19	22	24	27	30	33
11		1	4	7	9	12	15	18	22	25	28	31	34	37
12		2	5	8	11	14	17	21	24	28	31	35	38	42
13	0	2	5	9	12	16	20	23	27	31	35	39	43	47
14	0	2	6	10	13	17	22	26	30	34	38	43	47	51
15	0	3	7	11	15	19	24	28	33	37	42	47	51	56

Values of $U_{0.01}$

n_1 \ n_2	3	4	5	6	7	8	9	10	11	12	13	14	15
3							0	0	0	1	1	1	2
4				0	0	1	1	2	2	3	3	4	5
5			0	1	1	2	3	4	5	6	7	7	8
6		0	1	2	3	4	5	6	7	9	10	11	12
7		0	1	3	4	6	7	9	10	12	13	15	16
8		1	2	4	6	7	9	11	13	15	17	18	20
9	0	1	3	5	7	9	11	13	16	18	20	22	24
10	0	2	4	6	9	11	13	16	18	21	24	26	29
11	0	2	5	7	10	13	16	18	21	24	27	30	33
12	1	3	6	9	12	15	18	21	24	27	31	34	37
13	1	3	7	10	13	17	20	24	27	31	34	38	42
14	1	4	7	11	15	18	22	26	30	34	38	42	46
15	2	5	8	12	16	20	24	29	33	37	42	46	51

TABLE VIII Critical values of u[†]

Values of $u_{0.025}$

n_1 \ n_2	4	5	6	7	8	9	10	11	12	13	14	15
4		9	9									
5	9	10	10	11	11							
6	9	10	11	12	12	13	13	13	13			
7		11	12	13	13	14	14	14	14	15	15	15
8		11	12	13	14	14	15	15	16	16	16	16
9			13	14	14	15	16	16	16	17	17	18
10			13	14	15	16	16	17	17	18	18	18
11			13	14	15	16	17	17	18	19	19	19
12			13	14	16	16	17	18	19	19	20	20
13				15	16	17	18	19	19	20	20	21
14				15	16	17	18	19	20	20	21	22
15				15	16	18	18	19	20	21	22	22

Values of $u'_{0.025}$

n_1 \ n_2	2	3	4	5	6	7	8	9	10	11	12	13	14	15
2											2	2	2	2
3					2	2	2	2	2	2	2	2	2	3
4				2	2	2	3	3	3	3	3	3	3	3
5			2	2	3	3	3	3	3	4	4	4	4	4
6		2	2	3	3	3	3	4	4	4	4	5	5	5
7		2	2	3	3	3	4	4	5	5	5	5	5	6
8		2	3	3	3	4	4	5	5	5	6	6	6	6
9		2	3	3	4	4	5	5	5	6	6	6	7	7
10		2	3	3	4	5	5	5	6	6	7	7	7	7
11		2	3	4	4	5	5	6	6	7	7	7	8	8
12	2	2	3	4	4	5	6	6	7	7	7	8	8	8
13	2	2	3	4	5	5	6	6	7	7	8	8	9	9
14	2	2	3	4	5	5	6	7	7	8	8	9	9	9
15	2	3	3	4	5	6	6	7	7	8	8	9	9	10

[†] This table is adapted, by permission, from F. S. Swed and C. Eisenhart, "Tables for testing randomness of grouping in a sequence of alternatives," *Annals of Mathematical Statistics*, Vol. 14.

TABLE VIII   Critical Values of u (Continued)

Values of $u_{0.005}$

n_1 \ n_2	5	6	7	8	9	10	11	12	13	14	15
5		11									
6	11	12	13	13							
7		13	13	14	15	15	15				
8		13	14	15	15	16	16	17	17	17	
9			15	15	16	17	17	18	18	18	19
10			15	16	17	17	18	19	19	19	20
11			15	16	17	18	19	19	20	20	21
12				17	18	19	19	20	21	21	22
13				17	18	19	20	21	21	22	22
14				17	18	19	20	21	22	23	23
15					19	20	21	22	22	23	24

Values of $u'_{0.005}$

n_1 \ n_2	3	4	5	6	7	8	9	10	11	12	13	14	15
3										2	2	2	2
4						2	2	2	2	2	2	2	3
5				2	2	2	2	3	3	3	3	3	3
6			2	2	2	3	3	3	3	3	3	4	4
7			2	2	3	3	3	3	4	4	4	4	4
8		2	2	3	3	3	3	4	4	4	5	5	5
9		2	2	3	3	3	4	4	5	5	5	5	6
10		2	3	3	3	4	4	5	5	5	5	6	6
11		2	3	3	4	4	5	5	5	6	6	6	7
12	2	2	3	3	4	4	5	5	6	6	6	7	7
13	2	2	3	3	4	5	5	5	6	6	7	7	7
14	2	2	3	4	4	5	5	6	6	7	7	7	8
15	2	3	3	4	4	5	6	6	7	7	7	8	8

TABLE IX Values of $Z = \frac{1}{2} \cdot \ln \frac{1+r}{1-r}$

r	.00	.01	.02	.03	.04	.05	.06	.07	.08	.09
0.0	0.000	0.010	0.020	0.030	0.040	0.050	0.060	0.070	0.080	0.090
0.1	0.100	0.110	0.121	0.131	0.141	0.151	0.161	0.172	0.182	0.192
0.2	0.203	0.213	0.224	0.234	0.245	0.255	0.266	0.277	0.288	0.299
0.3	0.310	0.321	0.332	0.343	0.354	0.365	0.377	0.388	0.400	0.412
0.4	0.424	0.436	0.448	0.460	0.472	0.485	0.497	0.510	0.523	0.536
0.5	0.549	0.563	0.576	0.590	0.604	0.618	0.633	0.648	0.662	0.678
0.6	0.693	0.709	0.725	0.741	0.758	0.775	0.793	0.811	0.829	0.848
0.7	0.867	0.887	0.908	0.929	0.950	0.973	0.996	1.020	1.045	1.071
0.8	1.099	1.127	1.157	1.188	1.221	1.256	1.293	1.333	1.376	1.422
0.9	1.472	1.528	1.589	1.658	1.738	1.832	1.946	2.092	2.298	2.647

For negative values of r put a minus sign in front of the corresponding Z's, and vice versa.

TABLE X Binomial Coefficients

n	$\binom{n}{0}$	$\binom{n}{1}$	$\binom{n}{2}$	$\binom{n}{3}$	$\binom{n}{4}$	$\binom{n}{5}$	$\binom{n}{6}$	$\binom{n}{7}$	$\binom{n}{8}$	$\binom{n}{9}$	$\binom{n}{10}$
0	1										
1	1	1									
2	1	2	1								
3	1	3	3	1							
4	1	4	6	4	1						
5	1	5	10	10	5	1					
6	1	6	15	20	15	6	1				
7	1	7	21	35	35	21	7	1			
8	1	8	28	56	70	56	28	8	1		
9	1	9	36	84	126	126	84	36	9	1	
10	1	10	45	120	210	252	210	120	45	10	1
11	1	11	55	165	330	462	462	330	165	55	11
12	1	12	66	220	495	792	924	792	495	220	66
13	1	13	78	286	715	1287	1716	1716	1287	715	286
14	1	14	91	364	1001	2002	3003	3432	3003	2002	1001
15	1	15	105	455	1365	3003	5005	6435	6435	5005	3003
16	1	16	120	560	1820	4368	8008	11440	12870	11440	8008
17	1	17	136	680	2380	6188	12376	19448	24310	24310	19448
18	1	18	153	816	3060	8568	18564	31824	43758	48620	43758
19	1	19	171	969	3876	11628	27132	50388	75582	92378	92378
20	1	20	190	1140	4845	15504	38760	77520	125970	167960	184756

For $r > 10$ it may be necessary to make use of the identity $\binom{n}{r} = \binom{n}{n-r}$.

TABLE XI Random Numbers[†]

04433	80674	24520	18222	10610	05794	37515
60298	47829	72648	37414	75755	04717	29899
67884	59651	67533	68123	17730	95862	08034
89512	32155	51906	61662	64130	16688	37275
32653	01895	12506	88535	36553	23757	34209
95913	15405	13772	76638	48423	25018	99041
55864	21694	13122	44115	01601	50541	00147
35334	49810	91601	40617	72876	33967	73830
57729	32196	76487	11622	96297	24160	09903
86648	13697	63677	70119	94739	25875	38829
30574	47609	07967	32422	76791	39725	53711
81307	43694	83580	79974	45929	85113	72268
02410	54905	79007	54939	21410	86980	91772
18969	75274	52233	62319	08598	09066	95288
87863	82384	66860	62297	80198	19347	73234
68397	71708	15438	62311	72844	60203	46412
28529	54447	58729	10854	99058	18260	38765
44285	06372	15867	70418	57012	72122	36634
86299	83430	33571	23309	57040	29285	67870
84842	68668	90894	61658	15001	94055	36308
56970	83609	52098	04184	54967	72938	56834
83125	71257	60490	44369	66130	72936	69848
55503	52423	02464	26141	68779	66388	75242
47019	76273	33203	29608	54553	25971	69573
84828	32592	79526	29554	84580	37859	28504
68921	08141	79227	05748	51276	57143	31926
36458	96045	30424	98420	72925	40729	22337
95752	59445	36847	87729	81679	59126	59437
26768	47323	58454	56958	20575	76746	49878
42613	37056	43636	58085	06766	60227	96414
95457	30566	65482	25596	02678	54592	63607
95276	17894	63564	95958	39750	64379	46059
66954	52324	64776	92345	95110	59448	77249
17457	18481	14113	62462	02798	54977	48349
03704	36872	83214	59337	01695	60666	97410
21538	86497	33210	60337	27976	70661	08250
57178	67619	98310	70348	11317	71623	55510
31048	97558	94953	55866	96283	46620	52087
69799	55380	16498	80733	96422	58078	99643
90595	61867	59231	17772	67831	33317	00520
33570	04981	98939	78784	09977	29398	93896
15340	93460	57477	13898	48431	72936	78160
64079	42483	36512	56186	99098	48850	72527
63491	05546	67118	62063	74958	20946	28147
92003	63868	41034	28260	79708	00770	88643
52360	46658	66511	04172	73085	11795	52594
74622	12142	68355	65635	21828	39539	18988
04157	50079	61343	64315	70836	82857	35335
86003	60070	66241	32836	27573	11479	94114
41268	80187	20351	09636	84668	42486	71303

[†] Based on parts of *Tables of 105,000 Random Decimal Digits*. Interstate Commerce Commission, Bureau of Transport Economics and Statistics, Washington D.C.

TABLE XI   Random Numbers (*Continued*)

48611	62866	33963	14045	79451	04934	45576
78812	03509	78673	73181	29973	18664	04555
19472	63971	37271	31445	49019	49405	46925
51266	11569	08697	91120	64156	40365	74297
55806	96275	26130	47949	14877	69594	83041
77527	81360	18180	97421	55541	90275	18213
77680	58788	33016	61173	93049	04694	43534
15404	96554	88265	34537	38526	67924	40474
14045	22917	60718	66487	46346	30949	03173
68376	43918	77653	04127	69930	43283	35766
93385	13421	67957	20384	58731	53396	59723
09858	52104	32014	53115	03727	98624	84616
93307	34116	49516	42148	57740	31198	70336
04794	01534	92058	03157	91758	80611	45357
86265	49096	97021	92582	61422	75890	86442
65943	79232	45702	67055	39024	57383	44424
90038	94209	04055	27393	61517	23002	96560
97283	95943	78363	36498	40662	94188	18202
21913	72958	75637	99936	58715	07943	23748
41161	37341	81838	19389	80336	46346	91895
23777	98392	31417	98547	92058	02277	50315
59973	08144	61070	73094	27059	69181	55623
82690	74099	77885	23813	10054	11900	44653
83854	24715	48866	65745	31131	47636	45137
61980	34997	41825	11623	07320	15003	56774
99915	45821	97702	87125	44488	77613	56823
48293	86847	43186	42951	37804	85129	28993
33225	31280	41232	34750	91097	60752	69783
06846	32828	24425	30249	78801	26977	92074
32671	45587	79620	84831	38156	74211	82752
82096	21913	75544	55228	89796	05694	91552
51666	10433	10945	55306	78562	89630	41230
54044	67942	24145	42294	27427	84875	37022
66738	60184	75679	38120	17640	36242	99357
55064	17427	89180	74018	44865	53197	74810
69599	60264	84549	78007	88450	06488	72274
64756	87759	92354	78694	63638	80939	98644
80817	74533	68407	55862	32476	19326	95558
39847	96884	84657	33697	39578	90197	80532
90401	41700	95510	61166	33757	23279	85523
78227	90110	81378	96659	37008	04050	04228
87240	52716	87697	79433	16336	52862	69149
08486	10951	26832	39763	02485	71688	90936
39338	32169	03713	93510	61244	73774	01245
21188	01850	69689	49426	49128	14660	14143
13287	82531	04388	64693	11934	35051	68576
53609	04001	19648	14053	49623	10840	31915
87900	36194	31567	53506	34304	39910	79630
81641	00496	36058	75899	46620	70024	88753
19512	50277	71508	20116	79520	06269	74173

TABLE XI Random Numbers (*Continued*)

24418	23508	91507	76455	54941	72711	39406
57404	73678	08272	62941	02349	71389	45605
77644	98489	86268	73652	98210	44546	27174
68366	65614	01443	07607	11826	91326	29664
64472	72294	95432	53555	96810	17100	35066
88205	37913	98633	81009	81060	33449	68055
98455	78685	71250	10329	56135	80647	51404
48977	36794	56054	59243	57361	65304	93258
93077	72941	92779	23581	24548	56415	61927
84533	26564	91583	83411	66504	02036	02922
11338	12903	14514	27585	45068	05520	56321
23853	68500	92274	87026	99717	01542	72990
94096	74920	25822	98026	05394	61840	83089
83160	82362	09350	98536	38155	42661	02363
97425	47335	69709	01386	74319	04318	99387
83951	11954	24317	20345	18134	90062	10761
93085	35203	05740	03206	92012	42710	34650
33762	83193	58045	89880	78101	44392	53767
49665	85397	85137	30496	23469	42846	94810
37541	82627	80051	72521	35342	56119	97190
22145	85304	35348	82854	55846	18076	12415
27153	08662	61078	52433	22184	33998	87436
00301	49425	66682	25442	83668	66236	79655
43815	43272	73778	63469	50083	70696	13558
14689	86482	74157	46012	97765	27552	49617
16680	55936	82453	19532	49988	13176	94219
86938	60429	01137	86168	78257	86249	46134
33944	29219	73161	46061	30946	22210	79302
16045	67736	18608	18198	19468	76358	69203
37044	52523	25627	63107	30806	80857	84383
61471	45322	35340	35132	42163	69332	98851
47422	21296	16785	66393	39249	51463	95963
24133	39719	14484	58613	88717	29289	77360
67253	67064	10748	16006	16767	57345	42285
62382	76941	01635	35829	77516	98468	51686
98011	16503	09201	03523	87192	66483	55649
37366	24386	20654	85117	74078	64120	04643
73587	83993	54176	05221	94119	20108	78101
33583	68291	50547	96085	62180	27453	18567
02878	33223	39199	49536	56199	05993	71201
91498	41673	17195	33175	04994	09879	70337
91127	19815	30219	55591	21725	43827	78862
12997	55013	18662	81724	24305	37661	18956
96098	13651	15393	69995	14762	69734	89150
97627	17837	10472	18983	28387	99781	52977
40064	47981	31484	76603	54088	91095	00010
16239	68743	71374	55863	22672	91609	51514
58354	24913	20435	30965	17453	65623	93058
52567	65085	60220	84641	18273	49604	47418
06236	29052	91392	07551	83532	68130	56970

TABLE XI Random Numbers (*Continued*)

94620	27963	96478	21559	19246	88097	44926
60947	60775	73181	43264	56895	04232	59604
27499	53523	63110	57106	20865	91683	80688
01603	23156	89223	43429	95353	44662	59433
00815	01552	06392	31437	70385	45863	75971
83844	90942	74857	52419	68723	47830	63010
06626	10042	93629	37609	57215	08409	81906
56760	63348	24949	11859	29793	37457	59377
64416	29934	00755	09418	14230	62887	92683
63569	17906	38076	32135	19096	96970	75917
22693	35089	72994	04252	23791	60249	83010
43413	59744	01275	71326	91382	45114	20245
09224	78530	50566	49965	04851	18280	14039
67625	34683	03142	74733	63558	09665	22610
86874	12549	98699	54952	91579	26023	81076
54548	49505	62515	63903	13193	33905	66936
73236	66167	49728	03581	40699	10396	81827
15220	66319	13543	14071	59148	95154	72852
16151	08029	36954	03891	38313	34016	18671
43635	84249	88984	80993	55431	90793	62603
30193	42776	85611	57635	51362	79907	77364
37430	45246	11400	20986	43996	73122	88474
88312	93047	12088	86937	70794	01041	74867
98995	58159	04700	90443	13168	31553	67891
51734	20849	70198	67906	00880	82899	66065
88698	41755	56216	66852	17748	04963	54859
51865	09836	73966	65711	41699	11732	17173
40300	08852	27528	84648	79589	95295	72895
02760	28625	70476	76410	32988	10194	94917
78450	26245	91763	73117	33047	03577	62599
50252	56911	62693	73817	98693	18728	94741
07929	66728	47761	81472	44806	15592	71357
09030	39605	87507	85446	51257	89555	75520
56670	88445	85799	76200	21795	38894	58070
48140	13583	94911	13318	64741	64336	95103
36764	86132	12463	28385	94242	32063	45233
14351	71381	28133	68269	65145	28152	39087
81276	00835	63835	87174	42446	08882	27067
55524	86088	00069	59254	24654	77371	26409
78852	65889	32719	13758	23937	90740	16866
11861	69032	51915	23510	32050	52052	24004
67699	01009	07050	73324	06732	27510	33761
50064	39500	17450	18030	63124	48061	59412
93126	17700	94400	76075	08317	27324	72723
01657	92602	41043	05686	15650	29970	95877
13800	76690	75133	60456	28491	03845	11507
98135	42870	48578	29036	69876	86563	61729
08313	99293	00990	13595	77457	79969	11339
90974	83965	62732	85161	54330	22406	86253
33273	61993	88407	69399	17301	70975	99129

TABLE XII   Random Normal Numbers[†]

1.801	0.459	1.102	−1.072	−0.336	0.942	−0.290	−0.716	1.396	−0.466
−0.175	−0.754	−0.134	1.231	1.483	−0.149	0.555	1.401	−1.142	0.205
−0.861	−1.460	0.526	0.239	−0.206	2.021	0.313	−0.253	−0.891	1.135
−0.577	0.335	−0.820	0.140	−0.333	0.426	0.209	−0.024	0.323	1.223
0.827	0.802	−0.457	0.560	0.643	−0.729	−0.249	0.338	−0.281	−1.804
−1.344	0.949	−1.459	−1.210	1.016	−0.148	−1.737	0.069	−1.185	0.040
1.476	1.262	−1.428	0.489	−0.523	−0.646	1.721	0.749	0.179	−0.922
0.527	−1.045	0.877	0.646	2.957	−0.972	−1.796	0.309	2.224	−0.070
−0.645	0.117	0.059	−0.080	−1.637	−0.746	1.256	2.520	−0.673	0.994
−0.514	−1.510	−0.714	−1.581	0.905	1.745	1.767	0.682	−0.648	−1.742
−0.656	−0.217	0.287	0.114	1.175	0.791	−0.263	−0.695	−1.348	1.239
−0.778	1.177	0.180	1.156	0.458	1.089	0.339	1.304	0.402	−0.831
0.352	−1.829	−0.645	0.236	0.641	0.920	−1.287	−0.187	−2.339	−0.237
1.352	−0.076	−1.962	0.827	0.252	1.621	0.770	1.324	0.488	−0.037
0.017	0.030	0.211	2.276	0.693	−1.733	0.773	0.652	−0.947	0.148
−0.218	−1.060	−0.553	1.043	2.305	0.380	−0.794	−1.498	1.088	−0.689
1.118	0.816	0.713	0.485	0.185	0.318	−1.050	0.110	0.563	1.177
−1.622	0.436	0.481	0.021	2.070	−0.845	−0.257	−0.680	−0.565	0.024
−1.103	−0.210	−1.088	−0.033	−1.022	0.366	−0.531	2.022	0.210	1.037
−0.677	−0.737	−0.950	−1.517	1.148	0.377	−0.397	−1.902	−0.748	−1.753
1.110	1.120	1.163	1.577	−1.172	−0.133	−0.213	0.154	−0.435	0.218
−0.278	0.569	0.586	1.523	−0.244	−0.170	−1.274	0.874	−1.020	−0.809
0.178	1.314	0.462	−0.253	−0.122	0.108	−1.256	−0.137	1.043	−0.135
0.312	−2.287	−0.655	−1.459	0.075	−0.457	−0.206	−0.326	0.489	−0.149
0.469	−2.066	−0.973	−1.009	−1.410	0.505	0.459	−0.572	−1.186	0.978
−0.730	1.650	0.760	−0.520	−0.671	−0.122	−0.324	−0.202	0.411	−2.103
0.834	0.280	0.744	0.598	0.122	−0.460	−1.310	−1.271	−0.917	0.650
−1.397	−1.053	0.412	1.286	−0.820	−0.371	0.826	−0.666	0.505	0.733
0.238	−0.668	1.861	0.051	0.460	0.079	1.008	−0.487	0.306	−0.061
0.102	−0.907	−0.833	1.103	−0.921	0.145	−0.904	−0.401	0.553	−1.422
−0.160	0.567	−0.638	0.355	0.427	−0.695	−0.846	0.359	1.500	−0.926
0.496	1.179	−0.776	0.511	−1.325	0.275	−0.130	−0.123	1.175	−0.102
0.307	−0.328	−2.474	−0.121	1.371	0.266	1.235	1.827	−0.296	−2.715
−0.559	0.523	1.264	−0.018	−2.791	0.139	1.515	1.976	0.173	−1.728
0.658	−0.261	0.004	−1.296	0.568	−1.215	0.104	0.178	1.126	1.134
−0.856	−2.278	−0.140	−0.164	1.416	−0.043	0.243	−1.399	−0.448	0.120
2.778	0.245	0.282	0.301	−1.506	1.805	1.798	1.078	1.629	−0.648
0.543	0.761	−2.038	−0.533	−0.594	1.742	0.487	1.432	−0.210	−0.358
−0.008	−0.445	−2.551	0.935	1.961	−0.270	−1.557	−1.318	−0.744	−0.860
−1.147	−1.151	−0.522	−2.118	−0.667	0.906	0.639	1.005	−0.480	−1.354
−0.851	0.585	0.672	0.481	−0.888	−0.480	0.041	0.345	−0.537	−0.589
0.023	0.609	0.623	0.356	0.279	−0.051	0.158	−0.353	0.776	0.102
−0.257	0.152	−1.413	0.175	0.149	−1.354	0.286	1.794	−0.571	−0.202
−0.421	−0.344	−0.803	0.832	0.256	−1.296	−1.390	0.379	0.955	0.366
−1.681	2.444	−1.025	1.178	−0.827	−0.200	0.727	0.778	0.169	−1.363
0.717	−1.666	1.071	−2.061	−1.367	−0.450	−0.038	−1.004	−1.240	0.901
−1.266	0.256	−1.312	−0.582	−0.351	−1.002	0.648	0.873	0.015	0.641
0.350	0.552	−1.549	−1.680	1.417	−0.769	−0.514	−1.900	1.017	−1.222
−0.186	0.006	0.148	0.560	−1.081	−0.637	−1.968	−0.623	0.009	−0.369
1.359	1.027	0.740	−2.067	0.543	1.099	0.543	0.064	0.589	−0.016

[†] Reproduced by permission from RAND Corporation, *A Million Random Digits with 100,000 Normal Deviates* (New York: Macmillan Publishing Co., Inc., third printing, 1966).

TABLE XII   Random Normal Numbers (*Continued*)

0.048	1.040	−0.111	−0.120	1.396	−0.393	−0.220	0.422	0.233	0.197
−0.521	−0.563	−0.116	−0.512	−0.518	−2.194	2.261	0.461	−1.533	−1.836
−1.407	−0.213	0.948	−0.073	−1.474	−0.236	−0.649	1.555	1.285	−0.747
1.822	0.898	−0.691	0.972	−0.011	0.517	0.808	2.651	−0.650	0.592
1.346	−0.137	0.952	1.467	−0.352	0.309	0.578	−1.881	−0.488	−0.329
0.420	−1.085	−1.578	−0.125	1.337	0.169	0.551	−0.745	−0.588	1.810
−1.760	−1.868	0.677	0.545	1.465	0.572	−0.770	0.655	−0.574	1.262
−0.959	0.061	−1.260	−0.573	−0.646	−0.697	−0.026	−1.115	3.591	−0.519
0.561	−0.534	−1.730	−1.172	−0.261	−0.049	0.173	0.027	1.138	0.524
−0.717	0.254	0.421	−1.891	2.592	−1.443	−0.061	−2.520	−0.497	0.909
−2.097	−0.180	−1.298	−0.647	0.159	0.769	−0.735	−0.343	0.966	0.595
0.443	−0.191	0.705	0.420	−0.486	−1.038	−0.396	1.406	0.327	1.198
0.481	0.161	−0.044	−0.864	−0.587	−0.037	−1.304	−1.544	0.946	−0.344
−2.219	−0.123	−0.260	0.680	0.224	−1.217	0.052	0.174	0.692	−1.068
1.723	−0.215	−0.158	0.369	1.073	−2.442	−0.472	2.060	−3.246	−1.020
−0.937	1.253	0.321	−0.541	−0.648	0.265	1.487	−0.554	1.890	0.499
−0.568	−0.146	0.285	1.337	−0.840	0.361	−0.468	0.746	0.470	0.171
−1.717	−1.293	−0.556	−0.545	1.344	0.320	−0.087	0.418	1.076	1.669
−0.151	−0.266	0.920	−2.370	0.484	−1.915	−0.268	0.718	2.075	−0.975
2.278	−1.819	0.245	−0.163	0.980	−1.629	−0.094	−0.573	1.548	−0.896
−0.650	0.669	−0.761	0.154	0.872	0.914	−0.563	−1.434	−0.006	−0.975
−1.086	0.810	0.461	−0.528	2.130	−0.218	0.111	−0.412	−0.580	−1.487
−0.143	−1.196	−1.254	−0.133	0.937	−0.475	−2.348	0.618	−0.057	−0.710
−2.072	0.711	1.241	0.066	−0.341	0.356	1.220	0.431	0.263	−1.623
−0.394	−0.368	−2.108	0.605	0.485	2.068	0.687	−1.474	0.071	−1.196
0.174	−1.131	0.870	2.114	0.201	−0.373	−0.284	−0.234	−2.087	−1.304
0.020	0.102	−1.911	−1.132	1.267	0.420	0.791	1.548	−0.147	−0.453
0.297	0.449	−0.604	−0.858	−1.739	1.143	0.131	0.740	−1.596	0.165
1.160	0.253	0.716	−1.032	−0.595	−1.662	0.632	−0.315	−0.374	0.700
−0.351	−0.490	−0.632	−0.409	−0.116	−1.153	−0.266	−0.125	0.489	−0.366
−0.594	−0.214	−0.461	0.030	−0.595	−0.889	0.638	−0.488	0.418	−0.693
−1.882	1.890	−0.236	0.006	0.966	−0.723	0.229	−2.136	−1.017	−0.008
0.041	2.955	−1.526	2.114	−0.540	1.040	0.753	0.025	0.462	1.221
−0.403	1.237	−1.938	−1.704	−0.103	−0.346	1.214	0.826	0.336	−1.140
−0.068	0.599	0.192	1.503	−0.579	−1.485	−1.645	0.302	−1.348	0.553
−0.361	0.958	0.807	0.787	−0.547	−0.074	−1.378	−0.010	−1.096	0.789
−0.251	0.629	0.459	−0.165	0.016	0.489	−1.205	−0.260	−0.256	−0.399
−1.011	0.893	−0.741	−0.514	−0.576	−0.929	0.478	−0.374	1.950	−0.695
0.780	−2.464	−0.522	0.767	−1.657	−0.983	0.217	−0.529	−0.648	1.454
−0.712	−0.355	−0.564	1.052	−0.169	−0.410	1.543	−2.330	−0.008	−0.955
−0.612	−1.068	−0.644	−0.007	−0.835	0.623	0.093	0.105	−0.318	−0.228
−0.064	0.012	−0.676	0.349	0.303	1.539	0.792	−0.101	−0.344	−0.096
−0.379	1.504	2.375	0.498	−0.996	0.174	−1.268	−1.137	−0.618	0.173
1.145	−1.403	0.770	0.799	0.844	−1.361	−1.059	0.128	1.398	0.277
−0.117	0.585	−1.763	−0.632	0.239	−0.854	1.684	1.024	−0.067	−0.045
1.333	1.374	−0.515	−1.655	0.607	−0.885	−0.902	−1.010	−1.297	−0.139
−0.249	−0.747	1.044	−0.930	0.346	0.575	0.335	−1.159	−1.651	−1.642
−1.022	0.085	−1.441	−0.198	0.844	0.697	0.548	−0.080	0.656	0.443
−0.780	−0.534	−0.339	−0.642	−0.902	−0.827	0.071	−0.678	−0.359	−0.479
−0.687	−0.418	0.991	0.331	−1.003	0.061	−1.416	0.876	0.125	−2.246

TABLE XIII   Values of e^{-x}

x	e^{-x}	x	e^{-x}	x	e^{-x}	x	e^{-x}
0.0	1.000	2.5	0.082	5.0	0.0067	7.5	0.00055
0.1	0.905	2.6	0.074	5.1	0.0061	7.6	0.00050
0.2	0.819	2.7	0.067	5.2	0.0055	7.7	0.00045
0.3	0.741	2.8	0.061	5.3	0.0050	7.8	0.00041
0.4	0.670	2.9	0.055	5.4	0.0045	7.9	0.00037
0.5	0.607	3.0	0.050	5.5	0.0041	8.0	0.00034
0.6	0.549	3.1	0.045	5.6	0.0037	8.1	0.00030
0.7	0.497	3.2	0.041	5.7	0.0033	8.2	0.00028
0.8	0.449	3.3	0.037	5.8	0.0030	8.3	0.00025
0.9	0.407	3.4	0.033	5.9	0.0027	8.4	0.00023
1.0	0.368	3.5	0.030	6.0	0.0025	8.5	0.00020
1.1	0.333	3.6	0.027	6.1	0.0022	8.6	0.00018
1.2	0.301	3.7	0.025	6.2	0.0020	8.7	0.00017
1.3	0.273	3.8	0.022	6.3	0.0018	8.8	0.00015
1.4	0.247	3.9	0.020	6.4	0.0017	8.9	0.00014
1.5	0.223	4.0	0.018	6.5	0.0015	9.0	0.00012
1.6	0.202	4.1	0.017	6.6	0.0014	9.1	0.00011
1.7	0.183	4.2	0.015	6.7	0.0012	9.2	0.00010
1.8	0.165	4.3	0.014	6.8	0.0011	9.3	0.00009
1.9	0.150	4.4	0.012	6.9	0.0010	9.4	0.00008
2.0	0.135	4.5	0.011	7.0	0.0009	9.5	0.00008
2.1	0.122	4.6	0.010	7.1	0.0008	9.6	0.00007
2.2	0.111	4.7	0.009	7.2	0.0007	9.7	0.00006
2.3	0.100	4.8	0.008	7.3	0.0007	9.8	0.00006
2.4	0.091	4.9	0.007	7.4	0.0006	9.9	0.00005

TABLE XIV Logarithms

N	0	1	2	3	4	5	6	7	8	9
10	0000	0043	0086	0128	0170	0212	0253	0294	0334	0374
11	0414	0453	0492	0531	0569	0607	0645	0682	0719	0755
12	0792	0828	0864	0899	0934	0969	1004	1038	1072	1106
13	1139	1173	1206	1239	1271	1303	1335	1367	1399	1430
14	1461	1492	1523	1553	1584	1614	1644	1673	1703	1732
15	1761	1790	1818	1847	1875	1903	1931	1959	1987	2014
16	2041	2068	2095	2122	2148	2175	2201	2227	2253	2279
17	2304	2330	2355	2380	2405	2430	2455	2480	2504	2529
18	2553	2577	2601	2625	2648	2672	2695	2718	2742	2765
19	2788	2810	2833	2856	2878	2900	2923	2945	2967	2989
20	3010	3032	3054	3075	3096	3118	3139	3160	3181	3201
21	3222	3243	3263	3284	3304	3324	3345	3365	3385	3404
22	3424	3444	3464	3483	3502	3522	3541	3560	3579	3598
23	3617	3636	3655	3674	3692	3711	3729	3747	3766	3784
24	3802	3820	3838	3856	3874	3892	3909	3927	3945	3962
25	3979	3997	4014	4031	4048	4065	4082	4099	4116	4133
26	4150	4166	4183	4200	4216	4232	4249	4265	4281	4298
27	4314	4330	4346	4362	4378	4393	4409	4425	4440	4456
28	4472	4487	4502	4518	4533	4548	4564	4579	4594	4609
29	4624	4639	4654	4669	4683	4698	4713	4728	4742	4757
30	4771	4786	4800	4814	4829	4843	4857	4871	4886	4900
31	4914	4928	4942	4955	4969	4983	4997	5011	5024	5038
32	5051	5065	5079	5092	5105	5119	5132	5145	5159	5172
33	5185	5198	5211	5224	5237	5250	5263	5276	5289	5302
34	5315	5328	5340	5353	5366	5378	5391	5403	5416	5428
35	5441	5453	5465	5478	5490	5502	5514	5527	5539	5551
36	5563	5575	5587	5599	5611	5623	5635	5647	5658	5670
37	5682	5694	5705	5717	5729	5740	5752	5763	5775	5786
38	5798	5809	5821	5832	5843	5855	5866	5877	5888	5899
39	5911	5922	5933	5944	5955	5966	5977	5988	5999	6010
40	6021	6031	6042	6053	6064	6075	6085	6096	6107	6117
41	6128	6138	6149	6160	6170	6180	6191	6201	6212	6222
42	6232	6243	6253	6263	6274	6284	6294	6304	6314	6325
43	6335	6345	6355	6365	6375	6385	6395	6405	6415	6425
44	6435	6444	6454	6464	6474	6484	6493	6503	6513	6522
45	6532	6542	6551	6561	6571	6580	6590	6599	6609	6618
46	6628	6637	6646	6656	6665	6675	6684	6693	6702	6712
47	6721	6730	6739	6749	6758	6767	6776	6785	6794	6803
48	6812	6821	6830	6839	6848	6857	6866	6875	6884	6893
49	6902	6911	6920	6928	6937	6946	6955	6964	6972	6981
50	6990	6998	7007	7016	7024	7033	7042	7050	7059	7067
51	7076	7084	7093	7101	7110	7118	7126	7135	7143	7152
52	7160	7168	7177	7185	7193	7202	7210	7218	7226	7235
53	7243	7251	7259	7267	7275	7284	7292	7300	7308	7316
54	7324	7332	7340	7348	7356	7364	7372	7380	7388	7396

TABLE XIV Logarithms (*Continued*)

N	0	1	2	3	4	5	6	7	8	9
55	7404	7412	7419	7427	7435	7443	7451	7459	7466	7474
56	7482	7490	7497	7505	7513	7520	7528	7536	7543	7551
57	7559	7566	7574	7582	7589	7597	7604	7612	7619	7627
58	7634	7642	7649	7657	7664	7672	7679	7686	7694	7701
59	7709	7716	7723	7731	7738	7745	7752	7760	7767	7774
60	7782	7789	7796	7803	7810	7818	7825	7832	7839	7846
61	7853	7860	7868	7875	7882	7889	7896	7903	7910	7917
62	7924	7931	7938	7945	7952	7959	7966	7973	7980	7987
63	7993	8000	8007	8014	8021	8028	8035	8041	8048	8055
64	8062	8069	8075	8082	8089	8096	8102	8109	8116	8122
65	8129	8136	8142	8149	8156	8162	8169	8176	8182	8189
66	8195	8202	8209	8215	8222	8228	8235	8241	8248	8254
67	8261	8267	8274	8280	8287	8293	8299	8306	8312	8319
68	8325	8331	8338	8344	8351	8357	8363	8370	8376	8382
69	8388	8395	8401	8407	8414	8420	8426	8432	8439	8445
70	8451	8457	8463	8470	8476	8482	8488	8494	8500	8506
71	8513	8519	8525	8531	8537	8543	8549	8555	8561	8567
72	8573	8579	8585	8591	8597	8603	8609	8615	8621	8627
73	8633	8639	8645	8651	8657	8663	8669	8675	8681	8686
74	8692	8698	8704	8710	8716	8722	8727	8733	8739	8745
75	8751	8756	8762	8768	8774	8779	8785	8791	8797	8802
76	8808	8814	8820	8825	8831	8837	8842	8848	8854	8859
77	8865	8871	8876	8882	8887	8893	8899	8904	8910	8915
78	8921	8927	8932	8938	8943	8949	8954	8960	8965	8971
79	8976	8982	8987	8993	8998	9004	9009	9015	9020	9025
80	9031	9036	9042	9047	9053	9058	9063	9069	9074	9079
81	9085	9090	9096	9101	9106	9112	9117	9122	9128	9133
82	9138	9143	9149	9154	9159	9165	9170	9175	9180	9186
83	9191	9196	9201	9206	9212	9217	9222	9227	9232	9238
84	9243	9248	9253	9258	9263	9269	9274	9279	9284	9289
85	9294	9299	9304	9309	9315	9320	9325	9330	9335	9340
86	9345	9350	9355	9360	9365	9370	9375	9380	9385	9390
87	9395	9400	9405	9410	9415	9420	9425	9430	9435	9440
88	9445	9450	9455	9460	9465	9469	9474	9479	9484	9489
89	9494	9499	9504	9509	9513	9518	9523	9528	9533	9538
90	9542	9547	9552	9557	9562	9566	9571	9576	9581	9586
91	9590	9595	9600	9605	9609	9614	9619	9624	9628	9633
92	9638	9643	9647	9652	9657	9661	9666	9671	9675	9680
93	9685	9689	9694	9699	9703	9708	9713	9717	9722	9727
94	9731	9736	9741	9745	9750	9754	9759	9763	9768	9773
95	9777	9782	9786	9791	9795	9800	9805	9809	9814	9818
96	9823	9827	9832	9836	9341	9845	9850	9854	9859	9863
97	9868	9872	9877	9881	9886	9890	9894	9899	9903	9908
98	9912	9917	9921	9926	9930	9934	9939	9943	9948	9952
99	9956	9961	9965	9969	9974	9978	9983	9987	9991	9996

ANSWERS TO EXERCISES

1.2 (a) The interviewer is baiting the question by suggesting that the practice is unfair.

(b) There are many objections to this procedure, including the omission of unlisted telephones and the difficulty of reaching people who spend little time at home.

1.3 (a) Persons coming out of the building housing the national headquarters of a political party are more likely to support that party.

(b) December (pre-Christmas) spending is not typical of spending throughout the year.

1.4 (a) A systematic defect effecting (say) every tenth can might be completely missed.

(b) Graduates with lower than average incomes are less likely to return the question-naires.

(c) People are reluctant to give completely truthful answers regarding questions of personal hygiene.

1.5 (a) Since only 79 and 88 exceed 75, the conclusion follows from the data.

(b) If we can assume that the student received the grades in the given order, the conclusion follows from the data.

(c) Since there may have been other reasons, the conclusion goes beyond the data.

(d) Since $88 - 46 = 42$, the conclusion follows from the data.

1.6 (a) Since Mary sold 11 houses and Jean sold 9, the conclusion follows from the data.

(b) This conclusion goes beyond the data.

(c) This conclusion follows from the data.

(d) This conclusion goes beyond the data.

1.7 (a) and (b) are numeric descriptions of the data, while (c) and (d) make generalizations which go beyond the data.

1.8 (a) Follows directly from the data, while (b) is a generalization.

1.9 (a) Ordinal; (b) nominal;
(c) ordinal; (d) nominal.

1.10 (a) Nominal; (b) ratio;
(c) ratio; (d) ordinal.

1.11 Since IQ scores are rescaled z-scores (see Section 4.3), it is unrealistic to interpret three times a "smartness" difference.

1.12 The zero point on the calendar time scale is artificial. Athletic event times are ratio data.

2.1

15	8	Note: 15 | 8 means 158.
16	13667	
17	0122478	
18	2	
19	1	
20	1	

2.2

16	4	Note: 16 | 4 means 164.
17	0268	
18	0123668	
19	3348	
20	58	
21	0	
22	5	

2.3

3	29	Note: 3 | 2 means $320.
4	2456888	
5	0126	
6	12	

2.4 Assuming that the stem gives the tens digit, the values are 40, 42, 43, 51, 51, 58, 59, 62, 63, 63, 67, 67, 69, and 70.

2.5

16·	4	Note: 16· | 4 means 164.
16∗		17∗ | 6 means 176.
17·	02	
17∗	68	
18·	0123	
18∗	668	
19·	334	
19∗	8	
20·	58	
20∗		
21·	0	
21∗		
22·		
22∗	5	

2.6

6·	1234			Note: 6· | 1 means 61.
6∗	56666	78888	99	6∗ | 5 means 65.
7·	01333	344		
7∗	6779			
8·	0112			

2.7 160–169, 170–179, 180–189, 190–199, 200–209, 210–219, 220–229, 230–239, 240–249, 250–259, and 260–269.

2.8 148.0–149.9, 150.0–151.9, 152.0–153.9, 154.0–155.9, 156.0–157.9, 158.0–159.9, and 160.0–161.9. The class width is 2°, and the intervals begin at 148°, 150°, 152°, and so on.

2.9 225.00–249.99, 250.00–274.99, 275.00–299.99, 300.00–324.99, 325.00–349.99, 350.00–374.99, and 375.00–399.99. The class width is $25, and the intervals begin at $225, $250, $275, and so on.

2.10 (a) Yes; (b) no; (c) yes; (d) no; (e) no.

2.11 (a) Yes; (b) no; (c) no; (d) yes.

2.12 (a) 89; (b) 36; for the remaining parts, the exact numbers cannot be given.

2.13 (a) 15–29, 30–44, 45–59, 60–74, and 75–89;
(b) 14.5, 29.5, 44.5, 59.5, 74.5, and 89.5;
(c) 22, 37, 52, 67, and 82;
(d) 15.

2.14 (a) $0.00–49.99, $50.00–99.99, $100.00–149.99, $150.00–199.99, $200.00–249.99, $250.00–299.99, and $300.00 and over;
(b) −$0.005, $49.995, $99.995, $149.995, $199.995, $249.995, and $299.995. Note that the final class has no upper boundary;
(c) $24.995, $74.995, $124.995, $174.995, $224.995, and $274.995. The final class does not have a class mark;
(d) $50, except for the final class, which is unbounded.

2.15 (a) −0.5, 8.5, 17.5, 26.5, 35.5, and 44.5;
(b) 0–8, 9–17, 18–26, 27–35, and 36–44.

2.16 (a) −0.5, 12.5, 25.5, 38.5, and 51.5;
(b) 0–12, 13–25, 26–38, and 39–51.

2.17 There is no provision for 17 days. Also, the value 24 appears in two different classes.

2.18 The second and third classes overlap, in that values from $35.00 to $35.99 could be placed in either of these classes. Also, there is no provision for values from $49.91 to $49.99.

2.19

Grades	Frequency
30–39	3
40–49	2
50–59	2
60–69	10
70–79	12
80–89	7
90–99	4
Total	40

2.20

Grades	Percentage
30–39	7.5%
40–49	5.0%
50–59	5.0%
60–69	25.0%
70–79	30.0%
80–89	17.5%
90–99	10.0%
Total	100.0%

2.21

Grades	Frequency
less than 30	0
less than 40	3
less than 50	5
less than 60	7
less than 70	17
less than 80	29
less than 90	36
less than 100	40

2.22

Weights	Frequency
80– 89	2
90– 99	3
100–109	9
110–119	13
120–129	13
130–139	7
140–149	3
Total	50

2.23

Weights	Percentage
80– 89	4%
90– 99	6%
100–109	18%
110–119	26%
120–129	26%
130–139	14%
140–149	6%
Total	100%

2.24

Weights	Frequency
80 or more	50
90 or more	48
100 or more	45
110 or more	36
120 or more	23
130 or more	10
140 or more	3
150 or more	0

2.25

Number of customers	Frequency
40–44	4
45–49	9
50–54	17
55–59	34
60–64	38
65–69	13
70–74	3
75–79	2
Total	120

2.26 (a)

Number of customers	Percentage
40–44	3.33%
45–49	7.50%
50–54	14.17%
55–59	28.33%
60–64	31.67%
65–69	10.83%
70–74	2.50%
75–79	1.67%
Total	100.00%

(b)

Number of customers	Percentage
Less than 40	0.00%
Less than 45	3.33%
Less than 50	10.83%
Less than 55	25.00%
Less than 60	53.33%
Less than 65	85.00%
Less than 70	95.83%
Less than 75	98.33%
Less than 80	100.00%

2.27

Miles per gallon	Frequency
22.5–22.9	2
23.0–23.4	3
23.5–23.9	7
24.0–24.4	8
24.5–24.9	14
25.0–25.4	5
25.5–25.9	1
Total	40

2.28 (a)

Miles per gallon	Frequency
More than 22.4	40
More than 22.9	38
More than 23.4	35
More than 23.9	28
More than 24.4	20
More than 24.9	6
More than 25.4	1
More than 25.9	0

(b)

Miles per gallon	Percentage
More than 22.4	100.0%
More than 22.9	95.0%
More than 23.4	87.5%
More than 23.9	70.0%
More than 24.4	50.0%
More than 24.9	15.0%
More than 25.4	2.5%
More than 25.9	0.0%

2.29

Number of accidents	Frequency
0	6
1	10
2	9
3	11
4	7
5	4
6	2
7	1
Total	50

2.30

Number of mistakes	Frequency
0	25
1	19
2	10
3	4
4	2
Total	60

2.31

Number of mistakes	Frequency
Fewer than 1	25
Fewer than 2	44
Fewer than 3	54
Fewer than 4	58
Fewer than 5	60

2.32

Means of transportation	Frequency
Bus	8
Car	18
Plane	10
Train	4
Total	40

2.33 Many dresses are made from blends of fibers, and this distribution does not allow for these blends.

2.35

Amount in dollars	Frequency
Less than 20.00	22
Less than 40.00	69
Less than 60.00	135
Less than 80.00	170
Less than 100.00	191
Less than 120.00	200

2.37

Weight in pounds	Frequency
Less than 90	0
Less than 100	6
Less than 110	31
Less than 120	77
Less than 130	114
Less than 140	136
Less than 150	143
Less than 160	146
Less than 170	149
Less than 180	149
Less than 190	150

2.38 The central angles of the sectors are $82 \cdot 3.6 = 295.2°$, $13 \cdot 3.6 = 46.8°$, $2 \cdot 3.6 = 7.2°$, and $3 \cdot 3.6 = 10.8°$.

2.39 We can make the areas represent the class frequencies by dividing the height of the fourth rectangle by 2.

2.40 Since each dimension is doubled, we might make each dimension of the larger figure $\sqrt{2} = 1.41$ times that of the smaller figure.

2.41 The central angles of the six sectors are 209°, 52°, 35°, 10°, 46°, and 8°, rounded to the nearest degree.

2.45 15–21, 22–28, 29–35, 36–42, 43–49, and 50–56.

2.46 The class frequencies are 7, 10, 20, 22, 10, 2, and 1.

2.48 The cumulative "or less" frequencies are 7, 17, 37, 59, 69, 71, and 72.

2.49 There is no place for 19, and 34 can go in the fourth class or in the fifth class.

2.50 (a) No;   (b) yes;   (c) yes;   (d) no.

2.51 (a) 0.5, 2.5, 4.5, 6.5, and 8.5;
    (b) -0.5, 1.5, 3.5, 5.5, 7.5, and 9.5;
    (c) 2.

2.52 The "or more" cumulative percentages are 100, 73.33, 31.67, 10, 3.33, and 0.

2.53 (b) The central angles of the five sectors are 88.2°, 210.5°, 21.5°, 25.2°, and 14.7°.

2.54 A good choice would be 18,000–20,499, 20,500–22,999, 23,000–25,499, 25,500–27,999, 28,000–30,499, and 30,500–32,999.

2.55 (a) Bar charts and pie charts;
    (b) bar charts and pie charts;
    (c) histograms, bar charts, and pie charts.

2.56 6.0–7.9, 8.0–9.9, 10.0–11.9, 12.0–13.9, and 14.0–15.9.

2.57 The frequencies corresponding to never, rarely, occasionally, and frequently are 15, 21, 11, and 3.

2.58 (a) 1,253, 1,250, 1,254, 1,258, 1,257, 1,256, 1,256, and 1,255;
    (b) 3,467, 3,405, 3,419, and 3,448.
    (c) 11, 10, 18, 16, 17, 17, 22, 24, 20, 21, 21, 23, 29, 26, 27, 27, 33, 31, 32, 38, 35, and 42.

2.59 The frequencies corresponding to 2, 3, 4, 5, 6, 7, 8, and 9 are 2, 4, 8, 6, 5, 3, 1, and 1.

2.61

```
12 | 4            Note: 16 | 2 means 162.
13 | 05
14 | 1269
15 | 135689
16 | 225
17 | 37
18 | 2
19 |
20 | 4
```

2.62 Some may not have finished high school.

CHAPTER 3

3.1 (a) This would be regarded as a population if we are interested only in this single company.

(b) This would be regarded as a sample if we are interested generally in the expenses of computer support groups of many companies.

3.2 (a) We would regard these as a population if the candidates are running for a county office.

(b) We would regard these as a sample if the candidates are running for an office for a region larger than the county.

3.3 (a) These would be regarded as a population if the set of 23 department stores constitutes a single chain of stores run by one central management.

(b) These would be regarded as a sample if these were 23 stores out of a larger chain or if the 23 stores were completely unrelated.

3.4 (a) These would be regarded as a population if we have enumerated the entire student body of the only college in which we are interested.

(b) These would be regarded as a sample if these 848 students constitute a subset of the student body.

3.5 41.8.

3.6 (a) 12;   (b) $67.

3.8 58.1.

3.9 (a) 335.07.

(b) If you subtract 300 from each value, the average of what remains is $526/15 = 35.07$. Adding back 300 will yield 335.07.

3.10 The bridge is safe.

3.11 Not overloaded.

3.12 $6.90.

3.13 (a) 4.01 ounces;   (b) 0.075 ounce.

3.14 25.

3.15 (a) 87.5%;   (b) 2/3.

3.16 The deviations from average can be rather uncomfortable.

3.17 (a) 1.6;   (b) 4;   (c) 96 cases on the fourth day and 192 cases on the fifth day.

3.18 (c) 72.

3.19 The average grade is $84\frac{2}{3}$.

3.20 $31,042.

3.21 1.597.

3.22 0.336.

3.23 (a) 13th value;   (b) mean of the 16th and 17th values.

3.24 (a) 19th value;   (b) mean of the 32nd and 33rd values.

3.25 52.

3.26 65.

3.27 55.

3.28 2.

3.29 (a) $\bar{x} = 0.79$;   (b) 0.

3.30 (a) 140;   (b) 140.

3.31 68.5.

3.32 $\bar{x} = 110.5\%$; the median is 92%. The mean is influenced strongly by the single person at 353% of quota, and the median is a better indicator of "average" performance.

3.33 (a) The medians are 4, 5, 4, 3, 5, 2, 3, 5, 3, 2, 3, and 4, and the means are 4, $4\frac{1}{3}$, 4, $3\frac{1}{3}$, 4, 2, $2\frac{2}{3}$, 4, $3\frac{1}{3}$, 3, 3, and 4.

Medians	Frequency	Means	Frequency
1.5–2.5	2	1.5–2.5	1
2.5–3.5	4	2.5–3.5	5
3.5–4.5	3	3.5–4.5	6
4.5–5.5	3	4.5–5.5	0

3.35 (a) The manufacturers of car A can use the mean;

(b) the manufacturers of car B can use the median.

3.36 The manufacturers of car C can use the midrange.

3.37 (a) 65.5;   (b) the lower hinge is 50; the upper hinge is 75.

3.38 (a) 67;   (b) the lower hinge is 58; the upper hinge is 70.

3.39 130 minutes and 146 minutes.

3.40 Here are the positions:

	n	Median	Lower quartile	Upper quartile	Lower hinge	Upper hinge
(a)	40	20.5	10.5	30.5	10.5	30.5
(b)	41	21	10.5	31.5	11	31
(c)	42	21.5	11	32	11	32
(d)	43	22	11	33	11.5	32.5

Half-integers refer to the means of the values in the surrounding positions.

3.41 The lower quartile is 129. The upper quartile is 150.5.

3.42 The lower quartile and lower hinge are both 32. The upper quartile and upper hinge are both 84.5.

3.43 142 minutes.

3.44 Hinges are 4 and 12. Quartiles are 4 and 12.

3.47 7.

3.48 1.

3.49 (a) 146;   (b) 149;   (c) no mode.

3.50 Blue.

3.51 (a) 581.2, 591.5, and 514;

(b) 580, 590, and 480;

(c) 585, 600, and 600;

(d) the mean and median are altered only modestly by rounding. The mode can change radically when data are rounded.

3.52 (a) The mean and the median can both be determined.

(b) The mean cannot be determined because two of the classes are open; the median can be determined because it does not fall into one of the open classes.

(c) The mean cannot be determined because there is an open class; the median cannot be determined because it falls into the open class.

3.53 56.45.

3.54 (a) 59.0;

(b) $Q_1 = 41.3$ and $Q_3 = 72.4$.

3.55 (a) $D_1 = 25.8$ and $D_9 = 82.2$;

(b) $P_5 = 18.0$ and $P_{95} = 87.8$.

3.56 7.8.

3.57 (a) 6.8;

(b) $Q_1 = 3.0$ and $Q_3 = 12.0$.

3.58 (a) 3.7 and 10.6;

(b) $P_5 = 0.2$ and $P_{98} = 19.5$.

3.59 28.03.

3.60 (a) 27.23;

(b) 22.42 and 32.74.

3.61 24.36 and 30.43.

3.62 $117.5 - 117.88 = -0.38$.

3.63 $24.30 - 24.3075 = -0.0075$.

3.64 For 3.62, the class interval is 10, and the grouping error is -0.38. For 3.63, the class interval is 5, and the grouping error is -0.0075.

3.65 (a) $x_1 + x_2 + x_3 + x_4 + x_5 + x_6$;

(b) $y_1 + y_2 + y_3 + y_4 + y_5$;

(c) $x_1 y_1 + x_2 y_2 + x_3 y_3$;

(d) $x_1 f_1 + x_2 f_2 + x_3 f_3 + x_4 f_4 + x_5 f_5 + x_6 f_6 + x_7 f_7 + x_8 f_8$;

(e) $x_3^2 + x_4^2 + x_5^2 + x_6^2 + x_7^2$;

(f) $x_1 + y_1 + x_2 + y_2 + x_3 + y_3 + x_4 + y_4$.

3.66 (a) $x_1 + x_2 + x_3 + x_4 + x_5 + 5$;

(b) $3(y_1 + y_2 + y_3 + y_4)$;

(c) $3(x_1 + x_2 + x_3 + x_4)$.

3.67 (a) $\sum_{i=1}^{4} z_i$; (b) $\sum_{i=6}^{11} x_i$; (c) $\sum_{i=1}^{6} x_i f_i$; (d) $\sum_{i=1}^{5} y_i^2$; (e) $\sum_{i=1}^{6} 3x_i = 3\sum_{i=1}^{6} x_i$;

(f) $\sum_{i=1}^{4}(x_i - y_i) = \sum_{i=1}^{4} x_i - \sum_{i=1}^{4} y_i$; (g) $\sum_{i=1}^{3}(w_i - 5) = \sum_{i=1}^{3} w_i - 15$; (h) $\sum_{i=1}^{4} a_i b_i c_i$.

3.68 (a) 25; (b) 21; (c) 105; (d) 605.

3.69 (a) 6; (b) 30.

3.70 (a) 7; 4; -2; 10. (b) 4; 8; 7.

3.71 19.

3.73 Try $n = 2$ and $x_1 = 2$, $x_2 = 3$.

3.74 1.575 and 1.32.

3.75 The lower hinge is 1.34. The upper hinge is 1.645.

3.76 $2.80.

3.77 13.7.

3.78 (a) 14.17;

(b) the lower quartile is 10.75; the upper quartile is 16.73.

3.79 15.17.

3.80 (a) These could be viewed as a population in any situation in which they constitute all the commercials of interest.

(b) These could be viewed as a sample if our interest goes beyond the ten commercials.

3.81 (a) 16.

(b) Mean of the values in positions 40 and 41.

3.82 (a) 27.75; (b) 27.61.

3.83 The hinges are 25.8 and 30.5.

3.84 85.78.

3.85 Not possible.

3.86 Independent.

3.87 Median is in the twelfth position. Lower hinge is the mean of the values in the sixth and seventh positions. The upper hinge is the mean of the values in the seventeenth and eighteenth positions.

3.88 The median is the mean of the twelfth and thirteenth positions. Lower hinge is the mean of the values in the sixth and seventh positions. The upper hinge is the mean of the values in the seventeenth and eighteenth positions.

3.89 41.5.

3.90 The mean cannot be found.

3.91 (a) $Q_1 = 19.6$;   (b) $Q_3 = 29.5$;   (c) $D_2 = 18.1$;   (d) $P_{95} = 36.7$.

3.92 (a) 6;   (b) 4.

3.93 It is possible that the mean salary is higher at company B. Here is a very simple situation in which this can happen:

		Male Employees		Female Employees	
Company	Number	Mean Salary		Number	Mean Salary
A	1	$30,000		2	$24,000
B	2	$28,000		1	$23,000

4.1 8.

4.2 325.

4.3 $1\frac{1}{8}$ for stock A and $\frac{3}{8}$ for stock B.

4.4 8.

4.5 The hinges are 15 and 17; the interquartile range is 2.

4.6 34.

4.7 The lower quartile is 58. The upper quartile is 70. The interquartile range is 12.

4.8 68.

4.9 The lower hinge is 53. The upper hinge is 78. The semi-interquartile range is $\frac{1}{2}(78 - 53) = 12.5$

4.10 (a) 2.74;   (b) 2.74.

4.11 (a) 0.18;   (b) 0.18.

4.12 (a) 17.60;   (b) 17.60.

4.13 (a) 3.77;   (b) 3.77, identical with the result of part (a).

4.14 (a) 2.31;   (b) 54.77;   (c) 10.69.

4.15 (a) $s = 2$;   (b) $s = 2$;   (c) $s = 4.47$.

4.16 The range is 4, and the standard deviation is 2.06. The claim is reasonable.

4.17 The range is 12, and the standard deviation is 4.26. The range is roughly three times the standard deviation.

4.18 (b) 6.26.

4.19 (b) 6.78, and the standard deviation of the original data is $\dfrac{6.78}{10} = 0.68$.

4.20 (a) $\frac{35}{36}$;   (b) $\frac{143}{144}$.

4.21 (a) 93.75%;   (b) 98.77%.

4.22 (a) 96%;   (b) 98.4%;   (c) 99%;   (d) 99.75%.

4.23 (a) 75%;   (b) 96%.

4.24 (a) The thiamine content of $\frac{63}{64}$ of the slices must be between 0.220 and 0.300.
    (b) The thiamine content of 0.84 of the slices must be between 0.2475 and 0.2725.

4.25 The percentage is at least 88.89%. For a bell-shaped distribution, the percentage is about 99.7%.

4.26 91%.

4.27 The z-score for sirloin steak is -0.20, for chicken is -0.27, and for leg of veal is -0.53. The leg of veal is the best bargain.

4.28 $z = 2.2$ for the first man, and $z = 1.95$ for the second man.

4.29 The coefficients of variation are 0.44% for the first instrument and 0.49% for the second instrument. The first instrument is more precise.

4.30 The coefficients of variation are 7.55% for the first patient and 6.32% for the second patient. The first patient more variable.

4.31 For the chicken orders, the coefficient of variation is 18.7%. For the steak orders, the coefficient of variation is 21.0%. The numbers of order for steak are more variable.

4.32 9.4%.

4.33 (a) $\bar{x} = 24.5$ and $\tilde{x} = 24.21$;
    (b) 9.90.

4.34 (a) Following the half-cent suggestion, 150.68;
    (b) $13.79.

4.35 0.30. The distribution is moderately skewed, with the tail on the right.

4.36 20.62.

4.37 -0.37. The distribution is moderately skewed with the tail on the left.

4.38 1.03. The data are skewed, with a tail on the right.

4.39 The smallest value is 3, the lower hinge is 4, the median is 5, the upper hinge is 7, and the largest value is 10.

4.40 The smallest value is 35, the lower hinge is 50, the median is 65.5, the upper hinge is 75, and the maximum is 103.

4.41 The smallest value is 46, the lower hinge is 58, the median is 67, the upper hinge is 70, and the largest value is 80. There is modest negative skewness.

4.42 The smallest value is 0, the lower hinge is 0, the median is 2, the upper hinge is 3, and the maximum is 4. The data do not lend themselves to a judgment about a notion as sophisticated as skewness.

4.43 The smallest value is 113, the lower hinge is 130, the median is 140, the upper hinge is 146, and the largest value is 164. The data are fairly symmetrical.

4.44 The frequencies corresponding to 0, 1, 2, and 3 are 28, 17, 4, and 1; the data are highly skewed with the tail on the right.

4.45 The smallest value is 0, the lower hinge is 0, the median is 0, the upper hinge is 1, and the largest value is 3. The data are highly skewed with the tail on the right.

4.46 The frequencies corresponding to 0, 1, 2, 3, 4, and 5 are 24, 11, 4, 3, 2, and 16. The data are U-shaped.

4.47 The smallest value is 0, the lower hinge is 0, the median is 1, the upper hinge is 5, and the largest value is 5. There are no "whiskers" on either side; the data are U-shaped.

4.48 7.23.

4.49 55.56%

4.50 (a) 31.245, 79.995;   (b) 48.750;   (c) 43.8%.

4.51 The values show a very slight positive skewness.

4.52 19.73.

4.53 σ^2 is 2, and this is precisely $\dfrac{5^2 - 1}{12}$.

4.54 $s = 1.92$.

4.55 (a) 17.75:   (b) 5.085.

4.56 The median is 16.94, and SK = 0.48.

4.57 $\mu = 796.88$ and $\sigma = 111.56$.

4.58 (a) $\frac{35}{36}$;   (b) $\frac{63}{64}$;   (c) $\frac{224}{225}$.

4.59 (a) 22;   (b) 12 and 29.

4.61 (a) 55.6%;   (b) 75%;   (c) 84%.

4.62 (a) 0.5;   (b) 2;   (c) 2.5.

4.63 103.74.

4.64 Given $n = 10$ and $\sum x = 40$, the smallest possible value for $\sum x^2$ is 160.

CHAPTER 5

5.1 There are three ways.

5.2 There are 15 ways:
  (a) in 4 cases he will be $2 ahead (ending with $5);
  (b) he can never be $1 ahead (ending with $4);
  (c) in 6 cases he will be even (ending with $3);
  (d) he can never be $1 behind (ending with $2);
  (e) in 3 cases he will be $2 behind (ending with $1).

5.3 (a) There are three ways;   (b) there are four ways.

5.4 There are 16 ways.

5.5 There are 12 ways.

5.6 There are nine ways. In five of the nine ways, the officers are different sexes.

5.7 30.

5.8 60.

5.9 8.

5.10 (a) 25;   (b) 20;   (c) 20;   (d) 16.

5.11 96.

5.12 (a) 72;   (b) 18;   (c) 24;   (d) 6.

5.13 240.

5.14 1024.

5.15 1024.

5.16 (a) true;   (b) false;   (c) true;   (d) true.

5.17 (a) false;   (b) false;   (c) true;   (d) true.

5.18 720.

5.19 1,862,784.

5.20 7,920.

5.21 491,400.

5.22 504.

5.23 40,320.

5.24 9! = 362,880 different batting orders. If the pitcher must bat ninth, there are 8! = 40,320 different batting orders.

5.25 (a) 384;   (b) 1,152;   (c) 2,880;   (d) 1,152;   (e) 2,880.

5.26 (a) 60;   (b) 120;   (c) 120;

     (d) The duplicate objects cause overcounting by a factor of $r!$.

5.27 (a) 180;   (b) 60;   (c) 20;

     (d) $\dfrac{n!}{r_1! \cdot r_2! \cdot r_3! \cdot \ldots \cdot r_k!}$;

     (e) 34,650.

5.28 (a) There are 39 permutations, 9 with two e's, 24 with one e, and 6 with no e's.

     (b) There are 34 permutations, 1 with three b's, 9 with two b's, 18 with one b, and 6 with no b's.

5.29 (a) The position of the first object is irrelevant. Once this is in place, there are $(n-1)!$ ways that the other objects can be arranged relative to it.

     (b) 120;   (c) 5,040;   (d) 144.

5.30 (a) Each key will always be adjacent to each of the other two, so there is only one arrangement;   (b) 6;   (c) 120.

5.31 1,365.

5.32 120.

5.33 210.

5.34 (a) 495;   (b) 66.

5.35 3,003.

5.36 (a) 55;   (b) 165.

5.37 (a) 56;   (b) 56;   (c) 8.

5.38 (a) 28;   (b) 6;   (c) 6;   (d) 16.

5.39 18,900.

5.40 (a) 6;   (b) 30;   (c) 20;

     (d) Since $6 + 30 + 20 = 56$, we have verified the claim.

5.41 (a) $(a+b)^1 = \dbinom{1}{0}a + \dbinom{1}{1}b$;

     (b) $(a+b)^2 = \dbinom{2}{0}a^2 + \dbinom{2}{1}ab + \dbinom{2}{2}b^2$.

     (c) $(a+b)^3 = \dbinom{3}{0}a^3 + \dbinom{3}{1}a^2b + \dbinom{3}{2}ab^2 + \dbinom{3}{3}b^3$.

     (d) $(a+b)^4 = a^4 + 4a^3b + 6a^2b^2 + {} + 4ab^3 + b^4 = \dbinom{4}{0}a^4 + \dbinom{4}{1}a^3b + \dbinom{4}{2}a^2b^2 + \dbinom{4}{3}ab^3 + \dbinom{4}{4}b^4$.

     (e) The results agree with Table X.

5.42 (a) 18,564;   (b) 3,003;   (c) 816;   (d) 4,368.

5.43 (a) $\dbinom{12}{4} = 3 \cdot \dbinom{11}{3}$.

5.44 The next three rows are

$$
\begin{array}{ccccccccc}
1 & 6 & 15 & 20 & 15 & 6 & 1 & & \\
1 & 7 & 21 & 35 & 35 & 21 & 7 & 1 & \\
1 & 8 & 28 & 56 & 70 & 56 & 28 & 8 & 1
\end{array}
$$

5.45 (a) 1 7 21 35 35 21 7 1;   (b) 35.

5.46 (a) 35;   (b) 20;   (c) 15;   (d) $20 + 15 = 35$.

5.47 (a) $\frac{1}{26}$;   (b) $\frac{3}{13}$;   (c) $\frac{1}{2}$;   (d) $\frac{4}{13}$;   (e) $\frac{1}{4}$.

5.48 (a) $\frac{25}{102}$;   (b) $\frac{1}{17}$.

5.49 (a) $\frac{11}{850}$;   (b) $\frac{6}{5525}$;   (c) $\frac{39}{850}$.

5.50 (a) $\frac{1}{6}$;   (b) $\frac{3}{6} = \frac{1}{2}$.

5.51 (a) $\frac{1}{9}$;   (b) $\frac{1}{6}$;   (c) $\frac{1}{18}$;   (d) $\frac{2}{9}$.

5.52 $\frac{1}{16}$, $\frac{4}{16}$, $\frac{6}{16}$, $\frac{4}{16}$,   and $\frac{1}{16}$.

5.53 (a) $\frac{15}{72}$;   (b) $\frac{50}{72}$;   (c) $\frac{7}{72}$;   (d) $\frac{35}{72}$.

5.54 (a) $\frac{37}{75}$;   (b) $\frac{1}{5}$;   (c) $\frac{1}{5}$.

5.55 $\frac{1}{n}$.

5.56 (a) $\frac{24}{91}$;   (b) $\frac{45}{91}$;   (c) $\frac{2}{91}$.

5.57 (a) 0.837;   (b) 0.159;   (c) 0.004.

5.58 0.362.

5.59 (a) $\frac{2}{7}$;   (b) $\frac{1}{21}$.

5.60 $\frac{8}{11}$.

5.61 $\frac{226}{300}$.

5.62 $\frac{103}{150}$.

5.63 $\frac{143}{842}$.

5.64 The probability estimate would be unreasonable if the law enforcement agencies change the number of people who work at solving robberies. Many other changes would also make the estimate unreasonable.

5.65 $\frac{126}{278} = 0.453$.

5.66 $\frac{1,558}{2,050}$.

5.70 24

5.71 (a) 108;   (b) 110.

5.72 (a) true;   (b) true;   (c) false;   (d) true.

5.73 $\frac{102}{625}$.

5.74 There are 20 ways.

5.75 $\frac{5}{36}$.

5.76 $10! = 3,628,800$.

5.77 $4^{10} = 1,048,576$.

5.78 27,720.

5.79 (a) 2;   (b) 3;   (c) 0;   (d) 1.

5.80 560.

5.81 (a) 210;   (b) 1,260.

5.83 $\binom{52}{10} = 15,820,024,220$.

5.84 $P(0) = \frac{1}{32}$, $P(1) = \frac{5}{32}$, $P(2) = \frac{10}{32}$, $P(3) = \frac{10}{32}$, $P(4) = \frac{5}{32}$, $P(5) = \frac{1}{32}$.

5.85 $\frac{3}{7}$.

5.86 0.81.

5.87 (a) 32;   (b) 96.

5.88 (a) $\frac{143}{280}$;   (b) $\frac{39}{140}$;   (c) $\frac{39}{280}$;   (d) $\frac{1}{560}$.

5.89 $\frac{756}{1,200}$.

5.90 $24 \cdot 2^5 = 24 \cdot 32 = 768$.

5.91 There are six ways.

5.92 93,024 and 3,876.

6.1 (a) $\{a, c, d, f, g\}$ is the event that the scholarship is awarded to one of the female applicants.

(b) $\{a, b, c, e, f, g\}$ is the event that the scholarship is not awarded to Mrs. Daly.

(c) $\{e\}$ is the event that the scholarship is awarded to Mr. Earl.

6.2 They are not mutually exclusive.

6.3 (a) $C' = \{7, 8, 9, 10\};$ (b) $C \cap D = \{5, 6\}.$

6.4 $C' \cap D' = \{10\}.$

6.5 (a) $\{8\}$ is the event that a person chooses a color other than red, yellow, blue, green, brown, white, or purple.

(b) This is an empty set.

(c) $\{1, 2, 3, 4, 5\}$ is the event that a person chooses red, yellow, blue, green, or brown.

(d) $\{3, 4, 8\}$ is the event that a person chooses blue, green, and another color not red, yellow, brown, white, or purple.

6.6 A and B are not mutually exclusive, A and C are mutually exclusive, and B and C are not mutually exclusive.

6.7 (a) The first salesman sells one car.

(b) The second salesman sells more cars than the first salesman, or the first salesman sells no cars.

(c) The salesmen sell the same number of cars.

(d) The first salesman sells more cars than the second salesman, or the second salesman sells no cars.

6.8 N and O, O and P, O and Q, and P and Q.

6.9 (a) $(0, 2), (1, 1), (2, 0);$ (b) $(2, 0), (1, 0);$ (c) $(2, 0), (1, 0), (1, 1).$

6.10 (b) $T = \{(0, 3), (1, 2), (2, 1), (3, 0)\},$

$U = \{(0, 3), (1, 2), (1, 3), (2, 3)\},$

$V = \{(3, 0), (3, 1), (3, 2), (3, 3)\},$ and

$W = \{(2, 2), (3, 3)\}.$

6.11 T and U are not mutually exclusive, T and V are not mutually exclusive, T and W are mutually exclusive, U and V are mutually exclusive, U and W are mutually exclusive,   and V and W are not mutually exclusive.

6.12 (a) $\{(0, 3), (1, 3), (1, 2), (2, 3), (2, 2), (2, 1)\}$ is the event that she will not see three movies on the first day.

(b) $\{(0, 3), (1, 2)\}$ is the event that she will see none of the movies on the first day or one on the first day and two on the second day.

(c) $\{(2, 2)\}$ is the event that she will see two movies on both days.

6.13 (b) $K = \{(0, 2), (0, 3), (1, 2)\},$

$L = \{(1, 0), (2, 0), (2, 1), (3, 0)\},$ and

$M = \{(0, 3), (1, 2), (2, 1), (3, 0)\}.$

(c) K and L are mutually exclusive, K and M are not mutually exclusive,   and L and M are not mutually exclusive.

6.14 (a) $\{(0, 0), (0, 1), (1, 0), (1, 1), (2, 0), (2, 1), (3, 0)\}$ is the event that at most one boat is rented out for the day.

(b) {(2, 1), (3, 0)} is the event that at least two of the boats are in dry dock and all the boats not in dry dock are rented out for the day.

6.15 (a) {A, D}; (b) {C, E}; (c) {B}; (d) {A, C, E}.

6.16 (a) Not mutually exclusive; (b) mutually exclusive; (c) mutually exclusive; (d) not mutually exclusive; (e) not mutually exclusive.

6.17 Region 1 represents the event that a driver has both kinds of insurance; Region 2 represents the event that a driver has liability insurance but no collision insurance; Region 3 represents the event that a driver has collision insurance but no liability insurance; Region 4 represents the event that a driver has neither kind of insurance.

6.18 (a) Regions 1 and 2 together represent the event that a driver has liability insurance.
(b) Regions 2 and 4 together represent the event that a driver does not have collision insurance.
(c) Regions 1, 2, and 3 together represent the event that a driver has either or both kinds of insurance.

6.19 Region 1 represents the event that the flight leaves Denver on time and arrives in Houston on time;
Region 2 represents the event that the flight leaves Denver on time but does not arrive in Houston on time;
Region 3 represents the event that the flight does not leave Denver on time but arrives in Houston on time;
Region 4 represents the event that the flight does not leave Denver on time and does not arrive in Houston on time.

6.20 (a) Regions 1 and 3 together represent the event that the flight arrives in Houston on time.
(b) Regions 3 and 4 together represent the event that the flight does not leave Denver on time.
(c) Regions 2, 3, and 4 together represent the event that the flight either does not leave Denver on time or does not arrive in Houston on time.

6.21 60 of the players are men who do not use a two-handed backhand.

6.22 33.

6.23 (a) 3, 5, 6 and 8; (b) 2 and 7; (c) 1 and 3; (d) 7 and 8.

6.24 (a) Region 1 represents the event that the car needs an engine overhaul, transmission repairs, and new tires.
(b) Region 3 represents the event that the car needs transmission repairs and new tires, but not an engine overhaul.
(c) Region 7 represents the event that the car needs an engine overhaul, but neither transmission repairs nor new tires.
(d) Regions 1 and 4 together represent the event that the car needs an engine overhaul and new tires.
(e) Regions 2 and 5 together represent the event that the car needs transmission repairs but no new tires.
(f) Regions 3, 5, 6, and 8 together represent the event that the car does not need an engine overhaul.

6.25 (a) Region 5; (b) Regions 1 and 2 together; (c) Regions 3, 5, and 6 together; (d) Regions 1, 3, 4, and 6 together.

6.26 24; (b) 16.

6.28 (a) That there will not be enough capital for expansion;

(b) That there will not be adequate transportation;

(c) That there will be enough capital for expansion and/or adequate transportation;

(d) That there will be enough capital for expansion and adequate transportation.

6.29 (a) $P(C' \cap T)$;   (b) $P(C' \cap T')$.

6.30 (a) That a professor does not give difficult tests;

(b) That a professor gives difficult tests and/or is a rough grader;

(c) That a professor is not well versed in his field and gives difficult tests;

(d) That a professor is well versed in his field but not a rough grader.

6.31 (a) $P(R')$;   (b) $P(D' \cap R)$;   (c) $P(V' \cup D')$.

6.32 (a) Postulate 1;   (b) Postulate 2;   (c) Postulate 2;   (d) Postulate 3.

6.33 (a) The third rule;   (b) the first rule;   (c) the second rule;   (d) the first rule.

6.34 (a) $P(A \cup B) \geq P(A)$;   (b) $P(A \cap B) \leq P(A)$.

6.35 $0.23 + 0.17 \neq 0.38$.

6.36 Postulate 3 and Postulate 2.

6.37 Postulate 3 and Postulate 2.

6.38 Postulate 1.

6.39 (a) 0.69;   (b) 0.38;   (c) 0.93;   (d) 0.07.

6.41 (a) Use any example in which $A \cup B$ is not the whole space.

(b) Use any example in which $B = A'$.

6.42 (a) 0.43;   (b) 0.90;   (c) 0.10.

6.44 (a) 0.998;   (b) 0.007;   (c) 0.993.

6.46 (a) 3 to 2;   (b) 11 to 5;   (c) 7 to 2.

6.47 13 to 7.

6.48 9 to 4.

6.49 (a) $\frac{34}{55}$.   (b) $\frac{11}{13}$.   (c) $\frac{5}{6}$.

6.50 The probability that they will go down is $\frac{3}{4}$ and the probability that they will go up is $\frac{5}{6}$, and this is impossible since $\frac{3}{4} + \frac{5}{6} > 1$.

6.51 (a) $p = \frac{1}{4}$;   (b) $p < \frac{1}{4}$.

6.52 $\frac{3}{5} \leq p \leq \frac{2}{3}$.

6.53 The probability that he will run for the House is $\frac{1}{3}$; for the Senate is $\frac{1}{5}$; and for one or the other is $\frac{7}{12}$. Since running for the House and running for the Senate are mutually exclusive, his probabilities are inconsistent.

6.54 The probability for the $1,000 raise is $\frac{5}{12}$; for the $2,000 raise is $\frac{1}{12}$; and for either is $\frac{1}{2}$. Since $\frac{5}{12} + \frac{1}{12} = \frac{1}{2}$, the probabilities are consistent.

6.55 19 to 11.

6.56 The event in (b) has probability $\frac{1}{1,024}$, which is small but certainly not zero. Since commercial airline crashes in the United States occur several times a year, there is a very small, but non-zero, probability that three will occur on the same day. The events in (a) and (c) have incredibly small probabilities; you could reasonably assign them the value zero.

6.58 Bets (a) and (b) are attractive.

6.59 A possible bet might be 14.5 to 10.5, with Alan placing $10.50 on the Pythons and Ben placing $14.50 on the Rockets.

6.60 (a) 0.37;   (b) 0.77;   (c) 0.63.

6.61 0.23.

6.62 (a) 0.63;   (b) 0.23;   (c) 0.62.

6.63 $P(A) = \frac{15}{32}$.

6.64 (a) $\frac{4}{7}$;   (b) $\frac{24}{35}$;   (c) $\frac{1}{7}$.

6.65 $P(T) = \frac{2}{5}$, $P(U) = \frac{2}{5}$, $P(V) = \frac{2}{5}$, $P(W) = \frac{1}{5}$.

6.66 (a) 0.55;   (b) 0.53;   (c) 0.23;   (d) 0.80.

6.67 $\frac{1}{32}$, $\frac{5}{32}$, $\frac{10}{32}$, $\frac{10}{32}$, $\frac{5}{32}$,   and $\frac{1}{32}$.

6.68 0.95.

6.69 0.12.

6.70 (a) 0.32;   (b) 0.68.

6.71 $\frac{1}{16}$.

6.72 0.27;   (b) 0.40;   (c) 0.60.

6.73 7 to 3.

6.74 (a) $0.84 + 0.71 - 0.53 = 1.02 > 1$.
      (b) $0.48 + 0.36 + 0.12 = 0.96 < 1$.

6.75 (a) 0.68.   (b) 0.98.

6.76 (a) $P(Q|W)$;   (b) $P(W'|Q)$;   (c) $P(Q'|W')$.

6.77 (a) The probability that a worker who meets the production quota is well trained.
      (b) The probability that a worker who is well trained will not meet the production quota.
      (c) The probability that a worker who does not meet the production quota is not well trained.

6.78 (a) $P(N|I)$;   (b) $P(I'|A')$;   (c) $P(I' \cap A'|N)$.

6.79 (a) The probability that a student who rates high on the social adjustment scale will score high in intelligence.
      (b) The probability that a student who does not display neurotic tendencies will not rate high on the social adjustment scale.
      (c) The probability that a student who scores high in intelligence and rates high on the social adjustment scale will not display neurotic tendencies.

6.80 (a) $P(E|G)$;   (b) $P(G'|A')$;   (c) $P(A|E \cap G)$.

6.81 (a) The probability that an applicant who is employed will have the application approved.
      (b) The probability that an applicant who is not employed will not have the application approved.
      (c) The probability that an applicant who has the application approved will not be employed or not have a good credit rating.

6.82 (a) $\frac{36}{80}$;   (b) $\frac{20}{80}$;   (c) $\frac{24}{80}$;   (d) $\frac{8}{80}$;   (e) $\frac{24}{36}$;   (f) $\frac{8}{20}$;   (g) $\frac{8}{44}$;   (h) $\frac{72}{80}$.

6.83 (a) $\frac{24}{36}$;   (b) $\frac{8}{20}$.

6.84 (a) $\frac{3}{5}$;   (b) $\frac{1}{2}$;   (c) $\frac{2}{5}$;   (d) $\frac{1}{10}$;   (e) $\frac{2}{3}$;   (f) $\frac{1}{5}$.

6.85 (a) $\frac{2}{3}$;   (b) $\frac{1}{5}$.

6.86 (a) $\frac{2}{5}$;   (b) $\frac{1}{2}$.

6.87 $\frac{7}{10}$.

6.88 (a) 0.90;   (b) 0.15.

6.89 (a) 0.50;   (b) 0.50;   (c) 0.30;   (d) 0.30.

6.90 0.30.

6.91 0.36.

6.92 (a) $\frac{1}{16}$;   (b) $\frac{1}{17}$.

6.93 (a) $\frac{33}{59}$;   (b) $\frac{7}{118}$;   (c) $\frac{45}{118}$.

6.94 (a) $\frac{22}{145}$;   (b) $\frac{4}{25}$.

6.95 (a) 0.09;   (b) 0.20;   (c) 0.45;   (d) 0.36;   (e) 0.11;   (f) 0.55;
      (g) 0.44;   (h) $0.20 + 0.45 - 0.09 = 0.56$.

6.98 A and C are independent.

6.99 K and L are not independent.

6.100 M and N are not independent.

6.101 $\frac{1}{256}$.

6.102 $\frac{625}{1,296}$.

6.103 (a) $P(X \mid \text{Red, Deck } A) = \frac{5}{10} > P(X \mid \text{Green, Deck } A) = \frac{14}{30}$;
      (b) $P(X \mid \text{Red, Deck } B) = \frac{5}{25} > P(X \mid \text{Green, Deck } B) = \frac{2}{15}$.
      (d) For the combined deck,
             $P(X \mid \text{Red, Combined Deck}) = \frac{10}{35} < P(X \mid \text{Green, Combined Deck}) = \frac{16}{45}$.

6.104 (a) Baker's proportion was 0.60, Abel's was 0.55.
      (b) Baker's proportion was 0.375, Abel's was 0.20.
      (c) For the "season-to-date" Baker completed nine of 21 passes, for a proportion of
          0.429. Abel completed 12 of 25 passes, for a proportion of 0.48.

6.105 $\frac{1}{12}$.

6.106 $\frac{1}{91}$.

6.107 0.0288.

6.108 (a) 0.729;   (b) 0.144.

6.109 (a) 0.0616;   (b) 0.5325.

6.110 (a) 0.76;   (b) 0.4474.

6.111 (a) 0.242;   (b) 0.5372.

6.112 (a) $\frac{7}{16}$;   (b) $\frac{3}{7}$.

6.113 0.8051.

6.114 0.9167.

6.115 (a) 0.246;   (b) 0.999. Because the disease is rare, a positive test is only a weak
       suggestion that a person has the disease. Getting a negative test virtually guarantees
       that a person does not have the disease.

6.116 0.475.

6.117 0.568.

6.118 (a) 0.028;   (b) 0.0714;   (c) 0.2857.

6.119 Given the explosion the probabilities of the four causes are 0.229, 0.244, 0.183, and
       0.344. Based on these figures, the most likely cause is purposeful action.

6.120 (a) $P(M) = 5P(Y) - 2$;   (b) 0.12.

6.122 (a) 26 to 10 (or 13 to 5);   (b) 1 to 9;   (c) 42 to 22 (or 21 to 11).

6.124 (a) $\frac{1}{21}$;   (b) $\frac{5}{21}$;   (c) $\frac{15}{21}$;   (d) $\frac{6}{21}$;   (e) $\frac{6}{21}$;   (f) $\frac{15}{21}$.

6.125 (a) 0.076;   (b) 0.421.

6.126 (a) 0.72;   (b) 0.91;   (c) 0.09.

6.127 0.0046.

6.128 (a) The probability that a person who is qualified for the job will get the job.
      (b) The probability that a person who is not qualified for the job will not get the
          job.
      (c) The probability that the person who gets the job will be qualified for it.
      (d) The probability that a person who does not get the job will be qualified for it.

6.129 The two events are independent.

6.130 It is the amount of confidence we may have in a probability, and not the probability itself, that will increase if the probability is based on more information.

6.131 (a) $\frac{5}{24}$;   (b) $\frac{13}{16}$;   (c) $\frac{40}{77}$.

6.132 (a) $\{1, 6\}$ is the event that the program will be rated terrible or excellent.
(b) $\{2, 3, 4, 5, 6\}$ is the event that the program will not be rated terrible.
(c) $\{4, 5\}$ is the event that the program will be rated good or very good.
(d) $\{2, 3\}$ is the event that the program will be rated poor or fair.

6.133 0.68.

6.134 0.49.

6.135 (a) 0.38;   (b) 0.42;   (c) 0.50.

6.136 At least $\frac{3}{4}$, but less than $\frac{4}{5}$.

6.137 Any bet for which the probability on the 49er's is between 0.636 and 0.714 is plausible. Such a bet is 2 to 1 favoring the 49er's. In such a bet Joe would put $10 on the Rams and Walter would put up $20 on the 49ers.

6.138 (a) $\frac{15}{32}$;   (b) $\frac{13}{32}$;   (c) $\frac{5}{32}$;   (d) $\frac{23}{32}$;   (e) $\frac{8}{32} = \frac{1}{4}$;   (f) $\frac{9}{32}$.

6.139 (a) $\frac{5}{13}$;   (b) $\frac{8}{17}$.

6.140 (a) $\frac{18}{25}$;   (b) $\frac{2}{3}$;   (c) $\frac{4}{5}$;   (d) $\frac{20}{27}$;   (e) $\frac{11}{25}$;   (f) $\frac{10}{21}$.

6.141 (a) $\frac{94}{125}$;   (b) $\frac{4}{5}$;   (c) $\frac{4}{5}$;   (d) $\frac{40}{47}$;   (e) $\frac{11}{25}$;   (f) $\frac{20}{31}$.

6.142 (a) Fifty-fifty;   (b) fifty-fifty.

6.143 0.224.

6.144 Region 1 represents the event that the school's football team is rated among the top twenty by both AP and UPI.
Region 2 represents the event that the school's football team is rated among the top twenty by AP but not by UPI.
Region 3 represents the event that the school's football team is rated among the top twenty by UPI but not by AP.
Region 4 represents the event that the school's football team is rated among the top twenty by neither AP nor UPI.

6.145 The probability that the movie will be rated G is $\frac{1}{9}$, that it will be rated PG is $\frac{1}{6}$, and that it will be rated G or PG is $\frac{5}{18}$. The probabilities are consistent.

6.146 0.34.

6.147 Let R denote the event that the person is a security risk. Let F denote the event that this person fails the lie detector test:
(a) $P(R|F) = 0.187$.
(b) $P(R'|F') = 0.998$.

6.148 (a) Probabilities cannot be negative.
(b) The probability for a C cannot be $10(0.11) = 1.10$, which is greater than 1.
(c) $0.59 + 0.31 = 0.90$, which is less than 1.

6.149 (a) $\dfrac{1}{4,096}$;   (b) $\dfrac{1}{4,096}$.

6.150 Any example in which B is a subset of A will work.

6.151 (a) The person will visit Portugal, but neither Belgium nor England.
(b) The person will visit Belgium and England.
(c) The person will visit England, but not Belgium.

(d) The person will visit England and/or Portugal, but not Belgium.

(e) The person will visit neither Belgium nor England.

6.152 (a) 0.0435;   (b) 0.632.

6.153 (a) 0.48;   (b) 0.12;   (c) 0.08;   (d) 0.40.

6.154 (a) The death rate is higher at hospital A for each risk category.

(b) Hospital A has overall death rate 0.100. Hospital B has overall death rate of 0.138.

(c) You should prefer hospital B.

6.155 The odds that it will not win either race are 21 to 11.

CHAPTER 7

7.1 $0.20.

7.2 $0.23.

7.3 $5.00.

7.4 (a) $27,500.

(b) The player whose probability of winning is 0.60 has an expectation of $30,000, and the opponent has an expectation of $25,000.

(c) The player whose probability of winning is 0.70 has an expectation of $32,500, and the opponent has an expectation of $22,500.

7.5 Since the two expectations are $0.25 and $0.28, it would be smarter to draw a slip of paper from box 2.

7.6 (a) 25 cents.

(b) Since the expectation is less than the cost of the postage, it was not worth sending in the label.

7.7 $5\frac{13}{16}$.

7.8 $52.00.

7.9 $1.16.

7.10 The expected selling price is $22,560. His expected gross profit is $560.

7.11 1.81.

7.12 1.87.

7.13 (a) $p < 0.25$.   (b) $p > 0.15$.

7.14 $p = 0.20$.

7.15 He believes that $p < 0.571$.

7.16 $p > 0.40$.

7.17 $89.

7.18 $1,800.

7.19 Since the expectation is $3.50, it is not rational to pay $4.

7.20 The two expectations are $82,500 and $67,500.

7.21 Since the expectation is $120, it is not worth her time.

7.22 If the driver goes to the lumberyard which is 18 miles away, his expected distance is 46 miles. If the driver goes to the lumberyard which is 22 miles away, his expected distance is $44\frac{2}{3}$.

7.23 (a) If he goes first to the lumberyard which is 18 miles away, his expected distance is 44 miles. For the other option his expected distance is $45\frac{1}{3}$ miles.

(b) If he goes first to the lumberyard which is 18 miles away, his expected distance is 45 miles. If he goes first to the lumberyard which is 22 miles away, his expected distance is also 45 miles.

7.24 If they continue, the expected profit is $900,000, and if they do not continue, the expected profit is $-$675,000.

7.25 Since the expected profits are $-$300,000 and $-$300,000 it does not matter whether they continue the operation.

7.26 His expected profit if he stocks 4 pies is 3.10. His expected profit if he stocks 3 pies is 4.50. His expected profit if he stocks 2 pies is $5, and his expected profit if he stocks 1 pie is $3.40. His optimal strategy is to stock 2 pies.

7.27 The expected profits are $59,200 and $73,600. Delaying the expansion maximizes the expected profit.

7.28 (a) He should delay expansion; then he cannot do worse than $16,000.
(b) This criterion would lead him to stock zero pies!

7.29 (a) He should expand now, since this is the only strategy that has a chance of making $328,000.
(b) His maximax procedure is to stock 4 pies.

7.30 (a) The minimum profits are $-$2,700,000 and $-$1,800,000, and they will maximize the minimum profit if they do not continue the operation.
(b) The maximum profits are 4,500,000 and 450,000, and they will maximize the maximum profit if they continue the operation.

7.31 (a) The expected distance with perfect information is $21\frac{1}{3}$ miles, and the expected value of perfect information is 2 miles.
(b) The expected profit with perfect information is $2,475,000, and the expected value of perfect information is $1,575,000. It would be worthwhile to spend the $500,000.

7.32 (a) Unless he expands now he does not even have a chance of avoiding bankruptcy.
(b) He can avoid bankruptcy only by delaying the expansion.

7.33 (a) $132; (b) $192.

7.34 (a) the mode, 17; (b) the median, 18; (c) the mean, 19.

7.35 The midrange, which is $1,695.

7.37 His best strategy is to buy 63 papers.

7.38 1.1.

7.39 $p < 0.60$.

7.40 The expected profit if it grants the mortgage is $5,800. The best decision is to grant the mortgage.

7.41 If the credit manager accepts the applicant, the expected profit is $342. The credit manager should accept the applicant.

7.42 To minimize the maximum loss, the credit manager should turn down the applicant.

7.43 2.20.

7.44 $U = \$26.00$.

7.45 The mode, which is 40.

7.46 The mean, which is 40.8.

7.47 The probability is at least $\frac{2}{3}$, but less than $\frac{3}{4}$.

7.48 $5.00.

7.49 (a) $p > 0.10$; (b) $p < 0.10$; (c) $p = 0.10$.

7.50 At hotel A, the expected cost is $83.00; at hotel B, the expected cost is $81.80.

7.51 The expected cost with perfect information is $78.20; the expected value of perfect information is $3.60. It is worthwhile to spend the $2.40.

7.52 At hotel A, the expected cost is $82.40; at hotel B, the expected cost is $82.40.

7.53 At hotel A, the minimum is $80.00; at hotel B, the minimum is $72.80. Since $72.80 is smaller, she should make her reservation at hotel B.

7.54 $p = 0.16$.

7.55 The midrange, 41.5.

7.56 $p = \frac{5}{8}$. In general, $p = \dfrac{a}{a + b}$.

CHAPTER 8

8.1 (a) No; (b) no; (c) yes.

8.2 (a) No; (b) yes; (c) no.

8.3 (a) No; (b) yes.

8.4 (a) No; (b) yes.

8.5 0.311.

8.6 0.057.

8.7 0.278.

8.8 (a) 0.250; (b) 0.056; (c) 0.922.

8.9 0.235.

8.10 (a) 0.251; (b) 0.251.

8.11 (a) 0.115; (b) 0.115.

8.12 (a) 0.016; (b) 0.298; (c) 0.823.

8.13 (a) 0.175; (b) 0.228; (c) 0.588.

8.14 (a) 0.397; (b) 0.941; (c) 0.010.

8.15 (a) 0.448; (b) 0.129; (c) 0.672.

8.16 (a) 0.203; (b) 0.316; (c) 0.050.

8.17 0.000, 0.000, 0.001, 0.009, 0.037, 0.103, 0.200, 0.267, 0.233, 0.121, and 0.028.

8.18 (a) 0.401; (b) 0.349; (c) 0.107; (d) 0.028.

8.19 (a) 0.311; (b) 0.230; (c) 0.654.

8.20 (a) $\dfrac{n - x}{x + 1} \cdot \dfrac{p}{1 - p}$;

(b) $f(0) = \dfrac{729}{4{,}096}$, $f(1) = \dfrac{1{,}458}{4{,}096}$,

$f(2) = \dfrac{1{,}215}{4{,}096}$, $f(3) = \dfrac{540}{4{,}096}$,

$f(4) = \dfrac{135}{4{,}096}$, $f(5) = \dfrac{18}{4{,}096}$,

$f(6) = \dfrac{1}{4{,}096}$.

8.21 (a) 0.0864; (b) 0.0791; (c) 0.0630.

8.22 (a) 0.2969, so that 15 is not enough.

(b) 0.8867, so that 20 is not enough.

(c) 0.9998, so that 30 is more than what is needed.

(d) 21.

8.23 Using the binomial distribution with $p = 0.43$ and with varying values of n, we find that the smallest number she can purchase is 60.

8.24 0.293.

8.25 0.318.

8.26 (a) 0.045; (b) 0.409; (c) 0.545.

8.27 (a) 0.295; (b) 0.491; (c) 0.196; (d) 0.018.

8.28 (a) 0.833; (b) 0.455; (c) 0.773.

8.29 (a) Not satisfied; (b) satisfied; (c) satisfied; (d) not satisfied.

8.30 (a) 0.207450; (b) 0.203627; (c) -0.003823. In percentage terms, -1.84%.

8.31 (a) 0.206218; (b) 0.204800; (c) -0.001418. In percentage terms, -0.69%.

8.32 (a) 0.0667; (b) 0.16; (c) $f(1) = \dfrac{a}{a+b}$.

8.33 $f(x) = \dfrac{\binom{b}{x-2}\binom{a}{1}}{\binom{a+b}{x-1}} \cdot \dfrac{a-1}{a+b-(x-1)}$ for $x = 2, 3, \ldots, b, b+1, b+2$.

8.34 (a) Satisfied; (b) not satisfied; (c) not satisfied.

8.35 $f(3) = 0.192$.

8.36 0.1680.

8.37 0.879.

8.38 (a) 0.162; (b) 0.152.

8.39 (a) 0.165; (b) 0.297.

8.40 0.1771.

8.41 0.156.

8.42 (a) 0.116; (b) 0.170; (c) 0.351.

8.43 (a) 0.1367; (b) 0.1321; (c) 0.4754; (d) 0.7301.

8.44 (a) 0.0988; (b) 0.0034.

8.45 0.117.

8.46 (a) 0.049; (b) 0.056.

8.47 0.040.

8.48 0.029.

8.49 (a) 0.0659; (b) 0.0580; (c) 0.000148.

8.50 (a) 0.0586. The error of approximation is -0.0073.

(b) 0.0544. The error of approximation is -0.0036.

(c) 0.000162. The error of approximation is 0.000014.

8.51 (a) 1; (b) 1.

8.53 (a) $\mu = 1.65$; (b) $\sigma^2 = 1.6475$. Then $\sigma = 1.28$.

8.54 $\sigma^2 = 1.6475$.

8.55 $\mu = 1.25$.

8.56 $\sigma = 1.34$.

8.57 (a) $\mu = 2$; (b) $\mu = 2$.

8.58 (a) $\sigma^2 = 1$; (b) $\sigma^2 = 1$; (c) $\sigma^2 = 1$.

8.59 (a) $\mu = 4.805$ and $\sigma^2 = 1.909$; (b) $\mu = 4.8$ and $\sigma^2 = 1.92$.

8.60 (a) $\mu = 7.990$. $\sigma^2 = 1.6679$. Then $\sigma = 1.29$.

(b) $\mu = 8$ and $\sigma^2 = 1.60$, giving $\sigma = 1.26$.

8.61 $\mu = 16$ and $\sigma = 1.79$.

8.62 (a) $\mu = 242$ and $\sigma = 11$;

(b) $\mu = 120$ and $\sigma = 10$;

(c) $\mu = 180$ and $\sigma = 11.2$;

(d) $\mu = 24$ and $\sigma = 4.8$;

(e) $\mu = 520$ and $\sigma = 13.5$.

8.63 $\mu = 1.499$.

8.64 $\mu = 1.5$.

8.65 $\mu = 0.937$.

8.66 $\mu = 0.9375$.

8.67 (a) $f(0) = \frac{1}{70}$, $\quad f(1) = \frac{16}{70}$, $\quad f(2) = \frac{36}{70}$,

$f(3) = \frac{16}{70}$, $\quad f(4) = \frac{1}{70}$;

(b) $\mu = 2$;

(c) $\mu = 2$.

8.68 $\sigma = 0.756$.

8.69 (a) $\sigma = 0.756$; (b) $\sigma = 0.584$; (c) $\sigma = 0.747$.

8.70 (a) 0.7895; (b) 0.9697; (c) 0.9970; (d) 0.7035.

8.71 $\mu = 0.799$, and this is very close to $\lambda = 0.8$.

8.72 $\mu = 2.4 = \lambda$.

8.73 $\sigma = 0.8905$, which is very close to $\sqrt{0.8} = 0.8944$.

8.74 At least $\frac{15}{16}$.

8.75 (a) At least $\frac{143}{144}$; (b) at least $= 0.91$.

8.78 (a) 00–24, 25–74, and 75–99.

8.79 (a) 00–13, 14–40, 41–67, 68–85, 86–94, 95–98, and 99.

8.81 (a) 00–40, 41–77, 78–93, 94–98, and 99.

8.82 (a) 0000–2465, 2466–5917, 5918–8334, 8335–9462, 9463–9857, 9858–9968, 9969–9994, and 9995–9999.

8.83 (a) 00–04, 05–19, 20–44, 45–69, 70–89, and 90–99.

8.85 (a) $= 0.681$; (b) $= 0.279$.

8.86 The ratio is 1.39.

8.87 (a) 0.0232; (b) 0.0231.

8.88 (a) Their total is 0.98; (b) $f(4)$ is negative; (c) their total is 1.25.

8.89 (a) $\mu = 1.48$; (b) $\sigma^2 = 1.37$.

8.90 0.082.

8.91 (a) $= 0.396$; (b) $= 0.183$.

8.92 00–45, 46–72, 73–87, 88–95, and 96–99.

8.93 0.0245.

8.94 (a) Satisfied; (b) not satisfied; (c) not satisfied.

8.95 (a) $\mu = 4.5$; (b) $\mu = 4.5$.

8.96 (a) $\sigma^2 = 2.258$, so that $\sigma = \sqrt{2.258} = 1.503$; (b) $\sigma = 1.5$.

8.97 (a) 0.1985; (b) 0.8109; (c) $0.9684 - 0.0991 = 0.8693$.

8.98 (a) Satisfied.

(b) The probability by the Poisson approximation is 0.122138. The exact binomial probability is 0.122730. The error of approximation is -0.000592.

8.99 (a) At least $\frac{48}{49}$; (b) at least 0.84.

8.100 The probabilities are 0.48, 0.50, 0.51, 0.51, 0.50, and 0.49, and they are a maximum for $N = 11$ and $N = 12$.

8.101 (a) The hypergeometric probability is $\dfrac{\dbinom{1393}{48}\dbinom{35}{2}}{\dbinom{1428}{50}}$.

      (b) The binomial probability is 0.223551.

      (c) The Poisson probability is 0.220480.

8.102 (a) 0.326;   (b) 0.325.

8.103 $\mu = 2.092$ and $\sigma^2 = 1.836$.

8.104 (a) 0.270;   (b) 0.485;   (c) 0.221;   (d) 0.118.

8.105 (a) 0.382;   (b) 0.930;   (c) 0.079.

8.106 (a) It can be a probability distribution.

      (b) It cannot be a probability distribution.

      (c) It cannot be a probability distribution.

8.107 (a) 0.146;   (b) 0.957;   (c) 0.189.

8.108 (a) $f(0) = 0.091$;   (b) $f(0) + f(1) + f(2) = 0.571$.

8.109 $\mu = 162$ and $\sigma = 9$.

8.110 At least 0.9975.

CHAPTER 9

9.1 (a) The total area under the curve is <1.

     (b) $f(x) < 0$ for $1 \le x < \frac{7}{4}$.

9.2 (a) 0.625;   (b) 0.8.

9.3 $\frac{7}{16}$.

9.4 (a) $\frac{1}{8}$;   (b) $\frac{5}{9}$.

9.5 (a) The first area is bigger;

     (b) the first area is bigger;

     (c) the first area is bigger;

     (d) the two areas are equal;

     (e) the first area is bigger;

     (f) the second area is bigger;

     (g) the two areas are equal;

     (h) the first area is bigger.

9.6 (a) The first area is bigger;

     (b) the second area is bigger;

     (c) the first area is bigger;

     (d) the first area is bigger;

     (e) the two areas are equal;

     (f) the second area is bigger;

     (g) the two areas are equal;

     (h) the first area is bigger.

9.7 (a) 0.3078;   (b) 0.4515;   (c) 0.3156;   (d) 0.6064;

     (e) 0.9032;   (f) 0.2148.

9.8 (a) 0.1598;   (b) 0.1068;   (c) 0.6444.

9.9 (a) 0.5792;   (b) 0.1251;   (c) 0.5871;   (d) 0.3708;   (e) 0.1271;

     (f) 0.7764;   (g) 0.1804.

9.10 (a) 0.5;   (b) 0.87;   (c) -2.22;   (d) 1.44 or -1.44.

9.11 (a) $z = \pm 1.48$;   (b) $z = -0.74$;   (c) $z = 1.12$;   (d) $z = \pm 2.17$.

9.12 (a) 0.6826;   (b) 0.9544;   (c) 0.9974;   (d) 0.99994;   (e) 0.9999994.

9.13 (a) 2.575;   (b) 1.96.

9.14 (a) The first probability is bigger;
   (b) the second probability is bigger;
   (c) the first probability is bigger;
   (d) the first probability is bigger.

9.15 (a) The two probabilities are equal;
   (b) the second probability is bigger;
   (c) the first probability is bigger;
   (d) the second probability is bigger;
   (e) the first probability is bigger.

9.16 (a) 0.9332;   (b) 0.7734;   (c) 0.2957;   (d) 0.9198.

9.17 (a) 0.0351;   (b) 0.2912;   (c) 0.2645;   (d) 0.3182.

9.18 $\sigma = 20$.

9.19 $\mu = 73.3$ and the probability is 0.9332.

9.20 (a) 0.330;   (b) 0.200;   (c) 0.202.

9.21 (a) 0.699;   (b) 0.449;   (c) 0.074.

9.22 (a) 0.135;   (b) 0.050;   (c) 0.393.

9.27 Roughly, $\mu = 18.6$ and $\sigma = 5.4$.

9.28 Roughly, $\mu = 24.5$ and $\sigma = 10$.

9.29 Roughly, $\mu = 32$ and $\sigma = 6$.

9.30 Normal probability paper should show the points, at least approximately, on a straight line.

9.31 Points should show a pronounced curve on the normal probability paper.

9.32 (a) 0.1093;   (b) 0.5733;   (c) 0.6826.

9.33 18.37.

9.34 (a) 0.7881;   (b) 0.2743.

9.35 (a) 0.0154;   (b) 0.3071.

9.36 4.75.

9.37 (a) 0.0618;   (b) 0.3520;   (c) 0.5862.

9.38 0.9772.

9.39 0.2643.

9.40 (a) 0.0809;   (b) 0.3859;   (c) 0.6950.

9.41 (a) 0.0521;   (b) 0.2611;   (c) 0.7910.

9.42 0.0661.

9.43 (a) Not satisfied;   (b) satisfied   (c) not satisfied.

9.44 (a) Not satisfied;   (b) not satisfied;   (c) satisfied.

9.45 The probability is 0.2128. The error is 0.004.

9.46 0.2266.

9.47 0.0808.

9.48 0.0012.

9.49 0.6950.

9.50 0.0087.

9.51 The probability is 0.0078. The error is 0.0042.

9.52 0.975.

9.53 The error is -0.0024.

9.54 0.0764.

9.55 (a) We cannot use the normal approximation; we cannot use the Poisson approximation.

(b) We can use the normal approximation; we cannot use the Poisson approximation.

(c) We cannot use the normal approximation; we can use the Poisson approximation.

(d) We can use the normal approximation; we can use the Poisson approximation.

9.56 (a) 0.0198;   (b) 0.0327;   (c) 0.0079.

9.57 The error is 0.000525.

9.62 (a) 0.1379;   (b) 0.5255.

9.63 The probability is 0.1200. The error is 0.000, to three figures.

9.64 (a) 0.4664;   (b) 0.9938;   (c) 0.7389;   (d) 0.1075;   (e) 0.2389.

9.65 (a) 0.9332;   (b) 0.7734;   (c) 0.0934;   (d) 0.6514.

9.66 (a) 2.05;   (b) 1.28.

9.68 0.0618.

9.71 (a) $\frac{1}{2}$;   (b) $\frac{3}{4}$;   (c) $\frac{7}{8}$.

9.72 (a) 0.2090;   (b) 0.4933.

9.73 0.8944.

9.74 The proportion of the time that job A will take longer than the average job B is 0.0643. The proportion of the time that job B will take longer than the average job A is 0.0143.

9.75 (a) 0.8078;   (b) 0.8700.

9.76 (a) 0.4%;   (b) 0.08%.

9.77 (a) 0.3665;   (b) 0.4714;   (c) 0.0534;   (d) 0.1293;   (e) 0.8544.

9.78 (a) Satisfied;   (b) not satisfied;   (c) satisfied;   (d) not satisfied.

9.79 (a) 0.0683;   (b) 0.4129.

9.80 (a) 0.54;   (b) -1.18;   (c) 0.92;   (d) 0.53 or -0.53;   (e) 0.41 or -0.41.

9.81 (a) 0.4908;   (b) 0.9556.

10.1 (a) 15;   (b) 45;   (c) 300.

10.2 (a) 455;   (b) 4,060;   (c) 34,220.

10.3 (a) $\frac{1}{495}$;   (b) $\frac{1}{26,334}$.

10.4 0.60.

10.5 ab

ac   bc

ad   bd   cd

ae   be   ce   de

af   bf   cf   df   ef

ag   bg   cg   dg   eg   fg

ah   bh   ch   dh   eh   fh   gh

10.6 $\frac{1}{4}$.

10.7 (a) $\frac{1}{15}$;   (b) $\frac{1}{3}$.

10.8 607, 776, 679, 320, 495, 457, 040, 783, 756, and 314.

10.9 5190, 1250, 1377, 1312, 7648, 6367, 0796, 7900, 5223, 6686, 1543, 5872, 1586, 3357, and 5209.

10.10 69, 58, 03, 57, 61, 39, 40, 80, 27, and 10.

10.11 0.8568.

10.12 $\frac{1}{10}$.

10.15 (a) $\frac{1}{10}$;   (b) $\frac{1}{10}$;   (c) 0.0082;   (d) less than 0.01.

10.16 If you pick the first number, your sample will consist of the values in columns 1 and 6 (and so on).

10.17 Since the values are displayed six across the page, each possible sample consists of one of the six columns.

10.18 There is a definite cyclical pattern and each month's figures (for instance, the four December figures, 97, 108, 111, and 117) go into the same sample.

10.19 (a) The probability that the sample mean will differ from the population mean 160 by more than 5 is 8/15, about 53%.

　　(b) The probability that the sample mean will differ from the population mean 160 by more than 5 is 2/9, about 22%.

　　(c) The probability that the sample mean will differ from the population mean 160 by more than 5 is 2/3, about 67%.

　　(d) The stratification proposed in (c) was actually destructive.

10.20 (a) 2,295;   (b) 4,896.

10.21 (a) 17,598,672,000;   (b) 17,524,416,000.

10.23 $n_1 = 20$, $n_2 = 48$, $n_3 = 8$,   and $n_4 = 4$.

10.24 (a) 7, 2, and 2;   (b) 8, 2, and 2;   (c) 7, 3, and 2.

10.25 Since all the values are less than 155 or greater than 165, the probability is 1.

10.27 $n_1 = 20$ and $n_2 = 80$.

10.28 $n_1 = 50$, $n_2 = 24$,   and $n_3 = 10$.

10.30 (a) $\bar{x}_w = 216.47$;   (b) 216.47.

10.31 (b)

Mean	Probability
3	1/25
4	2/25
5	3/25
6	4/25
7	5/25
8	4/25
9	3/25
10	2/25
11	1/25

10.32 (a) $\frac{13}{25}$;   (b) $\frac{19}{25}$.

10.33 $\sigma_{\bar{x}}^2 = 4$ and $\sigma_{\bar{x}} = 2$.

10.34 (d) The probability that the sample mean differs from 13.5 by less than 3 is $16/20 = 0.80$.

10.36 The probability that the sample median differs from 13.5 by less than 3 is $12/20 = 0.60$.

10.37

Range	6	9	12	15
Probability	4/20	6/20	6/20	4/20

10.38 (a) It is divided by 2.   (b) It is divided by 1.5.   (c) It is divided by 3.
(d) It is multiplied by 4.

10.39 (a) 0.9771;   (b) 0.9909;   (c) 0.9977;   (d) 0.9511;   (e) 0.9499;
(f) 0.9488.

10.40 Interpolation yields 0.6745 corresponding to 0.2500.   (a) 1.74;   (b) $1.23.

10.41 Their standard deviation is 2.48, which is greater than $\sigma_{\bar{x}}$ for the means of ordinary random samples.

10.42 Their standard deviation is 5.72, which is greater than $\sigma_{\bar{x}}$ for the means of ordinary random samples.

10.43 (a) $\mu_{\bar{x}} = 160$ and $\sigma_{\bar{x}} = 21.0$;   (b) $\mu_{\bar{x}} = 160$ and $\sigma_{\bar{x}} = 7.1$.

10.44 50.

10.45 $\mu_{\bar{x}} = 160$ and $\sigma_{\bar{x}} = 35$.

10.46 $\sigma_{\bar{x}} = 1.5$.   (a) The probability is at least $\frac{8}{9}$.   (b) The probability is 0.9974.

10.47 $\sigma_{\bar{x}} = 0.48$.   (a) The probability is at least 0.84.   (b) The probability is 0.9876.

10.48 $\sigma_{\bar{x}} = 0.3$.   (a) The probability is 0.8384.   (b) The probability is 0.0124.

10.49 $\sigma_{\bar{x}} = 0.00625$. The probability is 0.8904.

10.50 The probability is 0.0918.

10.51 (a) 0.0735;   (b) 0.0146.   (c) The observations would not be independent if based on nine cars:

10.52 The smallest sample size that will work here is $n = 46$.

10.54 72 is the smallest number of sets.

10.55 $\sigma = 288.7$

10.56 Their mean is 15.7 compared to $\mu = 16$, and their standard deviation is 2.00 compared to $\sigma_{\bar{x}} = 4/\sqrt{5} = 1.79$.

10.57 Their standard deviation is 2.06 compared to the 2.00 we obtained for the means.

10.58 The medians have a greater dispersion than the means.

10.59 31 to 9 against.

10.60 $s = 1.26$ compared to $\sigma_s = 1.26$.

10.62 (a) The probability is at least 0.9506.
(b) The probability is in excess of 0.9999.

10.63 (a) 300,105,000;   (b) 1,736,410,000.

10.64 213, 534, 474, 173, 616, 336, 357, 442, 424, 560, 202, 315, 623, 137, and 074.

10.65 $\dfrac{1}{82,160}$.

10.66 Their mean is 15.7 compared to $\mu = 16$ and their standard deviation is 1.43 compared to $\sigma_{\bar{x}} = 1.26$.

10.67 (b)

Mean	Probability
2	1/6
3	1/6
4	2/6
5	1/6
6	1/6

(c) $\mu_{\bar{x}} = 4$ and $\sigma_{\bar{x}} = 1.29$.

10.68 (b)

Mean	Probability
1	1/16
2	2/16
3	3/16
4	4/16
5	3/16
6	2/16
7	1/16

(c) $\mu_{\bar{x}} = 4 = \mu$ and $\sigma_{\bar{x}}^2 = \frac{5}{2}$.

10.69 $n_1 = 100$, $n_2 = 48$, and $n_3 = 20$.

10.70 (a) $\mu_{\bar{x}} = 160$ and $\sigma_{\bar{x}} = 22.7$;   (b) $\mu_{\bar{x}} = 160$ and $\sigma_{\bar{x}} = 5.8$.

10.71 $\sigma_{\bar{x}} = 5.8$.

10.72 6.

10.73 (a) $= 0.870$;   (b) $= 0.981$.

10.74 (a) It is divided by 5.   (b) It is multiplied by 3.5.

10.75 (a) 0.0869;   (b) 0.0202.

10.76 21 persons is the maximum number.

10.77 (a) 3,060;   (b) 27,405;   (c) 3,921,225.

10.78 $\dfrac{1}{42,504}$.

10.79 (a) 0.9876;   (b) 0.7888.

11.1   4.24.

11.2   (67.96 seconds, 76.44 seconds).

11.3 $51.35 < \mu < 54.25$ mm.

11.4   1.45 mm.

11.5   0.59 minute.

11.6   0.77 minute.

11.7   (a) 24.15 ± 0.589;   (b) 24.15 ± 0.700;   (c) 24.15 ± 0.775;   (d) 24.15 ± 0.844.

11.8   9.6 minutes.

11.9   1.50. The sample has not yet been taken.

11.10 $\$216.93 < \mu < \$220.41$. The probability is 0.10 that such an interval will not contain the population mean.

11.11   243.1 to 267.5 drinks.

11.12   99% confident.

11.13   22.7.

11.14   1.408.

11.15 $237.96 < \mu < 243.64$ pounds.

11.16 $\$216.99 < \mu < \$220.35$.

11.17   The variance of $\bar{x}$ is 1.872468.   (a) 2.69 cm;   (b) 3.53 cm.

11.18 (a) A plausible value for σ could be around 0.8 ounce. The maximum error would then be 0.25 ounce.

(b) A pessimistic value for σ could be around 1.0 ounce; The maximum error would then be 0.31 ounce.

11.19 1.24 ounces.

11.20 $n = 153$. The assertion is made now, whereas in Exercise 11.19 the assertion will be made after the data have been collected.

11.21 47.

11.22 (a) 89.

(b) Here we have in mind a statement that will be made after seeing the data.

(c) The solution will be the same.

11.23 40.

11.24 62.

11.25 (a) 27;   (b) 26 of the 30 confidence intervals contain μ.

11.26 (a) (55.83, 60.61);   (b) (54.94, 61.50);   (c) 2.39;   (d) 3.28.

11.27 The 99% interval is longer by 37%.

11.28 (a) 43%;   (b) 36%;   (c) 34%.

11.29 (a) 0.0023;   (b) 0.0034.

11.30 0.69 microgram.

11.31 $1.79 < \mu < 2.73$ micrograms.

11.32 12,840.

11.33 (a) $24.61 < \mu < 30.05$;   (b) $23.49 < \mu < 31.17$.

11.34 $21.35 < \mu < 24.65$ minutes.

11.35 0.11 gram.

11.36 $1.35 < \mu < 4.65$ fillings.

11.37 0.2794.

11.38 $360.02.

11.39 The posterior mean is $360.02, and the posterior standard deviation is 11.44. The probability in this distribution between $360 and $400 is 0.5000, to four figures.

11.40 0.8985.

11.41 (a) $\mu_1 = 70.0$;   (b) $\sigma_1 = 0.92$. The probability is 0.0150.

11.42 (a) Use the alternative hypothesis $\mu < 20$ and make the modification only if the null hypothesis can be rejected.

(b) Use the alternative hypothesis $\mu > 20$ and make the modification unless the null hypothesis can be rejected.

11.43 (a) Use the alternative hypothesis $\mu_2 > \mu_1$ and switch to the radial tires only if the null hypothesis can be rejected.

(b) Use the alternative hypothesis $\mu_2 < \mu_1$ and switch to the radial tires unless the null hypothesis can be rejected.

(c) Use the alternative hypothesis $\mu_2 \neq \mu_1$.

11.44 (a) Type I error;   (b) they will not be making an error.

11.45 (a) Reject the null hypothesis (error);

(b) reject the null hypothesis (correct);

(c) reject the null hypothesis (error);

(d) reject the null hypothesis (correct).

11.46 (a) $\mu_1 = \mu_2$;   (b) $\mu_1 < \mu_2$;   (c) $\mu_1 > \mu_2$.

11.47 If it erroneously rejects the null hypothesis, the testing service commits a Type I error. If it erroneously accepts the null hypothesis, the testing service commits a Type II error.

11.48 We would commit a Type I error if we erroneously reject the null hypothesis that the device is effective. We would commit a Type II error if we erroneously accept the null hypothesis that the device is effective.

11.49 Use the null hypothesis that the antipollution device for cars is not effective.

11.50 0.0087.

11.51 0.2743.

11.52 (a) 0.0456; (b) 0.8413.

11.53 (a) 0.1056; (b) 0.7340; (c) 0.2660.

11.54 (a) 0.8212; (b) 0.5000; (c) 0.1762; (d) 0.0307.

11.55 (a) 0.1616.

(b) The probabilities are:

$$0.0096 \text{ for } \mu = 38.5 \text{ or } 46.5,$$
$$0.0808 \text{ for } \mu = 39.5 \text{ or } 45.5,$$
$$0.3192 \text{ for } \mu = 40.5 \text{ or } 44.5,$$
$$0.6712 \text{ for } \mu = 41.5 \text{ or } 43.5.$$

11.56 (a) 0.0414.

(b) The probability is

$$0.0207 \text{ for } \mu = 38.5 \text{ or } 46.5;$$
$$0.1539 \text{ for } \mu = 39.5 \text{ or } 45.5,$$
$$0.5000 \text{ for } \mu = 40.5 \text{ or } 44.5,$$
$$0.8450 \text{ for } \mu = 41.5 \text{ or } 43.5.$$

11.57 (a) 0.0192; (b) 0.3192; (c) 0.0808.

11.58 The term "statistically significant" applies to the difference between the sample means. For the population means, there either is a difference or no difference.

11.59 Not necessarily, as it does not take 60% of the vote to win.

11.60 It does not make sense since we are not dealing with sample data.

11.61 By chance, 2.5 of the persons can be expected to score that well, so we cannot conclude that they must have extraordinary powers.

11.62 By chance, 2 times can be expected, but 6 times is reason for concern.

11.63 If the null hypothesis is true, then we have observed an event whose probability is at most 0.001.

11.64 (a) The alternative should be $\mu \neq 3.4$. (b) Two-tailed test.

11.65 (a) We should presume that the research worker's only concern is that the reduction might be overstated, so that the alternative should be $\mu < 20$.

(b) One-tailed test.

11.66 The alternative hypothesis is $\mu > 6.2$ minutes.

11.67 By using a very small sample, the probability of Type II error is large, and he has essentially guaranteed the acceptance of the null hypothesis. The experiment shows virtually nothing.

11.68 $\bar{x} = 24.08$ is so close that we are unconcerned.

11.69 The results are shocking and should be publicized immediately.

11.70 $z = -3.54$ and the null hypothesis must be rejected.

11.71 $p < 0.0001$.

11.72 $z = 2.73$ and the null hypothesis must be rejected.

11.73 $z = 3.02$ and the null hypothesis must be rejected.

11.74 0.0026.

11.75 $z = 2.92$ and the null hypothesis must be rejected.

11.76 Since 0.005 is less than 0.01, the null hypothesis must be rejected.

11.77 $z = -1.44$ and the null hypothesis cannot be rejected.

11.78 0.0749.

11.79 (a) $z = 2.35$; reject the null hypothesis.

(b) $z = 2.35$; accept the null hypothesis.

11.80 0.0188.

11.81 (a) $n = 166$. The null hypothesis will be rejected for $\bar{x} \geq 551.24$.

(b) $n = 14$. The null hypothesis will be rejected for $|\bar{x} - 650| \geq 13.62$.

11.82 $t = -0.79$ and the null hypothesis cannot be rejected.

11.83 $t = 4.00$ and the null hypothesis must be rejected.

11.84 $t = -1.47$ and the null hypothesis must be rejected.

11.85 $t = 2.63$ and the null hypothesis cannot be rejected.

11.86 $t = 2.64$; reject the claim.

11.87 $t = -0.58$ and the null hypothesis must be accepted.

11.88 (a) $t = 2.45$ and the null hypothesis cannot be rejected.

(b) $t = 2.45$ and the null hypothesis must be rejected.

11.89 $t = 5.66$ and the null hypothesis must be rejected.

11.90 (a) 0.05;　(b) 9.75%.　(c) 0.8063.

11.91 (a) 0.8784;　(b) 0.6082.

11.92 0.0334.

11.93 0.3668.

11.94 (a) $t = -2.326$.

(b) $0.01 < p < 0.025$.

(c) The data support the claim that the new formulation is not significantly worse than the old.

(d) Selecting α after seeing the p-value is unscrupulous.

11.95 $z = 2.26$ and the null hypothesis must be rejected.

11.96 (a) $z = -2.12$ and the null hypothesis cannot be rejected.

(b) The tail probability is 0.0340, and the null hypothesis cannot be rejected when $\alpha = 0.01$.

(c) The null hypothesis cannot be rejected when $\alpha = 0.02$.

(d) The null hypothesis cannot be rejected when $\alpha = 0.03$.

(e) The null hypothesis will be rejected when $\alpha = 0.04$.

11.97 $z = 0.31$ and the null hypothesis must be accepted.

11.98 $t = -1.99$ and the null hypothesis must be rejected.

11.99 $z = 2.78$ and the null hypothesis must be rejected.

11.100 $z = 4.44$ and we reject the null hypothesis.

11.101 $30.60 < \delta < 46.68$.

11.102 $0.051 < \delta < 0.055$.

11.103 $t = -1.88$ and the null hypothesis cannot be rejected.

11.104 $t = 2.01$ and the null hypothesis cannot be rejected.

11.105 $t = 1.29$ and the null hypothesis cannot be rejected.

11.106 $t = 1.11$ and the null hypothesis cannot be rejected.

11.107 $t = 1.99$ and the null hypothesis must be rejected.

11.108 $t = -0.52$ and the null hypothesis cannot be rejected.

11.109 $-24.2 < \delta < 78.2$.

11.110 $z = 2.14$ and the null hypothesis must be rejected.

11.111 $t = 2.20$ and the null hypothesis cannot be rejected.

11.112 90.1% confidence.

11.113 $z = -28.9$ and the null hypothesis must be rejected.

11.114 (a) $H_0: \mu_A = \mu_B$ versus $H_A: \mu_A > \mu_B$.

   (b) $H_0: \mu_A = \mu_B$ versus $H_A: \mu_A \neq \mu_B$.

   (c) $H_0: \mu_A = \mu_B$ versus $H_A: \mu_A < \mu_B$. Your objective will be reached if you end up accepting H_0.

   (d) (c) and (d) have the same solutions.

   (e) In the framework of hypothesis testing, you cannot show that two parameters are equal.

11.115 $15.95 < \mu < 24.39$.

11.116 $d = 1.98$.

11.117 (a) $\bar{x} = 2.74$.

   (b) For $\mu = 2.50$, the probability is 0.5557. For $\mu = 5.00$, the probability is 0.0869. For $\mu = 7.50$, the probability is 0.0021.

11.118 (a) 0.6543;   (b) $\mu_1 = 4{,}721$ and $\sigma_1 = 109.1$;   (c) 0.9736.

11.119 (a) 0.4633;   (b) $\mu_1 = 4{,}677.36$ and $\sigma_1 = 110.60$;   (c) 0.9434;

   (d) The two analyses used different prior distributions and hence come up with different answers.

11.120 $n = 424$.

11.121 $t = 0.90$ and the null hypothesis cannot be rejected.

11.122 $z = -3.28$ and the null hypotheses must be rejected.

11.123 0.473 pound.

11.124 The values covered by the confidence interval correspond exactly to the set of μ_0 values which would have been accepted in the hypothesis test.

11.125 No, since we are dealing with actual votes and not sample data.

11.126 At most 0.0262.

11.127 $z = -2.55$ and the null hypothesis must be rejected.

11.128 The tail probability is 0.0054.

11.129 $z = 4.79$ and the null hypothesis must be rejected.

11.130 $z = 1.67$ and the null hypothesis cannot be rejected.

11.131 $z = -2.44$ and the null hypothesis must be rejected.

11.132 $t = -1.32$ and the null hypothesis cannot be rejected.

11.133 (a) The law firm should set up the null hypothesis that on the average he or she makes as many mistakes as all the secretaries, and the alternative hypothesis that on the average he or she makes more mistakes than all the secretaries.

   (b) The law firm should set up the null hypothesis that on the average he or she makes as many mistakes as all the secretaries, and the alternative hypothesis that on the average he or she makes fewer mistakes than all the secretaries.

   In (a) the secretary is let go if the null hypothesis can be rejected, and in (b) the secretary is let go unless the null hypothesis can be rejected.

11.134 (a) She should use the null hypothesis $\mu = 48$ and the alternative hypothesis $\mu > 48$, and she would prove her point if the null hypothesis can be rejected.

(b) She should use the null hypothesis $\mu \geqslant 48$ and the alternative hypothesis $\mu < 48$, and she would prove her point if the null hypothesis can be accepted.

12.1 $0.0105 < \sigma < 0.0210$.

12.2 $10.84 < \sigma < 23.34$.

12.3 $12,700 < \sigma < 39,100$.

12.4 (a) $3.60 < \sigma < 7.20$; (b) $0.0020 < \sigma < 0.0057$.

12.5 (a) $0.329 < \sigma < 1.490$; (b) $0.091 < \sigma < 0.366$.

12.6 (a) $4.504 < \sigma^2 < 42.680$; (b) $195.34 < \sigma^2 < 3,016.32$.

12.7 (a) $10.37 < \sigma < 16.72$; (b) $3.84 < \sigma < 5.44$.

12.8 (a) $2.92 < \sigma < 3.77$; (b) $13.60 < \sigma < 16.55$.

12.9 $0.065 < \sigma^2 < 0.133$.

12.10 38.89.

12.11 1.98, compared to $s = 1.79$.

12.12 0.0388, compared to $s = 0.0365$.

12.13 1.72, compared to $s = 1.58$.

12.14 $\chi^2 = 12.49$ and the null hypothesis cannot be rejected.

12.15 $\chi^2 = 6.66$ and the null hypothesis cannot be rejected.

12.16 $\chi^2 = 6.87$ and the null hypothesis cannot be rejected.

12.17 $\chi^2 = 1.88$ and the null hypothesis cannot be rejected.

12.18 $z = 1.87$ and the null hypothesis must be rejected.

12.19 0.0307.

12.20 $z = -2.43$ and the null hypothesis cannot be rejected.

12.21 $z = -1.68$ and the null hypothesis cannot be rejected.

12.22 $z = 1.99$ and the null hypothesis must be rejected.

12.23 $F = 2.86$ and the null hypothesis must be rejected.

12.24 $F = 1.96$ and the null hypothesis cannot be rejected.

12.25 $F = 1.81$ and the null hypothesis cannot be rejected.

12.26 $F = 2.57$ and the null hypothesis cannot be rejected.

12.27 $F = 1.80$ and the null hypothesis cannot be rejected.

12.28 0.355.

12.29 (a) 0.304; (b) 0.178.

12.30 $\chi^2 = 16.9$ and the null hypothesis cannot be rejected.

12.31 $F = 7.21$ and the null hypothesis must be rejected.

12.32 $1.37 < \sigma < 2.45$ grams.

12.33 (a) 4.21; (b) 5.56; (c) 3.68.

12.34 $167.50 < \sigma < 283.98$.

12.35 $F = 1.66$ and the null hypothesis cannot be rejected.

12.36 $z = -1.46$ and the null hypothesis cannot be rejected.

12.37 $\chi^2 = 24.07$ and the null hypothesis cannot be rejected.

12.38 $F = 5.44$ and the null hypothesis must be rejected.

13.1 (a) $0.02 < p < 0.98$; (b) $0.00625 < p < 0.99375$;
(c) $n > 125$; (d) $n > 250$.

13.2 (a) (0.676, 0.764).
(b) 0.044.

13.3 (a) (0.658, 0.782).
(b) (0.218, 0.342).
(c) The endpoints in (b) can be found directly from the endpoints in (a). $0.218 = 1 - 0.782$ and $0.342 = 1 - 0.658$.

13.4 (a) (0.3940, 0.4380).
(b) (0.4059, 0.4501).
(c) (0.1398, 0.1722).
(d) There are no simple relationships among the endpoints.

13.5 (a) (0.068, 0.199); (b) 0.0654.

13.6 $0.617 < p < 0.743$.

13.7 0.075.

13.8 $0.245 < p < 0.295$.

13.9 0.0625.

13.10 The confidence is in excess of 0.9999.

13.11 $0.642 < p < 0.758$.

13.12 0.0689.

13.13 $0.772 < p < 0.928$.

13.14 0.109.

13.15 94.88%.

13.16 $0.261 < p < 0.419$.

13.17 $0.182 < p < 0.418$.

13.18 (a) 2,401; (b) 2,185.

13.19 (a) 400; (b) 625; (c) 2,500.

13.20 (a) 156; (b) 625; (c) 2,500; (d) 10,000. This assumes 95% confidence and the approximation used in Exercise 13.19.

13.21 (a) 423; (b) 601; (c) 1,041.

13.22 (a) 2,172; (b) 1,825.

13.23 (a) 1,068; (b) 1,025.

13.24 0.256

13.25 (a) 0.50, 0.45, and 0.05; (b) 0.53, 0.44, and 0.03.

13.26 (a) 0.145, 0.053, 0.039, 0.105, and 0.658; (b) $\mu_1 = 3.08$.

13.29 (b) $40{,}768.32p^2 - 58{,}368.32p + 20{,}736 < 0$; (c) (0.6541, 0.7776).

13.30 The probability of 5 or more successes is 0.028 and the null hypothesis must be rejected.

13.31 The probability of 11 or fewer successes is 0.017 and the null hypothesis must be rejected.

13.32 The probability of 5 or more is 0.073 and the null hypothesis cannot be rejected.

13.33 The probability of 12 or more successes is 0.007 and the null hypothesis must be rejected.

13.34 The probability of 7 or more successes is 0.980 and the probability of 7 or fewer successes is 0.057 and the null hypothesis cannot be rejected.

13.35 The probability of 5 or fewer successes is 0.038 and the null hypothesis must be rejected.

13.36 (a) $x \leq 7$;   (b) $x \leq 7$;   (c) $x \leq 5$.

13.37 (a) $x \leq 4$ or $x \geq 13$;   (b) $x \leq 3$ or $x \geq 13$;   (c) $x \leq 2$ or $x \geq 15$.

13.38 $z = -0.89$ and the null hypothesis cannot be rejected.

13.39 $z = -0.71$ and the null hypothesis cannot be rejected.

13.40 (a) $z = -2.05$ and the null hypothesis must be rejected.

   (b) $z = -2.05$ and the null hypothesis must be accepted.

13.41 $z = 1.49$ and the null hypothesis cannot be rejected.

13.42 $z = -1.83$ and the null hypothesis must be rejected.

13.43 (a) The p-value is 0.11 and the null hypothesis cannot be rejected.

   (b) The p-value is 0.412 and the null hypothesis cannot be accepted. A perfect 100 percent success rate in the data is not sufficient to conclude that the success rate is not equal to 0.90.

13.44 $z = 1.33$ and the null hypothesis cannot be rejected.

13.45 $z = 0.81$ and the null hypothesis cannot be rejected.

13.46 $z = -3.21$ and the null hypothesis must be rejected.

13.47 $z = 2.72$ and the null hypothesis must be rejected.

13.48 $z = -1.22$ and the null hypothesis cannot be rejected.

13.49 $z = 2.55$ and the null hypothesis must be rejected.

13.50 $z = 1.24$ and the null hypothesis cannot be rejected.

13.51 $z = -3.44$ and the null hypothesis must be rejected.

13.52 (a) $z = 1.81$ and the null hypothesis must be rejected.

   (b) $z = 0.61$ and the null hypothesis cannot be rejected.

13.53 $z = 0.33$ and the null hypothesis cannot be rejected.

13.54 $-0.001 < p_1 - p_2 < 0.159$.

13.55 $0.009 < p_1 - p_2 < 0.221$.

13.56 $\chi^2 = 40.9$.

13.57 $\chi^2 = 1.06$.

13.58 The null hypothesis is that the probabilities of the four response categories (go down, remain the same, go up, can't tell) are the same for each of the three types of dealers.

13.59 H_0:   State of employment and family size are independent.

   H_A: State of employment and family size are not independent.

13.60 The null hypothesis is that the probability of favoring the tax increase is the same in each of the four cities.

13.61 $\chi^2 = 3.55$ and the null hypothesis cannot be rejected.

13.62 $\chi^2 = 25.5$ and the null hypothesis must be rejected.

13.63 $\chi^2 = 1.30$ and the null hypothesis cannot be rejected.

13.64 $\chi^2 = 88.9$ and the null hypothesis must be rejected.

13.65 $\chi^2 = 88.9$.

13.66 $\chi^2 = 1.41$ and the null hypothesis cannot be rejected.

13.67 $\chi^2 = 4.01$ and the null hypothesis cannot be rejected.

13.68 (a) 0.3045;   (b) 0.3205.

13.69 $\chi^2 = 14.92$ and the null hypothesis must be rejected.

13.70 $\chi^2 = 5.97$ and the null hypothesis cannot be rejected.

13.71 $\chi^2 = 0.97$ and the null hypothesis cannot be rejected.
13.72 $\chi^2 = 4.82$ and the null hypothesis cannot be rejected.
13.73 $\chi^2 = 16.55$ and the null hypothesis must be rejected.
13.76 $\chi^2 = 9.04$ and the null hypothesis must be rejected.
13.81 $\chi^2 = 9.48$ and the null hypothesis cannot be rejected.
13.82 $\chi^2 = 5.06$ and the null hypothesis cannot be rejected.
13.83 $\chi^2 = 13.8$ and the null hypothesis must be rejected.
13.84 $\chi^2 = 21.42$ and the null hypothesis must be rejected.
13.85 $\chi^2 = 17.0$ and the null hypothesis must be rejected.
13.86 $\chi^2 = 19.78$ on 1 degree of freedom, and the null hypothesis must be rejected.
13.87 $\chi^2 = 21.594$ on 7 degrees of freedom, and the null hypothesis must be rejected.
13.88 $\chi^2 = 1.77$ on 1 degree of freedom, and the null hypothesis cannot be rejected.
13.89 (a) The probabilities are 0.0158, 0.0810, 0.2332, 0.3291, 0.2371, 0.0868, 0.0155, and 0.0015.

 (b) The expected frequencies are 3.2, 16.2, 46.6, 65.8, 47.4, 17.4, 3.1, and 0.3. In what follows, the first two and the last three classes are combined.

 (c) $\chi^2 = 3.7$ on 3 degrees of freedom and the null hypothesis cannot be rejected.
13.90 $\chi^2 = 17.2$ on 2 degrees of freedom, and the null hypothesis must be rejected.
13.91 $0.605 < p < 0.795$.
13.92 $\chi^2 = 11.6$ and the null hypothesis must be rejected.
13.93 (a) 0.212;   (b) 0.500;   (c) 0.212.
13.94 $z = 2.88$ and the null hypothesis must be rejected.
13.95 $\chi^2 = 8.30$ and the null hypothesis must be rejected.
13.96 $z = 2.45$ and the null hypothesis must be rejected.
13.97 $0.0125 < p < 0.9875$.
13.98 $0.895 < p < 0.965$.
13.99 (a) 267;   (b) 225.
13.100 0.10.
13.101 (a) We talk about significant differences between sample data, or significant differences between sample data and population parameters, but not about significant differences between population parameters.

 (b) $z = 1.11$ and the null hypothesis cannot be rejected.
13.102 0.032.
13.103 The null hypothesis can be rejected at the 0.04 level of significance.
13.104 H_0: $p_{11} = p_{12}$, $p_{21} = p_{22}$,   and $p_{31} = p_{32}$
  H_A: $p_{i1} \neq p_{i2}$ for at least one value of i.
13.105 $\chi^2 = 4.24$ and the null hypothesis cannot be rejected.
13.106 $z = -7.83$ and the null hypothesis must be rejected.
13.107 $\chi^2 = 8.37$ and the null hypothesis must be rejected.
13.108 The null hypothesis is that the response to product A is independent of the response to product B. This chi-squared analysis does not ask about the relative merits of the two products.
13.109 $\chi^2 = 1.3$ on 2 degrees of freedom and the null hypothesis cannot be rejected.
13.110 (a) The probabilities are 0.04, 0.40, and 0.56

 (b) The three posterior probabilities are 0.60 for the purchasing agent, 0.13 for the head of the quality control department, and 0.27 for the chief engineer.

13.111 $0.160 < p < 0.290$.

13.112 $\chi^2 = 52.8$ and the null hypothesis must be rejected.

13.113 0.47.

13.114 $z = -0.81$ and the null hypothesis cannnot be rejected.

13.115 0.057.

13.116 $n = 271$.

13.117 (a) $x = 0$ or $x \geq 9$;   (b) $x \geq 8$;   (c) $x \leq 1$;   (d) $x \leq 1$ or $x \geq 8$.

13.118 $\chi^2 = 17.2$ and the null hypothesis must be rejected.

13.119 $z = -1.59$ and the null hypothesis cannot be rejected.

13.120 There are small expected frequencies in the second column.

13.121 Combining deaf and blind workers, $\chi^2 = 2.47$ and the null hypothesis cannot be rejected.

13.122 (a) 0.30, 0.60, and 0.10;   (b) 0.68, 0.30, and 0.02.

13.123 $\chi^2 = 58.23$ with 4 degrees of freedom. The null hypothesis should be rejected.

13.124 0.10.

CHAPTER 14

14.1 (a) $F = 0.64$;   (b) $F = 0.64$, and the null hypothesis cannot be rejected.

14.2 (a) $F = 2.25$;   (b) $F = 2.25$ and the null hypothesis cannot be rejected.

14.3 (a) $F = 6.54$;   (b) $F = 6.54$ and the null hypothesis must be rejected.

14.4 The standard deviations within the groups are very unequal, and this violates one of the assumptions.

14.5 The assumption that the populations have equal standard deviations seems untenable.

14.6 (a) It will be impossible to distinguish the effects of the balls from the effects of the golf pros.

   (b) It will be impossible to distinguish the effects of the balls from the effects of time sequence.

14.7 The three kinds of tulips should have been assigned at random to the twelve locations in the flower bed.

14.8 The probability that the five heaviest persons will all get diet 1 is $\dfrac{1}{3,003}$. The probability that they will all get the same diet is $\dfrac{3}{3,003}$, which is about 0.001.

14.9 This is controversial. Most practitioners will examine the results of their randomization to check for odd patterns. For this situation, it is likely that the analysis of variance will not be done.

14.10 $F = 0.64$ and the null hypothesis cannot be rejected.

14.11 $F = 2.25$ and the null hypothesis cannot be rejected.

14.12 $F = 6.54$ and the null hypothesis must be rejected.

14.13 $F = 0.68$ and the null hypothesis cannot be rejected.

14.14 $F = 10.79$ and the null hypothesis must be rejected.

14.15 $F = 11.70$ and we reject the null hypothesis.

14.16 $F = 11.05$ and the null hypothesis must be rejected.

14.17 $F = 3.18$ and the null hypothesis cannot be rejected.

14.19 $kn - 1$.

14.20 4.80.

14.22 $F = 1.235$ (or $t = 1.11$) and the null hypothesis cannot be rejected.

14.23 She might consider only programs of the same length, or she might use the program lengths as blocks and perform a two-way analysis of variance.

14.24 There appears to be a definite interaction.

14.25 $F = 6.62$ for treatments, and $F = 4.86$ for blocks and both null hypotheses must be rejected.

14.26 $F = 11.09$ for machines, and $F = 99.11$ for brands; both null hypotheses are rejected.

14.27 $F = 8.31$ for treatments, and $F = 0.58$ for blocks. The null hypothesis for treatments must be rejected; the null hypothesis for blocks cannot be rejected.

14.28 (a) $F = 0.08$. We must accept the null hypothesis.

(b) $F = 0.18$ for salespersons, and $F = 4.64$ for weeks. We must accept the null hypothesis for the salespersons, and reject the null hypothesis for the weeks.

14.29 $F = 4.42$ for factor A, and $F = 17.06$ for factor B. The null hypothesis for factor A cannot be rejected; the null hypothesis for factor B must be rejected.

14.30 $F = 0.35$ for factor A, and $F = 7.07$ for factor B. The null hypothesis for factor A cannot be rejected; the one for factor B must be rejected.

14.31 (b) 60.

14.32 160.

14.34 (a)

B	C	A
C	A	B
A	B	C

(b)

C	A	D	C
B	D	A	B
A	C	B	D
D	B	C	A

(c)

B	A	E	D	C
A	D	B	C	E
C	E	D	A	B
D	B	C	E	A
E	C	A	B	D

14.35 $F = 27.02$ for rows, $F = 2.56$ for columns, and $F = 94.17$ for treatments. The null hypothesis for rows (professional interest) must be rejected; the null hypothesis for columns (ethnic background) cannot be rejected; the null hypothesis for treatments (instructors) must be rejected.

14.36 $F = 15.64$ for rows, $F = 0.82$ for columns, and $F = 45.82$ for treatments. The null hypothesis for rows (golf pros) must be rejected; the null hypothesis for columns (drivers) cannot be rejected; the null hypothesis for treatments (designs) must be rejected.

14.37 The doctor who is a Westerner is a Republican.

14.38 A must appear with E on March 16.

F must appear with I on March 11.

H must appear with G on March 5.

G must appear with B and I on March 13.

D must appear with B and C on March 2 and with I on March 9.

14.39 2 must appear together with 5 and 7 on Thursday.

4 must appear together with 3 and 7 on Tuesday.

5 must appear together with 3 and 6 on Saturday.

14.40 Observed differences due to the designs may instead be due to the use of different tees.

14.41 (a) $n \cdot s_{\bar{x}}^2 = 28$, and $F = \dfrac{28}{9.56} = 2.93$.

(b) $F = 2.93$ and the null hypothesis cannot be rejected.

14.42 $F = 2.93$ and the null hypothesis cannot be rejected.

14.43 (b) There are two solutions:

Griffith	–	Dramatics
Anderson	–	Discipline
Evans	–	Tenure or salaries
Fleming	–	Salaries or tenure

14.44 (a) 5 df for treatments, 6 df for blocks, 30 df for error;

(b) 5 df for treatments, 9 df for blocks, 45 df for error;

(c) 4 df for treatments, 3 df for blocks, 12 df for error;

(d) 7 df for treatments, 6 df for blocks, 42 df for error.

14.45 $F = 15.2$ for factor A, and $F = 53.0$ for factor B. The null hypothesis for factor A (majors) must be rejected; the null hypothesis for factor B (instructors) must be rejected.

14.46 $F = 2.31$ for rows, $F = 8.24$ for columns, and $F = 31.28$ for treatments. The null hypothesis for rows (mechanics) cannot be rejected; the null hypothesis for columns (engines) must be rejected; the null hypothesis for treatments (fuels) must be rejected.

14.47 $F = 6.83$ and the null hypothesis must be rejected.

14.48 (b) four ways.

14.49 Rank.

14.50 Since all the male faculty members are economists and all the female faculty members are statisticians, any observed differences between the males and the females may actually be due to their being economists and statisticians.

14.51 $F = 7.75$ for treatments, and $F = 8.06$ for blocks. The null hypothesis for treatments (routes) must be rejected; the null hypothesis for blocks (days of the week) must be rejected.

CHAPTER 15

15.1 For the first line the sum of the squares of the errors is 8. For the second line the sum of the squares of the errors is 6. The second line provides a better fit.

15.2 $12 = 3a + 36b$
$120 = 36a + 120b$
By the method of elimination, $a = 8$ and $b = -\frac{1}{3}$.

15.3 (a) $107 = 6a + 36b$
$721 = 36a + 304b$
$b = 0.898$; $a = 12.445$;

(b) 19.6;

(c) 131.1. This is what happens when we make predictions for values very different from those in the original database.

15.4 (a) $\hat{y} = 6.226 - 0.079x$;

(b) 4.804 minutes;

(c) The value 4.9 minutes was an actual time in the database; the new figure in (b) is a prediction for a new week in which she trains for 18 hours.

15.5 The conclusion does not follow; the relationship applies only to the range of values of x on which the study is based.

15.6 (a) $\hat{y} = 2.039 - 0.102x$;

(b) 1.529;

(c) 1.223. The value 1.1 was part of the database actually observed, while this 1.223 is a fitted (or predicted) value.

(d) The readings would be dependent on each other, violating one of the assumptions.

15.7 $\hat{y} = 0.23 + 4.42x$.

15.8 (a) $\hat{y} = 24.92 + 0.69x$;   (b) $\hat{y} = 71.84$.

15.9 No. The couple will join a group of families whose average income is greater by $5,000, but this does not affect their actual income.

15.10 (b) $\hat{y} = 3.35 + 24.93x$;   (c) 177.86.

15.11 She will find $S_{xx} = 0$.

15.12 The sum of the squares of the errors is 20.94, which is less than either 44 or 26.

15.13 105.

15.14 (a) $\hat{y} = 0.491 + 0.272x$;   (b) 10.83.

15.15 $\hat{y} = 4.954 + 3.108x$. For $x = 10$, the predicted humidity is 36.034.

15.16 (a) $\hat{y} = 2.66 + 0.60x$. For $x = 3$, $\hat{y} = 2.66 + 0.60(3) = 4.46$.

(b) $\hat{y} = 2.95 + 0.242x$. For $x = 11$, $\hat{y} = \$5.6$ million.

15.17 (a) $\hat{y} = -5.07 + 0.92x$.

(b) For $x = 1$, $\hat{y} = -4.15$;   for $x = 10$, $\hat{y} = 4.13$.

(c) If the SAT ranks were assigned to the ten students at random, the expected rank for each student would be 5.5; the student ranked 1 on the PSAT test could expect his ranking to go down 4.5 points and the one ranked 10 could expect his ranking to go up 4.5 points.

15.18 (a) $a = 31.53$ and $b = 10.90$.

(b) $a = 24.91$ and $b = 0.69$.

15.19 $t = -1.21$ and the null hypothesis cannot be rejected.

15.20 $t = 3.70$ and the null hypothesis must be rejected.

15.21 (a) $\hat{y} = 106.026 + 14.927x$;

(b) $t = 8.197$ and the null hypothesis must be rejected.

15.22 (a) $\hat{y} = 1.259 + 1.483x$.

(b) $t = -1.36$ and the null hypothesis cannot be rejected.

15.23 $t = -2.76$ and the null hypothesis must be rejected.

15.24 $t = -0.52$ and the null hypothesis cannot be rejected.

15.25 (a) $\hat{y} = 257.19 - 14.16x$.

(b) $-18.14 < \beta < -10.18$.

15.26 $0.155 < \beta < 0.389$.

15.27 $1.160 < \beta < 1.806$.

15.28 $1.703 < \alpha < 2.101$.

15.29 $0.592 < \alpha < 1.926$.

15.30 $10.195 < \alpha < 52.871$.

15.31 (13.6, 22.6).

15.32 (170.68, 175.72).

15.33 4.01 and 4.44.

15.34 12.45 and 26.75.

15.35 (166.00, 180.40).

15.36 (a) True;   (b) false.

15.38 (a) $\hat{y} = -16{,}740 + 1{,}961x_1 + 5{,}976b_2$;   (b) \$83,643.

15.39 (a) $\hat{y} = 14.56 + 30.11x_1 + 12.16x_2$;

    (b) \$101.42;

    (c) \$105.06;

    (d) \$137.55;

    (e) \$141.20.

15.40 (a) $\hat{y} = 197.65 + 37.19x_1 - 0.120x_2$;

    (b) $\hat{y} = 70.8$.

15.41 (a) $\hat{y} = 69.73 + 2.975x_1 - 11.97x_2$;   (b) 71.2.

15.42 $\hat{y} = 101(0.96)^x$. For $x = 25$ we get $\hat{y} = 34.3\%$.

15.43 $\hat{y} = 68.9(1.23)^x$. For $x = 5$ we get $\hat{y} = 197$ thousand.

15.44 (a) $\hat{y} = 298(1.23)^x$;   (b) at $x = 9$, we get $y = 1{,}980$.

15.45 $\hat{y} = 27{,}200 \cdot x^{-2.36}$. For $x = 12$ we get $\hat{y} = 77.2$ thousand units.

15.46 $\hat{y} = 6{,}115 \cdot x^{-1.49}$. For $x = 15$ we get $\hat{y} = 108$ psi.

15.47 $\hat{y} = 14.789 + 38.892x - 17.559x^2$. For $x = 1.5$ we get $\hat{y} = 33.6$ pounds per square yard.

15.48 $\hat{y} = 19{,}715.56 - 828x - 227.99x^2$.

15.49 $\hat{y} = 3.143 + 0.15x + 0.057x^2$.

15.50 $\hat{y} = 384.39 - 36.00x + 0.896x^2$.

15.51 The parabola cannot be used for values of x beyond the range of values of x investigated, because it turns up again for larger values of x.

15.52 (a) $\hat{y} = 10.954 - 0.3326x$;   (b) $\hat{y} = 4.302$ hours.

15.53 $-0.652 < \beta < -0.013$.

15.54 $1.288 < \mu_{y|20} < 7.316$.

15.55 Use -2, -1, 0, 1, and 2 as the month numbers. The fitted equation is $\hat{y} = 56.66 - 5.1x_1 - 2.93x_2$.

15.56 (a) $\hat{y} = -0.627 + 0.0972 + 0.662$;   (b) \$29.05 thousand.

15.57 $\hat{y} = 28.3(1.02)^x$. For $x = 60$ we get $\hat{y} = 74.2$ mrem/year.

15.58 (a) $\hat{y} = -1.0695 + 1.2196x$;   (b) for $x = 2.5$, we get $\hat{y} = 1.98$.

15.59 $t = 2.47$ and the null hypothesis must be rejected.

15.60 (a) $\hat{y} = 50.001 + 7.439x$;   (b) for $x = 25$, we get $\hat{y} = 236.0$.

15.61 $t = -0.09$ and the null hypothesis cannot be rejected.

15.62 89.8 and 382.2.

15.63 (a) $\hat{y} = -32.702 + 3.819x_1 + 0.058x_2$.

    (b) 71.13;   94.04;   82.73;   105.64.

CHAPTER 16   **16.1** 0.92.

    16.2 -0.01.

    16.3 0.01%.

    16.4 No. It does not matter which variable is labeled x and which variable is labeled y.

    16.5 0.68.

    16.6 We still get $r = 0.68$.

16.7 No correction is needed.

16.8 It should not come as a surprise, since we can always draw a straight line passing through two distinct points.

16.9 $r = 0.704$.   49.6% of the variation in failure time is due to differences in resistance.

16.10 A correlation is symmetric in the two variables.

16.11 (a) Positive correlation.

  (b) Negative correlation.

  (c) Negative correlation.

  (d) No correlation.

  (e) Positive correlation.

16.12 (a) Positive correlation.

  (b) Positive correlation.

  (c) Positive correlation.

  (d) No correlation.

  (e) Negative correlation.

16.13 (a) $r = 0.199$.

  (b) $r = 0.199$.

16.14 (a) $r = 0.15$.

  (b) $r = -0.78$.

  (c) $r = 0.51$.

16.15 1.9988; the first relationship is twice as strong.

16.17 Both variables depend on other variables, such as population growth and economic conditions in general.

16.18 (a) 0.7976;   (b) 0.5413.

16.19 (a) $r = 0.13$;   (b) $r = 0.81$;   (c) $r = 0.61$;   (d) $r = -0.18$.

  (e) the effect is felt most strongly after two years, and after that it diminishes gradually.

16.20 $r = -0.01$, and $z = -0.030$ and the null hypothesis cannot be rejected.

16.21 $r = 0.704$, and $z = 3.03$ and the null hypothesis must be rejected.

16.22 (a) $r = 0.92$

  (b) $r = 0.92$, and $z = 4.767$ and the null hypothesis must be rejected.

16.23 (a) $r = 0.98$.   96.0% of the variation in the yield can be attributed to differences in the rainfall.

  (b) $r = 0.98$, and $z = 5.14$ and the null hypothesis must be rejected.

16.24 (a) $t = 1.83$ and the null hypothesis cannot be rejected.

  (b) $t = 3.35$ and the null hypothesis must be rejected.

16.25 (a) $t = 4.32$ and the null hypothesis must be rejected.

  (b) $t = 2.22$ and the null hypothesis cannot be rejected.

16.26 $t = -0.03$ and the null hypothesis cannot be rejected.

16.27 (a) 12.88;   (b) 12.88. These statistics are always identical.

16.29 $r = 0.73$, and $z = 0.93$ and the null hypothesis cannot be rejected.

16.30 $r = 0.94$, and $z = 2.42$ and the null hypothesis cannot be rejected.

16.31 $r = -0.64$, and $z = -1.735$ and the null hypothesis must be rejected.

16.32 (a) $0.49 < \rho < 0.93$;   (b) $-0.54 < \rho < 0.15$;   (c) $0.51 < \rho < 0.74$.

16.33 (a) $-0.96 < \rho < -0.60$;   (b) $-0.15 < \rho < 0.75$.

16.34 0.223.

16.35 Assign the categories scores -1, 0, and 1. The correlation is $r = 0.023$.

16.36 Use scores (say) -2, -1, 0, 1 for the columns.
  (a) Use scores -1, 0, and 1 for the three rows. $r = 0.733$.
  (b) Use scores 1, 2, and 4 for the three rows. $r = 0.652$.
  (c) The score spacings 1, 2, and 4 are equivalent to -1, 0, and 2;

16.37 $R^2 = 0.3320$ and $R = 0.576$.

16.38 0.9998.

16.39 0.959.

16.40 The multiple correlation coefficient based on both variables cannot be numerically less than the ordinary correlation coefficient based on only one of the two variables.

16.41 $r_{13.2} = -0.80$.

16.42 (a) $r_{13.2} = 1.00$;   (b) $r_{23.1} = 1.00$.

16.43 There are many reasons why these figures cannot all be correct. For instance, $r_{12.3} = 4.59$.

16.44 The second relation is three times as strong.

16.45 This would lead to a correlation coefficient of -1.03, which certainly indicates an error.

16.46 (a) $z = 1.64$ and we cannot reject the null hypothesis.
  (b) $t = 1.73$ and we cannot reject the null hypothesis.

16.47 -0.92.

16.48 $r = -0.92$, and $z = -2.75$ and the null hypothesis must be rejected.

16.49 6.25%.

16.50 (a) $0 < \rho < 0.74$;   (b) $-0.58 < \rho < 0$.

16.51 (a) Negative correlation.
  (b) Positive correlation.
  (c) No correlation.
  (d) Positive correlation.

16.52 0.68.

16.53 $r = -0.65$ and $z = -2.05$ and the null hypothesis cannot be rejected.

16.54 $r = 0.86$.

16.55 $r = 0.86$ and $z = 0.95$ and the null hypothesis cannot be rejected.

16.56 0.86.

16.57 $r_{12.3} = -0.74$.

CHAPTER 17

17.1 The probability of 8 or more $+$ signs is 0.194, so $p = 0.388$ and the null hypothesis cannot be rejected.

17.2 The sample size is only $n = 8$. The probability of 7 or more successes is 0.035 and the null hypothesis must be rejected.

17.3 The probability of 5 or more successes is 0.110 and the null hypothesis cannot be rejected.

17.4 The two values which equal 100.0 have to be discarded; the sample size is only $n = 12$. The probability of 9 or more successes is 0.073, and the probability of 9 or fewer successes is 0.981, and both exceed 0.025, so that the null hypothesis cannot be rejected.

17.5 The probability of 9 or more successes is 0.073 and the null hypothesis cannot be rejected.

17.6 Since one of the pairs, 75 and 75, has to be discarded, the sample size is only $n = 14$. The probability of 4 or fewer successes is 0.090 and the null hypothesis cannot be rejected.

17.7 Three pairs have to be discarded. Then $z = -2.67$, and the null hypothesis must be rejected.

17.8 Three of the values equal 23 and have to be discarded. $z = -2.47$ and the null hypothesis must be rejected.

17.9 The two values which equal 100.0 must be discarded. $z = 1.73$ and the null hypothesis cannot be rejected.

17.10 Four of the values equal 24.2 and must be discarded. $z = 2.3333$ and the null hypothesis must be rejected.

17.11 Four of the pairs must be discarded. $z = 2.75$ and the null hypothesis must be rejected.

17.12 Two of the pairs must be discarded. $z = 2.56$ and the null hypothesis must be rejected.

17.13 One of the pairs, 6 and 6, must be discarded. $z = -2.04$ and the null hypothesis must be rejected.

17.14 $U_1 = 34$, $U_2 = 2$, and $U = 2$. The null hypothesis must be rejected.

17.15 $U_1 = 5.5$, $U_2 = 42.5$, and $U = 5.5$. The null hypothesis must be rejected.

17.16 $U_1 = 13$, $U_2 = 36$, and $U = 13$. The null hypothesis cannot be rejected.

17.17 $U_1 = 30$, $U_2 = 6$, and $U = 6$. The null hypothesis cannot be rejected.

17.18 $U_1 = 88$, $U_2 = 92$, and $U = 88$. The null hypothesis cannot be rejected.

17.19 $U_1 = 109$ and $z = 2.14$; the null hypothesis must be rejected.

17.20 $U_1 = 24$ and $z = -1.71$ and the null hypothesis cannot be rejected.

17.21 $U_1 = 127.5$ and $z = 3.20$ and the null hypothesis must be rejected.

17.22 $U_1 = 88$ and $z = -0.10$ and the null hypothesis cannot be rejected.

17.23 $U_1 = 70$, and $z = 1.51$ and the null hypothesis cannot be rejected.

17.24 $R_1 = 64$, $R_2 = 59.5$, $R_3 = 47.5$, and $H = 0.85$. The null hypothesis cannot be rejected.

17.25 $R_1 = 53$, $R_2 = 68$, $R_3 = 30$, and $R_4 = 59$. $H = 4.51$ and the null hypothesis cannot be rejected.

17.26 $R_1 = 121$, $R_2 = 144$, and $R_3 = 86$. $H = 1.53$ and the null hypothesis cannot be rejected.

17.27 $R_1 = 107$, $R_2 = 93$, and $R_3 = 100$. $H = 0.245$ and the null hypothesis cannot be rejected.

17.28 $u = 15$ and the null hypothesis cannot be rejected.

17.29 $u = 4$ and the null hypothesis cannot be rejected.

17.30 $u = 5$ and the null hypothesis cannot be rejected.

17.31 $u = 17$ and the null hypothesis must be rejected.

17.32 $n_1 = 28$, $n_2 = 22$, $u = 23$, and $z = -0.77$; the null hypothesis cannot be rejected.

17.33 $n_1 = 38$, $n_2 = 12$, and $u = 17$. $z = -0.88$ and the null hypothesis cannot be rejected.

17.34 $n_1 = 13$, $n_2 = 12$, and $u = 15$. $z = 0.62$ and the null hypothesis cannot be rejected.

17.35 $n_1 = 20$, $n_2 = 36$, and $u = 28$. $z = 0.38$ and the null hypothesis cannot be rejected.

17.37 $n_1 = 12$, $n_2 = 12$, and $u = 8$. The null hypothesis cannot be rejected.

17.38 After the three 5's are discarded, we get $n_1 = 14$, $n_2 = 13$, and $u = 5$. $z = -3.73$ and the null hypothesis must be rejected.

17.39 $n_1 = 20$, $n_2 = 20$, and $u = 12$. $z = -2.88$ and the null hypothesis must be rejected.

17.40 $n_1 = 16$, $n_2 = 16$, and $u = 12$. $z = -1.80$ and the null hypothesis must be rejected.

17.41 $n_1 = 25$, $n_2 = 25$, and $u = 24$. $z = -0.57$ and the null hypothesis must be accepted.

17.42 After removing the value 4, only 21 remain. $u = 7$ and the null hypothesis must be accepted.

17.43 0.81.

17.44 $z = 3.03$ and the null hypothesis must be rejected.

17.45 0.65.

17.46 $z = 2.16$ and the null hypothesis must be rejected.

17.47 0.99.

17.48 $z = 2.97$ and the null hypothesis must be rejected.

17.49 $r_S = 0.893$.

17.50 $z = 1.79$ and the null hypothesis cannot be rejected.

17.51 0.75.

17.52 For judges A and B, $r_s = 0.61$.

For judges A and C, $r_s = -0.05$.

For judges B and C, $r_s = -0.18$.

(a) Judges A and B are most alike.

(b) Judges B and C differ the most.

17.53 $T = 19$ and the null hypothesis must be rejected.

17.54 $z = 1.07$ and the null hypothesis cannot be rejected.

17.55 $n_1 = 15$, $n_2 = 5$, and $u = 4$ and the null hypothesis must be rejected.

17.56 $R_1 = 56.5$, $R_2 = 41$, $R_3 = 37.5$, and $R_4 = 75$. $H = 5.03$ and the null hypothesis cannot be rejected.

17.57 $U_1 = 109$, $U_2 = 35$, and $U = 35$ and the null hypothesis must be rejected.

17.58 $U_1 = 35$, and $z = -2.14$ and the null hypothesis must be rejected.

17.59 $r_S = 0.39$. $z = 1.46$ and the null hypothesis cannot be rejected.

17.60 The probability of $x \leq 9$ is 0.995 and the probability of $x \geq 9$ is 0.032 and the null hypothesis cannot be rejected.

17.61 $z = -2.11$ and the null hypothesis must be rejected.

17.62 $x = 13$, $z = 1.34$ and the null hypothesis cannot be rejected.

17.63 (a) Use U_1 and reject the null hypothesis if $U_1 \leq 23$.

(b) Use U_2 and reject the null hypothesis if $U_2 \leq 23$.

(c) Use U and reject the null hypothesis if $U \leq 19$.

17.64 $n_1 = 24$, $n_2 = 12$, and $u = 14$. $z = -1.15$ and the null hypothesis cannot be rejected.

17.65 $r_S = 0.27$. $z = 1.32$ and the null hypothesis cannot be rejected.

17.66 The probability of $x \leq 4$ is 0.039 and the null hypothesis must be rejected.

17.67 $n_1 = 9$, $n_2 = 9$, and $u = 4$ and the null hypothesis must be rejected.

17.68 $U_2 = 18.5$ and the null hypothesis cannot be rejected.

INDEX

PROBLEMS OF ESTIMATION

Confidence Interval for

Mean (large sample, σ known or estimated by s)

$$\bar{x} - z_{\alpha/2} \cdot \frac{\sigma}{\sqrt{n}} < \mu < \bar{x} + z_{\alpha/2} \cdot \frac{\sigma}{\sqrt{n}}$$

Mean (small sample)

$$\bar{x} - t_{\alpha/2} \cdot \frac{s}{\sqrt{n}} < \mu < \bar{x} + t_{\alpha/2} \cdot \frac{s}{\sqrt{n}}$$

Proportion (large sample)

$$\hat{p} - z_{\alpha/2} \cdot \sqrt{\frac{\hat{p}(1-\hat{p})}{n}} < p < \hat{p} + z_{\alpha/2} \cdot \sqrt{\frac{\hat{p}(1-\hat{p})}{n}}$$

where $\hat{p} = \dfrac{x}{n}$

Maximum Error

Estimation of mean

$$E = z_{\alpha/2} \cdot \frac{\sigma}{\sqrt{n}}$$

Estimation of proportion

$$E = z_{\alpha/2} \cdot \sqrt{\frac{\hat{p}(1-\hat{p})}{n}}$$

Sample Size

Estimation of mean

$$n = \left[\frac{z_{\alpha/2} \cdot \sigma}{E} \right]^2$$

Estimation of proportion

$$n = p(1-p)\left[\frac{z_{\alpha/2}}{E} \right]^2$$